# Biomarkers of Disease

An Evidence-based Approach

This new publication takes a critical, evidence-based look at the efficacy of new diagnostic tests which are increasingly being used to evaluate organ damage and dysfunction. The use of biomarkers is growing, with a steady stream of new products being brought out by the diagnostics industry. Some of these assist in diagnosis, while others provide a means of monitoring the state of progression of disease and the effectiveness of therapeutic options. However, in many cases, the evidence which supports the use of these new methods as opposed to traditional biochemical tests has not yet been demonstrated, and it is intended that this volume will help clarify the strengths and weaknesses of using these biomarkers across a wide range of applications and in the various organs of the body. This approach will provide clinicians, pathologists, clinical biochemists and medical laboratory scientists with an invaluable overview of the diverse applications of biomarkers in medicine.

# Biomarkers of Disease

## An Evidence-based Approach

Edited by

Andrew K. Trull, Lawrence M. Demers

David W. Holt, Atholl Johnston, J. Michael Tredger

and Christopher P. Price

CAMBRIDGE UNIVERSITY PRESS
Cambridge, New York, Melbourne, Madrid, Cape Town, Singapore, São Paulo, Delhi

Cambridge University Press
The Edinburgh Building, Cambridge CB2 8RU, UK

Published in the United States of America by Cambridge University Press, New York

www.cambridge.org
Information on this title: www.cambridge.org/9780521811026



First published 2002
This digitally printed version 2008

*A catalogue record for this publication is available from the British Library*

ISBN 978-0-521-81102-6 hardback
ISBN 978-0-521-08860-2 paperback

---



---

# Contents

## Part 7 Biomarkers of cardiovascular disease and dysfunction



## Part 8 Biomarkers of neurological disease and dysfunction



## Part 9 Biomarkers in transplantation

# Contributors

**Victor W. Armstrong**
Abteilung Klinische Chemie
Zentrum Innere Medizin
Georg-August-Universität Göttingen
Robert-Koch-Strasse 40
Göttingen
Germany
varmstro@med.uni-goettingen.de

**John Azami**
Addenbrooke's Hospital
Hills Road
Cambridge
CB2 2QQ
UK

**Lina Badimon**
Cardiovascular Research Center
IIBB/CSIC-HSCSP-UAB
Jordi Girona, 18–26
08034 Barcelona
Spain
lbmucv@cid.csic.es

**Antonio Bayés-Genís**
Cardiovascular Research Center
IIBB/CSIC-HSCSP-UAB
Jordi Girona, 18–26
08034 Barcelona
Spain

**Katherine M. Beckett**
AstraZeneca
Macclesfield Site
Charterway
Silk Road Business Park
Macclesfield
Cheshire
SK10 2NA
UK

**Johannes G. Bode**
Institute of Biochemistry
RWTH Aachen
Pauweisstrasse 30
D-52074 Aachen
Germany

**Andy Bufton**
Abbott Diagnostics Division
Abbott House
Norden Road
Maidenhead
Berkshire
SL6 4XF
UK
andy.bufton@abbott.com

**Martin Burdelski**
Department of Paediatrics
University Hospital Hamburg Eppendorf
Martinistrasse 52

D-20246 Hamburg
Germany
burdelsk@uke.uni-hamburg.de

**Kelvin Cain**
MRC Toxicology Unit
Hodgkin Building
University of Leicester
Lancaster Road
Leicester
LE1 9HN
UK
kc5@leicester.ac.uk

**Jean Campbell**
Department of Pathology
University of Washington School of Medicine
Seattle
WA 98195
USA

**Stewart Campbell**
Department of Clinical Biochemistry
St Bartholomew's and The Royal London Hospitals
West Smithfield
London
EC1A 7BE
UK

**Julio E. Celis**
Department of Medical Biochemistry and Danish Centre for Human Genome Research
University of Aarhus
Ole Worms Allé
Building 170
DK-800 Aarhus C
Denmark
jec@biokemi.aau.di

**John C. Chambers**
National Heart and Lung Institute
Hammersmith Hospital
Du Cane Road
London
W12 0NN
UK
jchambers@eht.org.uk

**Paul J. Ciclitira**
Gastroenterology Unit
The Rayne Institute (KCL)
St Thomas' Hospital
Lambeth Palace Road
London
SE1 7EH
UK

**Paul O. Collinson**
St George's Hospital and Medical School
St George's Hospital
Cranmer Terrace
London
SW17 0BE
UK
poctrop@poctrop.demon.co.uk

**Juliet E. Compston**
Department of Medicine
Level 5 Box 157
Addenbrooke's Hospital
Cambridge
CB2 2QQ
UK
jec1001@cam.ac.uk

**Laurence M. Demers**
Pennsylvania State University College of Medicine
The M.S. Hershey Medical Center
500 University Drive
Hershey
PA 17033
USA

**Peter T. Donaldson**
Centre for Liver Research
The School of Clinical Medical Sciences
University of Newcastle
Newcastle upon Tyne
NE2 4HH
UK
p.t.donaldson@ncl.ac.uk

**Richard Eastell**
Division of Clinical Science (NGHT)
Section of Medicine
Bone Metabolism Group
Northern General Hospital
Herries Road
Sheffield
S5 7AU
UK
r.eastell@sheffeld.ac.uk

**H. Julia Ellis**
Gastroenterology Unit
The Rayne Institute (KCL)
St Thomas' Hospital
Lambeth Palace Road
London
SE1 7EH
UK
julia.ellis@kcl.ac.uk

**Vincent C. Emery**
Department of Virology
Royal Free and University College Medical School
Rowland Hill Street
London
NW3 2PF
UK

**Nelson Fausto**
Department of Pathology
University of Washington School of Medicine
Seattle
WA 98195
USA
nfausto@u.washington.edu

**Anthony J. FitzGerald**
Department of Histopathology
Hammersmith Hospital
Du Cane Road
London
W12 0NN
UK

**Jocelyn S. Fraser**
Gastroenterology Unit
The Rayne Institute (KCL)
St Thomas' Hospital
Lambeth Palace Road
London
SE1 7EH
UK

**Peter Fraunberger**
Institute of Clinical Chemistry
Klinikum Grosshadern
LM-University of Munich
Munich
Germany

**Patrick Garnero**
Unité INSERM 403
Hôpital E Herriot
Pavillion F
69437 Lyon Cedex 03
France
Patrick.garnero@synarc.com

**Lutz Graeve**
Institute of Biochemistry
RWTH Aachen
Pauweisstrasse 30
D-52074 Aachen
Germany

**Paul D. Griffiths**
Department of Virology
Royal Free and University College Medical School
Rowland Hill Street
London
NW3 2PF
UK

**Pavel Gromov**
Department of Medical Biochemistry
University of Aarhus
Ole Worms Allé
Building 170
DK-800 Aarhus C
Denmark

**Irina Gromova**
Department of Medical Biochemistry
University of Aarhus
Ole Worms Allé
Building 170
DK-800 Aarhus C
Denmark

**Jean-Pierre Grünfeld**
Service de Néphrologie
Hôpital Necker
161 rue de Sèvres
75743 Paris Cedex 15
France
jean-pierre.grunfeld@nck.ap-hp-paris.fr

**Serge Haan**
Institute of Biochemistry
RWTH Aachen
Pauweisstrasse 30
D-52074 Aachen
Germany

**Claus Hammer**
Institute of Surgical Research
Klinikum Grosshadern
LM-University of Munich
Munich
Germany
Hammer@icf.med.uni-muenchen.de

**Stephanie Hammer**
Institute of Surgical Research
Klinikum Grosshadern
LM-University of Munich
Munich
Germany

**Sheila Hart**
Division of Clinical Science (NGHT)
Section of Medicine
Bone Metabolism Group
Northern General Hospital
Herries Road
Sheffield
S5 7AU
UK

**Aycan F. Hassan-Walker**
Department of Virology
Royal Free and University College Medical School
Rowland Hill Street
London
NW3 2PF
UK
aycan@rfhsm.ac.uk

**Pekka Hayry**
Transplantation Laboratory
PO Box 21 (Haartmaninkatu 3)
00014 University of Helsinki
Helskinki
Finland
pekka.hayry@helsinki.fi

**Peter C. Heinrich**
Institute of Biochemistry
RWTH Aachen
Pauweisstrasse 30
D-52074 Aachen
Germany
heinrich@rwth-aachen.de

**Gudrun Höbbel**
Institute of Surgical Research
Klinikum Grosshadern
LM-University of Munich
Munich
Germany

**David W. Holt**
The Analytical Unit
St George's Hospital Medical School
London
SW17 0BE
UK

**Robin D. Hughes**
Institute of Liver Studies
Guy's, King's and St Thomas' School of Medicine
Bessemer Road
London
SE5 9PJ
UK
robin.hughes@kcl.ac.uk

**Humphrey J. F. Hodgson**
Royal Free and University College School of Medicine
Rowland Hill Street
London
NW3 2PF
UK
h.hodgson@rfc.ucl.ac.uk

**Tor Ingebrigtsen**
Department of Neurosurgery
Tromsø University Hospital
N-9038 Tromsø
Norway
tor.ingebrigtsen@rito.no

**Atholl Johnston**
Clinical Pharmacology
St Bartholomew's Hospital and the Royal London School of Medicine and Dentistry
Charterhouse Square
London
EC1M 6BQ
UK
A.Johnston@mds.qmw.ac.uk

**Richard Jones**
Directorate of Pathology
Leeds Teaching Hospitals NHS Trust and University of Leeds
Leeds
LS1 3EX
UK
r.g.jones@leeds.ac.uk

**Arja Jukkola**
Department of Clinical Chemistry, Oncology and Pathology
University of Oulu
PO Box 5000
FIN-90014 University of Oulu
Oulu
Finland

**Juan Carlos Kaski**
Coronary Artery Disease Unit
Department of Cardiological Sciences
St George's Hospital Medical School
Cranmer Terrace
London
SW17 0BE
UK
jkaski@sghms.ac.uk

**Saila Kauppila**
Department of Clinical Chemistry, Oncology and Pathology
University of Oulu
PO Box 5000
FIN-90014 University of Oulu
Oulu
Finland

**Bertrand Knebelmann**
Service de Néphrologie
Hôpital Necker
161 rue de Sèvres
75743 Paris Cedex 15
France

**Harry A. Kuiper**
RIKILT (National Institute for Quality Control of Agricultural Products)
Wageningen University and Research Centre
Wageningen
The Netherlands

**John M. Land**
Neurometabolic Unit
Institute of Neurology and National Hospital
Queen Square
London
UK
jland@ion.ucl.ac.uk

**Fin Stolze Larsen**
Department of Haematology
Rigshospitalet
Copenhagen University Hospital
Copenhagen
Denmark
stolze@post3.tele.dk

**Johannes Mair**
Department of Internal Medicine
Division of Cardiology
University of Innsbruck
Anichstrasse 35
A-6020 Innsbruck
Austria
Johannes.Mair@uibk.ac.at

**Michael P. Manns**
Department of Gastroenterology and Hepatology
Medical School of Hannover
Carl-Neuberg Strasse 1
D-30625 Hannover
Germany
manns.michael@mh-hannover.de

**Eevastiina Marjoniemi**
Department of Clinical Chemistry, Oncology and Pathology
University of Oulu
PO Box 5000
FIN-90014 University of Oulu
Oulu
Finland

**Astrid Martens**
Institute of Biochemistry
RWTH Aachen
Pauweisstrasse 30
D-52074 Aachen
Germany

**Basil F. Matta**
Perioperative Care Services
Addenbrooke's Hospital
Hills Road
Cambridge
CB2 2QQ
UK
basil@bmatta.demon.co.uk

**Bruno Meiser**
Department of Cardiothoracic Surgery
Klinikum Grosshadern
LM-University of Munich
Munich
Germany

**Jukka Melkko**
Department of Clinical Chemistry, Oncology and Pathology
University of Oulu
PO Box 5000
FIN-90014 University of Oulu
Oulu
Finland

**Carl E. Mogensen**
Medical Department M
Aarhus Kommunehospital
Aarhus University Hospital
DK-800
Aarhus
Denmark
akh.grpozs.com@aaa.dk

**R. Andrew Moore**
Pain Research and Nuffield Department of Anaesthetics
The Churchill
Oxford
OX3 7LJ
UK
andrew.moore@pru.ox.ac.uk

**Gerhard Müller-Newen**
Institute of Biochemistry
RWTH Aachen
Pauweisstrasse 30
D-52074 Aachen
Germany

**Ariane Nimmesgern**
Institute of Biochemistry
RWTH Aachen
Pauweisstrasse 30
D-52074 Aachen
Germany

**Hubert P. J. M. Noteborn**
RIKILT (National Institute for Quality Control of Agricultural Products)
Wageningen University and Research Centre
Wageningen
The Netherlands

**Mark Nutley**
Department of Surgery
University of Calgary and Foothills Medical Centre
Calgary
Alberta
Canada

**Petra Obermayer-Straub**
Department of Gastroenterology and Hepatology
Medical School of Hannover
Carl-Neuberg Strasse 1
D-30625 Hannover
Germany

**Michael Oellerich**
Abteilung Klinische Chemie
Zentrum Inner Medizin
Georg-August-Universtät
Göttingen
Germany
moeller@med.uni-goettingen.de

**Morten Østergaard**
Department of Medical Biochemistry
University of Aarhus
Ole Worms Allé
Building 170
DK-800 Aarhus C
Denmark

**Timo Paavonen**
Transplantation Laboratory
PO Box 21 (Haartmaninkatu 3)
00014 University of Helsinki
Helskinki
Finland

**Hildur Pálsdóttir**
Department of Medical Biochemistry
University of Aarhus
Ole Worms Allé
Building 170
DK-800 Aarhus C
Denmark

**Ad A. C. M. Peijnenburg**
RIKILT (National Institute for Quality Control of Agricultural Products)
Wageningen University and Research Centre
Wageningen
The Netherlands
A.A.C.M.Peijnenburg@rikilt.dlo.nl

**John Pernow**
Department of Cardiology
Karolinska Hospital
S-171 76 Stockholm
Sweden
jpw@cardio.ks.se

**Christopher P. Price**
Bayer Diagnostics Division
Stoke Court
Stoke Poges
Berks
SL2 4LY
UK
c.p.price@mds.qmw.ac.uk

**Robert G. Price**
Division of Life Sciences
King's College London
150 Stamford Street
London
SE1 8WA
UK
r.price@kcl.ac.uk

**Juha Risteli**
Department of Clinical Chemistry, Oncology and Pathology
University of Oulu
PO Box 5000
FIN-90014 University of Oulu
Oulu
Finland
juha.risteli@oulu.fi

**Leila Risteli**
Department of Clinical Chemistry, Oncology and Pathology
University of Oulu
PO Box 5000
FIN-90014 University of Oulu
Oulu
Finland

**Simon P. Robins**
Skeletal Research Unit
Rowett Research Institute
Bucksburn
Aberdeen
AB21 9SB
UK
S.Robins@rri.sari.ac.uk

**Bertil Romner**
Department of Neurosurgery
Lund University Hospital
S-22 185 Lund
Sweden
bertil.romner@neurokir.lu.se

**Marlene L. Rose**
National Heart and Lung Institute
Heart Science Centre
Harefield Hospital
Harefield
Middlesex
UB9 6JH
UK
marlene.rose@harefield.nthames.nhs.uk

**Fred Schaper**
Institute of Biochemistry
RWTH Aachen
Pauweisstrasse 30
D-52074 Aachen
Germany

**Jochen Schmitz**
Institute of Biochemistry
RWTH Aachen
Pauweisstrasse 30
D-52074 Aachen
Germany

**Markus J. Seibel**
University of Sydney
Department of Endochinology and Metabolism C64
Level 6 – Medical Centre
Concord Hospital
Sydney
2139 NSW
Australia
Markus.Seible@
email.cs.nsw.gov.au

**Linda D. Sharples**
MRC Biostatistics Unit
University Forvie Site
Robinson Way
Cambridge
CB4 3EU
UK
linda.sharples@mrc-bsu.cam.ac.uk

**Elmar Siewert**
Institute of Biochemistry
RWTH Aachen
Pauweisstrasse 30
D-52074 Aachen
Germany

**Vinijar Srikanthan**
Department of Clinical Biochemistry
St Bartholomew's and The Royal London Hospitals
West Smithfield
London
EC1A 7BE
UK

**Eero Taskinen**
Transplantation Laboratory
PO Box 21 (Haartmaninkatu 3)
00014 University of Helsinki
Helskinki
Finland

**Peter M. Timms**
Department of Clinical Biochemistry
St Bartholomew's and The Royal London Hospitals
West Smithfield
London
EC1A 7BE
UK
p.m.timms@mds.qmw.ac.uk

**J. Michael Tredger**
Institute of Liver Studies
Guy's, King's and St Thomas' School of Medicine
Bessemer Road
London
SE5 9PJ
UK
michael.tredger@kcl.ac.uk

**Andrew K. Trull**
Department of Pathology
Papworth Hospital
Papworth Everard
Cambridge
CB3 8RE
UK
Andrew.Trull@papworth-tr.anglox.nhs.uk

**Jonathan D. Tugwood**
AstraZeneca
Macclesfield Site
Charterway
Silk Road Business Park
Macclesfield
Cheshire
SK10 2NA
UK
jonathan.tugwood@astrazeneca.com

**Andre G. Uitterlinden**
Department of Internal Medicine/
Department of Epidemiology and Biostatistics
Erasmus University Medical School
Rotterdam
The Netherlands
uitterlinden@endov.fgg.eur.nl

**Nicholas A. Wright**
Histopathology Unit
Imperial Cancer Research Fund
Hammersmith Hospital
44 Lincoln Inn Fields
London
WC2A 3PX
UK
n.wright@ic.ac.uk

**Serdar Yilmaz**
Division of Transplantation
Department of Surgery
University of Calgary and Foothills Medical Centre
Calgary
Alberta
Canada

# Preface

Biological markers or 'biomarkers' of organ damage and dysfunction occupy a central position in the armamentarium of the clinician that is used for the screening, diagnosis and management of disease. Our knowledge of the pathophysiological basis of individual diseases continues to increase inexorably and the discoveries emanating from the Human Genome Project are set to enhance this knowledge immeasurably. Understanding the aetiopathogenesis of changes that take place in individual tissues, organs or compartments of the body can help in the search for markers that reflect these changes. Some of these changes may be directly related to the pathological abnormality while others might be a secondary consequence of the abnormality.

Basic research into the pathophysiology of a disease provides the foundation of knowledge that can lead to the discovery of valuable biomarkers. This foundation can also act as the starting point for the discovery of pharmaceutical interventions. Increasingly, with a more systematic approach to biomarker development and drug discovery, we are seeing the measurement of the biomarker playing a greater role in monitoring the efficacy and/or side effects of the therapeutic intervention. From a clinical standpoint, this can have a major benefit in assessing compliance with therapy, which is acknowledged to be one of the key determinants of efficacy, especially when there is no other ready means to judge the patient's response.

The discovery of a new biomarker is complemented by the development and validation of appropriate analytical technology. The appropriateness can be viewed in relation to a number of performance characteristics and the mode of delivery. Thus, the test may be delivered on an automated analytical platform from a centralized laboratory facility or by means of a point of care testing device. The ability to deliver the test at the point of care, with the immediacy of response, may be a key factor in ensuring the clinical utility of the test. The basic analytical characteristics include imprecision and inaccuracy which, together with the issues surrounding the biological variation and stability of the biomarker, constitute the core components of technical performance.

The diagnostic performance of the test must then be established by comparing biomarker measurements in appropriate populations of patients and controls in the relevant clinical setting. A diagnostic test is one element of a process which begins with a clinical question; the ideal test is one which provides an unequivocal answer to the question and enables a decision to be made on what action must be taken next. When effective action is then taken, a positive outcome or benefit can accrue to the patient, the healthcare system and society as a whole. Unfortunately, few tests are ideal in this respect and the medical practitioner usually has to appreciate the limited diagnostic performance of a test in order to make a balanced interpretation of the results in the context of the patient's clinical status.

The above description of the technical and clinical validation of a test applies to most research into diagnostic tests today. However, this is not the end of the process, as it is also necessary to consider the decision-making process associated with the introduction of a new test into the routine laboratory. In order to justify investment – from a clinical, operational and economic standpoint – it is not only important to prove that the test meets required standards of diagnostic performance but also that it delivers a positive clinical, operational or economic outcome – and, preferably, all three! Incorporating these outcome metrics into the evaluation of laboratory diagnostics provides a more rigorous and holistic standard of healthcare than is often embraced by laboratory medicine and illustrates the importance of integrating laboratory medicine fully into clinical practice – in terms of investment, delivery and quality management.

This book is the first attempt to review the current literature on biomarkers from an evidence-based and clinical outcomes perspective. It represents a distillation of the presentations made at the 'EMBODY 2000' conference held in Cambridge, England. It covers selected key areas – a comprehensive book would have necessitated a whole textbook! However, each of the authors, all clinical scientists of international repute, was asked to review the literature in their chosen field from the perspective of the impact of biomarkers on clinical outcome. We hope that this initiative, both the meeting and the book, represents a watershed in the literature on biomarkers, focusing more attention on clinical outcomes.

It is self-evident, but still worth stating, that a positive outcome will not be achieved unless the right test is requested in the first instance and the result acted upon with the relevant therapeutic strategy. This book, therefore, is not only relevant to professionals in laboratory medicine but also to the users of biomarkers.

**Andrew K. Trull**
**Laurence M. Demers**
**David W. Holt**
**Atholl Johnston**
**J. Michael Tredger**
**Christopher P. Price**

# Part 1

# Assessing and utilizing the diagnostic or prognostic power of biomarkers

1

# Evidence-based medicine: evaluation of biomarkers

R Andrew Moore

Pain Research and Nuffield Department of Anaesthetics, University of Oxford, Oxford, UK

## Evidence-based medicine

Evidence-based medicine (EBM) has been described as the '*conscientious, explicit and judicious use of current best evidence in making decisions about the care of individual patients*' [1]. Because there are so many biomedical journals (perhaps as many as 30000), the chance of any practitioner being aware of all the developments of interest is vanishingly small. The philosophy of EBM, therefore, extends into ways of summarizing information to make it understandable and useful. The key tool is the systematic review, and most work on systematic reviews, and indeed on EBM, has concentrated on treating disease.

### Systematic review

Reviews are called systematic when they include a thorough search for all published (and sometimes unpublished) information on a topic. Empirical observation in systematic reviews of treatment efficacy demonstrates several sources of bias occurring because of the architecture of study design. The ones we know of are:

| | |
|---|---|
| Randomization | Nonrandomized studies can overestimate treatment effects by up to 40%, or even change the conclusions of a review. Including only randomized studies is likely to be sensible for reviews of the effectiveness of treatments. |
| Blinding | Open (nonblinded) studies overestimate treatment effects by about 17%. |
| Quality | Studies of lower reporting quality overestimate treatment effects. |
| Quantity | Small studies can overestimate treatment effects. |

Duplication — Trials may be reported more than once. This may be legitimate, but is often incorrect and without cross-referencing. Unrecognized duplicate publications can lead to an overestimation in treatment effects of 20%.

Now, not all of these sources of bias will occur in each circumstance, but some will, and there may be others that are yet to be identified. What the systematic review process teaches us about trials of effectiveness is that there are many sources of potential bias, and we may not know all of them. It is notable is that every one we know of tends to overestimate the effects of treatment. There are other factors that may be important as potential sources of bias, particularly issues relating to the validity of experimental design in specific clinical situations.

Since systematic reviews concentrate on all the worthwhile published material on a topic, they provide the basis for a fresh look at where we are. One of their main results is to refresh the research agenda. A particular example is the increasing concentration on outcomes – the change in a disease state that is worthwhile for patients, their carers or the healthcare system. All too often, research concentrates on what is measurable, rather than what is meaningful. The large, simple, clinical trial with patient-defined outcomes may be one of the most important developments of EBM.

## Size

Clinical trials are performed in order to tell whether one treatment is better than another. The statistical power of the trial is calculated on the basis of being able to say with confidence that there is a difference. It is the direction of the effect that is being measured. However, most of the time what we really want to know is the magnitude of the effect of treatment. To do this, we need much more information – perhaps 10 times as many patients need to be studied.

Figure 1.1 shows the results of 56 meta-analyses of placebo in about 12000 patients in acute pain trials [2]. Overall, 18% of patients given placebo had more than 50% pain relief over 6 hours. All trials in the meta-analyses were randomized, all were double blind and all had the same outcomes measured in the same way. The variability with small samples is huge, from 0% to nearly 60%. Only when the sample is above 1000 patients given placebo is the true rate measured.

This is just one example of how small studies can be affected by random chance. This should not be surprising: calculating confidence intervals around small samples will demonstrate that uncertainty is large with small samples. However, it serves to illustrate the power of random variation with the use of small samples, and why it is dangerous to extrapolate from a single small trial to clinical practice.

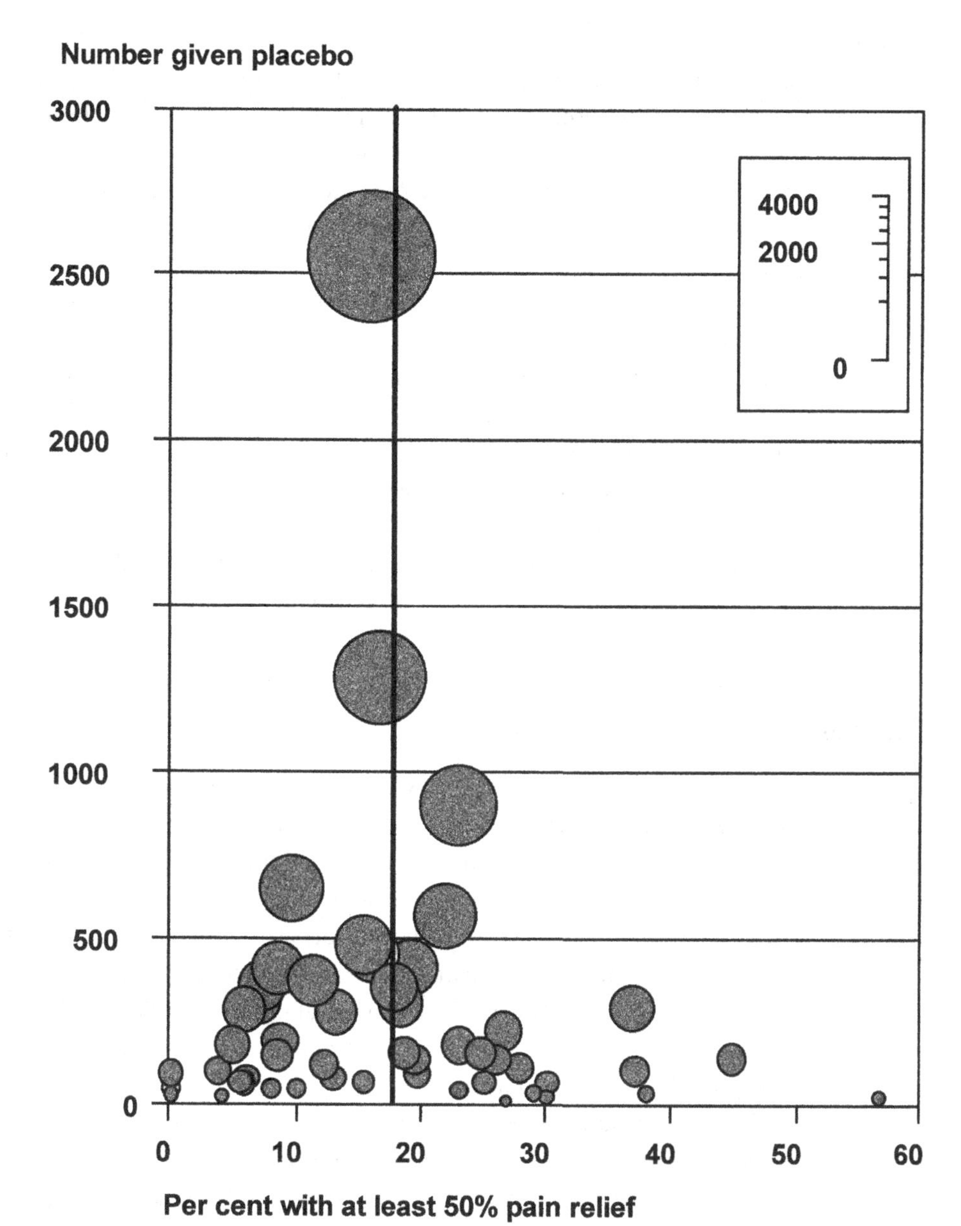


Figure 1.1 Per cent of patients with at least 50% pain relief from meta-analyses of acute pain studies. Each symbol represents one meta-analysis; all trials were randomized and double blind and with the same outcome measured over the same time (2). Size of the symbol is proportional to the number of patients included. The vertical line shows the overall average response (18%) from over 12 000 placebo patients.

### Expressing results

EBM has a real problem in how to express the results of research so that they can be understood and used. Statistical significance is in itself an unhelpful output, as are odds ratios, risk ratios, relative risks, weighted mean differences or effect sizes. The simple fact is that few people understand them and even fewer can use them.

What catapulted EBM into the real world was the use of the number-needed-to-treat (NNT). This is the inverse of the absolute risk reduction and describes the therapeutic effort required to produce one patient with the required clinical outcome [3]. It has proved particularly useful when there are many different treatments, as in analgesics for pain. By producing tables of NNTs for analgesics, choice can be made in terms of efficacy, harm and cost.

However, better understanding of the requirements of large samples to assess clinical outcomes accurately [4] is likely to lead to even simpler outcomes than the NNT. The future holds the prospect of being able to say, with confidence, that a given treatment in patients with a given disease and severity will lead to a successful outcome in x% – which would be understandable by doctors, patients and policy makers.

## Evidence-based laboratory medicine

There are various types of evidence we accept for laboratory tests and biomarkers: evidence about the analytical performance of an assay; evidence about quality control in the laboratory and quality assurance from external schemes; and evidence about issues like sensitivity and specificity in particular clinical circumstances. What we rarely have, though, is evidence that the use of a laboratory test can, for a given patient or group of patients, make a clinically relevant difference to the diagnosis or treatment. Evidence-based laboratory medicine (EBLM) has to encompass all of these types of evidence, of course, but the judgement will increasingly be made on clinical outcomes.

Whether systematic reviews will be helpful for EBLM, as they have been for treatments, is questionable, however. One description of levels of evidence commonly used for studies of diagnostic tests is shown in Table 1.1. The keys to good quality have been said to be independence, masked comparison with a reference standard and consecutive patients from an appropriate population. Lower quality comes from inappropriate populations and comparisons that are not masked or with different reference standards. Until recently, we lacked any empirical or theoretical evidence about the levels of bias that any of these study architectures can impart.

A new contribution from Holland [5] provides the missing link. The authors searched for and found 26 systematic reviews of diagnostic tests with at least five included studies. Only 11 could be used in their analysis, because 15 were either not

**Table 1.1.** Levels of evidence for studies of diagnostic methods

| Level | Criteria |
|---|---|
| 1 | An independent, masked comparison with reference standard among an appropriate population of consecutive patients |
| 2 | An independent, masked comparison with reference standard among nonconsecutive patients or confined to a narrow population of study patients |
| 3 | An independent, masked comparison with an appropriate population of patients, but reference standard not applied to all study patients |
| 4 | Reference standard not applied independently or masked |
| 5 | Expert opinion with no explicit critical appraisal, based on physiology, bench research or first principles |

systematic in their searching or did not report any sensitivity or specificity. Data from the remainder were subjected to mathematical analysis, to investigate whether the presence or absence of some item of proposed study quality made a difference to the perceived value of the test.

There were 218 individual studies, only 15 of which satisfied all eight criteria of quality that this analysis concerned. Thirty per cent fulfilled at least six of eight criteria. To evaluate bias, the authors calculated the relative diagnostic odds ratio by comparing the diagnostic performance of a test in those studies that failed to satisfy the methodological criterion with the performance of the test in studies that did meet this criterion. Overestimation of effectiveness (positive bias) of a diagnostic test was shown by a lower confidence interval for the relative diagnostic odds ratio of more than 1.

The results are shown in Table 1.2. Use of different reference tests, lack of blinding and lack of a description of either the test or the population in which the test was studied led to positive bias. However, the largest factor leading to positive bias was evaluation of a test in a group of patients already known to have the disease and a separate group of normal patients – called a case-control study in the paper [5].

There are also pointers to good practice in the publication of articles on diagnostic tests. The authors of a most important paper [6] set out seven methodological standards (Table 1.3). They then looked at papers published in the *Lancet*, *British Medical Journal*, *New England Journal of Medicine* and *Journal of the American Medical Association* from 1978 through 1993 to see how many reports of diagnostic tests meet these standards. Between 1978 and 1993, they found 112 articles, predominantly on radiological tests and immunoassays. Few of the standards were met consistently – ranging from 51% avoiding workup bias down to 9% reporting accuracy in subgroups (Table 1.3). While there was an overall improvement over

**Table 1.2.** Empirical evidence of bias in diagnostic test studies of different architecture

| Study characteristic | Relative diagnostic odds ratio (95% CI) | Description |
|---|---|---|
| Case-control | 3.0 (2.0–4.5) | A group of patients already known to have the disease compared with a separate group of normal subjects |
| Different reference tests | 2.2 (1.5–3.3) | Different reference tests used for patients with and without the disease |
| Not blinded | 1.3 (1.0–1.9) | Interpretation of test and reference is not blinded to outcomes |
| No description of test | 1.7 (1.1–1.7) | Test not properly described |
| No description of population | 1.4 (1.1–1.7) | Population under investigation not properly described |
| No description of reference | 0.7 (0.6–0.9) | Reference standard not properly described |

*Note:*
The relative diagnostic odds ratio indicates the diagnostic performance of a test in studies failing to satisfy the methodological criterion relative to its performance in studies with the corresponding feature [5].

time for reports to score on more standards, even in the most recent period studied only 24% met up to four standards, and only 6% up to six.

Most diagnostic test evaluations are structured to examine patients with a disease compared with those without the disease – a case-control design. Astonishingly, few studies are performed according to the highest standard in Table 1.1. The studies which have been published are seriously flawed, as Read et al. [6] have demonstrated. It must be questioned, therefore, whether any systematic review of diagnostic tests is worthwhile.

## Size

Just as large samples are needed to overcome the random effects of chance for treatments, so they are also needed for tests. An example is the controversy over falling sperm counts. A meta-analysis [7] collected 61 studies on sperm counts published between 1938 and 1990. Almost one-half of these studies (29/61) studied fewer than 50 men. The smallest number was nine and the largest 4435 men. Only 2% of the data on nearly 15000 men was collected before 1970, in small studies. Figure 1.2 shows the variability by size. The overall mean sperm count was 77 million/ml, but small individual studies recorded means from 40 to 140 million/ml. Only large studies correctly estimated the overall mean, and any temporal relationship is spurious because the old studies were small.

**Table 1.3.** Standards of reporting quality for studies of diagnostic tests

| Reporting standard | Background | Criteria | Per cent meeting standard |
|---|---|---|---|
| **Spectrum composition** | The sensitivity and specificity of a test depend on the characteristics of the population studied. Change the population and you change these indices. Since most diagnostic tests are evaluated on populations with more severe disease, the reported values for sensitivity and specificity may not be applicable to other populations with less severe disease in which the test will be used. | For this standard to be met, the report had to contain information on any three of these four criteria: age distribution, sex distribution, summary of presenting clinical symptoms and/or disease stage, and eligibility criteria for study subjects. | 27 |
| **Pertinent subgroups** | Sensitivity and specificity may represent average values for a population. Unless the condition for which a test is to be used is narrowly defined, then the indices may vary in different medical subgroups. For successful use of the test, separate indices of accuracy are needed for pertinent individual subgroups within the spectrum of tested patients. | This standard is met when results for indices of accuracy were reported for any pertinent demographic or clinical subgroup (for example, symptomatic versus asymptomatic patients). | 9 |
| **Avoidance of workup bias** | This form of bias can occur when patients with positive or negative diagnostic test results are preferentially referred to receive verification of diagnosis by the gold standard procedure. | For this standard to be met in cohort studies, all subjects had to be assigned to receive both the diagnostic test and the gold standard verification either by direct procedure or by clinical follow up. In case-control studies, credit depended on whether the diagnostic test preceded or followed the gold standard procedure. If it preceded, credit was given if disease verification was obtained for a consecutive series of study subjects regardless of their diagnostic test result. If the diagnostic test followed, credit was given if test results were stratified according to the clinical factors which evoked the gold standard procedure. | 51 |

**Table 1.3.** *(cont.)*

| Reporting standard | Background | Criteria | Per cent meeting standard |
|---|---|---|---|
| **Avoidance of review bias** | This form of bias can be introduced if the diagnostic test or the gold standard is appraised without precautions to achieve objectivity in their sequential interpretation – like blinding in clinical trials of a treatment. It can be avoided if the test and gold standard are interpreted separately by persons unaware of the results of the other. | For this standard to be met in either prospective cohort studies or case-control studies, a statement was required regarding the independent evaluation of the two tests. | 43 |
| **Precision of results for test accuracy** | The reliability of sensitivity and specificity depends on how many patients have been evaluated. Like many other measures, the point estimate should have confidence intervals around it, which are easily calculated. | For this standard to be met, confidence intervals or standard errors must be quoted, regardless of magnitude. | 12 |
| **Presentation of indeterminate test results** | Not all tests come out with a black or white, yes/no, answer. Sometimes they are equivocal, or indeterminate. The frequency of indeterminate results will limit a test's applicability, or make it cost more because further diagnostic procedures are needed. The frequency of indeterminate results and how they are used in calculations of test performance represent critically important information about the test's clinical effectiveness. | For this standard to be met, a study had to report all of the appropriate positive, negative or indeterminate results generated during the evaluation and whether indeterminate results had been included or excluded when indices of accuracy were calculated. | 26 |
| **Test reproducibility** | Tests may not always give the same result – for a whole variety of reasons of test variability or observer interpretation. The reasons for this, and its extent, should be investigated. | For this standard to be met in tests requiring observer interpretation, at least some of the tests should have been evaluated for a summary measure of observer variability. For tests without observer interpretation, credit was given for a summary measure of instrument variability. | 26 |

*Source:* From Read et al. [6].

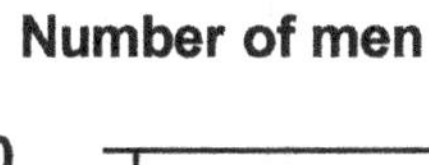

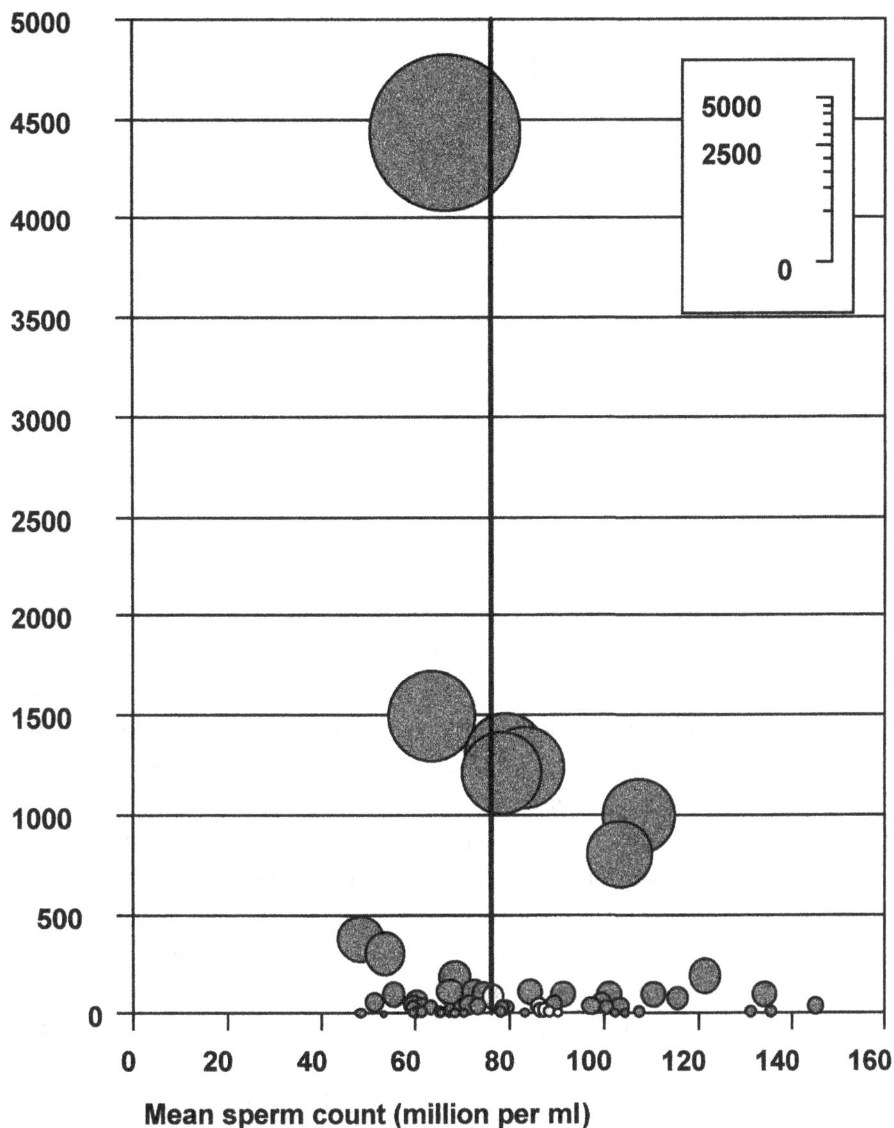


Figure 1.2 Mean sperm counts from individual studies in a meta-analysis (7). Each symbol represents one study. Size of the symbol is proportional to the number of patients included. The vertical line shows the overall mean (77 million/ml) from over 15 000 men.

**Table 1.4.** Frequency of use of methods of assessing test accuracy (50 physicians in each category)

| | Bayesian method | ROC curve | Likelihood ratios |
|---|---|---|---|
| Specialist physician | 5 | 1 | 1 |
| Generalist physician | 2 | 0 | 1 |
| Paediatrician | 1 | 1 | 0 |
| General surgeon | 0 | 1 | 0 |
| Family practice | 0 | 0 | 0 |
| Obstetrics/gynaecology | 0 | 0 | 0 |
| Overall percentage | 3% | 1% | 1% |

*Source:* From Read et al. [8].

## Expressing results

How do doctors use information about tests? A survey of US doctors [8] showed that almost none of them use these terms in any formal way. A stratified random sample of physicians in six specialties with direct patient care (at least 40% of time with patients) across the USA was determined by researchers at Yale. These physicians were then contacted, with a 10-minute telephone survey about their attitudes to formal methods of test use. There were 10 questions, reproduced in an appendix to the paper. A typical question was: *'Do you use test sensitivity and specificity values when you order tests or interpret test results?'*

There were 300 physicians in the final sample, 50 in each specialty. Few of them used formal methods of assessing test accuracy (Table 1.4). Bayesian methods were used by 3%, and receiver-operating characteristic (ROC) and likelihood ratio data by 1% each. Although as many as 84% said they used sensitivity and specificity at some time, from adopting the use of a new test to using them when interpreting a diagnostic test result, this was almost always done in an informal way.

The authors make a number of salient points. Firstly, information on test accuracy must be 'instantly available' when tests are ordered. Secondly, formal training needs to be improved. Thirdly, published information is mostly useless, because it usually fails to reflect the patient population in which it is being used. They might also have gone further and said that we need new and better ways of expressing the results of diagnostic tests. Sensitivity and specificity, or ROC curves or likelihood ratios are not understood or used by doctors. This remains a huge challenge.

## New paradigms for diagnostic test evaluation

When a diagnostic decision has to be made, physicians are often faced with a bewildering array of facts, which will include aspects of the unique biology of the individual patient, signs and symptoms from the clinical examination, and results from physiological, imaging and laboratory tests. What usually happens is that a diagnosis is made through a gestalt process where synthesis of all this information takes place, using weightings for individual pieces of information which are unformulated and unstated. How may this be changed?

We have some examples of where factors used to make a diagnosis have been evaluated retrospectively in order to determine statistical and clinical significance, and then tested prospectively in randomized controlled trials. The best of these are the Ottawa ankle and knee rules for determining whether a patient with an ankle or knee injury needs an X-ray [9, 10].

The way the knee rule was worked out was interesting. It started by having over 1000 knee injury patients examined for 23 standardized clinical findings. The variables found to be associated with a fracture were subjected to statistical analysis to assess the strength of the association. The five findings with the strongest association were then incorporated into a clinical decision rule.

Four hospitals in Ontario were chosen to test the rule. In two control hospitals, no intervention was made. In two intervention hospitals, the knee rule was introduced by means of a brief lecture, a pocket card and wall posters in the emergency department. Staff were given regular updates of progress and difficulties in implementing the decision rule. Eligible patients were those seen in the 12 months before and after the introduction of the rule.

About 4000 people were seen with knee injury in the four hospitals over the two periods. There was a 26% reduction in the proportion of patients sent for X-ray in the two intervention hospitals in the year after the implementation of the Ottawa knee rule compared with the previous year. Control hospitals showed no change. The knee rule correctly predicted all the fractures. No patient who did not have an X-ray was found subsequently to have a fracture. Time spent in the emergency department, time spent off work and overall medical costs were lower in those who did not have an X-ray than in those who did.

### CARE

One of the most exciting new developments in e-medical research is that on the Clinical Assessment of the Reliability of the Examination, which is a collaborative study of the accuracy and precision of the clinical examination (www.carestudy.com).

The all-too-common study of the accuracy and precision of the clinical examination comprises four experts examining 40 patients, the latter selected to confirm the biases and reputations of the former. The pioneering work of the US–Canadian Co-operative Research Group on the Clinical Examination reversed this trend, but even it has faced formidable problems in participation rates and patient numbers. A group of Canadians is trying to solve the problems of both numbers and clinical applicability [11]. Large (>100 clinicians enrolling >1000 patients), simple (<2 min per patient and <15 patients per participating clinician), fast (<2 weeks, with automatic data entry via the Internet) studies of the accuracy and precision of specific elements of the history and physical examination are in progress. Initial efforts led to over 160 clinicians from 20 countries joining CARE.

CARE works like this:

- Anybody, at any stage of training or experience, can join the enterprise just by signing up for it. The only prerequisites are an interest in the clinical examination and access to the Internet.
- Individuals in the collaboration nominate symptoms and signs they would like to validate (or debunk!) and broadcast them to the membership.
- Members who share an interest in this same topic come together electronically as Investigators, and proceed to design and debug the protocol and offer it to the entire collaboration.
- The membership-at-large vote with their precious time, enrolling just a few patients each and reporting their results electronically.
- Analyses are shared, *PowerPoint* summaries posted and papers published (with authorship by the Investigators, on behalf of CARE, and acknowledging every member who entered the requisite number of patients).

This could easily be the format for gaining the best evidence about laboratory tests and biomarkers, with occasional differences. Of vital importance is the potential to gather large amounts of information quickly and the combining of laboratory tests with demographic factors and signs and symptoms will be critical. Statistical analysis can sort those factors highly associated with the presence of disease, and can formulate decision rules deliverable through the electronic revolution when they are needed [12]. The effectiveness and cost-effectiveness of this can be tested in randomized controlled trials.

## Conclusions

The practice of EBM requires us to seek the best evidence with which to answer important questions about diagnosis, prognosis, therapy and other clinical and healthcare issues, and to appraise that evidence critically and consciously. The challenging task of introducing principles of EBM in diagnostic testing means a continuous process of identifying the diagnostic tests which offer the highest diagnostic efficiency and reliability and which cause fewer risks and less discomfort for patients. Compared with how much has been already achieved in the field of treatment strategies, there are vast unexplored areas in diagnostic testing.

It is uncomfortable to realize that so much of what we have done is of such poor quality. An old British political saying says: 'When you are in a hole, stop digging!' That is roughly where we are now, but the lessons of EBM and EBLM show us ways forward that can bring laboratory tests and biomarkers right to the forefront of modern medicine, through improvements in the diagnosis and management of disease.

## REFERENCES

1 Sackett, D.L., Rosenberg, W.M.C., Muir Gray, J.A., Haynes, R.B. and Richardson, W.S. (1996). Evidence-based medicine: what it is and what it isn't. *British Medical Journal*, **312**, 71–2.
2 Moore, R.A. (2000). Understanding clinical trials: what have we learned from systematic reviews? In *Proceedings of the 9th World Pain Congress.* Seattle: IASP Press, pp. 757–70.
3 McQuay, H.J. and Moore, R.A. (1997). Using numerical results from systematic reviews in clinical practice. *Annals of Internal Medicine*, **126**, 712–20.
4 Moore, R.A., Gavaghan, D., Tramèr, M.R., Collins, S.L. and McQuay, H.J. (1998). Size is everything – large amounts of information are needed to overcome random effects in estimating direction and magnitude of treatment effects. *Pain*, **78**, 209–16.
5 Lijmer, J.G., Mol, B.W. and Heisterkamp, S. (1999). Empirical evidence of design-related bias in studies of diagnostic tests. *JAMA*, **282**, 1061–6.
6 Read, M.C., Lachs, M.S. and Feinstein, A.R. (1995). Use of methodological standards in diagnostic test research: getting better but still not good. *JAMA*, **274**, 645–51.
7 Carlsen, E., Giwercman, A., Keiding, N. and Skakkebaek, N.E. (1992) Evidence for decreasing quality of semen during past 50 years. *British Medical Journal*, **305**, 609–13.
8 Read, M.C., Lane, D.A. and Feinstein, A.R. (1998). Academic calculations versus clinical judgements: practising physicians' use of quantitative measures of test accuracy. *American Journal of Medicine*, **104**, 374–80.
9 Stiell, I.G., Wells, G., Laupacis, A. et al. (1995). Multicentre trial to introduce the Ottawa ankle rules for use of radiography in acute ankle injuries. *British Medical Journal*, **311**, 594–7.
10 Stiell, I.G., Wells, G.A., Hoag, R.H. et al. (1997). Implementation of the Ottawa knee rule for the use of radiography in acute knee injuries. *JAMA*, **278**, 2075–9.
11 McAlister, F.A., Straus, S.E. and Sackett, D.L. (1999). Why we need large, simple studies of the clinical examination: the problem and a proposed solution. *Lancet*, **354**, 1721–4.
12 Jones, R.G. (1999). Informatics in point-of-care testing. In *Point-Of-Care Testing*, eds. C.P. Price and J.M. Hicks, pp. 175–96. Washington: AACC Press.

2

# Development of biomarkers: the industrial perspective

Andy Bufton

Abbott Diagnostics Division, Maidenhead, Berkshire, UK

One of the axioms by which successful sales organizations determine a decision maker among their potential clientele is to ask whether that person is the MAN. Not a politically incorrect measure of managerial position, the term MAN indicates that the individual has the Money, the Authority to spend it and the Need to do so. The diagnostics industry, prior to the spate of mergers and acquisitions of the recent decade, was made up of companies generally dedicated to the supply of products focused primarily on a particular analytical discipline, be it clinical chemistry, haematology, microbiology or histopathology. These serviced an expanding market-place where the sales teams would visit the MAN in the laboratory handling the particular discipline. Here, morale was high – the laboratory order book could be invoked, with authority to purchase, to meet the needs of professionals running a demand-led service.

During the following years, mergers and acquisitions led to a number of companies offering products across all disciplines, and, meanwhile, many laboratories reduced or removed the boundaries between disciplines. Throughout the world, the diagnostics industry continues to utilize the larger part of its sales and marketing resource in supporting and visiting the laboratory professional, and many companies have failed to recognize that decision making has moved away from the professionals working there. The money available has been reduced, and it is managed by others. The authority to make a purchasing decision lies outside of the laboratory and the need is to reduce workload, expenditure and, indeed, quality, from an excellent to an adequate level commensurate with reducing budgets and not insignificant deskilling of the laboratory workforce.

Market segmentation exercises frequently differentiate by descriptors, and the companies in the in vitro diagnostics (IVD) industry continue to describe their market segments in terms of size, according to workload, bed numbers in a hospital or academic status. Alternatively, the distinction is made between the different types of commercial and public institutions, and the scientific disciplines themselves continue to be cited as different market segments.

This article argues that the key differentiating factor between the segments of a market, however, is buyer motivation, not description. If this approach is applied to the laboratory professionals, we discover that the motivation has changed over the past two decades from promoting and developing their service, to cost reduction, consolidation and reduction of demand. The laboratory now acts as distributor of the IVD product to another MAN, the true end-user.

It is with the true end-users of the IVD medical device that this article is concerned, and the sites of use lead to recognizing user motivation.

## The laboratory

Hospital pathology laboratories provide a testing service, where short-term budgetary planning to reduce cost and concentration on the price of materials rather than value mitigate against the promotion of their service to outside organizations. It has been shown that, in the UK, there is a need for greater interaction between laboratory staff and the clinicians, whose knowledge of the availability and utility of IVDs is severely limited. In an effort to establish the current awareness of the role of IVDs, an audit was conceived and commissioned by a joint industry and professional executive group led by the British In Vitro Diagnostic Association (BIVDA), supported by the Association of Clinical Biochemists (ACB) [1].

The primary evaluation was carried out by postal audit of five stakeholder groups – laboratory scientists, general practitioners (GPs), hospital doctors, hospital managers and the commercial IVD manufacturers themselves. This was followed by a secondary phase, when the key outcomes were analysed and separate focus groups met to probe deeper into the reasons behind the recorded responses.

The strongest outcome and essentially the primary driver for the identified sub-optimal use of IVD testing was the low level of clinical laboratory testing knowledge among physicians, particularly those in a primary care role. One-half of both GPs and hospital doctors responded that they considered laboratory tests to be insufficiently accurate. This was particularly disturbing as many analytes are measured at levels of precision and accuracy far beyond those required for effective clinical management. The auditors' concerns as to the respondents' definition of 'inaccuracy' was pursued at the secondary phase, and was clarified by the hospital doctors as a surrogate expression which more appropriately represented concerns over the lack of adequate interpretation of and advice on the reported results.

More than one-half of the doctors indicated a desire for more 'definitive' tests, and showed an emerging desire for 'traffic light' laboratory testing, as opposed to its historical positioning as an adjunct in support of the doctor's diagnosis. The study clearly showed the physicians' inability to keep pace with the developments in IVD testing over the last 20 years. The study showed that physicians were seeking a service which converts a collection of numbers and statements into a precise diagnosis and prognosis, and, where appropriate, suggestions for consequent patient management. More than one-half of the responding doctors wanted a wider range of tests. Further probing resulted in the discovery of a huge gap in the awareness of GPs as to the availability of current routine tests. This gap is created, essentially, by the admission of the participating physicians, by an unstructured, unfocused and unsatisfactory education programme within the student medical syllabus. Unaware of new, disease-specific and definitive IVD tests, some, perhaps many, physicians continue to request 'old favourites' such as erythrocyte sedimentation rate, long replaced by newer, more disease-specific assays.

Costs associated with the carrying out of IVD tests were significantly overestimated by both primary and secondary care physicians. Cost was not, however, a barrier to use. There was a desire to be able to request tests by a single reference – for example, liver function tests – to allow the laboratory to include those tests it considered relevant to the pathology in question. Doctors felt that they suffered a suboptimal interpretation service from the laboratory, and there was a plea for greater interactive communication with the appropriate laboratory staff. Indeed, doctors selected the diagnostic laboratory staff as their overall primary choice as the source of learning about new IVD tests, and more than two-thirds of hospital doctors and one-half of the GPs wanted some form of access to interpretative advice from laboratory personnel on a 24-hour basis.

Physicians felt that laboratories should restructure their routine practices to reflect the changes in admission times, and the consequent therapeutic and presurgical management procedures resulting from changes in hospital practices designed to permit greater patient throughput. Hospital doctors felt that result turnaround time was less than satisfactory, wanting results from early morning samples before late afternoon.

Doctors regarded the construction of an interactive physician–laboratory communication platform as a key objective of any programme consequent on the survey's findings. Encouraged, laboratory-based participants not only accepted this but were anxious to assume the primary and co-ordinating role in providing this platform.

This author has found no evidence that these are circumstances specific to the UK. Ongoing work by the European Diagnostics Manufacturers Association [2] suggests this state of affairs is widespread across Europe.

## Testing near the patient

The benefits of rapid access to results and cost reductions are seen as the primary motivators for testing to be carried out near the patient. Many forecasts have been

**Table 2.1.** A representative list of self-testing products available on the Internet

| | |
|---|---|
| Cholesterol | Ovulation |
| Colorectal cancer | Pesticides |
| Drugs of abuse | Pregnancy |
| Glucose monitoring | Prostate status |
| Glycated haemoglobin | 'Genetic' testing |
| *Helicobacter pylori* | Inherited disease |
| Hepatitis B | Propensity for disease |
| HIV | 'DNA' testing |
| Osteoporosis | Paternity issues |

made for this market to expand rapidly, over the past two decades, based on these assumptions. The knowledge base of the attending clinician and his/her financial managers of the value and utility of the use of diagnostics are as much a defining factor at this site of use as in the laboratory. Consequently, much of the growth forecast for this area has been slow to materialize.

## Home testing

Providing convenience at a price, home testing also provides cost reductions to government-funded healthcare. In addition, it offers the opportunity for better disease management, and has been recognized in many countries by the partial or complete funding by government of products for the home monitoring of diabetes. Public awareness of IVDs is being improved by the Internet, where kits for many tests and conditions can be accessed (Table 2.1).

It has been said that the potential alliance of pressure from manufacturers and patients inevitably leads to too many tests being carried out (in the sense of an 'investigation carried out' rather than a 'procedure used to gather information about a patient to decide what action to take next') [3]. To ensure that the tests are indeed of the latter definition, the limited awareness of many clinicians as to the role and value of IVDs requires correction.

## Who is the customer?

Who, then, is the true customer of the IVD industry? Is it the laboratory staff purchasing the tests, or are these truly in the role of distributor of the end-product to another user? Is it the members of the public who buy the tests over the counter or from the Internet? Perhaps it is the physician who requests the tests, or hospital managers purchasing as part of their patient management frameworks. Or is it governments seeking healthier nations on limited healthcare budgets? It is all of these; these are the market segments, each with their own motivators. More importantly, for the industry, each has potential motivators which are poorly recognized but are to be found in the evidence base of both the peer-reviewed journals and the grey literature.

Many studies already exist as examples of the pivotal role IVDs and the laboratory professions can play in downstream economies for the healthcare budget and improvements in the quality of life for the individual members of society. In an announcement to the British Parliament on 31 March 1998, Health Secretary Frank Dobson said that one of the conditions which would get special attention in the new Health Action Zones was diabetes. 'Early diagnosis, advice, help and treatment could,' he said, 'enable patients to avoid the drastic consequences to which the disease can lead.' Responsible for over 8% of the total UK health expenditure [4], this condition remains undiagnosed in over 1 million citizens. For those who are diagnosed, the risk of sequelae can be reduced by between 40% and 76% by adequate monitoring [5, 6]. The Kings Fund and the British Diabetic Association estimate that the savings to be gained by adequate monitoring to be in excess of £500 million.

A similar argument can be supported for the early diagnosis of osteoporosis, a disease the results of which cost the American economy over US$10 billion annually. The risk of osteoporosis-related fracture is greater than the combined risks of breast, ovarian and uterine cancer, and 20% of women who suffer hip fractures die within a year [7]. With effective treatments emerging, the benefits of early prediction and prevention of this condition are clear.

Cost utility and quality of life benefits have been shown to support many other diagnostic procedures and the role of the professions associated with their use, leading to a proper recognition of the value of both to the healthcare system, and to society as a whole. The pursuit of evidence-based practice promises to provide the platform, but will ultimately succeed only if the opportunity is taken to fulfil the need and desire among medical staff to improve their knowledge of IVD through enhanced communication with their colleagues in the pathology laboratories. The pathology laboratories, in turn, are best placed to pursue, pragmatically and in a co-ordinated fashion, the evidence to support their value to healthcare.

To this end, a new paradigm is required for the collection of evidence in the case of the clinical laboratory – namely, the prospective or retrospective comprehensive audit of outcomes. Information would be collected on a large number of patients with defined clinical problems, and randomized trials utilized where the unit of randomization is based on geography, such as the location of a hospital or research group. Measurable outcomes might be in the realm of resource utilization, such as

**Table 2.2.** Emerging proprietary technologies

| | |
|---|---|
| Targeted screening | Autocyte – liquid-based prep for PAP smear slides, slide imaging technology |
| | Cytyc – liquid-based prep for PAP smear slides |
| | Digene – human papilloma virus testing in conjunction with cervical smears |
| | Visible genetics – DNA sequencing technology for genotyping, including cancer susceptibility |
| | Vysis – fluorescence in situ hybridization, a genetic analysis technology |
| Point of care | Amira medical – blood glucose metre technology |
| | Accumetrics – 2-minute monitor of platelet inhibition therapy |
| | Biosite – point of care (POC) platform, new cardiac product |
| | Diametrics – POC technology in critical care |
| | i-STAT – POC technology in critical care |

*Note:*
From [8].

length of stay per diagnosis, hospital days per 1000 of the population, and length of stay in intensive care units or cardiac care units. Intermediate outcomes might embrace the rate of screening per 1000 of the population at risk, and the rate of incidence of targeted conditions. Long-term outcomes would include patient satisfaction, the functional status of the individual, quality of life indices and rate of return to work.

The market segments offering the greatest potential future for the IVD medical device industry are, by these arguments, those motivated to use the real benefits of the products. These are governments, hospital managements, patients themselves and the clinicians whose intervention decisions can be better made in the light of the use of these products. It is necessary, however, for the industry itself, and the professions, to educate these segments as to their combined value and utility.

How does this influence the development of biomarkers by the IVD industry? It can be seen that a number of relatively small companies with proprietary technologies are arriving in those very areas where the wider and longer term benefits from their use have begun to be recognized (Table 2.2). These companies and their technologies are particularly interesting to the large IVD companies that are increasingly choosing to buy-in rather than innovate. The intellectual property rights associated with the development of assays for the prediction and prevention of disease, such as markers for the risk of coronary heart disease, are being jealously guarded by their owners. The pharmaceutical companies are recognizing that the ability to identify health risks, based on traditional or genetic-profile diagnostics, to enable the provision of tailored, preventative treatment provides an exciting opportunity for their industry [9]. The study and application of pharmacogenomics offers the opportunity to utilize drugs which have efficacy or toxicity according to the genetics of the patient, possibly releasing previously discarded compounds for use.

## The future – for the IVD industry and the professions it supplies

As industry comes to recognize the true market segments it services, development of biomarkers will be concentrated along the following lines. Firstly, for those associated with the prediction and prevention of disease, concentration will be on servicing the needs of governments with limited healthcare budgets, in order to avoid the expensive sequelae of late diagnosis. This will include markers for infection, noninvasive technologies for detection and control of diabetes, and markers to establish the risk of diseases such as osteoporosis, cancer and cardiac failure. Secondly, as the general public takes a more informed interest in its own health, the over-the-counter market will expand with new markers and technologies. It is worthy of note that, today, the two largest selling diagnostic products are over-the-counter pregnancy tests and, on prescription or reimbursed, glucose monitoring strips for home use. Thirdly, point of care technologies for use at the bedside or clinic, providing instant information on which to base medical intervention as part of standardized care pathways, will continue to develop, providing traditional markers on new platforms. Fourthly, the ability to tailor drug therapy to genetic predispositions to response through the study and implementation of pharmacogenomics draws the pharmaceutical industry into the decision-making process related to the priorities in marker research and development.

In order to harvest the full benefits of the use of these markers, both in the acute arena and in that of the prediction and prevention of disease, the professions traditionally associated with their use have an opportunity to place themselves at the centre of twenty-first-century healthcare. To do so, the stated need among physicians for an education and decision support programme developed and delivered by the laboratory professions needs to be met. The scientific endeavours of the laboratory community, traditionally communicated only among themselves, need to be brought to a wider audience, and include the cost utility (in the widest sense) of the discoveries and applications. The boundaries which resist the transfer of funding from the areas profiting from the use of these procedures back to the laboratory's budget must be strenuously questioned with health care management, politicians and patient interest groups. Already, associations, groups and working parties exist with these goals in mind, strongly supported, and often instigated, by the IVD industry.

## Conclusion

Analysts predict solid growth in the biomarker industry with the largest companies expected to outpace the overall market due to their developments or acquisitions of emerging technologies, such as nucleic acid probes, point of care and self-testing, and in situ hybridization. The key to the development of markers, however, is an understanding not only of the evidence supporting the clinical utility of their use, but that supporting their cost utility in prediction and subsequent prevention of the sequelae of occult conditions. In the acute arena, biomarkers allow hospitals to deliver more appropriate and timely care to the patient, avoiding unnecessary and more costly hospital treatment. It is these potential 'added value' components that biomarkers bring to total patient care, which in the past have been overshadowed by the pursuit of marginal advances in product performance, that require to be understood, developed and disseminated to all those that have the money, the authority and the need to maximize the effectiveness of healthcare provision.

## REFERENCES

1 Watson Biomedical (1997). *IVD in Health Care Survey*, Executive Report.
2 EDMA website, www.Edma-ivd.be
3 Kay, J.D. (1998). Review of the HTA report: near patient testing in primary care. *Bandolier*, 45, 8. (http://www.jr2.ox.ac.uk/bandolier/band45/b45-7.html)
4 King's Fund (1998). *Counting the Cost: The Real Impact of NonInsulin Dependent Diabetes*, Kings Fund Policy Institute Report.
5 The Diabetes Control and Complications Trial Research Group (1993). The effect of intensive treatment of diabetes on the development and prognosis of long-term complications in insulin-dependent diabetes mellitus. *New England Journal of Medicine* **329**, 977–86.
6 UKPDS Group (1998). Intensive blood glucose control with sulphonylureas or insulin compared with conventional treatment and risk of complications in patients with type 2 diabetes (UKPDS 33). *Lancet*, **352**, 837–53.
7 Anonymous (1993). Consensus development conference: diagnosis, prophylaxis and treatment of osteoporosis. *American Journal of Medicine*, **94**, 646–50.
8 Wilkin and Michelmore (1999). *In Vitro Diagnostics – Shaping up, a Leaner, Meaner IVD Industry*. SG Cowen Perspectives.
9 Gilham, R. (2000). Predictive medicine adding value to the Pharmaceutical Pipeline. In Proceedings of *Diagnostics in Pharmaceuticals* Seminar, Girton College, Cambridge.

3

# Statistical approaches to rational biomarker selection

Linda D Sharples

MRC Biostatistics Unit, Cambridge, UK

## Introduction

This chapter aims to present and discuss some statistical approaches to identifying and validating the relationship between diagnostic markers and their clinical end-points. Much of the work in this area has been developed in the context of clinical trials, although the principles extend naturally to cohort studies measuring exposure–end-point associations.

Informally, intermediate end-points or surrogates are laboratory measurements, X-ray results or other biological markers which are observed before the clinical end-point of interest occurs and which are useful in diagnosis, prognosis, monitoring patients or in the evaluation of the effects of an exposure or clinical intervention (E/I) to which patients are subjected (see Figure 3.1).

## Motivation for the use of intermediate end-points or surrogates

Ideally, when studying the effects of an E/I, the appropriate outcome measure is the true clinical end-point of interest (CE), which is usually survival or the incidence of a clinical event. However, the CE may require invasive tests, be expensive, occur rarely or be temporally distant from the E/I. For example, in trials of cholesterol-lowering drugs, the fundamental end-point of interest may be death or development of coronary artery disease (CAD). Such a study may require the recruitment of thousands of patients followed up for many years. Clinicians, patients and funding bodies are keen to assess the effectiveness of promising new therapies quickly. This leads investigators to explore markers that may serve as intermediate end-points or surrogate end-points (IE/Ss) of the CE and which are more frequently occurring and temporally closer to the E/I. Therefore, for example, it is of interest to know how useful a reduction in post-treatment cholesterol will be in predicting reductions in the incidence of CAD.

Fundamentally, an IE/S should be sufficiently closely related to the CE such that changes in the IE/S accurately predict changes in the CE. Implicit in this idea is the

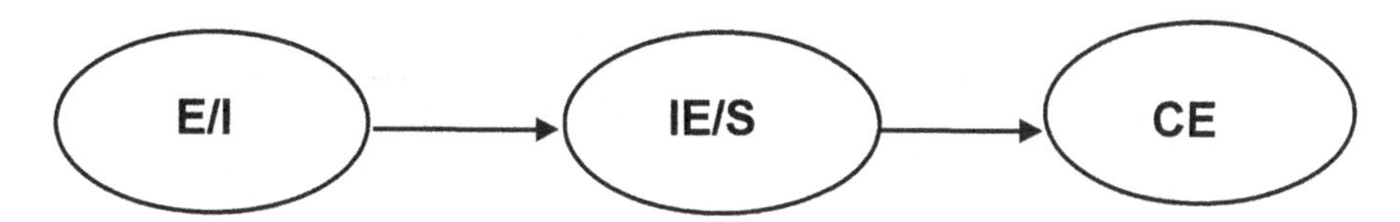


Figure 3.1 The position of an intermediate end-point or surrogate (IE/S) on the pathway between an exposure or intervention (E/I) and a clinical end-point of primary interest (CE).

condition that the CE should not be greatly influenced by factors independent of the chosen IE/S. Thus, all confounding factors should precede the measurement of the IE/S and act through the IE/S disease pathway. A poor choice of IE/S will overestimate the effects on the true CE, may underestimate confounding and side effects, and could lead to inappropriate treatment.

There may be other benefits in using IE/Ss. Since they are measured closer in time to the E/I, they may be easier to interpret. For example, in a study of surgical patients, deaths within 30 days may be easier to attribute to the procedure than deaths within 12 months, which may be influenced by many factors. A further benefit may be that identifying a useful IE/S adds to the understanding of the underlying disease process and, thus, potential intervention strategies may be suggested. A related point follows: if we can intervene based on the IE/S, we may prevent or attenuate the impact of the CE, thus addressing the underlying problem earlier in the disease process.

## Defining intermediate end-points or surrogates

Intuitively, a useful IE/S will correlate closely with the CE and will have a known, or at least plausible, biological mechanism. Although not universally accepted [1], the most commonly used statistical basis for the choice of a surrogate end-point in the context of clinical trials is the Prentice criterion [2]. He defines an IE/S as:

> a response variable for which a test of the null hypothesis of no relationship to the treatment groups under comparison is also a valid test of the corresponding null hypothesis based on the true endpoint.

This may be interpreted as follows. If we know the IE/S, we can directly predict the CE, independently of which treatment group the patient is in; that is, the intervention gives no extra information above that given by the IE/S. Schatzkin et al. [3] provide an illustrative hypothetical example of a cancer marker used in an intervention trial to prevent cancer onset (Table 3.1). We assume that 100 patients are

**Table 3.1.** Hypothetical data from an intervention study

| Group | Number of patients | Cancer proportion (%) |
|---|---|---|
| Intervention | | |
| Marker positive | 25 | 60% (15/25) |
| Marker negative | 75 | 12% (9/75) |
| Total | 100 | 24% (24/100) |
| Control | | |
| Marker positive | 50 | 60% (30/50) |
| Marker negative | 50 | 12% (6/50) |
| Total | 100 | 36% (36/100) |

*Note:*
Reproduced from Schatzkin et al. [3]

randomized to each treatment arm. In both groups, if the marker is positive, there is a 60% chance of developing cancer. If the marker is negative, the chance of developing cancer is 12%, irrespective of treatment group. Thus, the intervention works through reducing the proportion of marker positive cases, with a predictable effect on the development of cancer. It should be noted that the marker is a valid IE/S even though the difference in the proportion of marker positive cases (25% versus 50%) is not the same as the difference in proportion of cancer cases (24% versus 36%) and a hypothesis test based on the marker is more powerful than the one based on the CE.

The second point implicit in the Prentice criterion is that the IE/S is chosen in a way that depends on the interventions under comparison. For example, cholesterol reduction may serve as an IE/S for CAD in a trial of a cholesterol-lowering drug, but it is unlikely to be useful in a study of antihypertensive drugs. In addition, once validated, cholesterol may be a valid IE/S for trials of other cholesterol-lowering agents, but not for an E/I acting via a different pathway. Universal IE/Ss which apply for all E/Is are very rare. This point is directly related to the position of the IE/S on the disease pathway. Cholesterol changes may serve as an IE/S for CAD provided that the treatments under study do not differentially affect CAD via pathways that bypass the cholesterol-lowering effect – if they have a side effect of raising blood pressure, for example.

Extending the Prentice criterion to cohort studies is conceptually straightforward. An IE/S should be closely related to the CE independently of both the E/I and any other confounding variables. Therefore, if we consider the example in Table 3.1, the intervention might be exposure to tobacco smoking rather than a therapeutic intervention.

## Validating intermediate end-points or surrogates

Validation of IE/Ss should usually be done in large, prospective studies, which may be cohort studies or clinical trials. The statistical basis of validation is to examine the pathway represented in Figure 3.1. Freedman et al. [4] presented a practical strategy for validation and much of this section follows their methods.

The usual methodology for the study of E/I→IE/S→CE relationships is regression analysis. Here, the true CE, or a function of it, is determined by a function of the E/I and the IE/S. For example, if $p_{CAD}$ is the probability of developing CAD in patients receiving a lipid-lowering drug, we can, by the use of logistic linear regression, write:

$$\begin{aligned}\text{logit}(p_{CAD}) = &\ \text{average CAD rate}\\ &+ \text{cholesterol effect (IE/S)}\\ &+ \text{treatment effect (E/I)}\\ &+ \text{treatment} \times \text{cholesterol interaction}\end{aligned}$$

For cholesterol to be an effective IE/S, the cholesterol effect in this regression should be clinically and statistically significant. If the treatment effect is large, this indicates that the cholesterol-lowering drug alters the chance of acquiring CAD independently of cholesterol levels. Thus, cholesterol levels will not accurately predict the chance of CAD and it is likely to be an imperfect surrogate. Similarly, if there is a significant treatment × cholesterol interaction, this indicates that the treatment alters the chance of acquiring CAD differently for patients with different cholesterol levels. Again, it is more difficult to predict changes in CAD incidence based on changes in cholesterol and this relationship will depend on the cholesterol profile of the sample under study.

In practice, if either the E/I effect or the E/I × IE/S interaction is significant, the measurement is very unlikely to be a valid IE/S. If both are small and nonsignificant, and the IE/S effect is large and significant, then the data are consistent with the Prentice criterion.

In order to *quantify* the extent to which an IE/S is useful, we can estimate the proportion of the E/I effect which is explained by the IE/S from the attributable fraction (AF). In our example, AF is the proportion of the effect of cholesterol-lowering treatment which may be explained by the effect of cholesterol levels. Freedman et al. [4] suggested the estimate:

$$\text{AF} = 1 - \frac{\text{treatment effect adjusted for cholesterol}}{\text{treatment effect unadjusted for cholesterol}}$$

If the treatment effect adjusted for cholesterol is small compared with the unadjusted effect, i.e. cholesterol is an effective IE/S, the AF will be close to 1 and the

Prentice criterion is satisfied. Alternatively, if the adjusted treatment effect remains large, the AF will be close to zero and the Prentice criterion fails. In practical terms, an IE/S will be acceptable if the lower bound of the confidence interval for the AF is acceptable, say 75%. Further work in this area has provided methods for quantifying AF in applications of Cox regression [5].

Apart from sampling error, there may be several reasons why an IE/S is imperfect. Firstly, it is unlikely that a single measurement, possibly made several years before determination of the true CE, can fully represent the disease process. Secondly, the more variable an IE/S measurement is within patients, the less effective will its adjustment be in improving outcome. It is possible to design a study which uses repeated within-patient measures to gauge measurement error accurately. Finally, regression models are unlikely to be perfect representations of the disease process and should be chosen carefully, to reflect realistically both the disease process and all features of the study design, including adjustments for repeated within-patient measurements.

One attractive feature of this approach to the validation of IE/S is that there is a natural extension to composite measurements. Since validation is based on regression equations, the IE/S effect may be made up of a number of measurements which are added to the regression models as a group rather than individually. In practice, the use of composite measures can be problematic since there are often multiple dependencies between variables. Composite scores may need to be developed outside of the validation study. A further feature of this approach is that it is straightforward to choose between competing markers by means of standard regression analysis.

## Assessing markers for diagnosis

The special case of diagnosis is simpler in that we are only interested in the IE/S→CE link in Figure 3.1; that is, we have an IE/S and a CE, but they are not usually linked to an E/I. Consequently, we are only required to demonstrate the strength of the association between the IE/S and the CE. The most flexible methods for establishing these associations are regression-based models as described above.

In diagnosis, widely used measures of accuracy are the sensitivity and specificity of a marker, and their related quantities, i.e. the positive and negative predictive probabilities. The sensitivity of a marker is the number of true cases of a disease for which it is positive as a proportion of the total number of cases and this is often termed the true positive rate. The specificity is the number of noncases for which the marker is negative as a proportion of the total number of noncases and is often termed the true negative rate. Table 3.2 gives an example of the use of procalcitonin (PCT) as a marker of serious infection in transplant recipients. The sensitivity

**Table 3.2.** Procalcitonin as a marker for infection in transplant recipients

| Marker | Infection cases | No infection |
|---|---|---|
| PCT ≤0.125 (μg/l) | 70 | 216 |
| PCT>0.125 (μg/l) | 139 | 96 |
| Total | 209 | 312 |

of this marker was 67% (139/209) and its specificity was 69% (216/312). The greatest problem with this approach is the arbitrary threshold chosen for case definition. In this example, the average of the sensitivity and specificity was 68% over a wide range of thresholds from 0.125 μg/l to 0.275 μg/l PCT, and it is not clear whether to maximize sensitivity, specificity or the sum of the two. This may be especially problematic in radiology where summary measures often require subjective interpretation.

Predictive probabilities are derived from the sensitivity, specificity and prevalence of cases by the use of Bayes theorem. The positive predictive probability is the chance of being a true case given that the marker was positive. Similarly, the negative predictive probability is the chance of being a noncase given that the marker was negative. These values have a direct clinical interpretation, making them more useful in practice. However, since they are derived from the sensitivity and specificity, they again depend on the subjective choice of a diagnostic threshold.

One way around the problem of threshold selection is to summarize results on a receiver operating characteristic (ROC) curve (Figure 3.2). This technique plots sensitivity (the true positive rate) against 1−specificity (the false positive rate) for different thresholds and lends itself to discrete markers, such as those arising in radiology, as well as to continuous markers. The area under the ROC curve (AUC) is used as a summary measure of accuracy. Given two patients selected at random, one true case and one noncase, the AUC may be interpreted as the probability that the marker measurement is more positive for the true case than the noncase. Thus, an AUC of 1.0 indicates a perfect marker and a true case always has a more positive marker than a noncase, whereas an AUC of 0.5 indicates that the marker is of no diagnostic value. Confidence intervals can be calculated for estimates of the AUC and hypothesis tests of $AUC=0.5$ against $AUC>0.5$ are easy to perform. However, a marker with an AUC which is significantly greater than 0.5 may not be sufficiently accurate in diagnosing cases to be useful in clinical practice. The required level of accuracy is dependent on the clinical context and the consequences of both over- and underdiagnosis of true cases by the marker.

The principal advantages of ROC curve analysis are that it is simple to compute by means of parametric or nonparametric methods and that it is not dependent on

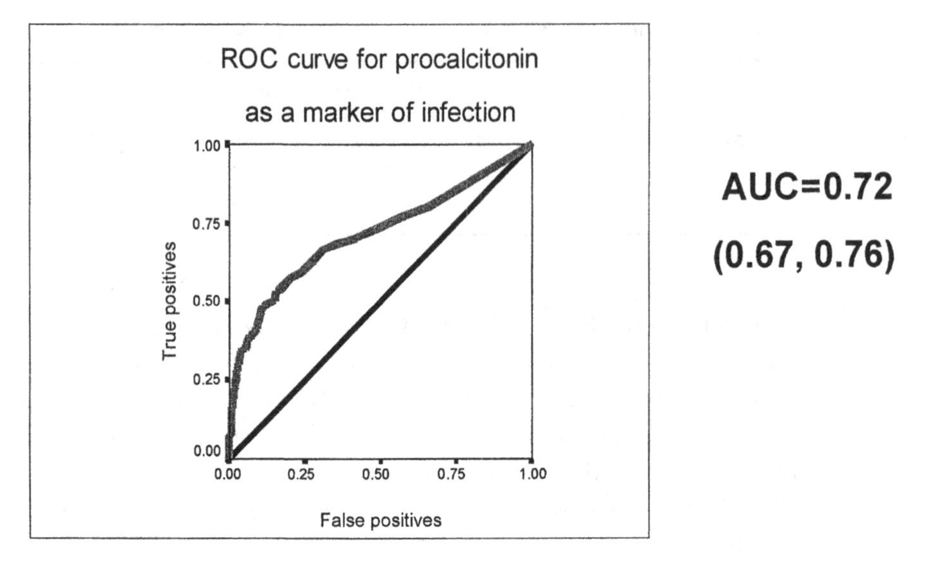


Figure 3.2 Receiver operating characteristic (ROC) curve for procalcitonin as a marker of infection.

a single threshold. However, data for ROC curve analysis often come from observational studies, which are subject to a number of avoidable and unavoidable biases. For example, the CE or gold standard may not always be measured, particularly if it involves an invasive and risky procedure. In this case, the gold standard will be missing but not at random, since patients with negative markers are less likely to undergo the gold standard test than those who are marker positive. Thus, systematic bias is introduced, which must be corrected for. Details of bias correction factors can be found elsewhere [6].

The relative value of competing markers can be assessed by comparing the summary measure AUC. Often, competing markers are measured on the same subjects and are correlated; that is, a patient with a high level of marker 1 is more likely to have a high level of marker 2 and, therefore, the two curves will be, in a sense, paired. Failure to take account of this design feature will underestimate differences in AUCs and resulting hypothesis tests will be insensitive. DeLong et al. [7], describe a nonparametric method for comparing AUCs from correlated ROC curves which is simple to compute.

One limitation of ROC curve analysis is the rather clumsy way in which composite tests are developed. In order to assess a second marker, undiagnosed cases based on the primary marker are used to construct a conditional or multi-ROC curve covering the reduced probability space [8]. The main flaw in this approach is that it requires specification of a threshold for the primary marker and the diagnostic accuracy of the secondary marker in the remaining undiagnosed true cases is conditional on this threshold. Thus, subjectively changing the diagnostic threshold of the primary marker will affect the diagnostic evaluation of the secondary marker. Clearly, a regression approach would use the actual value of each marker to assess the added value of the secondary marker and so would avoid these issues.

In addition to the papers cited in the text, researchers who are interested in this area are also directed to articles by Begg [9] and Wittes et al. [10].

## REFERENCES


1 Begg, B.B. and Leung, D.H.Y. (2000). On the use of surrogate end points in randomized trials. *Journal of the Royal Statistical Society*, **163**, 15–28.

2 Prentice, R.L. (1989). Surrogate endpoints in clinical trials: definition and operational criteria. *Statistics in Medicine*, **8**, 431–40.

3 Schatzkin, A., Freedman, L.S., Schiffman, M.H. and Sanford, M.D. (1990). Validation of intermediate end points in cancer research. *Journal of the National Cancer Institute*, **82**, 1746–52.

4 Freedman, L.S., Graubard, B.I. and Schatzkin, A. (1992). Statistical validation of intermediate endpoints for chronic diseases. *Statistics in Medicine*, **11**, 167–78.

5 Lin, D.Y., Fleming, T.R. and De Gruttola, V. (1997). Estimating the proportion of treatment effect explained by a surrogate marker. *Statistics in Medicine*, **16**, 1515–27.

6 Zhou, X-H. (1998). Correcting for verification bias in studies of a diagnostic test's accuracy. *Statistical Methods in Medical Research*, **7**, 337–53.

7 DeLong, E.R., DeLong, D.M. and Clarke-Pearson, D.L. (1988). Comparing the areas under two or more correlated receiver operating characteristic curves: a nonparametric approach. *Biometrics*, **44**, 837–45.

8 Shultz, E.K. (1995). Multivariate receiver-operating characteristic curve analysis: prostate cancer screening as an example. *Clinical Chemistry*, **41**, 1248–55.

9 Begg, C.B. (1991). Advances in statistical methodology for diagnostic medicine in the 1980's. *Statistics in Medicine*, **10**, 1887–95.

10 Wittes, J., Lakatos, E. and Probstfield, J. (1989). Surrogate endpoints in clinical trials: cardiovascular diseases. *Statistics in Medicine*, **8**, 415–25.

4

# Using intelligent systems in clinical decision support

Richard Jones

Leeds Teaching Hospitals NHS Trust and University of Leeds, Leeds, UK

## Introduction

The 1990s have seen two parallel technological revolutions, in bioscience and in information science, that are rapidly converging. Increasingly, biological concepts are being adopted by informaticians and chip technologies are being used for routine biological analysis. The emerging hybrid disciplines of bio-informatics and health informatics which they have spawned will be central to the effective clinical exploitation of the explosion of discoveries in the area of biomarkers.

Other chapters in this book describe many new biomarkers. These provide information about organ function and dysfunction to a degree of specificity and precision far beyond those in current use. However, the improvements in precision and specificity come at a price. The new tests have more focused functions. They require more discriminating use if they are to answer the discrete clinical questions for which they have been designed. For example, debate about the best 'cardiac enzyme' has been superceded by the need to select the best 'marker of risk in the acute coronary syndrome'. Given the relatively increased unit cost and limited health resources, their widespread introduction presents the twin challenges of how to select the best marker for any particular diagnostic problem and how to maximize the information provided by the test. Intelligent decision support systems may provide some of the answers.

## Clinical decision support systems

The practice of medicine is an inherently decision-based process. For individual patients, data gained from clinical examination and investigative procedures are combined by the physician and used to inform decisions about diagnosis or future management [1]. For policy definition, aggregation of data from individual patient episodes is used epidemiologically, both for a clinician's personal practice and for the institution. The synthesis of data from local practice together with that gained from clinical trials serves to inform strategic decision making on a regional, national or global basis.

To date, the main emphasis of laboratory information systems has been on control of analytical systems and administrative support of operational processes and decision making in test selection, and interpretation has generally exploited the personal skills of senior medical and scientific staff. The expanding information technology infrastructure of the modern hospital now provides fertile ground for the exploitation of clinical decision support systems [2] and the increasing interdependence of clinical and diagnostic services makes this area a particular focus for early implementation.

## System definitions

In considering clinical decision support, it is important to be clear about the definition and purpose of a decision support system [2, 3]. To many people, such systems are synonymous with so-called expert systems. However, this view is too restrictive as decision support tools need not be entirely computer based. The Harvard Business School Review [4] provides a good generic description of the qualities of a decision support system. It rightly emphasizes the powerful synergy between the power of the computer to handle data and the human to evaluate it:

> Decision support systems couple the intellectual resources of individuals with the capabilities of the computer to improve the quality of decisions. They are computer-based support systems for decision makers who deal with semi-structured problems. (Keen and Scott Morton, Harvard Business Review [4].)

In contrast, the following widely used definition of an expert system concentrates on the computational aspects alone and does not consider the entire context of use:

> An expert system embodies, in a computer, the knowledge based component of an expert's skill in such a form that the system can offer intelligent advice and on demand justify its own line of reasoning. (Donald Michie, 1985 [5])

Expert systems are widely used for decision support, but many of the best computer-based decision systems do not use this technology at all [3].

From a functional standpoint, a distinction can be drawn between systems providing *passive* as opposed to *active* decision support [2, 6]. Passive systems are ones which manipulate data or information in ways which make the appropriateness of decisions more apparent to human users but which in themselves do not *dictate* the action to be taken. In these terms, a computer-generated, paper-based graphical presentation of analytical results can be viewed as a decision-support aid. In contrast, active support systems are those that do dictate the outcome once the human expert has defined the rule set and the range of possible outcomes. Table 4.1

**Table 4.1.** Examples of different types of decision support systems in current clinical use

| | |
|---|---|
| *Examples of passive systems using open loop methodologies* | |
| Data presentation techniques | Sequential reporting for tumour marker data |
| Critiquing/modelling/reminder systems | HELP systems |
| Advisory systems | ECG diagnosis |
| *Examples of active systems using closed loop methodologies* | |
| Diagnostic support systems | Cascade testing rules in LIMS |
| | Autogenerated interpretative reports |
| Dose regulation modules | Anticoagulant clinic systems |
| | Artificial pancreas-regulating insulin dosage |
| | Intracranial pressure regulation |
| Intelligent medical devices | Implantable defibrillators |
| Robotic surgical tools | Hip replacement bone drills |

illustrates some examples of systems broken down in this way; those in the passive category can be considered 'open loop systems' whereas those in the active category generally are 'closed loop systems'. The importance of the distinction is that, since closed loop systems generally work without a human cross-check, they are required to perform to more exacting standards. The increase in performance achieved by transferring the responsibility for decision making to the system usually justifies the extra engineering effort involved. It should be noted that this does not necessarily imply that passive systems can be built to less exacting standards. Performance gains are just as important for both types of system and hence optimization is still a key development factor. More importantly, systems designed for passive use can become closed loop if the human user never questions the outcome. Potentially dangerous practices can be introduced, especially if unskilled users are unaware of system deficiencies. Passive systems should, therefore, be robustly developed and their clinical use rigorously controlled.

## Decision support in the diagnostic process

The diagnostic process is an iterative cycle in which clinicians form hypotheses based on clinical examination of the patient and perform investigations in order to rule in or rule out diagnoses. Information gained by each testing cycle feeds forward into the next cycle, informing the selection of subsequent tests by altering the balance of diagnostic possibilities. Many steps in the cycle are potential points for the deployment of decision support systems (Table 4.2).

**Table 4.2.** Steps in the diagnostic process which could be supported by decision aids

| Step | Decision aid |
|---|---|
| 1. Test selection | Diagnostic expert system<br>Diagnostic protocol selector |
| 2. Sample draw | Sample type selection aid<br>Automated tube labeller |
| 3. Laboratory accession | Automated aliquotting/sorting |
| 4. Laboratory analysis | Workflow optimizer<br>Reflex testing protocol |
| 5. Laboratory reporting | Rule-based data validation |
| 6. Clinical review | Automated test interpretation |

There are many good listings of existing decision support systems (see, for example, the web site of the Association of Clinical Biochemists: http://www.informatics–review.com/decision–support). However, few of the many hundreds of clinical decision support systems described in the medical literature have transferred into routine use for reasons explored below. It is noteworthy that almost all electrocardiograms (ECGs) taken in the developed world are now performed on instruments which provide automated, expert system reporting. By the mid1990s, more than 100 million ECGs per day were being routinely reported through decision support aids [7]. A small number are discussed below to illustrate applications in test selection and interpretation and to highlight issues in their development and successful deployment.

## Specific systems

### Function: test selection

#### Liver Unit Medical Protocol System

The Liver Unit Medical Protocol System (LUMPS) provides clinical protocol-driven control of test ordering. It works by creating a diary of tests for a patient based on their clinical diagnosis, which is modified interactively on the basis of test results or changes in clinical status [8]. In rigorous testing in the context of care of liver transplant patients, it produced benefits in quality and cost of care. The appropriateness of test use increased significantly and the need for *stat* (short turnaround tests) testing fell by 40%. With the increasing use of nonskilled staff and the increased complexity of test protocols, similar systems will become commonplace in acute centres to automate test ordering. A major challenge is to develop protocols with sufficient clinical sensitivity to target test use to areas of greatest benefit.

## Function: test interpretation

### Pathology Expert Interpretative Reporting System

(*Pathology Expert Interpretative Reporting System*) *PEIRS* is a system that automatically appends interpretative comments to pathology reports [9]. The knowledge base uses over 2300 '*ripple down rules*' rules accumulated from the analysis of routine, manually generated report comments. PEIRS can generate interpretative comments for a wide range of reports including thyroid function tests, arterial blood gases, glucose tolerance tests, human chorionic gonadotrophin (hCG), and catecholamines and a range of other hormones. Its strength is that it captures the pathologist's own behaviour in a self-correcting and adaptive way and thereby becomes an extremely personalized tool. The downside is that the rule set can never be formally proven which makes it difficult to transfer easily from site to site.

### Tumour marker interpretation

Many attempts have been made to apply mathematical techniques to tumour marker data, both to aid diagnosis and to provide earlier prediction of recurrence. These have generally been based on knowledge of the rates of tumour growth and biological clearance of markers following surgical intervention. One such system (MARKER) used regression analysis of falling hCG values in gestational trophoblastic tumours to predict time to baseline and compared this with the known cases [10]. Its elegance lay in its display technique that provided clinicians with a weighted comparison of outcomes on which to make their own clinical judgements (Figure 4.1). This particular approach could not be adopted for widespread use as the decay function of the hCG proved to be nonlinear. This illustrates a common reason for failure to progress systems in this area in that our knowledge and models of biological systems are as yet too crude and simplistic to apply in clinical practice.

## Function: diagnostic risk analysis

### TeleGastro

The TeleGastro system can be regarded as one of the milestone systems in medical decision support [11]. Though based on a relatively simple Bayesian algorithm, it was the first clinical decision support system to be subjected to a fully randomized, multicentre, controlled clinical trial. Indeed, it remains unique in this regard and is one of the seminal developments in the field alongside better known systems such as MYCIN [2] and INTERNIST [3]. The system aids the evaluation of acute abdominal pain, using simple laboratory and clinical data to provide a probability of positive diagnosis. In a trial of 16000 patient episodes, initial diagnostic accuracy improved from 45% to 65%, unnecessary operations and perforations were

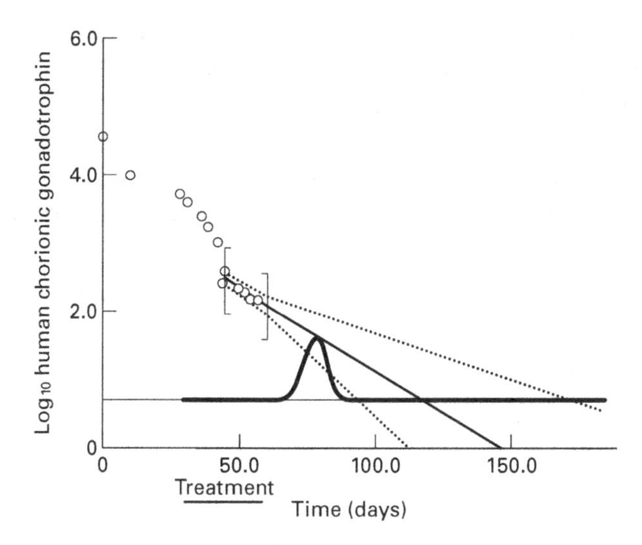


Figure 4.1 Graphical display of sequential hCG measurements (○) in a patient treated for gestational trophoblastic tumour. The display shows the distribution of the time taken for hCG values to fall to baseline for patients who survive following therapy and compares this with the predicted value for the patient based on an extrapolation from the individual measurements. Dotted lines indicate the confidence limits of the fitted regression line. The population curve is shown as the bold Gaussian curve above the X axis. In this case, the predicted time to baseline is to the right of the population curve indicating failure of disease control.

reduced by 15%, and 8000 bed-days of hospitalization were saved [11]. It also significantly reduced the number of avoidable deaths. This system is important in providing a model approach to decision support system development and deployment. In generic terms, it consists of a large database of cases with known outcomes coupled to an *a priori* set of rules based on known associations between the presence of pain, inflammation and acute abdominal pain. It uses Bayesian analysis to predict the likelihood of different diagnoses in unknown cases. Once outcomes are known, the data on these cases are added to the database, thereby strengthening the power of the system. The only biomarker used is the total white cell count, a crude and nonspecific marker of infection and inflammation. What makes TeleGastro so successful is the evaluation of this measurement in the fully documented context of a patient with abdominal pain of acute onset.

### Trisomy-21 risk analysis

Trisomy-21 risk analysis using biochemical markers is similar to the TeleGastro model. In place of a straightforward Bayesian analysis, individual cases are compared with a population model which combines a priori age-related risk with risks based on likelihood ratios for individual biological markers. Various adjustments can be made to the model to take account of twins or conditions such as diabetes. Interestingly, the first biochemical marker used (alpha-fetoprotein [AFP]) was discovered as a chance association from epidemiological data on neural tube screening programmes and the biological explanation is still controversial. The significance of these systems is that they do not provide a definitive diagnostic result but a probability of presence of abnormality such that a definitive amniocentesis test can be applied. They are successful only in the context of population screening since, by definition, a number of false positive and false negative results will be generated.

## Barriers to use

Identifying the barriers which have prevented the widespread implementation of existing systems should help direct future development towards more profitable approaches within the biomarker field [6].

### Availability of outcome data

It is noteworthy that, in both the case of trisomy-21 screening and TeleGastro, definitive outcomes are known in a short space of time and can be used to redefine and recalibrate the underlying models rapidly. The production of precise risk prediction systems should not be difficult in areas where outcomes are quickly available from definitive testing, e.g. from arteriography in acute cardiac states. For some biomarkers, such as those available for osteoporosis or prostatic cancer screening, robust outcome data may not be known for years. This will inhibit the speed at which risk calculation techniques can be applied. The research and development challenge is thus to capture rapidly the necessary outcome data and feed these back into the decision support tools. For trisomy-21 screening, co-operation between users and the diagnostics industry was important in obtaining large volumes of information from multicentre clinical trials. In other fields, the same approach may be necessary.

### Technical issues

Embedding decision support systems in conventional laboratory information systems (LIMS) has undoubtedly produced many good algorithms and evidence of utility. However, this approach has not led to the widespread use of even the most successful systems. Adopting a more modular approach, which separates the decision support element from the central LIMS, is more likely to succeed. Modern software technologies facilitate a 'plug and play scenario' where individual best of breed decision support systems, tailored for individual functions, can be attached to a LIMS, providing highly focused support for particular problems. The author's own group has developed a Down's Syndrome Screening calculation server. This can be called by client databases to estimate risks on the basis of clinical and biochemical data [12]. The communication is managed through an open database interface and the system can work on networks and even across the Internet. One obvious challenge is to create standard interfaces, along the lines of the HL7 analyser interface standard, to enable a truly plug and play approach. Given such interfaces, there would be no barrier to the rapid deployment of successful systems.

### Access to clinical information

All the examples given above require a combination of clinical and laboratory data in order to perform their calculations. A major barrier to the use of decision support tools in the laboratory has been the difficulty in acquiring adequate clinical information – for example, information on thyroxine dosages in thyroid test interpretation. The combination of the increasingly sophisticated clinical information systems and the rapid development of networks spanning the clinical and laboratory domains should overcome this. In the future, it may prove preferable to attach the decision support functions to clinical systems on the grounds that the volume of clinical data will outweigh the laboratory information. It will be a challenge for the pathologists and diagnostics industry to ensure that an appropriate transfer of knowledge and skills occurs and that the laboratory remains a stakeholder in the process.

### Professional needs and acceptance

In the late 1980s, professionals viewed intelligent systems with great suspicion, partly because of a perceived threat to professional status. If computers could be more accurate than doctors, then how could doctors still command respect and high salaries? With time, this view has changed. We do not expect planes to be flown without computer support and there is no reason why we should expect doctors to perform unsupported. The information explosion in medicine is now recognized and, with increasing demands on the time of medical professionals, any help is generally received in a positive light. However, there are still many problems to be solved with regard to the validation and accreditation of systems and the professional understanding of what constitutes acceptable performance. Computers are generally expected to be infallible, but all decision support systems are based at some level on statistical procedures. Hence, they will never be 100% accurate. It is more relevant to consider the marginal levels of improvement that can be achieved. We need to recognize that experts do not take most decisions in medicine and that even experts in some areas perform suboptimally. The TeleGastro abdominal pain diagnosis system was not perfect but it instantly brought performance standards of all novice users up to that of experts. The marginal improvements in performance were still highly significant.

### Evaluation and accreditation

There is a pressing need to develop methods to make comparative evaluations of decision support systems and to provide procedures for professional accreditation, quality assurance (QA) and licensing. Given a modular approach, laboratories are likely to be provided with a choice of which decision system to adopt for a particular biomarker evaluation, entirely analogous to the freedom to choose the best instrument for an analytical method. Of the alternative decision support aids, some will clearly be better engineered and will produce superior results. Robust methods of comparison will be needed, akin to those for analytical methods, in order to select those best suited to a particular purpose. For critical decision systems, a requirement for them to be accredited through a body such as the US Food and Drug Administration can be envisaged. Similarly, ongoing QA of system performance will be needed, especially where local customization or calibration is undertaken. The French Agency of Medicine recently embarked on a national scheme of screening for trisomy-21 and far-sightedly undertook a programme of accreditation of the calculators used in the risk assessment process. At present, many agencies lack the expertise to undertake this function, but the French approach could be an important indication of what the future might hold.

## Potential for development

There is no doubt that decision support systems will play an increasingly important role in the future of medicine. Enough evidence is currently available to show that they can significantly enhance the diagnostic and therapeutic performance of doctors and nurses without threatening their professional status [13]. Furthermore, they can do this in ways which can provide huge savings of scarce resources, both of time and material, and can thus contribute to alleviating the pressures on scarce health resources. However, there are many challenges in coupling this technology with the current systems of health provision since their use requires changes in many existing clinical behaviours and processes. Laboratories are among the most informatics literate of all healthcare organizations and are in the vanguard of use of decision support systems. They have great opportunities to extend their role into the implementation of decision support systems in the clinical world. In the area of the new biomarkers, intelligent systems will play an essential role in maximizing the utility of the analytical data. In some areas, as is already the case in trisomy-21 screening, the complete diagnostic marker system is likely to be a tightly coupled combination of analytical test and decision support system.

## REFERENCES


1 Jones, R.G. and Payne, R.B. (1997). *Clinical Investigation and Statistics in Laboratory Medicine.* London: ACB Venture Publications.

2 Coiera, E. (1997). *Guide to Medical Informatics, the Interent and Telemedicine.* London: Chapman and Hall Medical.

3 Wyatt, J. (1991). Computer-based knowledge systems. *Lancet*, **338**, 1431–6.

4 Keen, P.G.W. and Scott Morton, M.S. (1978). *Decision Support Systems: an Organizational Perspective.* Reading, MA and London: Addison-Wesley Publishing.

5 Michaelson, R.H., Michie, D. and Boulanger, A. (1985). The technology of expert systems. *Byte*, **10**, 303–12.

6 Heathfield, H.A. and Wyatt, J. (1993). Philosophies for the design and development of clinical decision-support systems. *Methods of Information in Medicine*, **32**, 1–8.

7 Drazen, E., Mann, N., Borun, R., Laks, M. and Bersen, A. (1988). Survey of computer-assisted electrocardiography in the United States. *Journal of Electrocardiology* **21** (Suppl), S98–104.

8 Peters, M. and Broughton, P.M. (1993). The role of expert systems in improving the test requesting patterns of clinicians. *Annals of Clinical Biochemistry*, **30**, 52–9.

9 Edwards, G., Compton, P., Malor, R., Srinivasan, A. and Lazarus, L. (1993). PEIRS: a pathologist maintained expert system for the interpretation of chemical pathology reports. *Pathology*, **25**, 27–34.

10 Leaning, M.S., Gallivan, S., Newlands, E.S. et al. (1992). Computer system for assisting with clinical interpretation of tumour marker data. *British Medical Journal*, **305**, 804–7.

11 Adams, I.D., Chan, M., Clifford, P.C. et al. (1986). Computer-aided diagnosis of acute abdominal pain: a multi-centre study. *British Medical Journal*, **293**, 800–4.

12 Jones, R.G. and Huntington, S. (1994). A client-server calculator for the assessment of Downs syndrome screening risks. *Clinical Chemistry*, **41**, S206.

13 Lundberg, G.D. (1998). Changing physician behavior in ordering diagnostic tests. *Journal of the American Medical Association*, **280**, 2036.

# Part 2

## Biomarkers of kidney disease and dysfunction

5

# Biomarkers in renal disease

Christopher P Price

St Bartholomew's and Royal London School of Medicine and Dentistry, London, UK

The kidneys play a central role in the homeostatic control of many substances, including water, and in the elaboration of urine in order to excrete water-soluble waste products (primarily nitrogenous) and toxins, thereby controlling the volume and composition of the extracellular fluid. In addition, the kidneys are responsible for some important endocrine functions, including haematopoiesis and vitamin D status. Typically, molecules under homeostatic control do not always provide sensitive and specific markers of pathological change, although they clearly represent a derangement of a homeostatic process. However, markers of functional change and structural damage can provide both sensitive and specific indications of pathological abnormality.

While a prerequisite for the utility of a biomarker is that its concentration alters as a consequence of structural or functional change, in itself this does not guarantee efficacy as a diagnostic test. Nevertheless, such information provides a valuable background to understanding the mechanisms of the underlying pathological change at a molecular, cellular or structural level. A diagnostic test can be used for either the screening, detection, diagnosis or prognosis of disease. The efficacy of a test should be judged by the ability of that test to enable a clinical decision to be made in the context of one or more of these situations. The implementation of the test then depends on its diagnostic performance and the potential clinical, operational or economic benefits that might accrue from its routine use [1]. This provides the foundation for the contribution of laboratory medicine to evidence-based clinical practice with the 'conscientious, explicit and judicious use of best evidence in the care of patients' [2].

Recognizing the fact that the availability of a diagnostic test underpins the ability to make a clinical decision and that effectiveness should determine the level of utilization, it is helpful to be aware of the pathophysiological basis for the choice of markers. In the case of biomarkers in renal disease, the pathophysiological origins of the test lie in changes in glomerular filtration rate, glomerular permeability, tubular function, tubular damage, urinary reflux, obstruction to urinary flow and deposition of collagen (fibrosis).

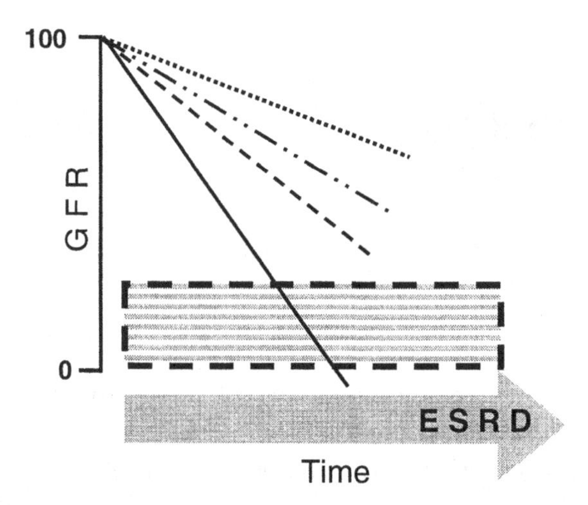


Figure 5.1 Schematic of the relationship between changes in glomerular filtration rate (GFR), appearance of symptoms (▤) and the progression to end-stage renal disease (ESRD). This pattern is applicable to different elements of renal function.

## Glomerular filtration rate

It is instructive to realize that clinical symptoms and signs are unremarkable until as little as 25% of the normal glomerular filtration rate (GFR) remains (Figure 5.1). Thus, changes in serum magnesium or phosphate do not generally occur until this reduction of GFR is reached. Furthermore, in patients with reduced renal function (as GFR), the rate of progression to end-stage renal failure is extremely variable from person to person. One of the reasons for this variability is thought to be glomerular hypertrophy, when remaining glomeruli can compensate for glomerular attrition and lead to the paradoxical stability of GFR. This can explain why diabetics often suddenly go into renal failure without any major deterioration in 'renal function' in the 10 years since their diagnosis was made. It may also explain why serum creatinine is such a poor prognostic marker. This said, it is important to have a sensitive, reliable measure of deteriorating renal function.

The reference procedure for the measurement of GFR is the clearance of an exogenous marker such as $^{57}$Cr ethylenediamine tetra-acetic acid or the contrast agent iohexol. However, these methods are not practical for frequent use and thus a variety of endogenous markers have been developed including measurements of urea, creatinine, urate and a range of low molecular weight proteins in serum, together with urinary clearance measurements, particularly of creatinine. The measurement of serum creatinine is probably the most commonly used test of renal dysfunction and yet it has many limitations, the circulating concentration being

**Table 5.1.** Potential use of cystatin C measurements in clinical decision making

| | |
|---|---|
| Early detection of renal dysfunction | Diabetes<br>Pre-eclampsia<br>Hypertension |
| Monitoring disease progression | Drug therapy<br>Graft rejection |
| Monitoring renoprotective strategies | Anti-hypertensive drugs |
| Assessing risk | Nephrotoxic drugs |

influenced by muscle mass, diet and tubular secretion, as well as there being certain methodological limitations.

The serum concentrations of several low molecular protein markers have been proposed as alternatives, including $\alpha_1$-microglobulin, $\beta_2$-microglobulin, retinol binding protein – and, more recently, cystatin C. The former markers have inherent disadvantages in that their circulating levels are influenced by factors other than changes in GFR (e.g. their rate of synthesis changes in an acute phase reaction). However, it appears that serum cystatin C may be a more viable alternative marker of GFR than serum creatinine. It is a protein that is freely filtered at the glomerulus, it is not an acute phase reactant and, apart from some rare autoimmune diseases, its circulating concentration does not appear to be influenced by pathological changes other than by an alteration in the GFR.

Experimental data have shown that serum cystatin C is better correlated with exogenous measures of GFR [3] than either serum creatinine or creatinine clearance. Furthermore, age-related changes in serum cystatin C are thought to be more accurate than changes in serum creatinine in reflecting changes in the GFR. While these data are important and provide good evidence of technical and diagnostic performance, they do not provide evidence of clinical (or cost) effectiveness. It has been shown that serum cystatin C increases before creatinine as the GFR falls, providing some indication of the potential clinical utility of the test and the benefits that might be gained from its routine use [4]. Indeed, Le Bricon et al. [5] have shown that cystatin C may be a more sensitive marker of graft rejection in renal transplant patients than creatinine. There is a clear implication for improved clinical outcome arising from the earlier detection of transplant rejection or dysfunction. It is possible to speculate that serum cystatin C may be a useful marker in other situations where renal function is impaired (Table 5.1) and where clinical decisions are required.

It has been shown that there is a significant deterioration in the GFR in women with pre-eclampsia, with little change in the serum creatinine but a significant increase in the albumin excretion rate [6]. Personal experience has shown that

**Table 5.2.** Serum cystatin C levels in 20 pregnant women with severe pre-eclampsia compared with matched pregnant women controls

| | Serum cystatin C (mg/l) | |
|---|---|---|
| | Mean | SD |
| Pre-eclampsia | 1.31 | 0.36 |
| Control | 0.76 | 0.17 |

serum cystatin C increases significantly in patients with pre-eclampsia, although there is no such change in the serum creatinine (see Table 5.2). Paternoster et al. [7] in an analysis of several markers showed that the albumin excretion rate was probably the best predictor of pre-eclampsia; they studied 108 hypertensive women and 104 controls, sampling at 28–30 weeks. Only albumin, uric acid and fibronectin could discriminate between those who developed pre-eclampsia and those who did not. Sibai et al. [8] studied births in 774 pregnant women and showed that clinical outcome was worse for babies born to women with proteinuria when compared with women with no evidence of proteinuria.

## Glomerular permeability

The development of proteinuria can be the result of many underlying pathologies, the most common being an increase in albumin excretion. Microalbuminuria can be due to increases in haemostatic pressure, increased glomerular and vascular permeability, tubular overload or reduced tubular reabsorption. The contribution of the latter two are probably quite common as the renal tubules normally operate at close to saturation of the albumin reabsorption capacity.

There is a clear diurnal variation in albumin excretion which disappears as blood pressure increases [9]. It has also been shown that the proportion of patients with so-called microalbuminuria increases with increased blood pressure – being 12–15% in patients with borderline hypertension to more than 50% in patients with severe hypertension [10]. Albumin excretion can be used to detect hypertension, although there are no comparative data to suggest that it is a more reliable indicator than the measurement of blood pressure itself. Albumin excretion also correlates with changes in blood pressure in patients treated with antihypertensive agents [11]. Indeed, this test can be used to compare the efficacy of different antihypertensive agents [11, 12] and it may prove to be useful in both the detection and monitoring of hypertension (Table 5.3).

**Table 5.3.** Potential use of the assessment of albumin excretion in clinical decision making

| |
|---|
| Detection of hypertension |
| Evolution of hypertension |
| Monitoring of hypertensive therapy |
| Prediction of outcome |

Increased albumin excretion is clearly associated with an increased incidence of cardiovascular disease [13]. Thus, Bigazzi et al. [14] demonstrated that the carotid artery wall thickness (a risk factor for cardiovascular disease) is thicker in patients with increased albumin excretion when compared with controls. Bigazzi et al. [15] also demonstrated that survival curves indicating the cumulative probability of developing cardiovascular complications were more favourable in patients with normal, when compared with those with increased, albumin excretion.

Increased albumin excretion in patients with diabetes mellitus is associated with the development of diabetic nephropathy. It is now accepted as an early predictor of disease progression. This subject has been reviewed on several occasions [16]. The value of early detection of increased albumin excretion is more apparent when it is appreciated that early treatment to reduce blood pressure through the use of angiotensin-converting enzyme inhibitors reduces the progress of the nephropathy, improving clinical outcome [17]. This observation provides an excellent example of the synergistic use of diagnostic test and pharmaceutical intervention to improve clinical outcome.

While increased albumin excretion may be a consequence of increased blood pressure and reduced renal function (e.g. due to altered glomerular membrane-sieving characteristics as well as saturated tubular reabsorption), it is also recognized that it may reflect increased systemic vascular permeability – not just involving the kidney [18]. McKinnon and colleagues [19] investigated the relationship between increased albumin excretion in patients admitted to an intensive care unit and their subsequent clinical outcome. By selecting a cut-off value of 3 mg albumin per mmol creatinine, the authors found positive and negative predictive values for mortality of 50% and 85%, respectively, when urine was collected at 6, 12 and 18 hours after admission to the unit.

## Tubular damage

Damage to the renal tubules will result in a deterioration or loss of function, together with the release of cellular contents into the urinary tract, either by active release (e.g. exocytosis) or by increased membrane permeability with release of intracellular proteins. Undoubtedly, one causative element in the increased excretion of albumin is failure of tubular reabsorption. However, this mechanism does not provide a suitable test of tubular damage due to the overpowering influences of altered glomerular and systemic permeability.

On the other hand, low molecular weight proteins that are freely filtered by the kidney may serve as markers of impaired tubular reabsorption. $\alpha_1$-microglobulin, retinol binding protein and $\beta_2$-microglobulin have all been evaluated as markers of tubular damage, although the latter protein has been discounted from having any practical value due to its instability in urine.

#### Nephrotoxicity

While tubular damage probably occurs in a wide range (if not all) of renal diseases, its detection is of no real diagnostic value in clinical decision making with the exception of its use in detecting the effects of nephrotoxic agents. The major classes of nephrotoxic agents are drugs and trace metals. The potential nephrotoxicity of therapeutic drugs is often avoided by monitoring circulating drug levels to guide dosage adjustment; examples of nephrotoxic drugs that are routinely monitored include aminoglycoside antibiotics, such as gentamicin, and the immunosuppressive calcineurin inhibitors, cyclosporin and tacrolimus. The measurement of individual proteins has been proposed for regular monitoring of workers who may be exposed to nephrotoxins, such as cadmium, lead and mercury [20].

An alternative approach to monitoring tubular damage is by the measurement of enzymes released from tubular cells, such as alkaline phosphatase, alanine aminopeptidase, N-acetyl-$\beta$-glucosaminidase, $\gamma$ glutamyl transferase and lactate dehydrogenase, to name but a few. This subject has been reviewed extensively by Price [21] and is discussed in Chapter 7.

#### Vesicouretic reflux

Reflux nephropathy is the second most common renal disease in children. It is responsible for 30% of advanced renal failure in children under 16 years of age and 15–20% of severe renal disease in adults under 50 years of age [22]. Several of the markers of tubular damage have been studied in children and adults with reflux. However, while increases in some markers have been described, the changes are predominantly associated with significant and longstanding reflux [23].

Pizzini and colleagues [24] studied 38 children aged between 1 and 12 years with at least two episodes of urinary tract infection (UTI); six patients were subsequently shown to have reflux. The levels of urinary N-acetyl-$\beta$-glucosaminidase and $\alpha_1$-microglobulin were both found to be higher in children with reflux when compared with levels found in both healthy controls and children with UTI and no reflux. However, in other studies of younger babies – in whom reflux has not yet led to scarring – no obvious increase in the urinary markers has been shown [23]. It is thought that most of the scarring as a result of reflux occurs in the first 2 years of life and new scarring is unusual after 5 years of age.

These statistics clearly identify the need for a marker to detect the early effects of reflux. Renal scarring occurs predominantly when the reflux has ascended well up into the nephrons and the ensuing inflammation affects both glomerular and tubular function. Thus, the diagnostic inaccuracies of previously studied markers are attributable to the fact that these tests predominantly reflect proximal tubular damage. The need exists for a marker to detect the effect of reflux on the collecting duct (or, less satisfactorily, the distal tubule). From a clinical standpoint, the ideal situation would be that the paediatrician is able to screen for the effects of reflux early in life. This will enable appropriate prophylactic action to be taken to reduce the possibility of renal scarring and repeated episodes of UTI, with a concomitant reduction in the risk of the development of renal failure.

## Matrix turnover and fibrosis

Changes in tissue architecture play an important role in the functional capability and capacity of any organ. However, normal function is generally well within the capacity of the organ and, consequently, tests of function do not always provide a sensitive measure of developing pathological change. This is particularly true in relation to the tissue's response to pathological damage and inflammation, with the development of tissue fibrosis. Especially in relation to the kidney, changes in the rate of matrix turnover are thought to play an important part in alterations in glomerular permeability–selectivity and glomerular sclerosis, which precede the development of renal failure.

Matrix turnover and the development of fibrotic tissue are strongly influenced by the matrix metalloproteinases (MMPs) and their inhibitors, the tissue inhibitors of metalloproteinases (TIMPs). There are currently 17 MMPs and four TIMPs that have been described which are widely distributed throughout the body (in most, but not in all, cases) – including in mesangial cells and monocytes (see Chapter 37). Their expression is regulated by a variety of agents and mechanisms, including cytokines and growth factors. The biochemistry of matrix turnover is presently a major field of research activity and there is a growing understanding of the mechanisms that underlie changes in particular disease states. Unfortunately, there has been little application of this knowledge in diagnostic testing and clinical decision making to date.

It has been shown by Anderson and colleagues [25] that glycation of matrix leads to an increase in the amount of collagen present, with a reduction in the expression of MMP2 and an increase in the expression of TIMP1. These observations led to the suggestion that glycation may alter mesangial function leading to an imbalance in mesangial matrix synthesis and degradation which contributes to the mesangial expansion seen in renal disease. It is, therefore, interesting to note that Ebihara and colleagues [26] demonstrated an increase in plasma MMP9 levels in diabetics before the appearance of microalbuminuria. These observations may provide an example of the detection of structural change before the demonstration of altered function, which is probably most important in organ systems where there is substantial functional reserve – including both the liver and kidney.

Akiyama and coworkers [27] have found the levels of MMPs 1, 2 and 3 and TIMPs 1 and 2 to be elevated in patients with different types of glomerulopathies including immunoglobulin A nephropathy, membranous nephropathy and lupus nephritis. The authors suggested that the varied changes in the MMPs and TIMPs in these diseases might reflect the differing stimuli to production and aetiologies of the diseases. Ebihara and colleagues [28] showed that MMP9 mRNA expression was increased in monocytes in patients with renal failure. The metalloproteinase MMP9 is thought to be the product of migratory cells. The MMP9 levels, however, were not increased in the serum. The authors suggested that this enhancement may play a role in the cardiovascular complications associated with chronic renal disease. Certainly, increased synthesis of MMPs has been demonstrated in regions of atherosclerotic plaque, and may be linked to plaque rupture.

## Biomarkers and complications of renal disease

There are a number of complications of chronic renal disease and, while it is beyond the scope of this chapter to cover them in detail, it is worth illustrating the use of biomarkers in a number of examples.

### Cardiovascular risk stratification

It is well known that there is a significant alteration in cholesterol metabolism in patients with renal failure which contributes to their increased risk of cardiovascular disease. More recently, it has been suggested that markers of cardiac muscle damage may be of prognostic value in patients on chronic haemodialysis. Ooi and colleagues [29] corroborated this for serum troponin T in their study of survival curves over a 60-week observation period.

### Metabolic bone disease

Renal osteodystrophy is a recognized complication in patients with chronic kidney disease as a consequence of their reduced ability to synthesize 1,25 dihydroxy cholecalciferol (calcitriol). The early detection of metabolic bone disease and the parathyroid gland response to the deficiency of calcitriol with the regular monitoring of parathyroid hormone (and, possibly, bone alkaline phosphatase status) is now

**Table 5.4.** Summary of use of biomarkers in renal disease

| |
|---|
| Early detection of disease |
| Assessing disease progression |
| Assessing disease remission |
| Assessing transplant rejection |
| Assessing nephrotoxic damage |
| Assessing reflux damage |
| Monitoring therapy |
| Prognosis |

standard practice in guiding the use of therapeutic interventions such as 1 $\alpha$-hydroxycholecalciferol [30].

### Anaemia and renal failure

Anaemia is an important complication of renal failure as a result of the loss of the capacity to synthesize erythropoietin. However, iron deficiency is also an important cause of erythropoietin resistance. While ferritin is a good marker of iron status, this is less true in patients with a chronic disease such as renal failure. An alternative proposed marker is soluble transferrin receptor, the level of which is influenced by iron status but not the coexistence of chronic disease.

Daschner and colleagues [31] have investigated the efficacy of both ferritin and soluble transferrin receptor in the monitoring of erythropoietin and iron therapy in 27 patients on dialysis (11 haemodialysis and 16 peritoneal dialysis). They studied the relationship between erythropoietin requirements and various parameters of erythropoiesis. A significant correlation was shown between the erythropoietin efficacy index (erythropoietin dose divided by haemoglobin concentration) and soluble transferrin receptor ($r=0.65$; $p=0.001$), while neither ferritin nor transferrin saturation were indicative of erythropoietin requirement. The authors suggested that intensified iron substitution in patients with elevated levels of soluble transferrin receptor (indicative of iron deficiency) might improve the cost-effectiveness of erythropoietin therapy.

## Conclusions

There is a wealth of evidence linking changes in biomarkers in patients with renal disease with the pathophysiology of the disease. There is also evidence to indicate how these markers might be used diagnostically (Table 5.4). There is, however, less evidence of the clinical, operational and economic benefits to be achieved from their use – with some notable exceptions. The early detection of the renal complications of diabetes mellitus illustrates both the opportunity and the complexity of assessing the benefits to be gained from the use of markers while also illustrating the synergy between the use of markers in both the detection and diagnosis of disease, together with the monitoring of the effectiveness of interventions.

## REFERENCES


1 Price, C.P. (2000). Evidence based laboratory medicine: supporting decision making. *Clinical Chemistry*, **46**, 1041–50.

2 Sackett, D.L., Rosenberg, W.M.C., Gray, J.A.M., Haynes, R.B. and Richardson, W.S. (1996). Evidence based medicine: what it is and what it isn't. *British Medical Journal*, **312**, 71–2.

3 Price, C.P. and Finney, H. (2000). Developments in the assessment of glomerular filtration rate. *Clinica Chimica Acta*, **297**, 55–66.

4 Newman, D.J., Thakkar, H., Edwards, R.G. et al. (1995). Serum cystatin C measured by automated immunoassay: a more sensitive marker of changes in GFR than serum creatinine. *Kidney International*, **47**, 312–18.

5 Le Bricon, T., Thervet, E., Benlakehal, M. et al. (1999). Changes in plasma cystatin C after renal transplantation and acute rejection in adults. *Clinical Chemistry*, **45**, 2243–9.

6 Lafayette, R.A., Druzin, M., Sibley, R. et al. (1998). Nature of glomerular dysfunction in pre-eclampsia. *Kidney International*, **54**, 1240–9.

7 Paternoster, D.M., Stella, A., Mussap, M. et al. (1999). Predictive markers of pre-eclampsia in hypertensive disorders of pregnancy. *International Journal of Gynecology and Obstetrics*, **66**, 237–43.

8 Sibai, B.M., Lindheimer, M., Hauth, J. et al. (1998). Risk factors for preeclampsia, abruptio placentae, and adverse neonatal outcomes among women with chronic hypertension. *New England Journal of Medicine*, **339**, 667–71.

9 Bianchi, S., Bigazzi, R., Baldari, G. et al. (1994). Diurnal variations of blood pressure and microalbuminuria in essential hypertension. *American Journal of Hypertension*, **7**, 23–9.

10 Hornych, A. and Asmar, R. (1999). Microalbuminurie et hypertension arterielle. *La Presse Medicale*, **28**, 597–604.

11 Bianchi, S., Bigazzi, R. and Campese, V.M. (1999). Microalbuminuria in essential hypertension: significance, pathophysiology, and therapeutic implications. *American Journal of Kidney Disease*, **34**, 973–95.

12 Redon, J. (1998). Renal protection by antihypertensive drugs: insights from microalbuminuria studies. *Journal of Hypertension*, **16**, 2091–100.

13 Calvino, J., Calvo, C., Romero, R. et al. (1999). Atherosclerosis profile and microalbuminuria in essential hypertension. *American Journal of Kidney Disease*, **34**, 996–1001.

14 Bigazzi, R., Bianchi, S., Nenci, R. et al. (1995). Increased thickness of the carotid artery in patients with essential hypertension and microalbuminuria. *Journal of Human Hypertension*, **9**, 827–33.

15 Bigazzi, R., Bianchi, S., Baldari, D. et al. (1998). Microalbuminuria predicts cardiovascular events and renal insufficiency in patients with essential hypertension. *Journal of Hypertension*, **16**, 1325–33.

16 Bennett, P.H., Haffner, S., Kasiske, B.L. et al. (1995). Screening and management of microal-

buminuria in patients with diabetes mellitus: recommendations to the Scientific Advisory Board of the National Kidney Foundation from an Ad Hoc Committee of the Council on Diabetes Mellitus of the National Kidney Foundation. *American Journal of Kidney Diseases*, **25**, 107–12.

17 Lewis, E.J., Hunsicker, L.G., Bain, R.P. et al. (1993). The effect of angiotensin-converting-enzyme inhibition on diabetic nephropathy. *New England Journal of Medicine*, **329**, 1456–62.

18 Gosling, P. (1995). Microalbuminuria: a sensitive indicator of non-renal disease? *Annals of Clinical Biochemistry*, **32**, 439–41.

19 MacKinnon, K.L., Molnar, Z., Lowe, D. et al. (2000). Use of microalbuminuria as a predictor of outcome in critically ill patients. *British Journal of Anaesthesia*, **84**, 239–41.

20 Chia, K.S., Jeyaratnam, J., Lee, J. et al. (1995). Lead-induced nephropathy: relationship between various biological exposure indices and early markers of nephrotoxicity. *American Journal of Industrial Medicine*, **27**, 883–95.

21 Price, R.G. and Whiting, P.H. (1992). Urinary enzymes and nephrotoxicity in humans. In *Urinary Enzymes in Clinical and Experimental Medicine*, eds. K. Jung, H. Mattenheimer and U. Burchardt, pp. 203–21. Berlin: Springer-Verlag.

22 Smellie, J.M., Prescod, N.P., Shaw, P.J. et al. (1998). Childhood reflux and urinary infection: a follow-up of 10–41 years in 226 adults. *Pediatric Nephrology*, **12**, 727–36.

23 Hanbury, D.C. and Calvin, J. (1992). Proteinuria and enzymuria in vesicoureteric reflux. *British Journal of Urology*, **70**, 603–9.

24 Pizzini, C., Mussap, M., Mangiarotti, P. et al. (1999). Urinary biomarkers in children with urinary tract infections with and without reflux on antibacterial prophylaxis with cefaclor. *Clinical Drug Investigation*, **18**, 461–6.

25 Anderson, S.S., Wu, K., Nagase, H. et al. (1996). Effect of matrix glycation on expression of type IV collagen, MMP-2, MMP-9 and TIMP-1 by human mesangial cells. *Cell Adhesion and Communication*, **4**, 89–101.

26 Ebihara, I., Nakamura, T., Shimada, N. et al. (1998). Increased plasma metalloproteinase-9 concentrations precede development of microalbuminuria in non-insulin-dependent diabetes mellitus. *American Journal of Kidney Disease*, **32**, 544–50.

27 Akiyama, K., Shikata, K., Sugimoto, H. et al. (1997). Changes in serum concentrations of matrix metalloproteinases, tissue inhibitors of metalloproteinases and type IV collagen in patients with various types of glomerulonephritis. Research Communications in *Molecular Pathology and Pharmacology*, **95**, 115–28.

28 Ebihara, I., Nakamura, T., Tomino, Y. et al. (1998). Metalloproteinase-9 mRNA expression in monocytes from patients with chronic renal failure. *American Journal of Nephrology*, **18**, 305–10.

29 Ooi, D.S., Veinot, J.P., Wells, G.A. et al. (1999). Increased mortality in hemodialyzed patients with elevated serum troponin T: a one-year outcome study. *Clinical Biochemistry*, **32**, 647–52.

30 Hruska, K.A. and Teitelbaum, S.L. (1995). Renal osteodystrophy. *New England Journal of Medicine*, **333**, 166–74.

31 Daschner, M., Mehls, O. and Schaefer, F. (1999). Soluble transferrin receptor is correlated with erythropoietin sensitivity in dialysis patients. *Clinical Nephrology*, **52**, 246–52.

6

# The genetics of renal disease

Jean-Pierre Grünfeld, Bertrand Knebelmann

Service de Néphrologie, Hôpital Necker, Paris, France

## Introduction

As in other fields of genetics, two approaches have been combined in the genetics of human kidney diseases in order to locate and identify loci and genes: positional cloning (previously termed reverse genetics) and the candidate gene approach. Both approaches have been successful in inherited monogenic renal diseases, often successively. Positional cloning is based on linkage analysis in families using polymorphic markers – mainly microsatellites. The candidate gene approach may either follow localization of the morbid locus by positional cloning or derive from an a priori hypothesis based on cell physiology/biology. After cloning of the gene of interest, mutations have to be found which segregate with the disease in families. Various techniques are used to identify mutations: Southern hybridization (to detect gross rearrangements), polymerase chain reaction (PCR)-based approaches, such as single strand conformational polymorphism (SSCP), denaturing gradient gel electrophoresis (DGGE) or direct DNA sequencing, and new DNA technology (see below). Either genomic or complementary DNA can be studied. Even if the gene expression is tissue specific, one can take advantage of the illegitimate transcription of any gene in any cell type in order to study accessible cells easily, such as lymphocytes. Abnormal transcripts can, therefore, be identified by using PCR amplification of lymphocyte mRNA.

Today, most of the genes implicated in inherited kidney disorders have been localized, and many of them have been cloned. We are entering the 'post-gene era'. The main questions are: what is/are the function(s) of these genes? And, what is, therefore, the mechanism of disease when the function of the gene is altered by mutation? These questions are raised for PKD genes involved in autosomal dominant polycystic kidney disease (ADPKD). New tools have to be used to answer these questions and to identify the molecular pathways downstream from the gene products: an in vitro approach using new DNA technology, such as DNA chip or microarray techniques, or an in vivo approach by generating transgenic animals, with hyperexpression or no expression (knock-out) of the gene. Answering these basic questions is now the challenge since this is a prerequisite to designing pharmacological interventions aimed at stopping or slowing the course of the disease. In some instances (e.g. in Anderson–Fabry disease), these successive steps have been completed and therapeutic trials in humans have been performed.

Another challenge is to approach multigenic diseases where predisposition to disease results from the addition of various minor polymorphisms of various genes [1]. These diseases or disease states include hypertension, diabetes, genetic predisposition to renal disease (such as diabetic nephropathy) or to renal failure, etc. [2, 3]. Minor polymorphisms in 'modifier genes' can also influence the rate of progression of inherited monogenetic diseases. This is well exemplified in ADPKD where, in a given family with a given PKD1 mutation, the rate of progression can be strikingly different from one affected subject to another. The polymorphisms may consist of either various DNA repeats in microsatellites or single nucleotide polymorphisms (SNPs) whose identification in the genome is in progress. This approach has been greatly facilitated by using new DNA technology, with the automated assays of many polymorphisms in large numbers of samples becoming increasingly feasible.

## Genetic diagnosis in inherited kidney disorders

There is a wide spectrum of inherited kidney disorders (Table 6.1). We will not review all of them. Our aim is, rather, to take selected examples and to put emphasis on the clinical use and relevance of biomarkers.

### Cystic kidney diseases

Autosomal dominant PKD is the prototype, and by far the most prevalent, inherited cystic kidney disease. Diagnosis is based mainly on renal ultrasonographic findings, in at-risk subjects belonging to ADPKD families [4]. Normal renal ultrasonography after 30 years of age excludes the diagnosis, but this is not true for younger subjects. Genetic diagnosis may therefore be required in rare instances: (i) when living related kidney donation is considered in at-risk subjects aged less than 30 years; (ii) in young at-risk subjects belonging to families with intracranial aneurysms; (iii) in young subjects who 'need-to-know' for various reasons, including family planning or professional choice; and (iv) in the very rare PKD families in which rapidly progressive ADKD appears in the first years of life and carries a high risk of recurrence in siblings. In these cases, linkage analysis is often the only method available. Indeed, the direct identification of mutations has been performed so far in a limited number of ADPKD families. The disease is genetically heterogeneous, with two different genes being involved: the PKD1 gene located on

**Table 6.1.** Classification of the main inherited kidney disorders

1. Cystic kidney diseases
    - Autosomal dominant polycystic kidney disease
    - Autosomal recessive polycystic kidney disease
    - Other cystic kidney diseases
2. Inherited diseases with glomerular involvement
    - Alport's syndrome and variants
    - Benign familial haematuria
    - Congenital nephrotic syndrome of the Finnish type
    - Nail-patella syndrome
    - Metabolic diseases with glomerular involvement
        - Fabry's disease
        - LCAT (lecithin: cholesterol acyltransferase)/deficiency
        - Type I glycogen storage disease
    - Familial primary glomerulonephritis
3. Inherited tubulointerstitial disorders
    - Juvenile nephronophthisis
    - Familial nephropathy with juvenile hyperuricaemia or gout
4. Genetic diseases with nephrolithiasis
5. Inherited tubular disorders
    - Cystinuria; Lowe's syndrome and other causes of Fanconi syndrome
    - Bartter's syndrome
    - Gitelman's syndrome
    - Liddle's syndrome
    - Nephrogenic diabetes insipidus (vasopressin resistant)

chromosome 16p and the PKD2 gene on 4q [5]. The structure of the PKD1 gene has been shown to be complex, with a number of duplicated regions, rendering detection of mutations rather tricky and tedious. The PKD2 gene is simpler and smaller in size, and, although the gene has been identified more recently, mutations have been detected more easily and numerously [5].

The prevalence of intracranial aneurysm is five times higher in ADPKD than in the general population. In addition, intracranial aneurysms (IAs) are approximately 2.5 times more frequent in families with a history of IA than in other families. Rupture is the main complication of IA, leading to subarachnoid haemorrhage. This complication has a very poor prognosis with a mortality rate of about 50%. Detection of occult, asymptomatic IA can be achieved today by noninvasive means such as magnetic resonance imaging–angiography. Treatment of IA can be considered, according to size and site, either by surgical clipping or by detachable coil-induced thrombosis performed by an interventional radiologist [6]. Thus, early identification of IA may be necessary in some families. In a retrospective European study, it was found that 10% of IA rupture accidents occurred at 20 years of age or less [7]. Such a strategy of early identification may imply genetic testing in young at-risk subjects.

Apart from ADPKD, other inherited cystic kidney diseases are known. For example, autosomal recessive PKD, which has a low prevalence when compared with ADPKD (1/40000 versus 1/500–1000 individuals) and becomes manifest in childhood, although patients may require renal replacement therapy later in life, is consistently associated with congenital hepatic fibrosis (interestingly, a few families with ADPKD and congenital hepatic fibrosis have been reported). Other examples of inherited cystic kidney disease include: (i) glomerulocystic kidney disease which may be dominantly inherited; (ii) X-linked oro-facio-digital syndrome type I with lethality in males; and (iii) autosomal dominant phakomatoses – namely, tuberous sclerosis and von Hippel–Lindau (VHL) disease which includes many extrarenal systemic abnormalities. Correct clinical diagnosis is the first step in genetics. However, this diagnosis may be occasionally difficult in cystic kidney diseases, especially when renal cysts are discovered in utero or early in life.

## Alport's syndrome

This syndrome is defined by the association of progressive haematuric nephritis and sensorineural hearing loss. Eye abnormalities (bilateral anterior lenticonus, retinal macular and perimacular flecks, and, occasionally, recurrent corneal erosions) are found in 35% of patients. Alport's syndrome is a disease of type IV collagen, a major component of the basement membranes [8]. In fact, it includes at least two diseases; the first is X-linked dominant, due to mutations involving COL4A5, the gene coding for the $\alpha5$ chain of type IV collagen ($\alpha5$[IV]). Affected males are hemizygotes and all of them progress to end-stage renal disease, (ESRD). Hearing defects are absent in some families. Carrier females are heterozygous and often have only slight or intermittent microhaematuria. However, 15% of the female heterozygotes progress to renal failure, usually later than males [9]. The second disease is autosomal recessive, due to mutations in the genes encoding $\alpha3$(IV) or $\alpha4$(IV). The phenotypic manifestations are similar to the classical Alport's syndrome, but: (i) homozygous patients may result from a consanguineous marriage, although many affected subjects are compound heterozygotes; (ii) the renal disease progresses to ESRD before 20–30 years of age, at a similar rate in women and in men; (iii) both parents are asymptomatic, with 50% of them having microhaematuria – an abnormality which if found in the father (in the absence of proteinuria and renal failure) is suggestive of autosomal recessive inheritance.

Immunofluorescence or immunoperoxidase staining is valuable for diagnosis.

Indeed, in the X-linked disease, $\alpha5$ is absent from the glomerular basement membrane (GBM) as well as $\alpha3/\alpha4$, which are not normally incorporated. The normal epidermal basement membrane (EBM) contains $\alpha5$ but not $\alpha3/\alpha4$ chains; in X-linked disease, $\alpha5$ is missing. In the autosomal recessive form, $\alpha3/\alpha4$ are absent from the GBM whereas $\alpha5$ cannot be integrated. In contrast, $\alpha5$ is normally present in the EBM. Thus, skin biopsy may be an interesting tool for differentiating X-linked and autosomal recessive forms. The absence of $\alpha5$ in the EBM is, to the best of our knowledge, specific for X-linked Alport's syndrome, but is found in only 75% of cases [10].

Gene testing may be clinically relevant in Alport's syndrome in order to provide genetic counselling, to identify heterozygous carriers in X-linked disease and to allow prenatal diagnosis if required by the parents. Unfortunately, so far, detection of mutations is successful in only 50–60% of families, after tedious laboratory work on the large genes involved [11].

## von Hippel–Lindau disease

This multisystemic and autosomal dominant disease encompasses retinal haemangioblastoma, central nervous system haemangioblastoma, pancreatic cysts and, more rarely, tumours, phaeochromocytoma (often bilateral, in 20% of kindreds), tumour of the endolymphatic sac (in less than 5% of the cases), and renal cysts and carcinomas, usually bilateral and multifocal in 70% of patients. Kidney involvement requires regular screening. Renal clear cell carcinomas need nephron-sparing surgery or nephrectomy, according to tumour size and surgical possibilities [12]. This multisystemic involvement does not occur concomitantly in all affected organs. In contrast, it develops successively, during life, until about 60 years of age. This explains why screening procedures should be repeated in affected subjects, and why it is important to differentiate early in life the subjects who carry the mutation and need screening from the subjects who do not carry the mutation, do not require follow up and have no risk of transmitting the mutated gene to their offspring.

The VHL gene has been cloned (it is a tumour suppressor gene) and it comprises three exons. Mutations are detected in 70–100% of the families, thus allowing early identification of the heterozygotes (affected subjects). This approach requires the correct diagnosis of VHL disease and information shared with families and individuals.

## Anderson–Fabry disease

This X-linked recessive disease, also called Fabry's disease, is characterized by $\alpha$-galactosidase A (lysosomal enzyme) deficiency, resulting in glycosphingolipid deposition, mainly in the cardiovascular and renal system. The first manifestations in hemizygotes are painful acroparesthesias, appearing in childhood, often prevented by the continuous administration of carbamazepine or phenytoin. Subsequently, angiokeratomas, anhidrosis and corneal deposits develop. Ischaemic cerebrovascular complications, cardiac valve abnormalities, myocardial deposition of glycolipids and coronary accidents are the most severe manifestations, along with renal involvement.

In the kidney, glycolipid deposition involves glomerular epithelial cells, tubular cells, and endothelial and smooth muscle cells of intrarenal arteries. The latter changes are responsible for progressive renal ischaemia. Renal disease is revealed by proteinuria at around 20 years, and then progresses to end-stage between 40 and 60 years of age, necessitating regular dialysis and/or kidney transplantation. Glycolipid deposition does not recur in the renal graft which contains normal $\alpha$-galactosidase activity [13].

Diagnosis is based on the determination of $\alpha$-galactosidase activity in leucocytes, serum, hair or fibroblasts. The gene encoding the enzyme has been cloned and mutations can be identified in affected families.

Enzyme therapy will possibly be available in the future. After transfection of cells by the human $\alpha$-GAL gene, two laboratories have been able to produce human recombinant $\alpha$-GAL A, and to initiate prospective randomized trials in Fabry hemizygous patients. Concomitantly, transgenic knock-out mice have been generated, mimicking Fabry's disease. These mice have no enzyme activity and they accumulate glycosphingolipids in lysosomes. However, they do not develop severe organ dysfunction. The biochemical changes are improved by the administration of recombinant $\alpha$-GAL. This remains to be confirmed in humans [14]. Other therapeutic approaches have been conceived and found to be successful in vitro or in animals, such as the stabilization of some mutant enzymes by chaperone molecules, inhibition of glycosphingolipid synthesis, bone marrow transplantation and gene therapy [15–19].

## Genetic counselling and prenatal diagnosis in inherited kidney diseases

Genetic counselling should be considered to be a partnership between an at-risk individual and a counsellor. The latter offers information, investigation, options and support, and the at-risk subject (or patient) makes the decisions. Genetic counselling should not be directive. It should first concentrate on the presentation of accurate facts and options. This explains why counselling should be shared, for the families with inherited kidney disorders, by a clinical geneticist and a nephrologist. It may also involve other partners, such as a nonmedical genetic counsellor, a genetic specialist nurse, social workers etc. The information provided by foundations or patient associations is also very useful for affected families.

Genetic counselling should be performed with enough empathy to assist the patient or couple to make their own choice. The presentation should be personalized, corresponding to the patient's education, emotional state, and religious, ethnic and personal beliefs. Patients respond differently to genetic risk and wish to make their own reproductive decisions. Genetic counselling in ethnic minority communities often requires the assistance of a knowledgeable member of the same community.

The first step in genetic counselling is to identify correctly the inherited disorder and its mode of inheritance, and to label correctly the at-risk subject. The second step in counselling is to provide information about the natural history and clinical management of the inherited disorder. Presymptomatic testing and prenatal diagnosis should be integrated into the process of genetic counselling. Regarding prenatal diagnosis, it should be stressed that its main goal is to allow the birth of unaffected babies in at-risk families. The aim is to cope with the needs and demands of the families. Most prenatal diagnosis is performed in the first trimester in order to permit termination of pregnancy if the foetus is affected or to reassure the parents if the foetus is not affected.

Biochemical markers have been used for several decades in prenatal diagnosis, but they are being progressively replaced by DNA testing. They may still be used in Fabry's disease ($\alpha$-galactosidase deficiency) or in the congenital nephrotic syndrome of the Finnish type (raised amniotic fluid $\alpha$-fetoprotein concentration) when DNA study is not informative. The demands for prenatal diagnosis occur mainly in the most severe autosomal recessive disorders such as autosomal recessive PKD, juvenile nephronophthisis, Finnish-type congenital nephrotic syndrome [20], Lowe's syndrome, cystinosis and type I primary hyperoxaluria. The demand can also be expressed in dominant disorders such as juvenile X-linked Alport syndrome and in autosomal dominant VHL disease. In any case, identification of the mutation in the family will facilitate prenatal diagnosis. If this information is not available, linkage analysis may be performed (provided that this has been prepared before pregnancy, several affected and nonaffected members of the kindred have accepted to participate, and DNA marker study is informative).

## Multigenic approach: from modifier genes to SNPs

It is generally assumed that diseases like hypertension and diabetes mellitus are the prototypes of multigenic (or polygenic) diseases where several gene defects or polymorphisms are involved and interact with environmental factors to trigger disease appearance and progression. Monogenic inherited diseases are not devoid of such interactions. An environmental factor such as smoking has a strong detrimental effect on pulmonary involvement in $\alpha$1-antitrypsin deficiency. We will take some examples in nephrology where a multigenic approach has been initiated, even in monogenic disorders.

Polymorphism (deletion/insertion) in intron 16 of the angiotensin converting enzyme (ACE) gene [21] has been extensively investigated and the DD genotype has been found to be associated with a more rapid renal progression in IgA nephropathy, nephroangiosclerosis, childhood urological diseases and also in ADPKD [22–25]. In an animal model of ARPKD, two major modifier genes have been identified which accelerate the development of renal cysts [26]. The chloride channel, CFTR, mutated in cystic fibrosis, may act as a modifier gene in ADPKD, slowing the rate of progression when it is mutated [27], but the evidence for this hypothesis is still very uncertain.

The genetic predisposition to diabetic nephropathy has been noted for several years [28] and has been further confirmed. In some families with type I diabetes mellitus, there is an aggregation of patients who have developed diabetic nephropathy whereas others are almost completely devoid of this complication after a similar follow up period. The search for the gene(s) predisposing to diabetic nephropathy is currently intensive [2] and some polymorphisms have been identified, e.g. in the gene coding for endothelial nitric oxide synthase [29].

Predisposition to renal failure may also be, in part, genetically determined. Familial aggregation of renal disease has been documented, mainly in African–Americans in the USA [30]. Loci controlling renal disease susceptibility have been recognized in rats [31]; similar loci have been sought in humans, but with disappointing results so far [32].

The search for these modifier genes and polymorphisms, and the understanding of their mechanisms, is crucial because they are targets for designing pharmacological tools and for conceiving new therapeutic interventions aimed at slowing or stopping progression of renal disease, inherited or not.

## REFERENCES


1 Risch, N. and Merikangas, K. (1996). The future of genetic studies of complex human diseases. *Science*, **273**, 1516–17.

2 Adler, S.G., Pahl, M. and Seldin, M.F. (2000). Deciphering diabetic nephropathy: progress using genetic strategies. *Current Opinion in Nephrology and Hypertension*, **9**, 99–106.

3 Nenov V.D., Taal, M.W., Sakharova, O.V. and Brenner, B.M. (2000). Multi-hit nature of chronic renal disease. *Current Opinion in Nephrology and Hypertension*, **9**, 85–97.

4 Ravine, D., Walker, R.G., Gibson R.N. et al. (1992). Phenotype and genotype heterogeneity in autosomal dominant polycystic kidney disease. *Lancet*, **340**, 1330–3.

5 Somlo, S. (1999). The PKD2 gene: structure, interactions, mutations, and inactivation. *Advances in Neophrology*, **29**, 257–87.

6 Pirson, Y. and Chauveau, D. (1996). Subarachnoid haemorrhage in ADPKD patients: how to recognize and how to manage? *Nephrology, Dialysis, Transplantation*, **11**, 1236–8.

7 Chauveau, D., Pirson, Y., Verellen Dumoulin, C., Macnicol, A., Gonzalo, A. and Grünfeld, J.P. (1994). Intracranial aneurysms in autosomal dominant polycystic kidney disease. *Kidney International*, **45**, 1140–6.

8 Kashton, C.E. (1999). Alport syndrome. An inherited disorder of renal, ocular, and cochlear basement membranes. *Medicine (Baltimore)*, **78**, 338–60.

9 Jais, J., Knebelmann, B., Giatras, I. et al. (2000). X-linked Alport syndrome. Natural history in 195 families and genotype-phenotype correlations in males. *Journal of American Society of Nephrology*, **11**, 649–57.

10 Pirson, Y. (1999). Nephrology forum: making the diagnosis of Alport's syndrome. *Kidney International*, **56**, 760–75.

11 Flinter, F. and Plant, K. (1998). Why are mutations in COL4A5 not detectable in all patients with Alport's syndrome? *Nephrology, Dialysis, Transplantation*, **13**, 1348–51.

12 Neumann, H.P.H., Laubenberger, J., Wetterauer, U. and Zbar, B. (1998). von Hippel-Lindau syndrome. In *Inherited Disorders of the Kidney*, eds. S.H. Morgan, J.P. Grünfeld, Oxford: Oxford University Press, pp. 535–61.

13 Desnick, R.J. and Eng, C.M. (1998). Fabry's disease and the lipidoses. In *Inherited Disorders of the Kidney*, eds. S. Morgan, J.P. Grünfeld. Oxford: Oxford University Press, pp. 355–83.

14 Schiffmann, R., Murray, G.J., Treco, D. (2000). Infusion of alpha-galactosidase A reduces tissue globotriaosylceramide storage in patients with Fabry disease. *Proceedings of the National Academy of Sciences*, USA, **97**, 365–70.

15 Abe, A., Radin, N.S., Shayman, J.A. et al. (1995). Structural and stereochemical studies of potent inhibitors of glucosylceramide synthase and tumor cell growth. *Journal of Lipid Research*, **36**, 611–21.

16 Abe, A., Arend, L.J., Lee, L., Lingwood, C., Brady, R.O. and Shayman, J.A. (2000). Glycosphingolipid depletion in Fabry disease lymphoblasts with potent inhibitors of glucosylceramide synthase. *Kidney International*, **57**, 446–54.

17 Fan, J.Q., Ishii, S., Asano, N. and Suzuki, Y. (1999). Accelerated transport and maturation of $\alpha$-galactosidase A in Fabry lymphoblasts by an enzyme inhibitor. *Nature Medicine*, **5**, 112–15.

18 Lee, L., Abe, A. and Shayman, J.A. (1999). Improved inhibitors of glucosylceramide synthase. *Journal of Biological Chemistry*, **274**, 14662–9.

19 Takenaka, T., Quin, G.J., Brady, R.O. and Medin, J.A. (1999). Circulating alpha-glactosidase A derived from transduced bone marrow cells: relevance for corrective gene transfer for Fabry disease. *Human Gene Therapy*, **10**, 1931–9.

20 Tryggvason, K. (1999). Unraveling the mechanisms of glomerular ultrafiltration: nephrin, a key component of the slit diaphragm. *Journal of the American Society of Nephrology*, **10**, 2440–5.

21 Rigat, B., Hubert, C., Alhenc-Gelas, F., Cambien, F., Corvol, P. and Soubrier, F. (1990). An insertion/deletion polymorphism in the angiotensin I-converting enzyme gene accounting for half the variance of serum enzyme levels. *Journal of Clinical Investigation*, **86**, 1343–6.

22 Pei, Y., Scholey, J., Thai, K., Suzuki, M. and Cattran, D. (1997). Association of angiotensinogen gene T235 variant with progression of immunoglobin A neophropathy in Caucasian patients. *Journal of Clinical Investigation*, **100**, 814–20.

23 Jardine, A.G., Padmanabhan, N. and Connell, J.M.C. (1998). Angiotensin converting enzyme gene polymorphisms and renal disease. *Current Opinion in Nephrology and Hypertension*, 7, 259–64.

24 Brock, J.W., Adams, M., Hunley, T., Wada, A., Trusler, L. and Kon, V. (1997). Potential risk factors associated with progressive renal damage in childhood urological diseases: the role of angiotensin-converting enzyme gene polymorphism. *Journal of Urology*, **158**, 1308–11.

25 Baboolal, K., Ravine, D., Daniels, J. et al. (1997). Association of the angiotensin I converting enzyme gene deletion polymorphism with early onset of ESRF in *PKD1* adult polycystic kidney disease. *Kidney International*, **52**, 607–13.

26 Woo, D.D.L., Nguyen, D.K.P., Khatibi, N. and Olsen, P. (1997). Genetic identification of two major modifier loci of polycystic kidney disease progression in pcy mice. *Journal of Clinical Investigation*, **100**, 1934–40.

27 Torres, V.E., Wilson, D.M., Hattery, R.R. and Segura, J.W. (1993). Renal stone disease in autosomal dominant polycystic kidney disease. *American Journal of Kidney Disease*, **22**, 513–19.

28 Seaquist, E.R., Goetz, F.C., Rich, S. and Barbosa, J. (1989). Familial clustering of diabetic kidney disease. Evidence for genetic susceptibility to diabetic nephropathy. *New England Journal of Medicine*, **320**, 1161–5.

29 Zanchi, A., Moczulski, D.K., Hanna, L.S., Wantman, M., Warram, J.H. and Krolewski, A.S. (2000). Risk of advanced diabetic nephropathy in type 1 diabetes is associated with endothelial nitric oxide synthase gene polymorphism. *Kidney International*, **57**, 405–13.

30 Lei, H.H., Perneger, T.V., Klag, M.J., Whelton, P.K. and Coresh, J. (1998). Familial aggregation of renal disease in a population-based case-control study. *Journal of the American Society of Neophrology*, **9**, 1270–6.

31 Brown, D.M., Provoost, A.P., Daly, M.J., Lander, E.S. and Jacob, H.J. (1996). Renal disease susceptibility and hypertension are under independent genetic control in the fawn-hooded rat. *Nature Genetics*, **12**, 44–51.

32 Freedman, B.I., Yu, H., Spray, B.J., Rich, S.S., Rothschild, C.B. and Bowden, D.W. (1997). Genetic linkage analysis of growth factor loci and end-stage renal disease in African Americans. *Kidney International*, **51**, 819–25.

7

# Early markers of nephrotoxicity for environmental and occupational monitoring

Robert G Price

Division of Life Sciences, King's College London, London, UK

## Introduction

Nephrotoxicity is a feature of many chemicals [1, 2], which find widespread use in industry and it is also a feature of a wide variety of drugs; particularly important are the heavy metals, hydrocarbons and solvents as well as insecticides, herbicides and fungicides (Table 7.1). Many important drugs are nephrotoxic, including antibiotics, immunosuppressive agents and some anticancer drugs together with nonprescription drugs including nonsteroidal anti-inflammatory drugs. Although nephrotoxicity does not preclude their use, it is important that it is recognized, the exposure minimized and the effects monitored. Initially, a nephrotoxin will affect a discrete region or cell type in the nephron, some initially attacking the glomerular cells, e.g. hydrocarbons and some solvents, while others affect the tubules, e.g. antibiotics. The kidney is particularly susceptible to circulating toxins because of its high blood flow, its ability to concentrate chemicals in the tubular fluid prior to tubular absorption, transcellular transport which results in maximal exposure of the cells to the active metabolic systems present in the tubular cells and because many chemicals can induce an immune response by the kidney. Changes in kidney structure and function are reflected in the profile of excretory products in urine; changes in serum profiles normally occur late in the damage process and are, therefore, not good markers of renal damage. Biomarkers can be divided broadly into three types for diagnostic purposes. The first are biomarkers of exposure which can be xenobiotic chemicals or metabolites which may have a long half-life, e.g. cadmium, or a short half-life, e.g. trichloroethylene. The second, biomarkers of effect, represent measurable cellular, physiological or biochemical alterations in the kidney and the most commonly used are albumin and N-acetyl-$\beta$-D-glucosaminidase (NAG). The third category includes the least well defined – the markers of susceptibility – and these indicate the individual's ability to resist the effects of exposure to a xenobiotic substance, e.g. genetic factors or pre-existing

**Table 7.1.** Common chemical agents that are nephrotoxic

| |
|---|
| Industrial and environmental compounds |
| Heavy metals |
| Organic solvents |
| Glycols |
| Insecticides, herbicides, fungicides |
| Drugs |
| Restricted under medical control |
| Antibiotics |
| Antibacterial agents |
| Antiviral agents |
| Antifungal agents |
| Immunosuppressive agents |
| Anticancer agents |
| Generally available to the public without prescription |
| Nonsteroidal anti-inflammatory agents |
| Analgesics |
| Chinese herbs |
| Illicit |
| Heroin, cocaine |

disease; this category is important in defining 'at-risk' populations in environmental and occupational studies.

## The nephrotoxic cascade

Exposure to a nephrotoxic substance initiates a cascade of events starting with the initial cellular response and then progressing via a series of potentially reversible functional events to structural damage and 'eventually' to a 'point of no return' and full-blown renal disease [3] which can progress to renal failure (Figure 7.1). The rate of progress through the cascade depends on the level of exposure and the susceptibility of the individual to the toxin. Most of the currently used clinical tests reflect late changes in the cascade, e.g. with the use of serum creatinine, and occur after progressive functional changes and structural damage have become established. The desired screening point would be when the functional change first occurs and prior to structural damage. Therefore, many investigators have concentrated their efforts on early markers of nephrotoxicity, but of equal importance is the identification of markers at each stage of the cascade. A wide range of established and novel biomarkers has been described in the literature, including: a range

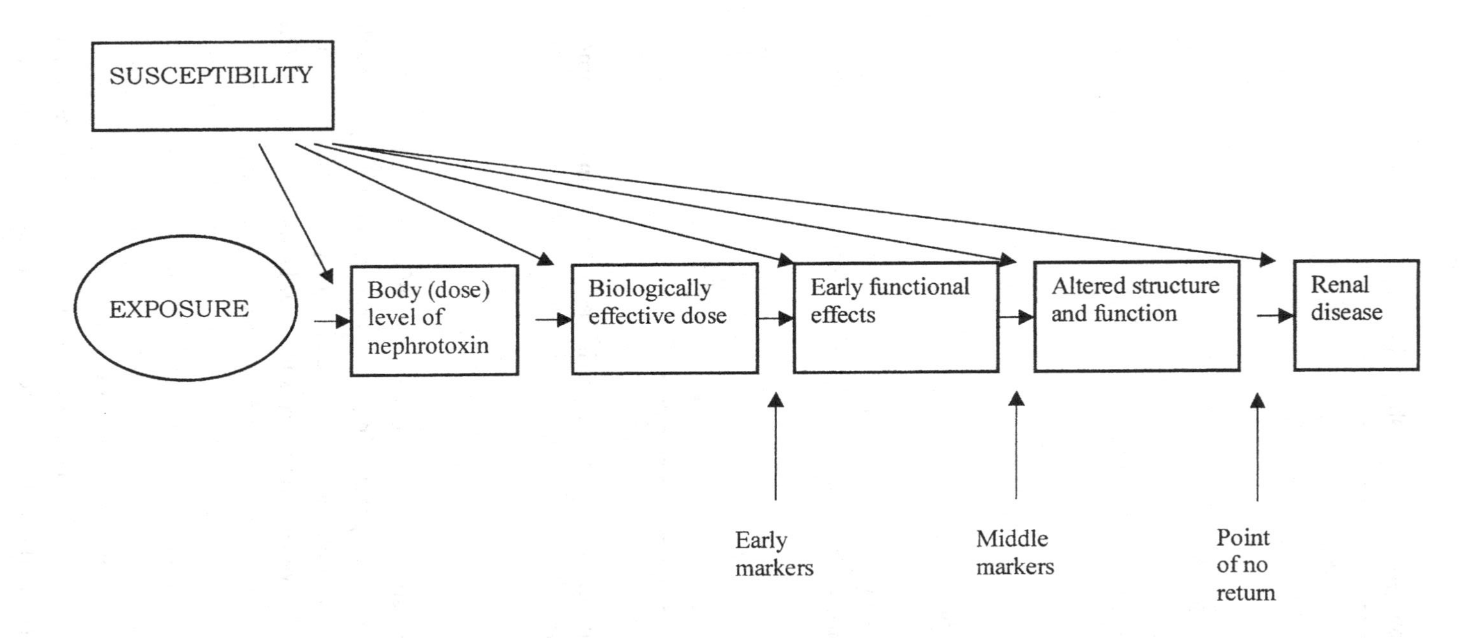


Figure 7.1 The nephrotoxic cascade. The boxes represent the progression of the damage sequence. Urinary biomarkers can reflect each step and individual susceptibility can influence each stage. The requirement is for early biomarkers which precede structural changes. Biomarkers that change later can be used to monitor progression.

of high and low molecular weight proteins; lysosomal and membrane-bound enzymes; antibody procedures that detect membrane fragments (derived from the brush border); prostaglandins; and structural components, e.g. extracellular matrix proteins. At the early stages of damage, the pattern of markers in the urine reflects the site of damage, but, as the effect becomes more severe, a mixture of marker abnormalities is present. Nephrotoxins can be distinguished by their initial effect, which can be on the glomerulus, tubules or the papillary region. As the number of segments affected increases, the profile of urinary markers changes and distinctive profiles have been observed depending on the extent of exposure, for example, to cadmium [4]. Different profiles have been reported with different agents and cohorts of individuals exposed to either lead, cadmium or mercury showed either tubular, glomerular or mixed profiles [5]. In addition to the location of the initial lesion, the subcellular site of damage also affects the urinary profile, and lysosomal (NAG) and brush border (alanine aminopeptidase) enzymes have proved, together with albumin, to be valuable early markers. Additional diagnostic information can be obtained by monitoring isoenzymes and the ratio of NAG A to NAG B changes following tubular damage.

## Application to environmental and occupational health studies

Investigations into the possible harmful effects of toxins encountered in the environment and in occupational medicine should be based on carefully prepared questionnaires for each of the individuals involved in the study. These should cover life style, drug intake and confounding factors, including smoking and alcohol consumption, and they should also seek to establish the level and length of exposure. The most valuable studies have been carried out on defined populations. A transeuropean study [5] investigated a wide range of established and novel biomarkers in workers exposed to heavy metals or solvents. A study of Italian dry-cleaners exposed to perchloroethylene demonstrated that it was possible to discriminate between the responses of different markers and to distinguish between glomerular and tubular effects – and, as a result, it was possible to establish the presence of glomerular dysfunction. Similar studies with workers exposed to lead, cadmium or mercury established that urinary changes were related to levels of exposure and that the marker profile was determined by the type of renal damage and the level of exposure. As a result of these and other studies, safe levels of exposure to metals [6] and solvents need to be redefined.

## Longitudinal studies

There is a lack of longitudinal studies into the application of biomarkers and these are necessary in order to determine the predictive value of biomarkers in general

**Table 7.2.** Potential new biomarkers for nephrotoxicity

| Biomarker | Associated structural/functional defect |
|---|---|
| Cystatin C | Glomerular filtration |
| Laminin, fibronectin | Markers of structural damage to the extracellular matrix |
| Epidermal growth factor (EGF), interleukins (IL6) | Cell response |
| $\alpha$- and $\pi$-isoenzymes of glutathione S-transferase | Located in defined regions of the nephron; provide comparison of distal and tubular effects |
| Various prostaglandins | Measure of metabolic and functional activity |

and particularly of those in common use. Populations exposed to cadmium have been monitored sequentially over an extensive period of time in Belgium, but the predictive value of commonly used biomarkers has not been clearly established. One of the difficulties with long-term studies is that, of necessity, the number of markers measured is limited, so that many possibly valuable markers are not measured. Another opportunity for longitudinal study may be to monitor exposure to immunosuppressive agents and some success has been found when NAG excretion is correlated with cyclosporine levels in the serum of individual patients.

## New biomarkers

The European study [5] incorporated a number of new markers and among the most promising were the matrix markers laminin fragments, fibronectin, the prostaglandins and intestinal alkaline phosphatase. These markers reflected both functional and structural damage. In addition, a number of other new markers are now available (Table 7.2) and, of these, cystatin C, which is used to measure glomerular function, and the isoenzymes of glutathione S-transferase, which are region specific, are particularly promising. Markers of cellular response have great promise and epidermal growth factor and the interleukins may well prove useful. The identification of new markers can only come from fundamental studies on the direct toxic effect on cells. However, the products of reactions, which may be characteristic of progression of cell damage following exposure to a toxin, will not necessarily be useful as markers of toxicity. A new marker should reflect a defined change in structure or function, preferably in a well-defined segment of the kidney. An essential criterion for all new markers is that they should be stable in urine, particularly when stored at low temperatures, and easy to assay, and the availability of kit procedures would be an advantage. The assays are best if carried out directly on the urine sample and if they avoid prior manipulation of the sample. All results should be standardized and expressed against creatinine concentration or collected over a defined time period. The second sample of the day is normally used and any abnormal results should always be confirmed.

## Cellular effects of nephrotoxins

Environmental effects on health are more difficult to study than those in potentially hazardous occupations. The effects of exposure in the general population are likely to be less severe and studies of these effects should take account of the movement of different populations. Recent studies have demonstrated that children are more susceptible to low levels of exposure than adults [7], which may reflect their increased risk due to the fact that their organ systems are still developing and they have a greater life expectancy.

Other studies have demonstrated that, if the exposure is in a restricted location (housing or near waste disposal sites), the whole population can be affected. In these situations, the assay of biomarkers is particularly valuable. The profile of excretion of biomarkers by children may not be identical to that of adults.

## Choice of biomarker

The biomarkers chosen for a study should be appropriate for the expected pattern of nephrotoxicity of the drug or toxicant under study. Since individual nephrotoxins cause different patterns of damage to the kidney, and it would be expected that the urinary biomarker profile would change, it is recommended that for a new study a small battery of tests should be used initially, and the core of tests can then be modified or expanded depending on the initial findings (Figure 7.2). The core group of tests recommended in the USA includes serum creatinine, which is viewed in Europe as insensitive. However, the core also includes albumin, which is considered to reflect glomerular function when low molecular weight proteinuria is normal, NAG and a membrane-bound enzyme alanine aminopeptidase and these, together with a low molecular weight protein (retinol binding protein), reflect aspects of tubular damage. In addition, microscopic investigation of the sediment should be carried out and a measure of osmolality and urinary creatinine taken in order to assess urine concentration. This approach is essentially similar to that adopted in the author's laboratory. A second stage would be to look in more detail at the function of individuals who give indications of abnormality in the first stage when the core battery of tests is applied. Glomerular function could be assessed by measuring transferrin and cystatin C or creatinine clearance. Tubular changes

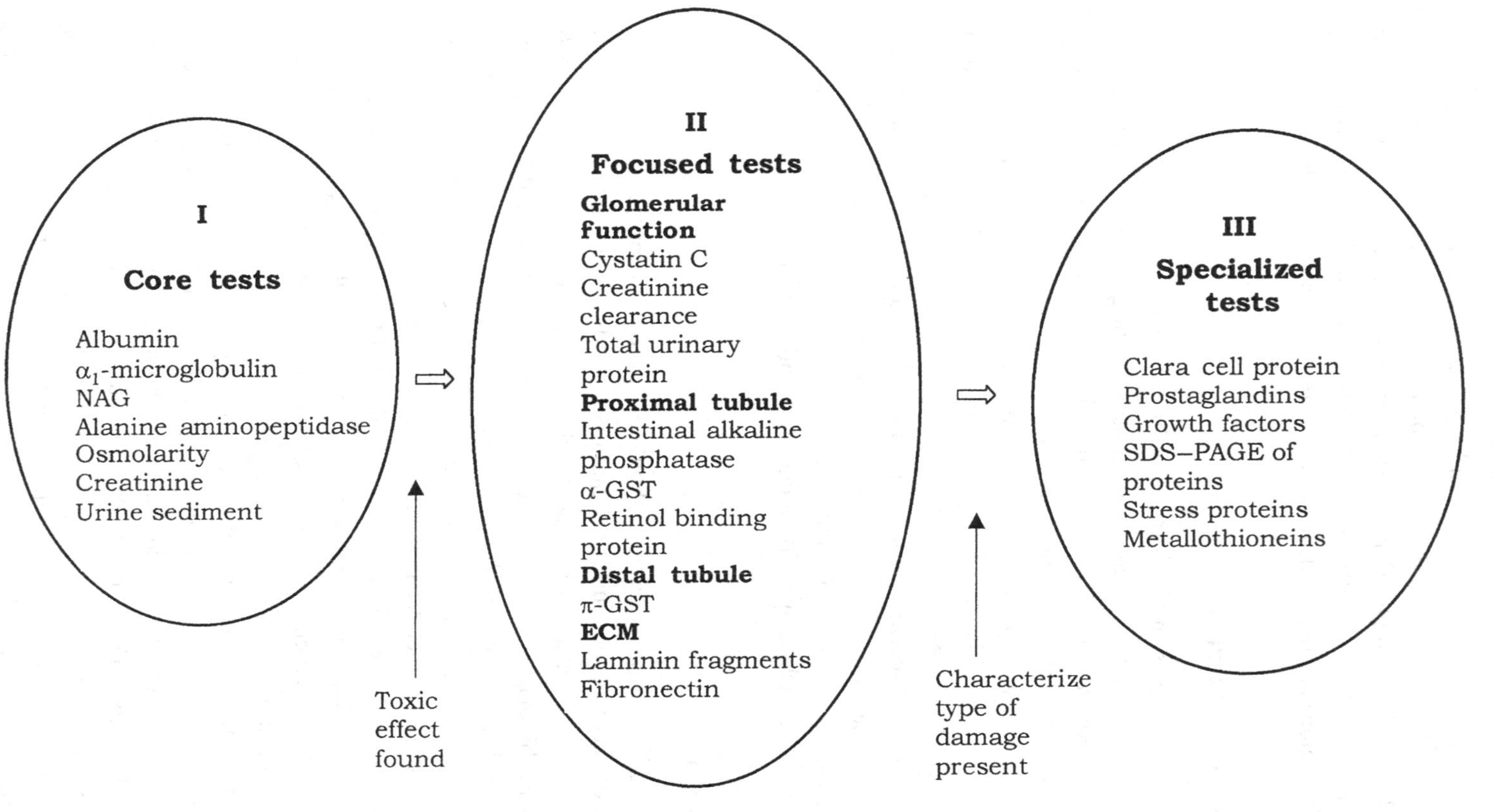


Figure 7.2 Tiered approach to using biomarkers for assessing nephrotoxicity. α-GST is α-glutathione s-transferase; ECM is extracellular matrix; SDS-PAGE is sodium dodecyl-sulphate polyacrylamide gel electrophoresis.

could be further assessed by measuring intestinal alkaline phosphatase, which is found in the third segment of the tubule, specific tubular proteins including ($\alpha_1$-microglobulin [8] and isoenzymes of glutathione-S-transferase which is a marker for the S3-segment [9]. The final stage would be to incorporate some of the newer tests outlined in Table 7.2 and, of these, the markers for the extracellular matrix and the prostaglandins are particularly promising.

## Difficulties in screening populations

One of the difficulties in screening at-risk populations is to determine which individuals are most susceptible to toxic exposure and, although the presence of pre-existing renal disease can be readily established, susceptibility due to genetic factors remains a problem. Although a number of candidate genes which are linked to end-stage real failure are known, including various growth factors and components of the renin–angiotensin system and sodium transport system, no defined link has been established. A similar situation exists in patients with diabetes where up to 40% of the patients can develop diabetic nephropathy even when blood pressure and glucose control are maintained. With the completion of the Human Genome Project, a more rational approach to the problem using susceptibility genes for nephrotoxicity should be possible and the results of these studies will aid in the monitoring of at-risk groups. A variety of in vitro methods are now available to study the direct effect of nephrotoxins on cells, and primary renal cell cultures are particularly useful. Acute models can be studied after 24 hours of exposure allowing metabolism, enzyme activities, respiration, transmembrane transport and metabolite concentrations to be measured. A chronic model would involve single or multiple exposures of up to 10 days allowing gene expression, protein, DNA and RNA synthesis, possible mutagenic effects and DNA repair to be measured. The chronic model would also give an opportunity to study acute toxicity. These types of studies would allow a whole range of potential new markers to be assessed; among them would be the cytokines, lipid mediators and growth and transcription factors. The effect of cellular production of extracellular matrix collagens and glycoprotein could also be measured. While cell adhesion molecules and heat shock proteins play a role in the mechanistic changes in cells following exposure to a nephrotoxin, they also provide a potential source of new markers. The immediate effect of exposure would be on the cell signalling system, which would result in a nuclear response leading to cellular and organ dysfunction. In turn, the structural dysfunction would eventually result in structural alterations leading, ultimately, to irreversible damage. The presence of structural damage would elicit a nuclear response and also affect the cell signalling system and, as a result, the rate of the damage cascade would accelerate towards renal failure. In order to assess the behaviour of new markers in people with various kinds of exposure, there is a need for longitudinal studies defining the various steps in the cascade of responses (effects) that result from exposure. The cascade includes different stages that interact with each other. This approach should lead to the identification of early, middle and late markers of the extent of damage and, as a result, appropriate remedial action could be taken.

## Future studies

The relationship between biomarker excretion and the damage cascade should be established alongside the evaluation of the new markers discussed above. It is essential that the appearance of a biomarker can be linked to a cellular event. This can be achieved by establishing more precisely the nephrotoxic mechanism of toxicity of different agents at the cellular level using cell culture techniques. These studies will aid in the identification of candidate genes which define the susceptibility of individuals to nephrotoxic exposure. Acceptable exposure levels should be agreed for known nephrotoxins and drugs and, where new nephrotoxins are identified, safe levels should be determined. Information on new nephrotoxins should be disseminated more widely than at present. The small battery of core tests used in many laboratories should be generally adopted and a tiered approach to environmental and occupational health monitoring adopted. This approach would allow at-risk populations to be clearly defined.

## REFERENCES


1 World Health Organization (1991). IPCS (International Programme on chemical safety). Principles and methods for the assessment of nephrotoxicity associated with exposure to chemicals. *Environmental Health Criteria*, **119**, 1–266.

2 National Research Council (1995). *Biomarkers in Urinary Toxicology*. Washington, DC, National Academy Press, pp. 16–185.

3 Mueller, P.W., Lash, L., Price, R.G., Stolte, H., Maack, T. and Berndt, W.O. (1997). Urinary biomarkers to detect significant effects of environmental and occupational exposure to nephrotoxins. I. Categories of tests for detecting effects of nephrotoxins. *Renal Failure*, **19**, 505–21.

4 Fels, L.M., Herbot, C., Pergrande, M. et al. (1994). Nephron target sites in chronic exposure to lead. *Nephrology, Dialysis, Transplantation*, **9**, 1740–6.

5 Price, R.G., Taylor, S.A., Chivers, I. et al. (1996). Development and validation of new screening tests for nephrotoxic drugs. *Human and Experimental Toxicology*, **15**, S15–19.

6 Roels, H.A., Hoet, P. and Lison, D. (1999). Usefulness of biomarkers of exposure to inorganic mercury, lead, or cadmium in controlling occupational and environmental risks of nephrotoxicity. *Renal Failure*, **21**, 251–62.

7 Price, R.G., Patel, S., Chivers, I., Milligan, P. and Taylor, S.A. (1999). Early markers of nephrotoxicity: detection of children at risk from environmental pollutants. *Renal Failure*, **21**, 303–8.
8 Guder, W.G., Ivandic, M. and Hofman, W. (1998). Physiopathology of proteinuria and laboratory diagnostic strategy based on single protein analysis. *Clinical and Chemical Laboratory Medicine*, **36**, 935–9.
9 Usuda, K., Kono, K., Dote, T. et al. (1998). Urinary biomarkers monitoring for experimental fluoride nephrotoxicity. *Archives of Toxicology*, **72**, 104–9.

8

# The early detection of renal impairment in diabetes mellitus. The case for microalbuminuria and other biomarkers

Carl E Mogensen

Aarhus Kommunehospital/Aarhus University Hospital, Aarhus, Denmark

Several chronic diseases may be without major symptoms or signs for many years; complications in diabetes and hypertension, e.g. early diabetic nephropathy or renal dysfunction, are excellent examples of this important feature [1, 2]. Nevertheless, renal disease may often, without intervention, progress to end-stage renal failure with an eventual need for dialysis or transplantation [3–5]. In fact, diabetes is now one of the major causes of end-stage renal failure in the USA and in Europe, as well as in the rest of the world. In many dialysis units, 30–50% of patients are diabetics [6]. The number of patients with type 2 diabetes developing end-stage renal disease is clearly increasing, as recently reviewed by Ritz and Rychlik [7]. Uraemia in type 1 diabetes is also a major concern. On the other hand, complications of diabetes may be postponed, or hopefully eventually prevented, by early and more effective management [1, 8, 9].

Since there are only a few symptoms or signs of the early stage of the disease, biomarkers should serve to identify those patients at risk of developing renal disease. Together with other organ biomarkers, these should also be used for the evaluation of disease progression, especially with respect to defining the appropriate time-point for intervention. In addition, biomarkers may also be used to evaluate treatment effect [1]. Usually, long-term follow ups are needed to define clear benefits with respect to preventing organ damage and, more specifically, end-stage renal disease and mortality (Figure 8.1).

Earlier studies indicated that about 30–40% of patients would develop end-stage renal disease and that the remaining would not. Therefore, it was argued that important susceptibility factors could be present, possibly related to genetic factors that could potentially identify patients at risk.

This chapter will survey a number of biomarkers proposed over the years. Obviously, the basis for the disease is its pathology: diabetic glomerulopathy, a slowly and gradually progressive disorder starting in glomeruli with subtle changes

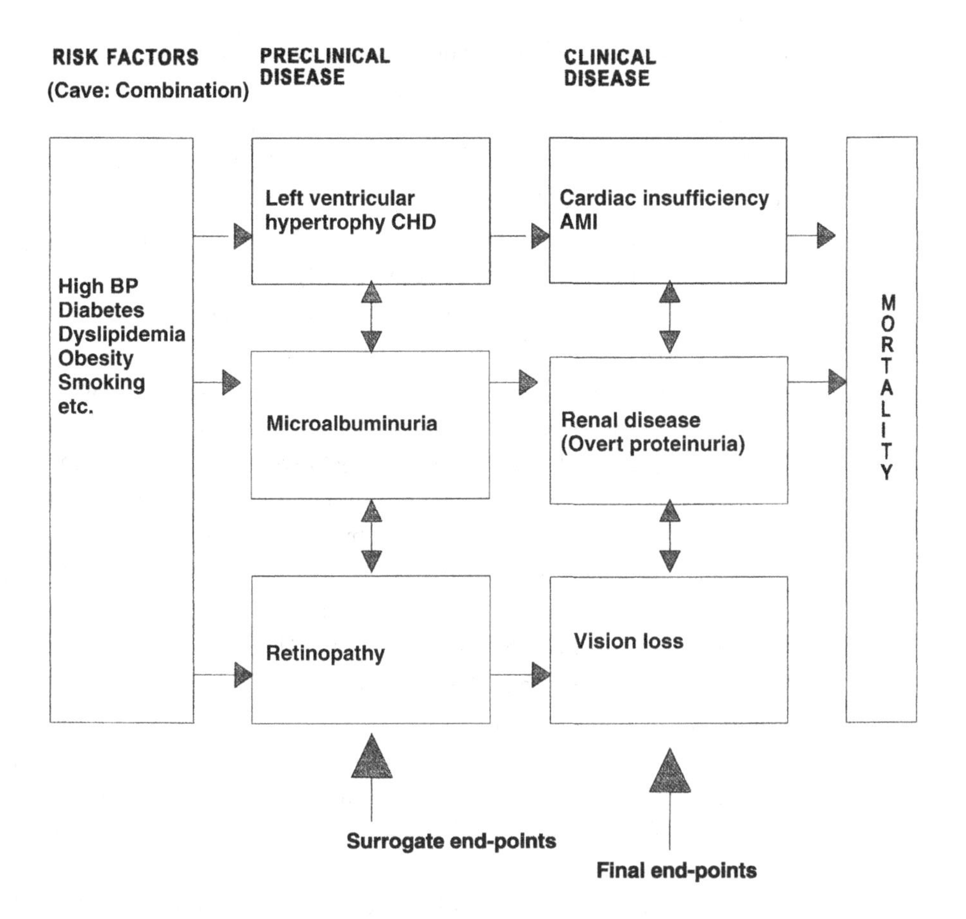


Figure 8.1 Risk factors and risk markers of diabetes and its complications. CHD is coronary heart disease and AMI is acute myocardial infarction.

which are only detectable by morphometry but ending up in total sclerotic and occluded glomeruli. Other parts of the nephron and the kidney may also be involved, but to a lesser and possibly secondary extent [10, 11].

## The structural basis for renal disease

Renal biopsies are progressively less used in the diagnosis of diabetic renal disease because the clinical implications of biopsy readings are increasingly limited. Some focus should be on the detection of nondiabetic renal disease possibly found in diabetic patients, but this is, in fact, quite rare and clinical management is usually not affected by biopsy descriptions [1, 12, 13]. In fact, this author has never seen a diabetic patient in whom treatment was changed due to renal biopsy readings.

By contrast, renal biopsies are indeed extremely useful when describing the natural history in research projects [10, 11, 14–22], as well as when defining the effect of treatment in controlled clinical trials [14].

In type 1 diabetes, structural lesions are not present at the time of diagnosis of nephropathy, although nephromegaly and glomerulomegaly may be quite common [11, 23, 24] These phenomena are, however, not signs of diabetic renal disease but may be predictive of late structural lesions and clinical nephropathy [23]. In type 1 diabetes, the structural changes, e.g. increase in basement membrane thickness and mesangial expansion, develop a few years after the onset of diabetes, and progress is thereafter related to glycaemic control and blood pressure [16]. Another risk factor is obviously duration of diabetes or exposure time to risk factors. Initially, the changes are only seen by morphometry, but, later on, clear-cut changes are found as initially described by Kimmelstiel and Wilson [25]. Regression of a lesion may or may not be seen with glycaemic normalization and stabilization with antihypertensive treatment [14]. Earlier in the history of diabetology, it was suggested that changes could be found at the time of diagnosis depending on 'genetic elements' of the patients [26].

In type 2 diabetes, the picture is not so clear since patients may have proteinuria or microalbuminuria without any significant changes and the biopsy may be read as 'near-normal structure' [13]. However, such readings do not exclude a rather rapid progression of diabetic renal disease, especially in the presence of profound proteinuria, which is the major risk marker [1]. More nonspecific changes can be present in these patients due to, for example, atherosclerosis or hypertension [21]. One should also remember that lesions may be located in the interstitium and in tubular cells. Indeed, changes in the vas-efferent and vas-afferent vessels are characteristic of diabetic renal disease [11]. Again, however, the clinical value of biopsy specimens, also in type 2, is quite limited. Some studies have described a rather high percentage of patients with nondiabetic renal lesions in type 2 diabetes with proteinuria, but more recent studies found a few, maybe 10% [13], and, in microalbuminuric type 2 patients, nondiabetic renal lesions are exquisitely rarely found, just as in type 1 diabetes [1]. Since diabetes may remain undiagnosed for several years in this type of diabetes, it is not surprising that lesions may be present when diabetes is clinically diagnosed – just as in the case of retinopathy.

To conclude, renal biopsies are very rarely indicated in patients with either type of diabetes, and the clinical implications are practically nonexistent. However, exact readings from renal biopsies by morphometry are key issues in research [11].

**Table 8.1.** Diagnostic limits of normo-, micro- and macroalbuminuria

| Condition | 24-hour urinary albumin excretion rate | Overnight urinary albumin excretion rate | Albumin: creatine ratio |
|---|---|---|---|
| Macroalbuminuria (overt nephropathy) | >300 mg/day | >200 μg/min | >25 mg/mmol |
| Microalbuminuria | 30–300 mg/day | 20–200 μg/min | 2.5–25.0 mg/mmol (for men)<br>3.5–25.0 mg/mmol (for women) |
| Normoalbuminuria | <30 mg/day | <20 μg/min | <2.5 mg/mmol (for men)<br><3.5 mg/mmol (for women) |

## Glycaemic control, the main metabolic basis

Good glycaemic control is obviously of paramount importance in preventing nephropathy [27–31]. Historically, it is strange and ironic that this message only came many years after the discovery of insulin and oral agents for the treatment of diabetes [1]. Continuous glycated haemoglobin (HbA1c) measurement is fundamental in this respect to all diabetes care [32–35].

## Normoalbuminuria, microalbuminuria and macroalbuminuria

Measurement of albumin in small concentrations is a necessary tool to detect microalbuminuria [1, 36–38]. Microalbuminuria is usually defined as a urinary excretion rate of albumin between 20 and 200 μg/min under standard conditions [39]. Values below this rate are termed normoalbuminuria and, values above, macroalbuminuria or clinical proteinuria (Table 8.1).

Microalbuminuria is an important risk marker for future nephropathy in type 1 and type 2 diabetes [1]. The progression rate in microalbuminuria is best described in type 1 diabetes and is 10–20% per year, influenced by the level of blood pressure and HbA1c and the level of albuminuria in itself [1]. In longitudinal studies, about 8% of patients per year may progress to clinical proteinuria – of course, mostly those in the upper range of microalbuminuria [1, 3, 31]. It should be emphasized that albuminuria is a continuous variable and that also patients with 'high normoalbuminuria' are at higher risk for progression compared with patients with very low values.

There is some correlation between albuminuria and structural lesions, although the correlation is not very precise [11, 15]. This is particularly true in type 2 diabetes and there is a large overlap between normo-, micro- and macroalbuminuria with respect to structural lesions [40]. Since it is known that microalbuminuria clearly predicts overt disease, measurement of microalbuminuria should be more sensitive than biopsies in detecting patients at risk. This is another good reason to abstain from biopsies in such patients since microalbuminuria is an ideal biomarker for damage and dysfunction – it can be monitored longitudinally both easily and cost-effectively with a minimum of inconvenience for the patients, just like HbA1c.

In addition, microalbuminuria predicts early mortality both in type 1 and type 2 diabetes [1, 41–48], and indeed also in the background population [41]. Therefore, detection of microalbuminuria is important in order to identify persons at risk in clinical trials such as the Heart Outcomes and Prevention Evaluation (HOPE) Study [1]. In the latter, microalbuminuric patients also had the greatest benefit from treatment with the angiotensin converting enzyme (ACE) inhibitor.

It can be concluded that the measurement of albuminuria is a key marker in the monitoring of patients – in general using an early morning urine and measuring the albumin to creatinine ratio (Table 8.1). For treatment purposes, this is the ideal marker for the detection of disease and follow up of disease progression. Several studies show that microalbuminuria usually indicates that the glomerular filtration rate (GFR) is still well preserved in type 1 diabetes at a supranormal level [1, 49–52]. A fall in GFR usually takes place in the phase of overt diabetic renal disease with proteinuria. Now, studies document preservation of GFR by ACE inhibition [53]. Abnormal albuminuria is often found in patients with uncontrolled hypertension [54–60]. Microalbuminuria is also found in patients with cardiovascular disease [61, 62].

## Estimation of GFR and renal function

It is characteristic for diabetic patients with normo- or microalbuminuria for the GFR to be increased above normal, the so-called hyperfiltration phenomenon [1, 49–51, 63] although this is not so pronounced in type 2 diabetes. Hyperfiltration may be predictive of late nephropathy [50, 51], as nephromegaly is also [23]. Therefore, in early diabetic renal disease, it can be expected that serum creatinine or other similar serum or plasma markers may be decreased, reflecting the glomerular hyperfiltration. With overt nephropathy, GFR starts to decrease, correlated to blood pressure level and HbA1c [1, 64]. Without any intervention, GFR typically decreases around 10 ml/min/year in type 1 diabetes and to a similar extent in type 2 diabetes in poorly controlled individuals [65]. However, this rate of decline can clearly be reduced by antihypertensive treatment and improved glycaemic control – the rate can be most clearly reduced in type 1 diabetes [1, 64].

In the clinical management of patients, GFR is usually not measured by precise techniques, with clinicians relying on the serum creatinine or reciprocal serum creatinine; markers relating serum substances to GFR (The Cockroft Gould Equation) may also be used. Reciprocal serum creatinine is interesting in the routine follow up of patients with early nephropathy in order to assess whether intervention is effective with respect to preservation of GFR.

In several trials, a doubling or 50% increase of serum creatinine has been used as a parameter for effective treatment [1]. However, more precise procedures for GFR determination may be interesting in studies with fewer patients, such as described earlier in the first trials documenting the beneficial effects of antihypertensive treatment in patients with type 1 diabetes [1].

It can be concluded that the measurement of serum creatinine or related parameters is important in the management of patients to detect early deterioration in GFR. However, accurate measurements of GFR are rarely performed in the clinic and are probably not needed for clinical management. More exact and practical markers should be sought. Obviously, in the follow up of patients, a careful evaluation of electrolytes is important, especially of serum potassium.

Regarding nephromegaly, it has been known for many years that the kidney volume is increased in patients with diabetes. This is more pronounced with poor metabolic control – for instance, before the treatment of diabetes. After initial insulin treatment, the kidney size is usually reduced [1]. In patients who are poorly treated with respect to glycaemia, the kidneys are often enlarged and this may be predictive of later nephropathy [23]. However, measurement of renal size is rarely done in such patients, although it may nevertheless be a marker of prolonged poor metabolic control.

## Genetic markers, family history and birth weight

Family clustering of diabetic renal disease suggests that genetic factors could be important in determining so-called susceptibility to nephropathy [66–73]. In some studies, diabetic siblings of probands with type 1 diabetes had a prevalence of diabetic nephropathy. Taking this fact into consideration, along with a cumulative incidence of diabetic renal disease in the siblings of diabetic patients that is as high as 70% versus 25% in siblings of probands without renal disease, again suggested a possible genetic background for renal disease in diabetes [73].

Several genes have been proposed as biologically interesting candidates for the so-called 'nephropathy gene'. Polymorphisms of the angiotensin converting enzyme, the so-called I-D-genotype, have been proposed, and the angiotensinogen gene and angiotensin II type 1 receptor gene have also been suggested. The I-D genotype has been examined in various populations in addition to other genes, e.g. the aldoreductase gene and lipid-related genes [66, 74].

So far, examination of these candidate genes in case-controlled studies has produced very conflicting results. However, it may well be that larger studies using new strategies such as family-based association studies of linkage analysis of sibpairs might give better information, although this is still unknown. In fact, a key point may be that the main determinants of diabetic renal disease are still poor glycaemic control and increased blood pressure. These two factors especially, considered together, seem to be major determinants of diabetic renal disease. So far, genetic analysis has found no place in the clinical management of diabetic patients and cannot be used as a biomarker for renal disease. The role of low birth weight and related parameters is also uncertain [1, 75, 76].

## Blood pressure level, the main haemodynamic basis

A rise in blood pressure is usually seen in parallel with microalbuminuria, at least in type 1 diabetes [77–82]. However, there is clearly an overlap in blood pressure between normo-, micro- and macroalbuminuria even after careful matching for age, sex and duration of diabetes [65].

In type 2 diabetes, blood pressure is usually increased with microalbuminuria, but again there is a huge overlap, and many patients with type 2 diabetes have elevated blood pressure even with normoalbuminuria and at the clinical diagnosis of diabetes [83]. This may be related to the metabolic syndrome or to simple obesity.

Ambulatory 24-hour blood pressure recordings give precise and detailed information [77], and it appears that blood pressure is increased in patients with renal complications. In follow up studies, there is an increase in blood pressure by about 3–4 mmHg per year in patients with microalbuminuria as compared with normoalbuminuric patients where the increase in blood pressure is usually 1 mmHg per year. In overt nephropathy, the increase is around 5–6 mmHg per year [1].

The best evidence for the importance of the need to control blood pressure stems from clinical trials. Several clinical trials in type 1 diabetes have shown that effective treatment of elevated blood pressure is important in slowing down progression. Even the treatment of blood pressure within the normal range appears to be beneficial, especially when using ACE inhibitors in microalbuminuric patients [58, 83–89].

To conclude, there is no clear-cut correlation between blood pressure level and development of renal disease; rather, the two abnormalities are characterized by a considerable overlap. However, a low blood pressure level is usually associated with a good prognosis. Careful monitoring of blood pressure is obviously of paramount importance in the management of diabetic patients.

## Cation cell transporters

Several studies have been performed based on the finding that increased erythrocyte sodium–lithium counter transport may be seen in patients with essential hypertension. Many studies have been published in this area but with conflicting results. A few papers suggest that there may be an increased number of cation cell transporters in patients with complications or at risk of complications, but in other studies this has not been confirmed. There are also methodological problems, as described in several papers, and certainly at this time results on cations transporters need to be confirmed in more reliable and large-scale studies. So far, measurements can by no means be used as a reliable clinical biomarker [90–97].

## Renin and prorenin and the renin angiotensin system

In cross-sectional studies, plasma prorenin has been shown to be elevated in patients with microalbuminuria. There are also results suggesting that prorenin may increase before the development of microalbuminuria in type 1 diabetes, but certainly there is a great overlap between progressors and nonprogressors [98–101].

The renin angiotensin system is important in diabetes, but mainly with respect to treatment with ACE inhibitors that are important in reducing renal complications. This effect may be related to blood pressure, but there seem also to be elements of intrinsic beneficial effects that are independent of reduction in blood pressure.

The renin angiotensin system is quite complex and there are related enzymes such as bradykinins that may be involved, along with other enzyme systems [102, 103]. To date, however, no discriminant function of any parameter of the renin angiotensin system has been described as predicting the risk of diabetic renal disease.

## Endothelial dysfunction and homocysteinaemia

There is some evidence that widespread endothelial dysfunction may result from endothelial damage that could be of importance in determining renal and cardiovascular disease [104, 105]. Some studies suggest that endothelial dysfunction is present in patients with microalbuminuria, especially circulating levels of von Willebrand factor and related markers. Other studies also suggest that concentrations of the von Willebrand factor may increase before the development of microalbuminuria. However, it can be concluded that more work needs to be done in this area and, in the clinical situation, measurement of endothelial dysfunction is presently not relevant in the management of patients with diabetes at risk for diabetic renal disease [106–110].

## Insulin resistance

Both type 1 and type 2 diabetics may be characterized by some degree of insulin resistance, but this may in fact be unrelated to microalbuminuria [1]. For example, in some studies in type 2 diabetes, no difference has been found in characteristic parameters related to insulin resistance – e.g. using the euglycaemic insulin clamp technique. Most studies are rather small and it seems unlikely that the measurement of insulin resistance will ever become useful in the prediction of diabetic renal disease or in the prediction of any complications related to diabetes, although insulin resistance may be related to blood pressure elevation and dyslipidaemia. It remains much more useful to focus on the direct and careful monitoring of blood pressure and lipid abnormalities, which are key elements of the insulin resistance syndrome [111–114].

## Lipid abnormalities

Lipid abnormalities may occur early in the course of renal disease, both in type 1 and type 2 diabetes. This may be the case not only for total cholesterol but also low-density lipoprotein (LDL)-cholesterol and triglycerides. Several studies have been performed in this area and there may be some association between nephropathy and increased serum lipids [12].

Despite some evidence, it seems clear that lipid concentrations are not useful in the prediction of diabetic renal disease as these abnormalities lack not only sensitivity but also specificity. Obviously, measurements of dyslipidaemia are important, but mainly for the more effective treatment of dyslipidaemia per se, which may be an important predictor of cardiovascular complications in diabetes [115–118].

## Other concomitant microvascular lesions and cardiovascular disease

A number of studies suggest that retinopathy, neuropathy and nephropathy are often found in the same patients. There are, however, important exceptions, especially in patients with type 2 diabetes and, more specifically, in nephropathy patients with type 2 diabetes who may not have retinopathy [1]. However, similar risk factors may be important for the development of retinopathy and nephropathy and the same treatment modalities also seem relevant [119, 120].

Therefore, although the presence of other microvascular complications should alert clinicians that patients are at risk of diabetic renal disease, they are not useful as real clinical markers for nephropathy; rather, evaluation of retinopathy and other microvascular lesions is important in its own right, especially as regards the early treatment of diabetic retinopathy. Certainly, evaluation of retinopathy, especially retinal photographs, is part of the package of early evaluation for microvascular complications in diabetes, with the option of laser treatment.

## Smoking

Several studies have suggested that cigarette smoking is a risk factor associated with microalbuminuria, both in type 1 and type 2 diabetes [121–123]. There is some evidence that smoking could exacerbate endothelial changes and damage and, therefore, induce microalbuminuria. However, there is too great an overlap between smokers and nonsmokers, and the relationship is weak and not really useful in predicting those patients at risk. Smoking should, however, be discouraged.

## Treatment options: based on biomarkers and clinical course

The main purpose in identifying risk factors is to define the clinical course and, primarily, to indicate treatment options. In this respect, evaluation of microalbuminuria is an ideal marker for the early identification of diabetic renal disease and several studies have been performed even in patients with normal blood pressure and microalbuminuria [1].

A total of 10 clinical trials with a duration of 2 years or more have been completed [1]. Practically all studies show a reduction of microalbuminuria and two important studies suggest an effect on stronger end-points such as preservation of GFR and preservation of glomerular structure [14]. Therefore, measurement of microalbuminuria is a key issue not only with respect to defining those patients at risk and determining the clinical course but also when aiming to identify those patients who are in need of better glycaemic control, as shown in the DCCT (Diabetes Control of Complications Trial), postDCCT study and other studies, for example [29–31, 88, 124–127]. However, on the whole, all patients need the best glycaemic control attainable, although this may be more pertinent to patients with microalbuminuria, because they are at an even greater risk. The main focus should be early treatment with ACE inhibitors in patients with persistent microalbuminuria. There is good evidence that this will ameliorate the course of renal disease and also the course of cardiovascular disease, as shown in the recent HOPE and MICRO-HOPE studies [1]. According to new evidence, angiotensin receptor antagonist should be first choice in patients with type 2 diabetes and proteinuria [128, 129] and possibly microalbuminuria [102, 130].

## Conclusions

Prediction of overt diabetic renal disease, characterized by proteinuria, decreased GFR and hypertension, is fairly simple. Patients should be carefully screened for

**Table 8.2.** Alternative diagnostic and treatment strategies in diabetes mellitus

| Strategy | Helpful in diagnosis | Helpful in treatment |
|---|---|---|
| Analysis of genetic factors | No, but studies are needed | At-risk patients cannot yet be found<br>No genetic modulation possible |
| Familial factors | To some extent (early diagnosis of hypertension) | No |
| Antiglycaemic treatment | Yes ($HbA_{1C}$ monitoring) | Yes, clearly demanding and sometimes not feasible |
| Various types of ACE inhibitors | No | Some renal studies stopped; other studies started |
| Aldose reductase inhibitor | No | No, renal studies stopped (?) |
| Growth factor inhibition | No | Needs investigation |
| Protein kinase C inhibitors | No | Needs investigation |
| Antihypertensive treatment | Yes, often along with microalbuminuria and albuminuria or high BP | Yes, mainly ACE inhibitors as basis for combination therapy |
| ACE inhibitors | No (genotyping not useful) | Yes, profoundly, especially in microalbuminuric patients |
| Lipid-lowering | Yes, dyslipidaemia | Probably, but needs further confirmation |
| Aspirin | No | Probably (only for macrovascular disease) |
| C-peptide | Sometimes | Further studies planned |
| Endothelial and endopeptidase inhibitors | No | Under investigation |
| Glycosaminoglycans | No | Under investigation |
| Metalloproteinase | No(?) | Under investigation |

microalbuminuria in a longitudinal fashion. Those developing consistent microalbuminuria should be treated with ACE inhibitors, not only with small doses but probably with medium to high doses. Diuretics may be indicated as well as other antihypertensive agents if blood pressure is not sufficiently reduced. For instance, beta-blockers are useful in many type 2 diabetic patients with microalbuminuria [89] and, indeed, combination therapy is also effective. New studies suggest that dual blockade of the renin angiotensin system is important in reducing blood pressure more efficiently in patients with type 2 diabetes and microalbuminuria [102]. There is a good theoretical background for this treatment because complete blockade of the system may not be obtained with ACE inhibitors alone, and, indeed, using receptor blockers will not lead to the elevation of bradykinins – which is a property of these agents that can also be useful in reducing blood pressure in patients with diabetes.

It can be concluded that a considerable number of pathogenetic factors or biomarkers has been proposed. The focus today, however, is on the main biomarker, albuminuria, irrespective of its level; the main determinants of disease progression are hyperglycaemia and elevated blood pressure, local or systemic. Alternatives in diagnosis and treatment strategies are listed in Table 8.2 [1]. A main focus is to circumvent the deleterious effect of elevated blood and tissue glucose by so-called nonglycaemic intervention, where reduction of blood pressure is the only successful measure so far. Normoglycaemia in most diabetic patients is still an elusive goal, although this is a main focus in treatment as well as in research programmes. New antihyperglycaemic treatment strategies are also important to improve diabetes care.

## REFERENCES

1 Mogensen, C.E. (2000). Microalbuminuria, blood pressure and diabetic renal disease: origin and development of ideas. In *The Kidney and Hypertension in Diabetes Mellitus*, ed. C.E. Mogensen. Boston, Dordrecht and London: Kluwer Academic Publishers, pp. 655–706.

2 Thomas, S.M. and Viberti, G.C. (2000). Is it possible to predict diabetic kidney disease? *Journal of Endocrinological Investigation*, **23**, 44–53.

3 Mogensen, C.E. and Christensen, C.K. (1984). Predicting diabetic nephropathy in insulin-dependent patients. *New England Journal of Medicine*, **311**, 89–93.

4 Parving, H.H., Oxenboll, B., Svendsen, P.A., Christiansen, J.S. and Andersen, A.R. (1982). Early detection of patients at risk of developing diabetic nephropathy. A longitudinal study of urinary albumin excretion. *Acta Endocrinologica (Copenhagen)*, **100**, 550–5.

5 Viberti, G.C., Hill, R.D., Jarrett, R.J., Argyropoulos, A., Mahmud, U. and Keen, H. (1982). Microalbuminuria as a predictor of clinical nephropathy in insulin-dependent diabetes mellitus. *Lancet*, **1**, 1430–2.

6 Hostetter, T.H. (2001). Prevention of end-stage renal disease due to type 2 diabetes. *New England Journal of Medicine*, **345**, 910–12.

7 Ritz, E. and Rychlik, I. (eds.) (1999). *Nephropathy in Type 2 Diabetes. Oxford Clinical Nephrology Series*. New York: Oxford University Press.

8 Ibrahim, H.A.A. and Vora, J.P. (1999). Diabetic nephropathy. *Baillière's Clinical Endocrinology and Metabolism*, **13**, 239–64.

9 Mogensen, C.E., Christensen, C.K. and Vittinghus, E. (1983). The stages in diabetic renal disease. With emphasis on the stage of incipient diabetic nephropathy. *Diabetes*, 2 (Suppl 32) 64–78.

10 Osterby, R. (1965). A quantitative estimate of the peripheral glomerular basement membrane in recent juvenile diabetes. *Diabetologia*, **1**, 97–100.

11 Osterby, R. (1996). Lessons from kidney biopsies. *Diabetes Metabolism Reviews*, **12**, 151–74.

12 Olsen, S. and Mogensen, C.E. (1996). How often is type 2 diabetes mellitus complicated with non-diabetic renal disease? An analysis of renal biopsies and the literature. *Diabetologia*, **39**, 1638–45.
13 Olsen, S. (2000). Light microscopy of diabetic glomerulopathy: the classic lesions. In *The Kidney and Hypertension in Diabetes Mellitus*, ed. C.E. Mogensen. Boston, Dordrecht and London: Kluwer Academic Publishers, pp. 201–10.
14 Rudberg, S., Østerby, R., Bangstad, H.-J., Dahlquiest, G. and Persson, B. (1999). Effect of angiotensin converting enzyme inhibitor or beta blocker on glomerular structural changes in young microalbuminuric patients with type 1 (insulin-dependent) diabetes mellitus. *Diabetologia*, **42**, 589–95.
15 Bangstad, H.J., Osterby, R., Dahl-Jorgensen, K. et al. (1993). Early glomerulopathy is present in young type 1 (insulin-dependent) diabetic patients with microalbuminuria. *Diabetologia*, **36**, 523–9.
16 Bangstad, H.-J., Rudberg, S. and Østerby, R. (2000). Renal structural changes in patients with type 1 diabetes and microalbuminuria. In *The Kidney and Hypertension in Diabetes Mellitus*, ed. C.E. Mogensen: Boston, Dordrecht and London: Kluwer Academic Publishers, pp. 211–24.
17 Bertani, T., Gambara, V. and Remuzzi, G. (1996). Structural basis of diabetic nephropathy in microalbuminuric NIDDM patients. a light microscopy study. *Diabetologia*, **39**, 1625–8.
18 Chavers, B.M., Bilous, R.W., Ellis, E.N., Steffes, M.W. and Mauer, M. (1989). Glomerular lesions and urinary albumin excretion in type 1 diabetes without overt proteinuria. *New England Journal of Medicine*, **320**, 966–70.
19 Chavers, B.M., Mauer, S.M., Ramsay, R.C. and Steffes, M.W. (1994). Relationship between retinal and glomerular lesions in IDDM patients. *Diabetes*, **43**, 441–6.
20 Fioretto, P., Matier, M., Brocco, E. et al. (1996). Patterns of renal injury in NIDDM patients with microalbuminuria. *Diabetologia*, **39**, 1569–76.
21 Fioretto, P., Stehouwer, C.D., Mauer, M. et al. (1998). Heterogeneous nature of microalbuminuria in NIDDM. Studies of endothelial function and renal structure. *Diabetologia*, **41**, 233–6.
22 Walker, J.D., Close, C.F., Jones, S.L. et al. (1992). Glomerular structure in type-1 (insulin-dependent) diabetic patients with normo- and microalbuminuria. *Kidney International*, **41**, 741–8.
23 Baumgartl, H.J., Banholzer, P., Sigl, G., Haslbeck, M. and Standl, E. (1998). On the prognosis of IDDM patients with large kidneys. The role of large kidneys for the development of diabetic nephropathy. *Nephrology, Dialysis, Transplantation*, **13**, 630–4.
24 Bernard, C. (1849). *Compte Rendu de la Société du Biologie*, **1**, 80–1.
25 Kimmelstiel, P. and Wilson, C. (1936). Intercapillary lesions in the glomeruli of the kidney. *American Journal of Pathology*, **12**, 83–95.
26 Siperstein, M.D., Unger, R.H. and Madison, L.L. (1968). Studies of muscle capillary basement membranes in normal subjects, diabetic and prediabetic patients. *Journal of Clinical Investigation*, **47**, 1973–99.
27 Keiding, N.R., Root, H.F. and Marble, A. (1952). Importance of control of diabetes in prevention of vascular complications. *Journal of the American Medical Association*, **150**, 964–9.

28 Pirart, J. (1978). Diabetes mellitus and its degenerative complications: a prospective study of 4,400 patients observed between 1947 and 1973. *Diabetes Care*, **1**, 168–88.

29 The Diabetes Control and Complications Trial Research Group (1993). The effect of intensive treatment of diabetes on the development and progression of long-term complications in insulin-dependent diabetes mellitus. *New England Journal of Medicine*, **329**, 977–86.

30 The Diabetes Control and Complications Trial Research Group (1996). The absence of a glycemic threshold for the development of long-term complications. The perspective of the Diabetes Control and Complications Trial. *Diabetes*, **45**, 1289–98.

31 The Diabetes Control and Complications Trial Research Group (2000). Retinopathy and nephropathy in patients with type 1 diabetes four years after a trial of intensive therapy. *New England Journal of Medicine*, **342**, 381–9.

32 Krolewski, A.S., Laffel, L.M., Krolewski, M., Quinn, M. and Warram, J.H. (1995). Glycosylated hemoglobin and the risk of microalbuminuria in patients with insulin-dependent diabetes mellitus. *New England Journal of Medicine*, **332**, 1251–5.

33 Little, R.R. and Goldstein, D.E. (1994). Measurements of glycated haemoglobin and other circulating glycated proteins. In *Research Methodologies in Human Diabetes, Part 1*, eds. C.E. Mogensen and E. Standl. Berlin, New York: Walter de Gruyter.

34 Tanaka, Y., Atsumi, Y., Matsuoka, K., Onuma, T., Tohjima, T. and Kawamori, R. (1998). Role of glycemic control and blood pressure in the development and progression of nephropathy in elderly Japanese NIDDM patients. *Diabetes Care*, **21**, 116–20.

35 Warram, J.H., Scott, L.J., Hanna, L.S. et al. (2000). Progression of microalbuminuria to proteinuria in type 1 diabetes. Non-linear relationship with hyperglycemia. *Diabetes*, **49**, 94–100.

36 Berggård, I. and Risinger, C. (1961). Quantitative immunochemical determination of albumin in normal human urine. *Acta Societatis Medicorum Upsaliensis*, **66**, 217–29.

37 Dati, F. and Lammers, M. (1989). Immunochemical methods for determination of urinary proteins (albumin and α1-microglobin) in kidney disease. *Journal of the International Federation of Clinical Chemistry*, **1**, 68–77.

38 Keen, H. and Chlouverakis, C. (1963). An immunoassay method for urinary albumin at low-concentration. *Lancet*, **ii**, 913–14.

39 Mogensen, C.E., Chachati, A., Christensen, C.K. et al. (1985–6). Microalbuminuria: an early marker of renal involvement in diabetes. *Uremia Investigation*, **9**, 85–95.

40 Nosadini, R. and Fioretto, P. (1999). Renal involvement in type 2 diabetes mellitus: prognostic role of proteinuria and morphological lesions. *Journal of Nephrology*, **12**, 329–46.

41 Damsgaard, E.M., Froland, A., Jorgensen, O.D. and Mogensen, C.E. (1992). Eight to nine year mortality in known non-insulin dependent diabetics and controls. *Kidney International*, **41**, 731–5.

42 Jarrett, R.J., Viberti, G.C., Argyropoulos, A., Hill, R.D., Mahmud, U. and Murrells, T.J. (1984). Microalbuminuria predicts mortality in non-insulin-dependent diabetics. *Diabetic Medicine*, **1**, 17–19.

43 Macleod, J.M., Lutale, J. and Marshall, S.M. (1995). Albumin excretion and vascular deaths in NIDDM. *Diabetologia*, **38**, 610–16.

44 Mattock, M.B., Morrish, N.J., Viberti, G.C., Keen, H., Fitzgerald, A.P. and Jackson, G.

(1992). Prospective study of microalbuminuria as predictor of mortality in NIDDM. *Diabetes*, **41**, 736–41.

45 Mogensen, C.E. (1984). Microalbuminuria predicts clinical proteinuria and early mortality in maturity-onset diabetes. *New England Journal of Medicine*, **310**, 356–60.

46 Neil, A., Hawkins, M., Potok, M., Thorogood, M., Cohen, D. and Mann, J. (1993). A prospective population-based study of microalbuminuria as a predictor of mortality in NIDDM. *Diabetes Care*, **16**, 996–1003.

47 Schmitz, A. and Vaeth, M. (1988). Microalbuminuria: a major risk factor in non insulin dependent diabetes. A 10 year follow up study of 503 patients. *Diabetic Medicine*, **5**, 126–34.

48 Viberti, G.C. and Thomas, S. (2000). Microalbuminuria and cardiovascular disease. In *The Kidney and Hypertension in Diabetes Mellitus*, ed. C.E. Mogensen. Boston, Dordrecht and London: Kluwer Academic Publishers, pp. 39–54.

49 Hansen, K.W., Mau Pedersen, M., Christensen, C.K., Christiansen, J.S. and Mogensen, C.E. (1992). Normoalbuminuria ensures no reduction of renal function in type 1 (insulin-dependent) diabetic patients. *Journal of Internal Medicine*, **232**, 161–7.

50 Mogensen, C.E. (1986). Early glomerular hyperfiltration in insulin-dependent diabetics and late nephropathy. *Scandinavian Journal of Clinical and Laboratory Investigation*, **46**, 201–6.

51 Mogensen, C.E. (1994). Glomerular hyperfiltration in human diabetes. *Diabetes Care*, **17**, 770–5.

52 Yip, J.W., Jones, S.L., Wiseman, M.J., Hill, C. and Viberti, G.C. (1996). Glomerular hyperfiltration in the prediction of nephropathy in IDDM: a 10-year follow-up study. *Diabetes* 45; 1729–33.

53 Mathiesen, E.R., Hommel, E., Hansen, H.P., Smidt, U.M. and Parving, H.H. (1999). Randomised controlled trial of long-term efficacy of captopril on preserved kidney function in normotensive patients with insulin dependent diabetes. *British Medical Journal*, **319**, 24–5.

54 Bianchi, S., Bigazzi, R. and Campese, V.M. (1997). Microalbuminuria in essential hypertension. *Journal of Nephrology*, **10**, 216–19.

55 Bigazzi, R., Bianchi, S., Baldari, D. and Campese, V.M. (1998). Microalbuminuria predicts cardiovascular events and renal insufficiency in patients with essential hypertension. *Journal of Hypertension*, **16**, 1325–33.

56 Campese, V.M., Bigazzi, R. and Bianchi, S. (2000). Microalbuminuria in essential hypertension. Significance for the cardiovascular and renal systems. In *The Kidney and Hypertension in Diabetes Mellitus*, ed. C.E. Mogensen. Boston, Dordrecht and London: Kluwer Academic Publishers, pp. 575–86.

57 Jensen, J.S., Feldt-Rasmussen, B., Borch-Johnsen, K., Clausen, P., Appleyard, M. and Jensen, G. (1997). Microalbuminuria and its relation to cardiovascular disease and risk factors. A population-based study of 1254 hypertensive individuals. *Journal of Human Hypertension*, **11**, 727–32.

58 Viberti, G.C., Keen, H. and Wiseman, M.J. (1987). Raised arterial pressure in parents of proteinuric insulin dependent diabetics. *British Medical Journal*, **295**, 515–17.

59 Christensen, C.K., Krussel, L.R. and Mogensen, C.E. (1987). Increased blood pressure in diabetes: Essential hypertension or diabetic nephropathy? *Scandinavian Journal of Clinical and Laboratory Investigation*, **47**, 363–70.

60 Mogensen, C.E. (1994). Systemic blood pressure and glomerular leakage with particular reference to diabetes and hypertension. *Journal of Internal Medicine*, **235**, 297–316.
61 Earle, K.A, Walker, J., Hill, C. and Viberti, G.C. (1992). Familial clustering of cardiovascular disease in patients with insulin-dependent diabetes and nephropathy. *New England Journal of Medicine*, **326**, 673–7.
62 Mattock, M.B., Keen, H., Viberti, G.C. et al. (1988). Coronary heart disease and urinary albumin excretion rate in type 2 (non-insulin-dependent) diabetic patients. *Diabetologia*, **31**, 82–7.
63 Cambier, P. (1934). Application de la théorie de Rehberg a l'etude clinique des affections rénales et du diabete. *Annales Médicine*, **35**, 273–99.
64 Parving, H.-H. (1998). Renoprotection in diabetes: genetic and non-genetic risk factors and treatment. *Diabetologia*, **41**, 745–59.
65 Mogensen, C.E., Damsgaard, E.M. and Froland, A. (1992). GFR-loss and cardiovascular damage in diabetes: A key role for abnormal albuminuria. *Acta Diabetologica*, **29**, 201–13.
66 Bain, S.C. and Chowdhury, T.A. (2000). Genetics of diabetic nephropathy and microalbuminuria. *Journal of the Royal Society of Medicine*, **93**, 62–6.
67 Borch-Johnsen, K., Norgaard, K., Hommel, E. et al. (1992). Is diabetic nephropathy an inherited complication? *Kidney International*, **41**, 719–22.
68 Marre, M., Jeunemaitre, X., Gallois, Y. et al. (1997). Contribution of genetic polymorphism in the renin angiotensin system to the development of renal complications in insulin-dependent diabetes: Genetique de la Nephropathie Diabetique (GENEDIAB) study group. *Journal of Clinical Investigation*, **99**, 1585–95.
69 Parving, H.H., Tarnow, L. and Rossing, P. (1996). Genetics of diabetic nephropathy. *Journal of the American Society of Nephrology*, **7**, 2509–17.
70 Quinn, M., Angelico, M.C., Warram, J.H. and Krolewski, A.S. (1996). Familial factors determine the development of diabetic nephropathy in patients with IDDM. *Diabetologia*, **39**, 940–5.
71 Tarnow, L., Cambien, F., Rossing, P. et al. (1996). Angiotensin-II type 1 receptor gene polymorphism and diabetic microangiopathy. *Nephrology, Dialysis, Transplantation*, **11**, 1019–23.
72 Tarnow, L., Pociot, F., Hansen, P.M. et al. (1997). Polymorphisms in the interleukin-1 gene cluster do not contribute to the genetic susceptibility of diabetic nephropathy in Caucasian patients with IDDM. *Diabetes*, **46**, 1075–6.
73 Seaquist, E.R., Goetz, F.C., Rich, S. and Barbosa, J. (1989). Familial clustering of diabetic kidney disease. Evidence for genetic susceptibility to diabetic nephropathy. *New England Journal of Medicine*, **320**, 1161–5.
74 Fogarty, D.G., Harron, J.C., Hughes, A.E., Nevin, N.C., Doherty, C.C. and Maxwell, A.P. (1996). A molecular variant of angiotensinogen is associated with diabetic nephropathy in IDDM. *Diabetes*, **45**, 1204–8.
75 Vestbo, E., Damsgaard, E.M., Froland, A. and Mogensen, C.E. (1996). Birth weight and cardiovascular risk factors in an epidemiological study. *Diabetologia*, **39**, 1598–602.
76 Yudkin, J.S. and Stanner, S. (1998). Prenatal exposure to famine and health in later life. *Lancet*, **351**, 1361–2.

77 Hansen, K.W., Poulsen, P.L. and Ebbehøj, E. (2000). Blood pressure elevation in diabetes: the results from 24-h ambulatory blood pressure recordings. In *The Kidney and Hypertension in Diabetes Mellitus*, ed. C.E. Mogensen. Boston, Dordrecht and London: Kluwer Academic Publishers, pp. 339–62.
78 Mathiesen, E.R., Oxenboll, B., Johansen, K., Svendsen, P.A. and Deckert, T. (1984). Incipient nephropathy in type I (insulin-dependent) diabetes. *Diabetologia*, **26**, 406–10.
79 Mathiesen, E.R., Ronn, B., Jensen, T., Storm, B. and Deckert, T. (1990). Relationship between blood pressure and urinary albumin excretion in development of microalbuminuria. *Diabetes*, **39**, 245–9.
80 Microalbuminuria Collaborative Study Group United Kingdom (1993). Risk factors for development of microalbuminuria in insulin dependent diabetic patients. A cohort study. Microalbuminuria Collaborative Study Group, United Kingdom. *British Medical Journal*, **306**, 1235–9.
81 Microalbuminuria Collaborative Study Group (1999). Predictors of the development of microalbuminuria in patients with type 1 diabetes mellitus: a seven year prospective study. *Diabetic Medicine*, **16**, 918–25.
82 Ramsay, L.E., Williams, B., Johnston, G.D. et al. (1999). British Hypertension Society Guidelines for hypertension management 1999: Summary. *British Medical Journal*, **319**, 630–5.
83 Cooper, M., McNally, P.G. and Boner, G. (2000). Antihypertensive treatment in NIDDM, with special reference to abnormal albuminuria. In *The Kidney and Hypertension in Diabetes Mellitus*, ed. C.E. Mogensen. Boston, Dordrecht and London: Kluwer Academic Publishers, pp. 441–60.
84 Gæde, P., Vedel, P., Parving, H.-H. and Pedersen, O. (1999). The Steno Type 2 Study: intensive multifactorial intervention delays the progression of micro- and macroangiopathy in microalbuminuric type 2 diabetic patients. *Lancet*, **353**, 617–22.
85 Hansen, K.W., Poulsen, P.L. and Mogensen, C.E. (2000). Scientific basis for the new guidelines for the treatment of hypertension in type 2 diabetes. In Mogensen, C.E (ed). In *The Kidney and Hypertension in Diabetes Mellitus.* Kluwer Academic Publishers, Boston, Dordrecht and London. pp. 611–22.
86 Mogensen, C.E., Keane, W.F., Bennett, P.H. et al. (1995). Prevention of diabetic renal disease with special reference to microalbuminuria. *Lancet*, **346**, 1080–4.
87 UK Prospective Diabetes Study Group (1998). Intensive blood-glucose control with sulphonylureas or insulin compared with conventional treatment and risk of complications in patients with type 2 diabetes (UKPDS 33). UK Prospective Diabetes Study (UKPDS) Group. *Lancet*, **352**, 837–53.
88 UK Prospective Diabetes Study Group (1998). Tight blood pressure control and risk of macrovascular and microvascular complications in type 2 diabetes UKPDS 38. *British Medical Journal*, **317**, 703–13.
89 Trocha, A.K., Schmidtke, C., Didjurgeit, U. et al. (1999). Effects of intensified antihypertensive treatment in diabetic nephropathy: mortality and morbidity results of a prospective controlled 10-year study. *Journal of Hypertension*, **17**, 1497–1503.
90 Jensen, J.S., Mathiesen, E.R., Nørgaard, K. et al. (1990). Increased blood pressure and erythrocyte sodium/lithium countertransport activity are not inherited in diabetic nephropathy. *Diabetologia*, **33**, 619–24.

91 Koren, W., Koldanov, R., Pronin, V.S., Postnov, V.S. et al. (1998). Enhanced erythrocyte $Na^+/H^+$ exchange predicts diabetic nephropathy in patients with IDDM. *Diabetologia*, **41**, 201–5.

92 Monciotti, C.G., Semplicini, A., Morocutti, A. et al. (1997). Elevated sodium-lithium countertransport activity in erythrocytes is predictive of the development of microalbuminuria in IDDM. *Diabetologia*, **40**, 654–61.

93 Ng, L.L. and Davies, J.E. (1992). Abnormalities in $Na^+/H^+$ antiporter activity in diabetic nephropathy. *Journal of the American Society of Nephrology*, **3** (Suppl 4), S50–5.

94 Ng, L.L., Davies, J.E., Siczkowski, M. et al. (1994). Abnormal $Na^+/H^+$ antiporter phenotype and turn-over of immortalized lymphoblasts from type 1 diabetic patients with nephropathy. *Journal of Clinical Investigation*, **93**, 2750–7.

95 Rutherford, P.A. and Thomas, T.H. (1994). Methodologies in sodium-lithium countertransport as related to other cation transport systems in diabetes. In *Research Methodologies in Human Diabetes, Part 1*, eds. C.E. Mogensen and E. Standl. Berlin, New York: Walter de Gruyter.

96 Trevisan, R., Cipoilina, M.R., Duner, E., Trevisan, M. and Nosadini, R. (1996). Abnormal $Na^+/H^+$ antiport activity in cultured fibroblasts from NIDDM patients with hypertension and microalbuminuria. *Diabetologia*, **39**, 717–24.

97 Trevisan, R. and Viberti, G.C. (1997). Sodium-hydrogen antiporter: its possible role in the genesis of diabetic nephropathy. *Nephrology, Dialysis, Transplantation*, **12**, 643–5.

98 Allen, T.J., Cooper, M.E., Gilbert, R.E., Winikoff, J., Skinni S. I. and Jerums, G. (1996). Serum total renin is increased before microalbuminuria in diabetes. *Kidney International*, **50**, 902–7.

99 Daneman, D., Crompton, C.H., Balfe, J.W. et al. (1994). Plasma prorenin as an early marker of nephropathy in diabetic (IDDM) adolescents. *Kidney International*, **46**, 1154–9.

100 Deinum, J., Rønn, B., Mathiesen, E., Derkx, F.H.M., Hop, W.C.J. and Schalekamp, M.A.D.H. (1999). Increase in serum prorenin precedes onset of microalbuminuria in patients with insulin-dependent diabetes mellitus. *Diabetologia*, **42**, 1006–10.

101 Franken, A.A., Derkx, F.H., Blankestijn, P.J. et al. (1992). Plasma prorenin as an early marker of microvascular disease in patients with diabetes mellitus. *Diabetes Metabolisme*, **18**, 137–43.

102 Mogensen, C.E., Neldam, S., Tikkanen, I. et al. for the CALM Study Group (2000). Randomised controlled trial of dual blockade of renin-angiotension system in patients with hypertension, microalbuminuria and non-insulin dependent diabetes: the candesartan and lisinopril microalbuminuria. *British Medical Journal*, **9**, 1440–4.

103 Schmitt, F., Natov, S., Martinez, F., Lacour, B. and Hannedouche, T.P. (1996). Renal effects of angiotensin I-receptor blockade and angiotensin convertase inhibition in man. *Clinical Science*, **90**, 205–13.

104 Stehouwer, C.D. (2000). Dysfunction of the vascular endothelium and the development of renal and vascular complications in diabetes. In *The Kidney and Hypertension in Diabetes Mellitus*, ed. C.E. Mogensen. Boston, Dordrecht and London: Kluwer Academic Publishers, pp. 179–92.

105 Stephenson, J.M., Fuller, J.H., Viberti, G.C., Sjolie, A.K. and Navalesi, R. (1995). Blood pressure, retinopathy and urinary albumin excretion in IDDM: the EURODIAB IDDM Complications Study. *Diabetologia*, **38**, 599–603.

106 Greaves, M., Malia, R.G., Goodfellow, K. et al. (1997). Fibrinogen and von Willebrand factor in IDDM: relationships to lipid vascular risk factors, blood pressure, glycemic control and urinary albumin excretion rate: the EURODIAB IDDM Complications Study. *Diabetologia*, **40**, 698–705.

107 Gruden, G., Cavallo-Perin, P., Bazzan, M., Stella, S., Vuolo, A. and Pagano, G. (1994). PAI-1 and factor VII activity are higher in IDDM patients with microalbuminuria. *Diabetes*, **43**, 426–9.

108 Gruden, G., Pagano, G., Romagnoli, R., Frezet, D., Olivetti, O. and Cavallo-Perin, P. (1995). Thrombomodulin levels in insulin-dependent diabetic patients with microalbuminuria. *Diabetic Medicine*, **12**, 258–60.

109 Hofmann, M.A., Kohl, B., Zumbach, M.S. et al. (1997). Hyperhomocyst(e)inemia and endothelial dysfunction in IDDM. *Diabetes Care*, **20**, 1880–6.

110 Hofmann, M.A., Kohl, B., Zumbach, M.S. et al. (1998). Hyperhomocyst(e)inemia and endothelial dysfunction in IDDM. *Diabetes Care*, **21**, 841–8.

111 Ekstrand, A.V., Groop, P.H. and Gronhagen-Riska, C. (1998). Insulin resistance precedes microalbuminuria in patients with insulin-dependent diabetes mellitus. *Nephrology, Dialysis, Transplantation*, **13**, 3079–83.

112 Groop, P.H., Elliott, T., Ekstrand, A. et al. (1996). Multiple lipoprotein abnormalities in type 1 diabetic patients with renal disease. *Diabetes*, **45**, 974–9.

113 Hauerslev, C.F., Vestbo, E., Frøland, A., Mogensen, C.E. and Damsgaard, E.M. (2000). Normal blood pressure and preserved diurnal variation in offspring of type 2 diabetic patients characterised by the features of the metabolic syndrome. *Diabetes Care*, **23**, 283–9.

114 Jager, A., Kostense, P.J., Nijpels, G., Heine, R.J., Bouter, L.M. and Stehouwer, C.D.A. (1998). Microalbuminuria is strongly associated with NIDDM and hypertension but not with the insulin resistance syndrome: the Hoorn Study. *Diabetologia*, **41**, 694–700.

115 Jones, S. I., Close, C.F., Mattock, M.B., Jarrett, R.J., Keen, H. and Viberti, G.C. (1989). Plasma lipid and coagulation factor concentrations in insulin dependent diabetics with microalbuminuria. *British Medical Journal*, **298**, 487–90.

116 Lahdenpera, S., Groop, P.H., Tilly-Kiesi, M. et al. (1994). LDL subclasses in IDDM patients. Relation to diabetic nephropathy. *Diabetologia*, **37**, 681–8.

117 Niskanen, L., Uusitupa, M., Sarland, H. et al. (1990). Microalbuminuria predicts the development of serum lipoprotein abnormalities favouring atherogenesis in newly diagnosed type 2 (non-insulin-dependent) diabetic patients. *Diabetologia*, **33**, 237–43.

118 Uusitupa, M.I., Niskanen, L.K., Sütonen, O., Voutilainen, E. and Pyorala, K. (1993). Ten-year cardiovascular mortality in relation to risk factors and abnormalities in lipoprotein composition in type 2 (non-insulin-dependent) diabetic and non-diabetic subjects. *Diabetologia*, **36**, 1175–84.

119 Gilbert, R.E., Tsalamandris, C., Allen, T.J., Colville, D. and Jerums, G. (1998). Early nephropathy predicts vision-threatening retinal disease in patients with type 1 diabetes mellitus. *Journal of the American Society of Nephrology*, **9**, 85–9.

120 Mogensen, C.E., Vigstrup, J. and Ehlers, N. (1985). Microalbuminuria predicts proliferative diabetic retinopathy. *Lancet*, **II**, 1512–13.

121 Ikeda, Y., Suehiro, T., Takamatsu, K., Yamashita, H., Tamura, T. and Hashimoto, K. (1997).

Effect of smoking on the prevalence of albuminuria in Japanese men with non-insulin-dependent diabetes mellitus. *Diabetes Research and Clinical Practice*, **36**, 57–61.

122 Sawicki, P.T. (1998). Smoking and diabetic nephropathy. In *The Kidney and Hypertension in Diabetes Mellitus*, ed. C.E. Mogensen. Boston, Dordrecht and London: Kluwer Academic Publishers, pp. 209–16.

123 Sinha, R.N., Patrick, A.W., Richardson, L., Wallymahmed, M. and MacFarlane, I.A. (1997). A six-year follow-up study of smoking habits and microvascular complications in young adults with type 1 diabetes. *Postgraduate Medical Journal*, **73**, 293–4.

124 Feldt-Rasmussen, B. (2000). The course of incipient and overt diabetic nephropathy: the perspective of more optimal treatment, including UKPDS-perspective. In *The Kidney and Hypertension in Diabetes Mellitus*, ed. C.E. Mogensen. Boston, Dordrecht and London: Kluwer Academic Publishers, pp. 461–72.

125 Fleming, G.A. (2000). Regulatory considerations in the development of therapies for diabetic nephropathy and related conditions. In *The Kidney and Hypertension in Diabetes Mellitus*, ed. C.E. Mogensen. Boston, Dordrecht and London: Kluwer Academic Publishers, pp. 623–44.

126 Mogensen, C.E. (1998). Combined high blood pressure and glucose in type 2 diabetes: double jeopardy (editorial). *British Medical Journal*, **317**, 693–4.

127 Schichiri, M., Kishikawa, H., Ohkubo, Y. and Wake, N. (2000). Long-term results of the Kumamoto Study on optimal diabetes control in type 2 diabetic patients. *Diabetes Care*, 23(Suppl 2): B21–30.

128 Brenner, B.M., Cooper, M.E., de Zeeuw, D. et al. for the Reduction of End-Points in NIDDM with the Angiotensin II Antagonist Losartan (RENAAL) Study Investigators. (2001). Effects of losartan on renal and cardiovascular outcomes in patients with type 2 diabetes and nephropathy. *New England Journal of Medicine*, **345**, 861–9.

129 Lewis, E.J., Hunsicker, L.G., Clarke, W.R. et al. (2001). Renoprotective effect of the angiotensin-receptor antagonist irbesartan in patients with nephropathy due to type 2 diabetes. *New England Journal of Medicine*, **345**, 851–60.

130 Parving, H.H., Lehnert, H., Bröchner–Mortensen, J. et al. (2001). The effect of irbesartan on the development of diabetic nephropathy in patients with type 2 diabetes *New England Journal of Medicine*, **345**, 870–8.

# Part 3

# Biomarkers of bone disease and dysfunction

9

# Bone turnover markers in clinical practice

Richard Eastell, Sheila Hart

Northern General Hospital, Sheffield, UK

## Introduction

Bone turnover markers are an established tool in the diagnosis and treatment of metabolic bone disease such as Paget's disease of bone. The measurements for these markers have improved considerably in the last few years such that their use can be considered for patients with osteoporosis. In this chapter, we will review the most recent evidence for using bone turnover markers in clinical practice, and give recommendations where they may prove helpful in the management of osteoporosis.

The markers currently in use in clinical practice are serum osteocalcin, serum bone alkaline phosphatase (BAP), urinary N-terminal telopeptide of type 1 collagen (NTx), urinary C-terminal telopeptide of type 1 collagen (CTx) and free deoxypyridinoline (Dpd). The introduction of automated analysers for these markers means that the assays can be performed reliably and should be available in any clinical chemistry laboratory.

### Osteocalcin

Osteocalcin is the most abundant noncollagenous protein in bone. It has a high affinity for hydroxyapatite and its formation is vitamin K dependent. The serum sample should be separated and transported to the laboratory within 2 hours, but measurement of the intact and large N-terminal midfragment improves the stability of this marker [1].

### Serum bone alkaline phosphatase

Bone alkaline phosphatase has a molecular weight of approximately 140000 Dal and is found in the membrane of osteoblasts. It is released into the circulation during bone formation. This marker is very stable and is not affected by haemolysis. Assays have been developed which can detect the bone isoform of alkaline phosphatase [2].

### Crosslinks

Type 1 procollagen is secreted from the osteoblast and undergoes cleavage of the amino and carboxyl terminals. The collagen fibrils then link together at both the C- and N-nonhelical terminal portions (telopeptides) via covalent crosslinks. Most crosslinks are pyridinoline (Pyr) or Dpd. When collagen is broken down during bone resorption, these crosslinks cannot be degraded and are therefore excreted and filtered by the kidney. Forty to fifty per cent of these collagen crosslinks are released into the circulation as free Pyr and Dpd. Fifty to sixty per cent are released bound to type 1 collagen fragments, linked to the CTx or the NTx telopeptides. The advantages of using pyridinium crosslinks over hydroxyproline are that they are not metabolized or absorbed from the diet [3]. Recently, assays for CTx and NTx in serum have been introduced [4–6]. This may improve the interindividual variability observed with the urine marker assays.

The use of bone turnover markers for diagnosing osteoporosis has many limitations [7]. However, interest in the use of bone turnover markers other than for the diagnosis of osteoporosis falls into the areas given below.

## Predicting rate of bone loss

If bone turnover markers could reliably predict the rate of bone loss, they would provide a cheap and easy way of screening women at high risk of developing osteoporosis. Unfortunately, there are major methodological problems that arise when trying to examine the association of bone markers with bone mineral density (BMD) in longitudinal studies [8]. For example, the magnitude of the error associated with BMD measurements over time (in the region of a few per cent) is similar to the annual changes that are seen in BMD. Therefore, it is difficult to make a valid assessment of the relationship between the rate of bone turnover and the subsequent rate of bone loss in individual postmenopausal women. Studies have previously given conflicting results [8–11], but two recent studies suggest that bone markers cannot be used to predict the rate of bone loss. Yoshimura et al. [12] examined eight bone markers and their relationship with BMD change at the hip and femoral neck over 3 years in 400 subjects. There was no relationship between any of the markers measured and the subsequent bone loss observed at the femoral neck. However, a significant relationship was noted between osteocalcin levels and bone loss at the lumbar spine in women. The authors of this chapter have found that, while there may be a strong relationship between bone marker levels and spinal bone loss, it is not possible to use these levels to predict bone loss in an individual [13]. The ability of bone markers to predict the rate of bone loss remains controversial and using them to predict bone loss in an individual may give inconsistent results.

## Predicting risk of fracture

Several studies have reported either no change, or an increase, in bone turnover markers within a few days of sustaining a fracture [14–17]. It is very difficult to determine from these retrospective studies whether the elevation of bone marker level is related to the underlying rate of bone turnover leading to the fracture, or to the acute changes associated with the fracture [18]. In order to predict the risk of fracture with bone markers, it is necessary to relate baseline bone turnover markers to the subsequent risk of fracture in a prospective study, similar to the multitude of studies that have identified the relationship between BMD and fracture risk [19]. Such studies need to have a large number of subjects and a sufficiently long follow-up period due to the rare occurrence of osteoporotic fractures.

One such study [20] found that, in women followed for 2 years, the formation markers (serum osteocalcin and BAP) were not predictive of hip fracture. However, baseline CTx and Dpd were higher in those patients who had subsequent hip fractures than in age-matched, nonfracture controls. This study showed that CTx and Dpd levels above the normal range were associated with a two-fold increase in hip fracture risk. A more recent study [21] also observed a significant relationship between CTx and fracture risk. In contrast to the earlier study, the authors of this study found BAP to be predictive of fracture as well. There is some disagreement as to the predictive value of bone markers. Tromp et al. [22] did not find any relationship between urinary NTx and osteoporotic fracture risk among 348 postmenopausal women over a 5-year period. Delmas [23] commented recently on the use of bone turnover markers in the assessment of fracture risk. He stated that the 'combination of a bone resorption marker and hip BMD measurement can enhance the specificity of hip fracture prediction without a loss of sensitivity. The practical outcome of such a strategy is that the number of women who need to be treated in order to avoid one hip fracture is significantly reduced.' The combination of bone markers and BMD may be the approach for the future, but more studies are required to support this.

## Predicting response to treatment

The value of bone marker levels in predicting treatment response in an individual is disputed for several reasons, including the considerable biological variation that the markers exhibit and because of reports that bone marker levels during treatment may only account for around 16% of BMD changes after 12 months of therapy [24]. With factors other than bone turnover status contributing significantly to BMD changes, prediction of response based solely on bone markers may be unreliable. The ability of bone markers to predict response to treatment in a relatively short period of time, however, would be a great advance over the current use of BMD measurements. Hesley et al. [25] found that those subjects identified as having a high bone turnover (baseline Dpd levels greater than two standard deviations above the group mean) showed a greater response to therapy, with the gain in lumbar spine BMD (2.8% increase) being significantly greater than the lumbar spine BMD increase for those individuals classified as low turnover (0.7% increase). A second study [26] found that baseline NTx levels in women treated with alendronate were correlated with alkaline phosphatase and lateral spine BMD at 12–18 months ($r=0.28$ and $r=0.27$, respectively; $p<0.05$).

Chesnut et al. [24] carried out a detailed analysis of the effect of hormone replacement therapy (HRT) on urine NTx levels and its value in predicting response to treatment. They observed a significant correlation between baseline NTx levels and lumbar spine BMD at 1 year. The percentage change in NTx at 6 months was also associated with an increase in spinal BMD. Furthermore, they demonstrated that the response to HRT in those subjects who had a low NTx at baseline (NTx level in the lowest quartile, which gave a value of less than 39 nmol bone collagen equivalents [BCE]/mmol creatinine) was not significantly different from the placebo group. In contrast, in those subjects who had a high baseline NTx (NTx level in the highest quartile, greater than 69 nmol BCE/mmol creatinine), the response to treatment was much greater, and, if no treatment was offered, these subjects had a 17.3 times greater risk of bone loss compared with the low turnover subjects. The authors suggest that the NTx assay used could be utilized to predict who would benefit from HRT, i.e. those subjects with an NTx level greater than 39 nmol BCE/mmol creatinine at baseline would benefit from therapy. How reliable this would be on an individual basis, however, is still questionable.

Several recent studies provide more evidence to support the use of bone markers in predicting response to treatment [27, 28]. Dresner-Pollak et al. [27] concluded that the change in BAP after 6 months of HRT could be used to predict the change in spinal BMD at 2 years. Again, the value of BAP at the individual level is not clear.

A further problem that arises when trying to predict response to treatment is that, if an individual is classified as a high, or low, turnover patient, one assumes that this turnover status does not change with time. Bone turnover markers may be useful to predict treatment response in a population, but studies are needed to determine if the change in bone turnover is predictive of long-term changes in BMD in individuals.

## Monitoring response to therapy

There are a multitude of studies which have shown that bone turnover markers decrease with anti-resorptive therapy [25, 29–36]. It has been shown for HRT that the bone formation markers tend to decrease slowly, reaching a nadir around 6 months of therapy, whereas the resorption markers decrease immediately and generally reach a steady state within 3 months of treatment [29]. Bone turnover markers may be useful in assessing response and increasing compliance in individuals as they are a source of feedback for the patient and can indicate a positive response to treatment as early as 6 months after starting the therapy.

Hesley et al. [25] have investigated the effects of treatment with 17–$\beta$ oestradiol (0.05 mg/day) on urinary Dpd excretion in 91 women who had recently undergone a surgical menopause. Despite the low number of subjects in this study, the results are in agreement with other studies that have shown that short-term changes in Dpd response to antiresorptive treatment are inversely correlated with long-term changes in BMD [37, 38]. In this study, the changes in Dpd at just 6 months were correlated ($r=0.47$, $p<0.05$), with changes in lumbar spine BMD at 6 months in the subgroup of patients on the 0.05 mg therapy. Additionally, urinary Dpd levels decreased into the premenopausal range in 95% (59/62) of those subjects taking HRT (dose ranged from 0.025 to 0.1 mg/day). Delmas et al. [39] investigated the change in BAP, osteocalcin and CTx in response to different doses of transdermal 17 $\beta$-oestradiol patches in 569 women. They found that all markers decreased significantly after 3 and 6 months of treatment and that these changes were predictive of BMD change at 2 years. These results support the monitoring of BMD response to HRT with bone markers.

Similar findings have been reported for other bone turnover markers with either HRT or bisphosphonate therapy [40, 41]. Greenspan et al. [26] studied the effects of alendronate treatment on bone turnover markers in 120 elderly women (65 years or older). At 6 months, there were significant decreases in all the markers in women on alendronate treatment: osteocalcin decreased by 20%, BAP by 24%, NTx by 53% and Dpd by 10%. These decreases in bone turnover marker levels persisted for the 2.5 years of the study. They also showed in this study that subjects who exhibited the largest decrease in NTx levels (greater than 65%) had a greater gain in BMD at various sites (total hip, trochanter and anterior–posterior spine) when compared with those individuals who experienced a smaller decrease (less than 45%) in their NTx level in response to treatment. It can therefore be hypothesized that those individuals with the greatest percentage decrease in their bone marker level in response to treatment would have a lower risk of fracture. Clinical studies however, are needed to confirm this.

Ravn et al. [42] demonstrated a dose-dependent decrease in urinary NTx and CTx in response to 1 year of alendronate therapy. The suppression in the bone marker level was not maintained after the alendronate was withdrawn. Kress et al. [43] also observed a significant decrease in BAP after 3 months of alendronate therapy in 74 postmenopausal women. Fink et al. [44] reported changes in free Dpd and NTx in patients treated with alendronate and concluded that these bone markers were useful for monitoring individual response to treatment despite the inter-individual variability these markers exhibit.

## Day to day variability of bone turnover markers

The advantages of using bone markers to monitor response to therapy are that they are inexpensive, noninvasive and suitable for serial measurement, and that they reveal changes in bone turnover sooner than can be revealed by serial BMD measurements. Unfortunately, however, bone markers are not without their disadvantages. They provide no information about cellular activity and are influenced by nonosseous sources, which may give misleading results. Many of the markers show considerable biological variation and this may limit their use in the clinical setting.

The day to day variability seen with bone markers is predominantly due to biological variation and is more pronounced for markers of bone resorption (reported to be in the region of 16% for Dpd, 23% for NTx and CTx [45, 46]) than for markers of bone formation (usually less than 10% [47]). Rosen et al. [32] examined the value of bone markers for monitoring response to treatment. They highlight the fact that a cut-off point is needed when interpreting bone marker results. Below this cut off, it can be said that the change in a bone marker level is the result of the day to day and biological variation, a change greater than the cut off indicates a significant change in the marker level (the least significant change or LSC). In their study, it was noted that, while fasting urinary NTx demonstrated the greatest decline in response to treatment (58.2%), the LSC was relatively high (54%) due to the magnitude of variation exhibited by this marker. Only 57% of patients treated with bisphosphonates showed a response in their NTx marker levels which was greater than the LSC.

If bone markers are to be used successfully in the clinical setting, an understanding of the sources of variability, and the impact that this variation can have on individual results, is crucial to the appropriate interpretation of bone marker monitoring in the clinical setting.

## Circadian variation

The circadian variation in bone turnover has been extensively investigated [48–51]. A detailed knowledge of the cyclical variation in bone marker levels throughout the day is essential and requires standard sample collection times if its influence is to be minimized. Stone et al. [52] investigated the circadian variation of collagen (pyridinoline and Dpd) and elastin (desmosine and isodesmosine) crosslinks. Dpd was found to be 61% higher in women in an evening 2-hour collection of urine compared with a 2-hour morning collection. A similar, but nonsignificant change was seen in men. The information on the circadian rhythm of collagen crosslinks in this study is limited by the study technique. Aoshima et al. [53] have reported that the urinary resorption markers, Dpd and CTx, were 37–55% higher at night compared with the day in nine premenopausal women and nine postmenopausal osteoporotic women. There are a few conflicting results in the literature, which are usually the consequence of variations in the study design. The general consensus, however, is that most markers tend to be higher in the early morning hours, with peak levels reported to occur between 05:00 and 08:00 [49, 53].

## Growth and ageing

During childhood, when the skeleton is growing, bone turnover is high and, consequently, bone turnover markers are at greater levels than those seen in normal adults [54, 55]. With increasing age, there are many factors which influence bone turnover and it has been reported that bone markers increase with increasing age [56–59].

Mora et al. [60] observed a significant increase in BAP and osteocalcin and urinary Dpd during the initial stages of puberty, with a decline in the latter stages. Once the pubertal peak has passed, there remains a gradual change in bone turnover with age. Hoshino et al. [61] reported that urinary CTx levels increased with each decade in 246 healthy females. The lowest values were seen in those women less than 30 years of age. After this period, the levels increased. A 55% increase was seen between the fourth and fifth decades, followed by a 44% rise between the fifth and sixth decades. After the age of 60, the changes were less dramatic. In another cross-sectional study, Gallagher et al. [62] reported that serum osteocalcin was significantly correlated with age in men ($r=0.015$, $p<0.05$) and women ($r=0.10$, $p<0.05$). Twenty-four-hour urine NTx/creatinine levels also increased significantly with age in men ($r=0.23$, $p<0.001$), but not in women ($r=0.07$). Conversely, Orwoll et al. [63] found no significant changes in NTx levels with age in 206 men aged 31–86 years.

## Menstrual cycle

The variation in bone markers during the menstrual cycle must be taken into account when assessing a woman's bone turnover status, although it is not normally an issue in the clinical setting since most of the patients being evaluated are postmenopausal. From the results of several studies, it appears that, during the luteal phase of the menstrual cycle, bone formation markers are elevated, while resorption markers are suppressed [64, 65]. Gorai et al. [66] observed a significant increase ($0.0265 > p < 0.0373$) in BAP and a significant decrease ($0.0008 > p < 0.0347$) in CTx and free Dpd during the follicular phase of the menstrual cycle.

## Menopause

As the majority of patients who are suspected of having osteoporosis are close to the menopause or postmenopausal, it is necessary to have a clear understanding of how the menopause affects bone marker levels. The results of several studies suggest an increase of bone turnover at the time of menopause. Okano et al. [67] observed that bone markers are already elevated during the menopausal transition period, when menstruation is irregular and when rapid bone loss has already started. The results from this study suggested that changes in bone turnover may commence while menstruation is still regular, although the change is too small to reflect changes in spinal BMD. Perimenopausal changes have also been noted by others [68–70]. Hoshino et al. [71] reported that bone turnover was increased up to 4 years before the onset of menopause and increased significantly at the time of menopause. Interestingly, it is the peptide-bound crosslinks that appear to show the best discriminating power between pre- and postmenopausal women, with increases of NTx in the range of 110–171% [72, 73] and 71–114% for CTx [74, 75], compared with increases of 48–104% for BAP [37, 74].

## Drugs

One of the most common causes of drug-induced osteoporosis is corticosteroid therapy [76]. It is generally accepted that bone formation is depressed during corticosteroid treatment, while the change in bone resorption, which is not so well defined, is thought to be minimal. This finding was documented over 10 years ago [77], as determined by the suppression of serum osteocalcin. It has also been shown that there is no suppression of bone resorption markers during steroid treatment [78]. More recently, Ebeling et al. [79] found that, in women taking steroids, the levels of BAP and osteocalcin were 1.7 and 0.54 standard deviations below the level of normal premenopausal women, with no change in urinary Dpd. Pearce et al. [80] studied nine men with no systemic illness who were being treated with corticosteroids (up to 50 mg/day) to reduce antisperm antibodies. The bone formation markers were suppressed, with serum osteocalcin and BAP showing a $28.5 \pm 15.5\%$ and a $24.2 \pm 8.6\%$ decrease, respectively. The bone resorption marker, CTx, showed no elevation. Hughes et al. [81] observed similar changes in bone markers with inhaled corticosteroids. Serum osteocalcin increased by 14.3–16.9% over 1 year of treatment depending on the type of inhaled steroid used. There was no significant change in free Dpd and NTx after 1 year of treatment. Despite these changes in the bone markers, there was no evidence of a decrease in BMD during the year of steroid therapy. These studies add further support to the theory that corticosteroid-induced bone loss is due to a decrease in bone formation with unaltered but ongoing bone resorption.

## Bed rest and immobility

A knowledge of the effects of bed rest and immobility on bone turnover markers, while not as significant in the clinical setting, is important in order to preserve bone mass in patients who are bedridden for long periods of time. The results of several studies suggest that immobility leads to an increase in bone resorption (Dpd and NTx increased by 58% and 49%, respectively, after 12 weeks of bed rest) [82, 83]. The effects of bed rest on bone formation are less consistent, although the markers suggest a decrease in bone formation. This imbalance of bone turnover may lead to increased bone loss.

## Fracture healing

It has been shown that BAP and serum osteocalcin remain elevated by as much as 20% over baseline at 52 weeks' postfracture [84]. Dpd and NTx are also elevated following a fracture (18–35%), although levels returned to baseline by 1 year after the fracture. Several recent studies have documented an increase in bone turnover and a decrease in BMD following a fracture [85, 86]. A detailed fracture history must therefore be available when assessing a patient's bone marker levels.

This brief description of some of the sources of variation in bone turnover markers illustrates the importance of using the markers carefully in clinical practice. Whichever bone markers are used, biological variation in their levels must be considered.

## Conclusion

The best evidence for the use of bone turnover markers in clinical medicine is in the monitoring of the treatment for osteoporosis (Figure 9.1). This approach is useful if care is taken to minimize interindividual variability. This can be done if the blood and urine samples are collected in the morning. For example, the blood sample should be taken between 08:00 and 10:00 in the fasting state, and the urine collection between 07:00 and 09:00. Variability can be further reduced by obtaining markers on two occasions before starting therapy. The average of the two marker results is then calculated and forms a baseline to calculate the change in markers after treatment is started. Follow-up measurements can then

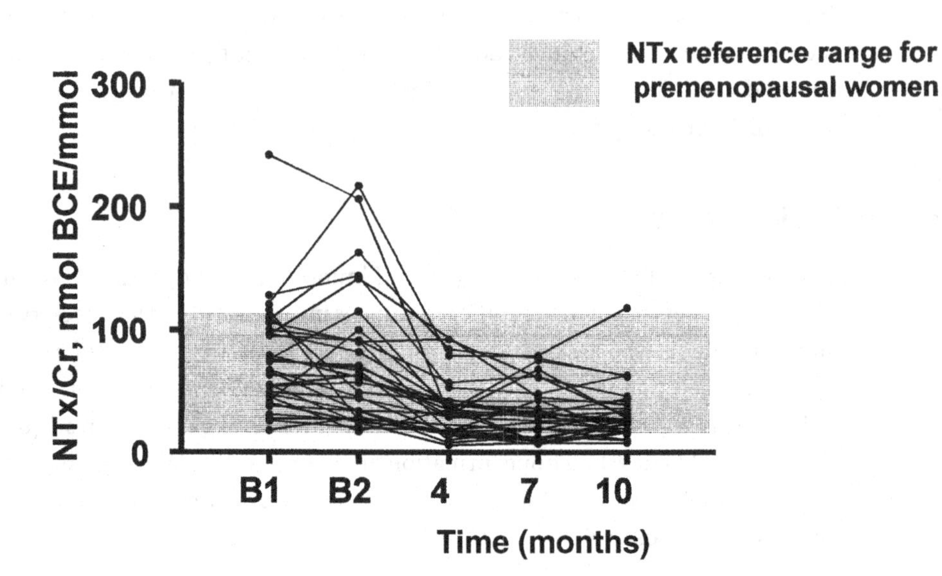


Figure 9.1 NTx was measured in the second morning void urine sample in 49 patient with osteoporosis (Bainbridge et al., 2000 [87]) on two occasions at baseline (B1 and B2) and then after 4, 7 and 10 months on treatment with HRT or bisphosphonates. Note that NTx does vary before starting treatment, and that there is an early and marked decrease in NTx after starting treatment such that the maximum effect is seen by 4 months in most patients. Note that NTx is in the lower half of the premenopausal reference range in most subjects after 10 months of treatment.

be made 3–6 months after starting therapy, and the average of these taken to compare with the baseline value. A decrease of more than 20–30% in a serum marker or of more than 40–60% in a urine marker would then indicate a response. It is also desirable to suppress the bone marker level into the normal reference range. Figure 9.1 shows the results of antiresorptive therapy in patients with osteoporosis. Note that there is variability in the two baseline measurements. The effect of treatment is rapid and the bone turnover level decreases to the lower half of the reference range for premenopausal women in most cases. This early sign of response assures a positive outcome in over one-half of patients. This test is useful when carried out in conjunction with follow-up measurements of lumbar spine BMD.

It is likely that bone turnover markers will eventually be used to predict future fracture risk so that patients who require treatment can be better managed. The best way to use markers for predicting fracture risk, however, has not yet been determined.

## REFERENCES


1 Garnero, P., Grimaux, M., Seguin, P. and Delmas, P.D. (1994). Characterisation of immunoreactive forms of human osteocalcin in vivo and vitro. *Journal of Bone and Mineral Research*, **9**, 255–64.

2 Garnero, P. and Delmas, P.D. (1993). Assessment of the serum levels of bone alkaline phosphatase with a new immunoradiometric assay in patients with metabolic bone disease. *Journal of Clinical Endocrinology and Metabolism*, **77**, 1046–53.

3 Delmas, P.D. (1995). Biochemical markers for the assessment of bone turnover. In *Osteoporosis: Etiology, Diagnosis, and Management*, eds. B.L. Riggs and L.J. Melton III. Philadelphia and New York: Lippincott-Raven Publishers, pp. 319–33.

4 Gertz, B.J., Clemens, J.D., Holland, S.D., Yuan, W. and Greenspan, S. (1998). Application of a new serum assay for type I collagen cross-linked N-telopeptides: assessment of diurnal changes in bone turnover with and without alendronate treatment. *Calcified Tissue International*, **63**, 102–6.

5 Christgau, S., Rosenquist, C., Alexandersen, P. et al. (1998). Clinical evaluation of the serum crosslaps one step ELISA. A new assay measuring the serum concentration of bone-derived degradation products of type I collagen C-telopeptides. *Clinical Chemistry*, **44**, 2290–300.

6 Pedrazzoni, M., Girasole, G., Giuliani, N., Passeri, G. and Passeri, M. (1998). Serum C-telopeptide of type I collagen closely reflects the intermittent inhibition of bone resorption during the cyclic administration of alendronate. *Bone*, **23**(Suppl), S628.

7 Dominguez Cabrera, C., Sosa Henríquez, M., Traba, M.L., Alvarez Villafañe, E. and de la Piedra, C. (1998). Biochemical markers of bone formation in the study of postmenopausal osteoporosis. *Osteoporosis International*, **8**, 147–51.

8 Blumsohn, A., Hannon, R.A. and Eastell, R. (1995). Long-term retest reliability of biochemical markers of bone turnover in healthy postmenopausal women. *Journal of Bone and Mineral Research*, **10**(Suppl), S182.

9 Hansen, M., Kirsten, O., Riss, B. and Christiansen, C. (1991). Role of peak bone mass and bone loss in postmenopausal osteoporosis: 12 year study. *British Medical Journal*, **303**, 961–4.

10 Ross, P. and Knowlton, W. (1998). Rapid bone loss is associated with increased levels of biochemical markers. *Journal of Bone and Mineral Research*, **13**, 297–302.

11 Garnero, P., Sornay-Rendu, E., Duboeuf, F. and Delmas, P.D. (1999). Markers of bone turnover predict postmenopausal forearm bone loss over 4 years: the OFELY study. *Journal of Bone and Mineral Research*, **14**, 1614–21.

12 Yoshimura, N., Hashimoto, T., Sakata, K., Morioka, S., Kasamatsu, T. and Cooper, C. (1999). Biochemical markers of bone turnover and bone loss at the lumbar spine and femoral neck: the Taiji study. *Calcified Tissue International*, **65**, 198–202.

13 Rogers, A., Hannon, R.A. and Eastell, R. (2000). Biochemical markers as predictors of rates of bone loss after menopause. *Journal of Bone and Mineral Research*, **15**, 1398–404.

14 McClung, M., Faulkner, K., Ravn, P. et al. (1996). Inability of biochemical markers to predict bone density changes in early postmenopausal women. *Journal of Bone and Mineral Research*, **11**(Suppl 1), S127.

15 Thompson, S., White, D., Hosking, D., Wilton, T. and Pawley, E. (1989). Changes in osteocalcin after femoral neck fracture. *Annals of Clinical Biochemistry*, **26**, 487–91.
16 Delmi, M., Rapin, C., Bengoa, J. et al. (1990). Dietary supplementation in elderly patients with fractured neck of femur. *Lancet*, **335**, 1013–16.
17 Akesson, K., Vergnaud, P., Gineyts, E., Delmas, P.D. and Obrant, K.J. (1993). Impairment of bone turnover in elderly women with hip fracture. *Calcified Tissue International*, **53**, 162–9.
18 Garcia-Unzueta, M.T., Amado, J.A. and Gonzalez-Macias, J. (1998). Osteocalcin following tibial shaft fracture (letter). *Annals of Clinical Biochemistry*, **35**, 324.
19 Melton, J.L. III, Atkinson, E.J., O'Connor, M.K., O'Fallon, W.M. and Riggs, B.L. (1998). Bone density and fracture risk in men. *Journal of Bone and Mineral Research*, **13**, 1915–23.
20 Garnero, P., Hausherr, E., Chapuy, M.C. et al. (1996). Markers of bone resorption predict hip fracture in elderly women: the EPIDOS prospective study. *Journal of Bone and Mineral Research*, **11**, 1531–8.
21 Ross, P.D., Kress, B.C., Parson, R.E., Wasnich, R.D., Armour, K.A. and Mizrahi, I.A. (2000). Serum bone alkaline phosphatase and calcaneus bone density predict fractures: a prospective study. *Osteoporosis International*, **11**, 76–82.
22 Tromp, A.M., Ooms, M.E., Popp-Snijders, C., Roos, J.C. and Lips, P. (2000). Predictors of fractures in elderly women. *Osteoporosis International*, **11**, 134–40.
23 Delmas, P.D. (1998). The role of markers of bone turnover in the assessment of fracture risk in postmenopausal women. *Osteoporosis International*, Suppl 1, S32–6.
24 Chesnut, C.H. III, Bell, N.H., Clark, G.S. et al. (1997). Hormone replacement therapy in postmenopausal women: urinary N-telopeptide of type I collagen monitors therapeutic effect and predicts response of bone mineral density. *American Journal of Medicine*, **102**, 29–37.
25 Hesley, R.P., Shepard, K.A., Jenkins, D.K. and Riggs, B.L. (1998). Monitoring estrogen replacement therapy and identifying rapid bone losers with an immunoassay for deoxypyridinoline. *Osteoporosis International*, **8**, 159–64.
26 Greenspan, S.L., Parker, R.A., Ferguson, L., Rosen, H.N., Maitland-Ramsey, L. and Darpf, D.B. (1998). Early changes in biochemical markers of bone turnover predict the long-term response to alendronate therapy in representative elderly women: a randomised clinical trial. *Journal of Bone and Mineral Research*, **13**, 1431–8.
27 Dresner-Pollak, R., Mayer, M. and Hochner-Celiniker, D. (2000). The decrease in serum bone-specific alkaline phosphatase predicts bone mineral density response to hormone replacement therapy in early postmenopausal women. *Calcified Tissue International*, **66**, 104–7.
28 Bjarnason, N.H. and Christiansen, C. (2000). Early response in biochemical markers predicts long-term response in bone mass during hormone replacement therapy in early postmenopausal women. *Bone*, **26**, 561–9.
29 Hannon, R., Blumsohn, A., Naylor, K. and Eastell, R. (1998). Response of biochemical markers of bone turnover to hormone replacement therapy: impact of biological variability. *Journal of Bone and Mineral Research*, **13**, 1–10.
30 Taga, M., Uemura, T. and Minaguchi, H. (1998). The effect of hormone replacement therapy in postmenopausal women on urinary C-telopeptide and N-telopeptide of type I collagen, new markers of bone resorption. *Journal of Endocrinological Investigation*, **21**, 154–9.

31 Fariney, A., Kyd, P., Thomas, E. and Wilson, J. (1998). The use of cyclical etidronate in osteoporosis: changes after completion of 3 year treatment. *British Journal of Rheumatology*, **37**, 51–6.
32 Rosen, H.N., Moses, A.C., Garber, J., Ross, D.S., Lee, S.L. and Greenspan, S.L. (1998). Utility of biochemical markers of bone turnover in the follow-up of patients treated with bisphosphonates. *Calcified Tissue International*, **63**, 363–8.
33 Lane, N.E., Sanchez, S., Modin, G.W., Genant, H.K., Pierini, E. and Arnaud, C.D. (1998). Parathyroid hormone treatment can reverse corticosteroid-induced osteoporosis: results of a randomised controlled clinical trial. *Journal of Clinical Investigation*, **102**, 1627–33.
34 Scopacasa, F., Horowitz, M., Wishart, J.M. et al. (1998). Calcium supplementation suppresses bone resorption in early postmenopausal women. *Calcified Tissue International*, **62**, 8–12.
35 Cosman, F. and Lindsay, R. (1998). Is parathyroid hormone a therapeutic option for osteoporosis? A review of the clinical evidence. *Calcified Tissue International*, **62**, 475–80.
36 Cosman, F., Nieves, J., Woelfert, L., Shen, V. and Lindsay, R. (1998). Alendronate does not block the anabolic effect of PTH in postmenopausal osteoporotic women. *Journal of Bone and Mineral Research*, **13**, 1051–5.
37 Garnero, P., Shih, W.J., Gineyts, E., Karpf, D.B. and Delmas, P.D. (1994). Comparison of new biochemical markers of bone turnover in late postmenopausal osteoporotic women in response to alendronate treatment. *Journal of Clinical Endocrinology and Metabolism*, **97**, 1693–700.
38 Garnero, P. and Delmas, P.D. (1996). New developments in biochemical markers for osteoporosis. *Calcified Tissue International*, **59**(Suppl), S1–9.
39 Delmas, P.D., Hardy, P., Garnero, P. and Dain, M. (2000). Monitoring individual response to hormone replacement therapy with bone markers. *Bone*, **26**, 553–60.
40 Ravin, P., Christensen, J., Baumann, M. and Clemmesen, B. (1998). Changes in biochemical markers and bone mass after withdrawal of ibandronate treatment: prediction of bone mass changes during treatment. *Bone*, **22**, 559–64.
41 Borderie, D., Cherruau, B., Dougados, M., Ekindjian, O.G. and Roux, C. (1998). Biochemical markers as predictors of bone mineral density changes after GnRH agonist treatment. *Calcified Tissue International*, **62**, 21–5.
42 Ravn, P., Weiss, S.R., Rodriguez-Portales, J.A. et al. (2000). Alendronate in early postmenopausal women: effects on bone mass during long-term treatment and after withdrawal. *Journal of Clinical Endocrinology and Metabolism*, **85**, 1492–7.
43 Kress, B.C., Mizrahi, I.A., Armour, K.W., Marcus, R., Emkey, R.D. and Santora, A.C. (1999). Use of bone alkaline phosphatase to monitor alendronate therapy in individual postmenopausal osteoporotic women. *Clinical Chemistry*, **45**, 1009–17.
44 Fink, E., Cormier, C., Steinmetz, P., Kindermans, C., Le Bouc, Y. and Souberbielle, J.C. (2000). Differences in the capacity of several biochemical bone markers to assess high bone turnover in early menopause and response to alendronate therapy. *Osteoporosis International*, **11**, 295–303.
45 Ju, H.J., Leung, S., Brown, B. et al. (1997). Comparison of analytical performance and biological variability of three bone resorption assays. *Clinical Chemistry*, **43**, 1570–6.
46 Eastell, R., Mallinak, N., Weiss, S. et al. (2000). Biological variability of serum and urinary N-

telopeptides of type I collagen in postmenopausal women. *Journal of Bone and Mineral Research*, **15**, 594–8.

47 Rosen, C.J. and Tenenhouse, A. (1998). Biochemical markers of bone turnover. *Postgraduate Medicine*, **104**, 101–14.

48 Eastell, R., Calvo, M.S., Burritt, M.F. et al. (1992). Abnormalities in circadian patterns of bone resorption and renal calcium conservation in type 1 osteoporosis. *Journal of Clinical Endocrinology and Metabolism*, **74**, 487–94.

49 Pedersen, B.J., Schlemmer, A., Hassager, C. and Christiansen, C. (1995). Changes in the carboxyl-terminal propeptide of type 1 procollagen and other markers of bone formation upon five days of bed rest. *Bone*, **17**, 91–5.

50 Gertz, B.J., Shao, P., Hanson, D.A. et al. (1994). Monitoring bone resorption in early postmenopausal women by an immunoassay for cross-linked collagen peptides in urine. *Journal of Bone and Mineral Research*, **9**, 135–41.

51 Schlemmer, A., Hassager, C., Pedersen, B.J. and Christiansen, C. (1994). Posture, age, menopause, and osteopenia do not influence the circadian variation in the urinary excretion of pyridinium cross-links. *Journal of Bone and Mineral Research*, **9**, 1883–8.

52 Stone, P.J., Beiser, A. and Gottlieb, D.J. (1998). Circadian variation of urinary excretion of elastin and collagen crosslinks. *Proceedings of the Society for Experimental Biology and Medicine*, **218**, 229–33.

53 Aoshima, H., Kushida, K., Takahashi, M. et al. (1998). Circadian variation of urinary type 1 collagen cross-linked C-telopeptide and free and peptide-bound forms of pyridinium cross-links. *Bone*, **22**, 73–8.

54 Beardsworth, L.J., Eyre, D.R. and Dickson, I.R. (1990). Changes with age in the urinary excretion of lysyl- and hydroxylysylpyridinoline, two new markers of bone collagen turnover. *Journal of Bone and Mineral Research*, **5**, 671–6.

55 Mora, S., Prinster, C., Proverbio, M.C. et al. (1998). Urinary markers of bone turnover in healthy children and adolescents: age-related changes and effect of puberty. *Calcified Tissue International*, **63**, 369–74.

56 Epstein, S., Poser, J., McClintock, R., Johnston, C., Bryce, G. and Hui, S. (1984). Differences in Serum bone Gla protein with age and sex. *Lancet*, **1**, 307–10.

57 Delmas, P.D., Stenner, D., Wahner, H.W., Mann, K.G. and Riggs, B.L. (1983). Increase in serum bone $\alpha$-carboxyglutamic acid protein with ageing in women. *Journal of Clinical Investigation*, **71**, 1316–21.

58 Duda, R.J. Jr, O'Brien, J.F., Datzmann, J.A., Peterson, J.M., Mann, K.G. and Riggs, B.L. (1988). Concurrent assays of circulating bone Gla-protein and bone alkaline phosphatase: effects of sex, age and metabolic bone disease. *Journal of Clinical Endocrinology and Metabolism*, **66**, 951–7.

59 Kushida, K., Takahashi, M., Kawana, K. and Inoue, T. (1995). Comparison of markers for bone formation and resorption in premenopausal and postmenopausal subjects, and osteoporosis patients. *Journal of Endocrinological Investigation*, **80**, 2447–50.

60 Mora, S., Pitukcheewanont, P., Kaufman, F.R., Nelsnon, J.C. and Gilsanz, V. (1999). Biochemical markers of bone turnover and the volume and density of bone in children at different stages of sexual development. *Journal of Bone and Mineral Research*, **14**, 1664–71.

61 Hoshino, H., Takahashi, M., Kushida, K., Ohishi, T. and Inoue, T. (1998). Urinary excretion of type I collagen degradation products in healthy women and osteoporotic patients with vertebral and hip fractures. *Calcified Tissue International*, **62**, 36–9.
62 Gallagher, J.C., Kinyamu, H.K., Fowler, S.E., Dawson-Hughes, B., Dalsky, G.P. and Sherman, S.S. (1998). Calciotropic hormones and bone markers in the elderly. *Journal of Bone and Mineral Research*, **13**, 475–82.
63 Orwoll, E.S., Bell, N.H., Nanes, M.S. et al. (1998). Collagen N-telopeptide excretion in men: the effects of age and intrasubject variability. *Journal of Clinical Endocrinology and Metabolism*, **83**, 3930–5.
64 Nielsen, H.K., Brixen, K., Bouillon, R. and Mosekilde, L. (1990). Changes in biochemical markers of osteoblastic activity during the menstrual cycle. *Journal of Clinical Endocrinology and Metabolism*, **70**, 1431–7.
65 Gorai, I., Chaki, O., Nakayama, C. and Minaguchi, H. (1995). Urinary biochemical markers for bone resorption during the menstrual cycle. *Calcified Tissue International*, **57**, 100–4.
66 Gorai, I., Taguchi, Y., Chaki, O. et al. (1998). Serum soluble interleukin-6 receptor and biochemical markers of bone metabolism show significant variations during the menstrual cycle. *Journal of Clinical Endocrinology and Metabolism*, **83**, 326–32.
67 Okano, H., Mizunuma, H., Soda, M. et al. (1998). The long-term effect of menopause on postmenopausal bone loss in Japanese women: results from a prospective study. *Journal of Bone and Mineral Research*, **13**, 303–9.
68 Taga, M., Shirashu, K. and Minaguchi, H. (1998). Changes in urinary excretion of type I collagen cross-linked C-telopeptide and N-telopeptide in perimenopausal women. *Hormone Research*, **49**, 86–90.
69 Knapen, M.H.J., Nieuwenhuijzen Kruseman, A.C., Wouters, R.S.M.E. and Vermeer, C. (1998). Correlation of serum osteocalcin fractions with bone mineral density in women during the first 10 years after menopause. *Calcified Tissue International*, **63**, 375–9.
70 De Leo, V., Ditto, A., la Marca, A., Lanzetta, D., Massafra, C. and Morgante, G. (2000). Bone mineral density and biochemical markers of bone turnover in peri- and postmenopausal women. *Calcified Tissue International*, **66**, 263–7.
71 Hoshino, H., Kushida, K., Takahashi, M. et al. (2000). Changes in levels of biochemical markers and ultrasound indices of Os calcis across the menopausal transition. *Osteoporosis International*, **11**, 128–33.
72 Gorai, I., Taguchi, Y., Chake, O., Nakayama, M. and Minaguchi, H. (1997). Specific changes of urinary excretion of cross-linked N-telopeptides of type 1 collagen in pre- and postmenopausal women: correlation with other markers of bone turnover. *Calcified Tissue International*, **60**, 317–22.
73 Ebeling, P.R., Atley, L.M., Guthrie, J.R. et al. (1996). Bone turnover markers and bone density across the menopausal transition. *Journal of Clinical Endocrinology and Metabolism*, **81**, 3366–71.
74 Garnero, P., Gineyts, E., Riou, J.P. and Delmas, P.D. (1994). Assessment of bone resorption with a new marker of collagen degradation on patients with metabolic bone disease. *Journal of Clinical Endocrinology and Metabolism*, **79**, 780–5.
75 Bonde, M., Qvist, P., Fledelius, C., Riis, B.J. and Christiansen, C. (1995). Applications of an

enzyme immunoassay for a new marker of bone resorption (crosslaps): follow-up on hormone replacement therapy and osteoporosis risk assessment. *Journal of Clinical Endocrinology and Metabolism*, **80**, 864–8.

76 Lukert, B.P and Raisz, L.G. (1990). Glucocorticoid-induced osteoporosis: pathogenesis and management. *Annals of Internal Medicine*, **112**, 352–64.

77 Ekenstam, E., Stalenheim, G. and Hallgren, R. (1988). The acute effect of high dose corticosteroid treatment on serum osteocalcin. *Metabolism*, **37**, 141–4.

78 Prummel, M.F., Wiersinga, W.M., Lips, P., Sanders, G.T.B. and Sauerwein, H.P. (1991). The course of biochemical parameters of bone turnover during treatment with corticosteroids. *Journal of Clinical Endocrinology and Metabolism*, **72**, 382–6.

79 Ebeling, P.R., Erbas, B., Hopper, J.L., Wark, J.D. and Rubinfeld, A.R. (1998). Bone mineral density and bone turnover in asthmatics treated with long-term inhaled or oral glucocorticoids. *Journal of Bone and Mineral Research*, **13**, 1283–9.

80 Pearce, G., Tabensky, A., Delmas, P., Gordon Baker, H. and Seeman, E. (1998). Corticosteroid-induced bone loss in men. *Journal of Clinical Endocrinology and Metabolism*, **83**, 801–6.

81 Hughes, J.A., Conry, B.G., Male, S.M. and Eastell, R. (1999). One year prospective open study of the effect of high dose inhaled steroids, fluticason propionate, and budesonide on bone markers and bone mineral density. *Thorax*, **54**, 223–9.

82 Smith, S.M., Nillen J. L., Leblanc, A. et al. (1998). Collagen cross-link excretion during space flight and bed rest. *Journal of Clinical Endocrinology and Metabolism*, **83**, 3584–91.

83 Zerwekh, J.E., Ruml, L.A., Gottschalk, F. and Pak, C.Y.C. (1998). The effects of twelve weeks of bed rest on bone histology, biochemical markers of bone turnover, and calcium homeostasis in eleven normal subjects. *Journal of Bone and Mineral Research*, **13**, 1594–1601.

84 Ingle, B.M., Hay, S.M., Bottjer, H.M. and Eastell, R. (1999). Changes in bone mass and bone turnover following distal forearm fracture. *Osteoporosis International*, **10**, 399–407.

85 Karlsson, M.K., Josefsson, P.O., Nordkvist, A., Akesson, K., Seeman, E. and Obrant, K.J. (2000). Bone loss following tibial osteotomy: a model for evaluating post-traumatic osteopenia. *Osteoporosis International*, **11**, 261–4.

86 Emami, A., Larsson, A., Petren-Mallmin, M. and Larsson, S. (1999). Serum bone markers after intramedullary fixed tibial fractures. *Clinical Orthopaedics and Related Research*, **368**, 220–9.

87 Bainbridge, P. and Eastelel, R. (2000). The clinical utility of bone turnover markers: nurse monitoring clinic. *Arthritis and Rheumatism*, **42**(Suppl 1), S388.

10

# Biomarkers of bone formation

Juha Risteli, Saila Kauppila, Arja Jukkola, Eevastiina Marjoniemi, Jukka Melkko, Leila Risteli

University of Oulu, Oulu, Finland

## Introduction

The biochemical markers of bone metabolism are substances that can be measured in a body fluid – typically serum or urine – and which are derived as a result of bone formation or bone resorption and thus act as indices of the activity of the bone remodelling process. Several markers of bone formation have been introduced in recent years, and it has become quite evident that they do not necessarily behave similarly under the different physiological and pathological situations that affect bone turnover.

Each marker of bone formation reflects one of the three different phases of bone formation: matrix synthesis, matrix maturation or mineralization (Table 10.1). The carboxy- and aminoterminal propeptides of type I procollagen (PICP and PINP, respectively) are liberated during the first phase. The second phase is reflected by bone-specific alkaline phosphatase (BAP) and the third by osteocalcin (OC). Since the organic matrix of bone is mostly type I collagen, the best biochemical marker of new bone formation should be the one that is able to estimate reliably the rate of synthesis of type I collagen.

## Knock-out experiments and human genetic diseases

Knock-out experiments have elucidated the role of proteins as markers of bone formation. Since type I collagen is essential for life, it is not possible to generate knock-out mice with a total lack of type I collagen. However, mice missing the $\alpha 2$-chain of type I collagen are available. These animals produce a variant form of type I collagen, called $\alpha 1$-homotrimer collagen [1]. Homozygous *oim*/*oim* mice have skeletal fractures, limb deformities and generalized osteopenia. In humans, the disease osteogenesis imperfecta is caused by genetic defects in type I collagen and many different mutations usually lead to this brittle bone disease. Knock-out of the tissue-nonspecific alkaline phosphatase (TNAP) gene leads to several abnormalities both in soft tissues and bones. The latter include impaired growth, abnormal

**Table 10.1.** Biochemical markers of bone formation

| Phase of bone formation | Events | Markers |
|---|---|---|
| Matrix synthesis | Procollagen synthesis | PINP and PICP |
| Matrix maturation | Propeptide cleavage and collagen crosslinking: initiation of mineralization | Alkaline phosphatase |
| Matrix mineralization | Calcium deposition | Osteocalcin |

bone mineralization, morphological changes in osteoblasts and spontaneous fractures, which increase with advancing age [2]. Also a genetic disease in humans, hypophosphatasia, with decreased activity of alkaline phosphatase, is associated with defective bone mineralization. Knock-out of osteocalcin has given surprising results: the bones are larger and better mineralized than in wild-type animals [3]. Thus, osteocalcin is not necessarily required for bone formation. However, it seems to function as a regulatory protein for bone mineral turnover. No human disease has been yet described with a lack of osteocalcin.

## An ideal assay for bone formation

Several questions should be considered when chemical analytes are described as clinical bone markers and compared with each other. The biochemical basis of the assay is very important. How does a certain analyte reflect the rate of bone formation? Is there a stoichiometric relationship between the amount of the analyte released and the amount of bone formed, or does the analyte reflect the enzymatic activity level of osteoblasts? The homogeneity or heterogeneity of the analyte will affect the technical behaviour of the assay, e.g. standardization and expression of the results. Also, the route of elimination and the metabolic clearance rate should be known. In the following synopsis of the bone markers, all of these aspects are considered for the individual assays.

## Markers of bone collagen biosynthesis

As type I collagen comprises more than 90% of the organic matrix of bone and is responsible for its mechanical strength, any biological process that has a significant effect on the amount of bone formed must affect type I collagen. There is also a stoichiometric relationship between the collagen deposited in the tissue and the procollagen propeptides released into the circulation.

Type I procollagen is the biosynthetic precursor of type I collagen. In comparison with the final product, the precursor contains large peptidal domains at both ends of the otherwise rod-like molecule. These parts are cleaved off *en bloc* by two specific enzymes and, at least when originating from bone tissue, reach the circulation relatively unchanged. The free parts are known as procollagen propeptides, although they are in fact proteins of reasonable size. Specific assays have been available for longer for the carboxyterminal propeptide, PICP, than for the aminoterminal, PINP. PICP is cleaved off immediately after the secretion of the procollagen, whereas the cleavage of the PINP is somewhat delayed and occurs in the osteoid during the matrix maturation phase. Although assays for the intact forms of PINP and PICP in principle reflect the same process, these two propeptides are quite different as proteins, and their circulating concentrations also behave differently during the course of certain clinical situations [4]. In particular, the percentage changes in the circulating concentrations of intact PINP are larger than those of PICP in clinical situations such as Paget's disease (active versus not active) and breast carcinoma (progressive disease versus stable). The exact reasons for this are not known. The activities of the scavenger and mannose receptors in the liver endothelial cells that clear PINP and PICP propeptides from the circulation, respectively, seem to be regulated differently [5].

In human serum, PICP antigenicity is homogeneous and of the same size as the authentic propeptide. However, PINP antigenicity is split into two forms, the major one having the size of the intact propeptide and the other likely having a single $\alpha1$-chain of the propeptide. It has recently been shown that the intact form of PINP can be denatured in vitro by maintaining it for an extended period at +37 °C [6]. However, it is not known if this denaturation is also important in vivo to produce the single $\alpha1$-chains of the propeptide found in the circulation. A similar appearance of single $\alpha1$(III)-chain fragments is even more prominent in the case of the analogous propeptide of type III procollagen (PIIINP), despite the fact that PIIINP does not undergo denaturation because of the stabilizing effects of interchain disulphide bridges [7]. Thus, in the case of PIIINP, the small antigenic forms are believed to originate from the degradation of the tissue forms of the type III procollagen (so-called pN-collagen). This explanation also seems plausible for the occurrence of similar fragments in the case of PINP. Furthermore, the thermal instability of the PINP does not seem to decrease the superior clinical utility of the intact PINP assay when compared with the total PINP assay [8]. The interpretation of the results of the total PINP assay is also more complex, since, in addition to the liver, the kidneys should be taken into account, since the single chain antigenic forms are eliminated into the urine. Impairment of liver function only seems to affect the clearance of the propeptides in advanced cases of liver disease. Thus, the increased concentration of the PIIINP, which is cleared via the same receptor as PINP [9], is prognostic for death in patients with primary biliary cirrhosis [10].

## Alkaline phosphatase

Alkaline phosphatase is the marker used most frequently to monitor bone formation. The bone-specific enzyme is a product of the TNAP gene. The function of alkaline phosphatase in bone tissue is unknown, although its activity seems to be needed for the initiation of the mineralization of the osteoid. It is a membrane protein of the osteoblasts and can be released into the extracellular space after cleavage by phospholipases C or D. The half-life of alkaline phosphatase in the circulation is several days [11] and the route of elimination is unknown, but believed to be via metabolism in the liver.

Alkaline phosphatase is present in several isoenzyme forms, the two most abundant – the liver and bone variants – being products of the same gene (TNAP). Differences in glycosylation have made it possible to separate these forms and several methods – e.g. based on electrophoresis, isoelectric focusing, enzyme inhibition and thermal stability – have been described. Recently, immunological methods have been introduced that measure BAP directly. Unfortunately, the immunoassay-based methods for the bone isoenzyme form still show significant cross-reactivity with the liver enzyme [12].

## Osteocalcin

Osteocalcin is a small protein (49 amino acids) produced by osteoblasts that is incorporated into the bone matrix during the mineralization process. However, the content of osteocalcin is not the same in different parts of the skeleton. Osteocalcin appears in the circulation, where the immunoreactive forms may in principle come via leakage from the osteoblasts or as degradation products of the osteocalcin found in the mineralized bone matrix. It has been estimated that about 10% of the osteocalcin synthesized is released into the circulation, although there is no biochemical basis to ascertain that this proportion remains stable in different clinical disease states. In many disease and normal bone remodelling conditions, most of the intact osteocalcin found in the circulation is believed to result from its synthesis by osteoblasts. Because of the small size of the protein, osteocalcin is cleared from the circulation by the kidneys.

Only a small proportion of the circulating osteocalcin is intact; in fact, several different types of fragments are found in the serum [13] and the large number of published immunoassays (up to 50 different methods, about one-half of them commercially available) measure both the intact form and the different fragments with different degrees of specificity. The different immunoassays also differ with respect to their sensitivity for the gamma-carboxylation of glutamic acid, a post-translational modification that is necessary for the proper function of the protein during the mineralization process. Lack of internationally recognized standards for osteocalcin also makes it difficult to compare the results obtained with the many different assays that are available [14].

## Changes in the bone formation markers with disease and therapy

The alterations in circulating levels of the bone formation markers differ depending on the type of bone disease present and the type and duration of therapy the patient receives. For example, the acute administration of large doses of glucocorticoids elicits a decrease in the serum concentrations of PICP and osteocalcin within 3 hours, whereas no changes are observed in the serum concentration of BAP [11]. The effect of oestrogen therapy can be monitored with just about any bone formation marker assay, although the response of intact PINP is about twice that of PICP [8]. In this clinical situation, the intact PINP assay seems to be superior to any other marker of either bone formation or resorption, since the extent of the biological variation is only about one-half of that of the treatment response [15]. In contrast, bone-specific and even total serum alkaline phosphatase are superior markers in Paget´s disease of bone, in which disease PICP and osteocalcin are of limited value. Surprisingly, serum intact PINP is as effective as BAP for monitoring disease activity in Paget´s disease [16]. In prostate cancer, intact PINP and BAP are adequate markers of bone formation used for confirming the presence of bone metastases [12].

## Future developments with bone formation markers

The Sp1 binding polymorphism of the gene for the $\alpha$1-chain of type I collagen (COLIA1) is believed to lead to the synthesis of the $\alpha$1-homotrimer variant of type I collagen, whose presence in bones may explain the observed increase in fracture risk in patients with this variant of type I collagen [17, 18]. The authors of this chapter recently purified from human ascitic and pleural fluid a variant of intact PINP which lacks the $\alpha$2-chain of the propeptide [19]. The homotrimer variant of PINP can be separated from the heterotrimer PINP by DEAE (diethylaminoethyl)-chromatography at low pH (5.0) conditions, since the former contains 50% more negatively charged phosphate groups. After the separation of these two forms, their concentrations can be estimated with the intact PINP assay, which measures both PINP variants equally. Unfortunately, the overall concentration of intact PINP in serum is too low to allow for DEAE separation by HPLC (high performance liquid chromatography). There is, however, a clear need for a direct assay for the homotrimer variant of PINP.

## Conclusions

The three biochemical assays for bone formation include assays for the type I procollagen propeptides, BAP and osteocalcin. The newest assay for intact PINP has turned out to be quite reliable and shows much more of a dynamic swing with bone disease than PICP. It has recently been shown that bone resorption is not a simple process but involves at least two routes of degradation leading to different degradation products. It is possible that no single analyte developed to date can reliably reflect bone resorption. Since bone formation and resorption are tightly coupled, it is possible that an assay for intact PINP may prove to be the most reliable marker of bone formation and a good marker for monitoring changes in the rate of bone turnover with metabolic bone disease.

## REFERENCES

1 Chipman, S.D., Sweet, H.O., McBride, D.J. et al. (1993). Defective pro $\alpha$2(I) collagen synthesis in a recessive mutation in mice: a model of human osteogenesis imperfecta. *Proceedings of the National Academy of Sciences USA*, **90**, 1701–5.

2 Narisawa, S., Frohlander, N. and Millan, J.L. (1997). Inactivation of two alkaline phosphatase genes and establisment of a model of infantile hypophosphatasia. *Developmental Dynamics*, **208**, 432–46.

3 Ducy, P., Desbois, C., Boyce, B. et al. (1996). Increased bone formation in osteocalcin-deficient mice. *Nature*, **382**, 448–52.

4 Risteli, J. and Risteli, L. (1999). Products of bone collagen metabolism. In *Dynamics of Bone and Cartilage Metabolism: Principles and Clinical Applications*, eds. M.J. Seibel, S.P. Robins and J.P. Bilezikian, pp. 275–87. London: Academic Press.

5 Toivonen, J., Tähtelä, R., Laitinen, K., Risteli, J. and Välimäki, M.J. (1998). Markers of bone turnover in patients with differentiated thyroid cancer on and off thyroxine suppressive therapy. *European Journal of Endocrinology*, **138**, 667–73.

6 Brandt, J., Krogh, T.N., Jensen, C.H., Frederiksen, J.K. and Teisner, B. (1999). Thermal instability of the trimeric structure of the N-terminal propeptide of human procollagen type I in relation to assay technology. *Clinical Chemistry*, **45**, 47–53.

7 Risteli. J. and Risteli, L. (1995). Analysing connective tissue metabolites in human serum. Biochemical, physiological and methodological aspects. *Journal of Hepatology*, **22** (Suppl 2), 77–81.

8 Suvanto-Luukkonen, E., Risteli, L., Sundström, H., Penttinen, J., Kauppila, A. and Risteli, J. (1997). Comparison of three serum assays for bone collagen formation during postmenopausal estrogen-progestin therapy. *Clinica Chimica Acta*, **266**, 105–16.

9 Melkko, J., Hellevik, T., Risteli, L., Risteli, J. and Smedsrød, B. (1994). Clearance of $NH_2$-terminal propeptides of type I and III procollagen is a physiological function of the scavenger receptor in liver endothelial cells. *Journal of Experimental Medicine*, **179**, 405–12.

10 Niemelä, O., Risteli, L., Sotaniemi, E.A., Stenbäck, F. and Risteli, J. (1988). Serum antigens reflecting basement membrane and type III procollagen metabolism in primary biliary cirrhosis. *Journal of Hepatology*, **6**, 307–14.

11 Peretz, A., Moris, M., Willems, D. and Bergmann, P. (1996). Is bone alkaline phosphatase an adequate marker of bone metabolism during acute corticosteroid treatment? *Clinical Chemistry*, **42**, 102–3.

12 Diaz-Martin, M.A., Traba, M.L., De La Piedra, C., Guerrero, R., Mendez-Davila, C. and De La Pena, E.G. (1999). Aminoterminal propeptide of type I collagen and bone alkaline phosphatase in the study of bone metastases associated with prostatic carcinoma. *Scandinavian Journal of Clinical and Laboratory Investigation*, **59**, 125–32.

13 Garnero, P., Grimaux, M., Seguin, P. and Delmas, P.D. (1994). Characterization of immunoreactive forms of human osteocalcin generated in vivo and in vitro. *Journal of Bone and Mineral Research*, **9**, 255–64.

14 Delmas, P.D., Price, P.A. and Mann, K.G (1990). Validation of the bone gla protein (osteocalcin) assay. *Journal of Bone and Mineral Research*, **5**, 3–11.

15 Hannon, R., Blumsohn, A., Naylor, K. and Eastell, R. (1998). Response of biochemical markers of bone turnover to hormone replacement therapy: impact of biological variability. *Journal of Bone and Mineral Research*, **13**, 1124–33.

16 Alvarez, L., RicOs, C., Peris, P. et al. (2000). Components of biological variation of biochemical markers of bone turnover in Paget′s bone disease. *Bone*, **26**, 571–6.

17 Grant, S.F., Reid, D.M., Blake, G., Herd, R., Fogelman, I. and Ralston, S.H. (1996). Reduced bone density and osteoporosis associated with a polymorphic Sp1 binding site in the collagen type I $\alpha$1 gene. *Nature Genetics*, **14**, 203–5.

18 Uitterlinden, A.G., Burger, H., Huang, Q. et al. (1998). Relation of alleles of the collagen type I $\alpha$1 gene to bone density and the risk of osteoporotic fractures in postmenopausal women. *New England Journal of Medicine*, **338**, 1016–21.

19 Kauppila, S., Jukkola, A., Melkko, J. et al. (in press). Aminoterminal propeptide of the $\alpha$1-homotrimer variant of human type I procollagen (hotPINP) in malignant pleural effusion. *Anticancer Research*, **21**.

11

# Biochemical markers of bone resorption

Simon P Robins

Rowett Research Institute, Bucksburn, Aberdeen, UK

## Introduction

The majority of biochemical markers of bone resorption are based on the measurement of fragments of collagen type I, the fibrillar component which constitutes over 90% of the protein in bone. Other resorption marker assays that are based either on noncollagenous proteins or specific enzyme activities have received more attention since they are available as serum assays and, until recently, most bone resorption markers were those available as urine assays. The aim of this chapter is to give an account of the recent progress in the development of bone markers, with particular emphasis on the biochemical basis for each assay, along with a review of the criteria necessary for properly interpreting the results.

Before 1990, the principal method available clinically for the assessment of bone resorption was urinary hydroxyproline. It was recognized, however, at the time that this method lacked specificity and sensitivity for a number of reasons. In addition to its known release from dietary sources, a major problem with measuring hydroxyproline is the presence of this amino acid in all connective tissues as well as some rapidly metabolized serum components such as C1q. A second major drawback with urinary hydroxyproline was the fact that this assay measures not only the degradation of insoluble tissue collagen but also the release of hydroxyproline from biosynthetic intermediates including degradation of precursors intracellularly, a process which could account for more than 15% of the total collagen synthesized. The N-terminal propeptides of procollagen type I also contain a short helical section with hydroxylated proline residues so that normal collagen synthesis will lead to a significant release of hydroxyproline.

The newer resorption assays have been developed in order to improve both the sensitivity and specificity of these assays but also to provide for more convenient assays (Table 11.1). Measurements of galactosyl-hydroxylysine, for example, give some improvements over hydroxyproline in terms of bone specificity and independence from dietary interference [1], but they lack simplicity as a method which has hampered the wider application of this assay. For urine, the main choices currently

**Table 11.1.** Markers of bone resorption

| Marker | Fluid | Method | Specificity |
|---|---|---|---|
| Hydroxyproline, total and dialyzable (Hyp, OHP) | Urine | Colorimetric, HPLC | All fibrillar collagens and collagen-like proteins, including C1q; present in newly synthesized and mature collagen |
| Hydroxylysine glycosides | Urine | HPLC | Most collagens; glucosylgalactosyl-hyl is present in higher proportions in soft tissues and C1q; galactosyl-hyl is the major glycoside in skeletal collagens |
| Pyridinoline (Pyd; Pyr; HP) | Urine Serum | HPLC, EIA | Collagens, with highest concentrations in cartilage and bone; absent from skin; present in mature collagen only |
| Deoxypyridinoline (Dpd, d-Pyr, LP) | Urine | HPLC, EIA | Collagens, with highest concentration in bone; absent from cartilage or skin; present in mature collagen only |
| Carboxyterminal crosslinked telopeptide of type I collagen (ICTP) | Serum | RIA | Collagen type I, with highest contribution probably from bone<br>More sensitive to pathological bone resorption than normal remodelling |
| Carboxyterminal crosslinked telopeptide of collagen type I ($\alpha$-, $\beta$-CTx) | Urine ($\alpha$, $\beta$) Serum ($\beta$ only) | EIA, RIA | Collagen type I, with highest contribution probably from bone<br>Ratios of $\alpha$ and $\beta$ forms have limited use as markers of collagen turnover rate |
| Aminoterminal crosslinked telopeptide of collagen type I (NTx) | Urine Serum | EIA | Collagen type I, with highest contribution probably from bone |
| Bone sialoprotein (BSP) | Serum | RIA, EIA | Synthesized by active osteoblasts<br>Appears to reflect osteoclast activity for reasons as yet unclear |
| Tartrate-resistant acid phosphatase (TRAP) | Plasma, serum | Colorimetric, RIA | Osteoclasts, platelets, erythrocytes; recent assays are isotype specific |

*Note:*
EIA is enzyme immunoassay; RIA is radio immunoassay; HPLC is high performance liquid chromatography.

available are based on the pyridinium crosslinks of collagen or on peptide assays in which crosslinked regions of the N- or C-terminal telopeptides are measured; the biochemical basis of these assays will be considered separately.

## Pyridinium crosslinks

The pyridinium crosslinks, pyridinoline (Pyd) and deoxypyridinoline (Dpd), are products of a collagen crosslinking pathway involving telopeptide hydroxylysine. The hydroxylation of telopeptide lysines is a crucial step in controlling the tissue specificity of collagen crosslinking, and is now known to involve a separate enzyme to that which hydroxylates lysines destined to be in the helix [2]. The tissue specificity of crosslinking helps to provide bone specificity for pyridinium crosslink measurements because of the restricted distribution of these compounds and their absence from skin. However, the fact that bone remodelling occurs constantly and throughout life, whereas the skin turnover process is very slow, ensures that, even though skin may constitute a slightly greater proportion of the total body pool of collagen type I, any component of fibrillar collagen will preferentially be derived from bone by virtue of its higher turnover rate. Dpd has been proposed as a more specific bone marker than Pyd because of its restricted tissue distribution. Dpd is, however, present in other tissues including intramuscular collagen, vascular tissue and ligaments [3], although the much lower turnover rates of these soft tissues again ensures that their contribution to Dpd excretion is negligible.

As the formation of pyridinium crosslinks occurs extracellularly during the final stage of maturation of the collagen fibril, these compounds reflect only the degradation of mature collagen and not any biosynthetic intermediates. Their application is also facilitated by the lack of any requirements for dietary restrictions [4].

### Development of urinary crosslink assays

Following studies in the early 1980s, in which the presence of collagen crosslinks in urine was established, the main interest in crosslink assays came with the development of an HPLC (high performance liquid chromatography) method applicable to measurement of pyridinium crosslinks in urine [5]. Development of an automated HPLC system including an internal standard improved the precision of the assay at least 3-fold [6]. Initially, measurements of total crosslinks were performed after acid hydrolysis of the urine. The validity of this method for assessing bone resorption was confirmed by the observed close correlation with bone histomorphometry and with radioisotopic methods [7].

The HPLC assay was used to show that about 40% of the pyridinium crosslinks were present in urine in a free form that could be measured without the need for hydrolysis [8]. The proportion of free crosslinks was found to be relatively constant in both healthy individuals and in patient groups with osteoporosis, thyroid disorders, hyperparathyroidism and arthritic diseases. These observations paved the way for the development of direct immunoassays that initially measured both Pyd and Dpd along with small molecular weight substances ($M_r < 1000$). This was followed by the commercial development of specific monoclonal antibody-based assays that measure the more bone-specific crosslink, Dpd [9]. More recently, immunoassays for free urinary Dpd have become widely available on several types of automated immunoassay analyzers in reference laboratories and clinical laboratories.

## Telopeptide assays

Several assays have been developed that are based not on the crosslinks themselves but on the peptides that originate in the vicinity of the crosslinks found in type I collagen fibrils. The NTx assay [10] is based on a monoclonal antibody that recognizes an amino acid sequence comprising part of the N-terminal telopeptides of type I collagen (NTx). Although the antibody was shown not to react with the linear sequence in the telopeptide, the presence of the pyridinium crosslink is not necessary for reaction with the antibody [10]. The presence of sequences from the $\alpha 2$(I)-chain telopeptide in the epitope that is recognized by the NTx antibody was stated to confer specificity for bone collagen. The assay procedure has been standardized in terms of 'bone collagen equivalents' by reaction with peptides derived from a digest of a known quantity of bone. Experiments in vitro, however, have suggested that the NTx assay appears to react generally with collagen type I peptides rather than being specific for a neoepitope created only as a result of bone collagen degradation [11]. As discussed previously, this does not detract greatly from assay specificity because of the relatively high turnover rate in bone. A modified assay applicable to serum measurements has recently been developed [12], although relatively few clinical data have been reported.

As an alternative approach, assays based largely on the C-terminal telopeptides have been developed. Antibodies to an 8.5-kD peptide derived from the collagenase digestion of bone collagen form the basis of the ICTP (carboxy-terminal peptide of type I collagen) assay in serum [13]. This assay has been shown to be relatively insensitive to changes in normal bone turnover, a fact which is now understood to result from cleavage of the epitope recognized by the antibody during resorption by the osteoclastic enzyme, cathepsin K; the ICTP assay does, however, appear to be sensitive in monitoring pathological changes in bone degradation brought about mainly by matrix metalloproteinases [14].

By raising polyclonal antibodies against a synthetic octapeptide encompassing the (hydroxy)lysine residue at position 17 of the C-terminal telopeptide of collagen type I (generically referred to as CTx), Bonde et al. [15] developed an inhibition-based enzyme-linked immunosorbent assay (ELISA) which showed statistically

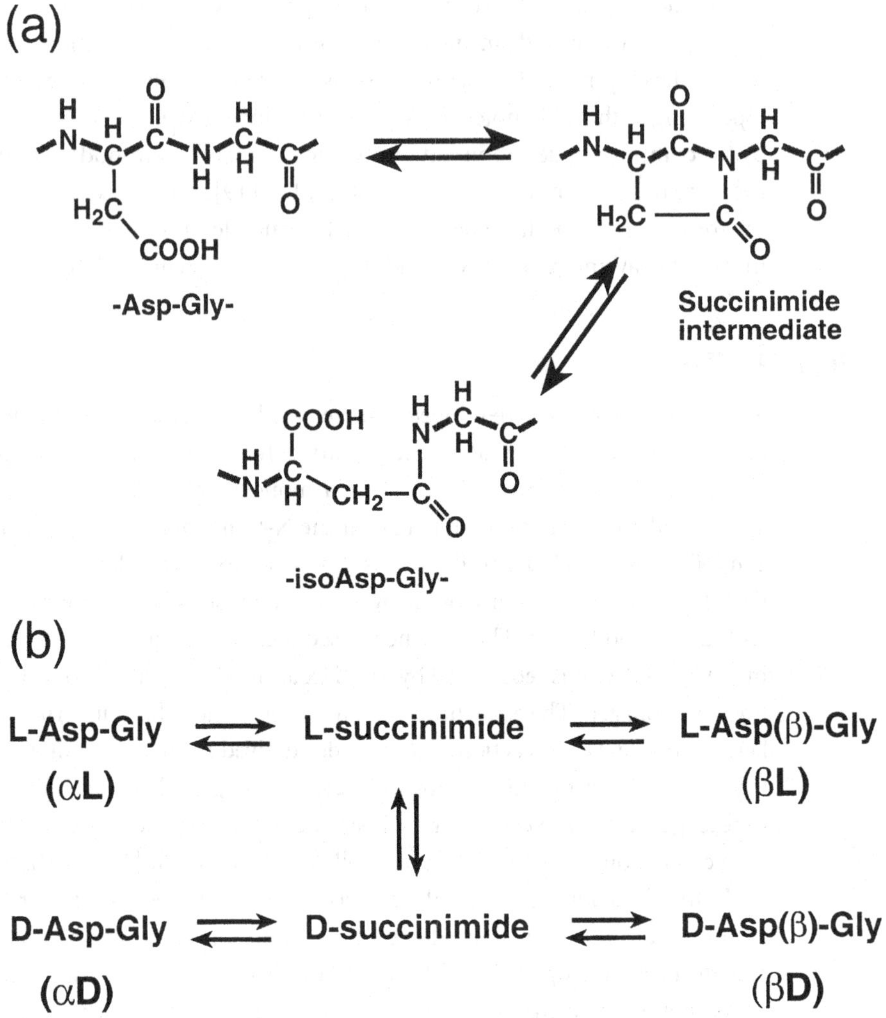

Figure 11.1 Isomerization and racemization of aspartyl residues in collagen telopeptides: (a) Asp-Gly bonds may isomerize to the $\beta$-aspartyl or isoaspartyl form via a succinimide intermediate; (b) racemization of the intermediates leads to the production of multiple forms of telopeptide, each of which may be measured with specific antibodies.

significant correlations with pyridinium crosslink measurements when applied to urine. Subsequent analysis has shown that the assay (termed 'Crosslaps') in fact recognizes only a form of the peptide containing an isoaspartyl (or $\beta$-aspartyl) peptide bond [16]. The transformation of aspartyl to isoaspartyl residues in proteins (Figure 11.1) is well known and, for extracellular proteins, the process is thought to be time dependent as well as a function of the particular environment of the susceptible bond. More recently, assays have been developed for all four forms of CTx ($\alpha$L, $\alpha$D, $\beta$L and $\beta$D) based on isomerization and racemization of the aspartyl residue (Figure 11.1). These assays have the potential to discriminate between fragments of collagen with different turnover rates. In practice, however, the equilibration rate within bone for the different forms of C-telopeptide aspartate is too rapid to allow useful data to be obtained, except in conditions with very high turnover rates such as Paget's disease of bone [17]. It has been shown that an Asp–Gly bond within the $\alpha$2(I) N-terminal telopeptide also undergoes isomerization [18], although the rate of this process in comparison with the C-terminal peptide has not been assessed.

The plethora of new assays for the C-terminal telopeptides has led to some complications in terms of the interpretation of results in metabolic bone disease, and further data are necessary to clarify the full clinical utility of these assays. Because each collagen molecule contains two identical $\alpha$1(I) chains, crosslinks from the C-terminal telopeptides can involve both $\alpha$- and $\beta$-aspartyl residues. In a recently introduced serum CTx assay, however, measurement of crosslinked peptides from more mature collagen is assured by using a sandwich assay format in which the two primary antibodies have identical specificity for the $\beta$-peptide [19]. Both the NTx and certain forms of the CTx immunoassays are, as with the Dpd assay, being developed for use in automated clinical immunoassay analysers.

## Comparison of crosslink and telopeptide assays

In general, studies have shown that there is excellent agreement between measurements in urine of free pyridinium crosslinks and the telopeptide assays. In certain instances, however, some disparity between urinary free crosslinks and telopeptides has been noted, particularly in patients treated with bisphosphonates. The proportion of free Dpd has been shown to increase during treatment for 4 weeks with an aminobisphosphonate [20], resulting in apparent responses that were lower than those obtained by measuring total crosslinks. Acute administration of bisphosphonates has also been shown to lead to a decreased response in free crosslink concentrations whereas the same study showed much larger per cent changes during treatment using both the NTx and CTx telopeptide assays [21]. These results, however, do not suggest that the telopeptides have a higher specificity for bone. The concept has been advanced that some treatments for high bone resorption states may not only affect true bone resorption but also the pattern of free crosslinks and peptides released into the urine [11]. These effects may be at the osteoclast level in bone or at other sites of the body such as the liver or kidney. The most reasonable explanation for the discrepant results in the presence of bisphosphonate therapy is that the free crosslink assays underestimate the true changes in bone resorption whereas the telopeptides assays most likely overestimate these effects. The latter point cannot be documented directly as it cannot be determined whether the urinary peptides measured by these assays constitute a consistent proportion of the bone collagen degraded. An important corollary to these inferences is that the measurement of total crosslinks, which gives intermediate rates of bone turnover, is likely to be most accurate as the uncertainties caused by changes in degradative metabolism are removed.

It has been suggested that changes in the proportions of the free crosslinks are simply a function of the rate of bone collagen resorption and not related to any particular form of treatment. Some evidence for this was provided by an observed inverse relationship between the proportion of free Dpd crosslink and the total amount of Pyd excreted [21]. The author's experience from a wide range of results from both healthy volunteers and patients with metabolic bone disease not receiving treatment is that the weak relationship between the proportion of free Dpd and the total Dpd excreted is unlikely to contribute significantly to the observed effects of treatment.

## Standardization of crosslink assays

Although the crosslink and telopeptide assays generally compare well, there are problems in attempting to correlate the results of the assays in absolute terms, because of difficulties in standardization. For the Dpd assay, the analyte is well defined and criteria have been established for the quantification of standard solutions using primarily ultraviolet extinction coefficients [22]. Also, Dpd with similar characteristics to the isolated material has now been chemically synthesized in several centres[23–25]. The monoclonal antibody used in the assay was shown to react only with Dpd in urine [9]. As the concentrations of Dpd in bone change little with age [26], the results of total urinary output can be related directly to an equivalent mass of bone being resorbed. By contrast, with the peptide assays, the exact nature of the analytes is unclear, making it difficult to generate pure calibrators. For the NTx assay, standards are prepared from a collagenase digest of bone [10] and the results calculated in terms of bone collagen equivalents, even though the patterns of peptides being analysed may be distinctly different from those of the prepared calibrators. The situation for the various forms of CTx assay is increasingly complex and calibrators may be based on incompletely characterized peptides isolated from urine.

## Tartrate-resistant acid phosphatase

Human acid phosphatases form a heterogeneous group of at least five electrophoretically distinct isoenzymes that are found in the prostate, various bloods cells and osteoclasts. All acid phosphatases are inhibited by tartrate, except for the band 5 isoenzymes, the subtype 5b of which appears to be found in and secreted by osteoclasts. Serum or plasma tartrate-resistant acid phosphatase (TRAP) activity has been suggested as a marker of osteoclast activity in bone resorption. Circulating levels of TRAP activity may be determined by electrophoretic or spectrophotometric procedures [27], although artificially high values are often obtained in haemolysed blood samples because of the release of TRAP from erythrocytes. Recent advances in the development of TRAP immunoassays include a direct two-site assay in serum which has been pretreated with a chelating agent to disperse a high $M_r$ complex (>250 kD) containing TRAP [28]. The complex, which appears to disperse on storage of the serum at 4 °C for 24 hours, does not affect activity assays, but immunoassays require the EDTA treatment to detect the 30-kD TRAP enzyme. The need to disperse the complex seems to explain the pronounced thermal instability of the enzyme which had been noted previously and which has, so far, precluded this marker from wider clinical application. Serum TRAP activity has been shown to be elevated in a variety of metabolic bone diseases associated with increased bone resorption, such as Paget's disease of bone, hyperparathyroidism, multiple myeloma and metastatic bone disease.

## Bone sialoprotein

Bone sialoprotein (BSP) is a product of active osteoblasts and odontoblasts and accounts for up to 10% of the noncollagenous matrix of bone. The expression of BSP is restricted to mineralized tissue such as bone and dentin and to the interface of calcifying cartilage. The intact molecule has a protein core of approximately 33 kD but is heavily glycosylated with a full molecular mass of about 80 kD. BSP contains an Arg–Gly–Asp (RGD) integrin recognition sequence, and appears to play a role in the supramolecular organization of the extracellular matrix of mineralized tissues.

Immunoassays for BSP in serum have been described [29, 30] and elevated concentrations have been noted in patients with high bone turnover diseases such as Paget's disease of bone, primary hyperparathyroidism, active rheumatoid arthritis, metastatic bone disease or multiple myeloma [31]. Based on clinical data, and an observed rapid decrease in serum BSP concentrations following intravenous bisphosphonate treatment, immunoreactive BSP appears to constitute a marker which reflects processes related to bone resorption or osteoclast activity. Further studies are necessary in order to determine whether the assays currently available measure intact protein in serum or some degraded forms released during bone resorption.

## Conclusion

An increased awareness of the prevalence and consequences, both social and economic, of osteoporosis has provided a major impetus for research into the development of biochemical markers of bone turnover. The newly developed assays for bone resorption provide physicians with additional tools to augment the clinical assessment and management of patients with metabolic bone disease. Thus, the measurement of bone markers provides a simple and cost-effective approach to assist in assessing the risk of osteoporotic bone fracture. This can help target preventative therapy more appropriately and could minimize the large healthcare costs associated with the treatment of fractures.

## Acknowledgement

I am indebted to the Scottish Executive Environment and Rural Affairs Department for support.

## REFERENCES

1 Moro, L., Modricky, C., Stagni, N., Vittur, F. and de Bernard, B. (1984). High-performance liquid chromatographic analysis of urinary hydroxylysyl glycosides as indicators of collagen turnover. *Analyst*, **109**, 1621–2.

2 Bank, R.A., Robins, S.P., Wijmenga, C. et al. (1999). Defective collagen crosslinking in bone, but not in ligament or cartilage, in Bruck syndrome: indications for a bone-specific telopeptide lysyl hydroxylase on chromosome 17. *Proceedings of the National Academy of Sciences USA*, **96**, 1054–8.

3 Seibel, M.J., Robins, S.P. and Bilezikian, J.P. (1992). Urinary pyridinium crosslinks of collagen: specific markers of bone resorption in metabolic bone disease. *Trends in Endocrinology and Metabolism*, **3**, 263–70.

4 Colwell, A., Russell, R.G. and Eastell, R. (1993). Factors affecting the assay of urinary 3-hydroxy pyridinium crosslinks of collagen as markers of bone resorption. *European Journal of Clinical Investigation*, **23**, 341–9.

5 Black, D., Duncan, A. and Robins, S.P. (1988). Quantitative analysis of the pyridinium crosslinks of collagen in urine using ion-paired reversed-phase high-performance liquid chromatography. *Analytical Biochemistry*, **169**, 197–203.

6 Pratt, D.A., Daniloff, Y., Duncan, A. and Robins, S.P. (1992). Automated analysis of the pyridinium crosslinks of collagen in tissue and urine using solid-phase extraction and reversed-phase high-performance liquid chromatography. *Analytical Biochemistry*, **207**, 168–75.

7 Eastell, R., Colwell, A., Hampton, L. and Reeve, J. (1997). Biochemical markers of bone resorption compared with estimates of bone resorption from radiotracer kinetic studies in osteoporosis. *Journal of Bone and Mineral Research*, **12**, 59–65.

8 Robins, S.P., Duncan, A. and Riggs, B.L. (1990). Direct measurement of free hydroxy-pyridinium crosslinks of collagen in urine as new markers of bone resorption in osteoporosis. In *Osteoporosis*, eds. C. Christiansen and K. Overgaard. Copenhagen: Osteopress ApS, pp. 465–8.
9 Robins, S.P., Woitge, H., Hesley, R., Ju, J., Seyedin, S. and Seibel, M.J. (1994). Direct, enzyme-linked immunoassay for urinary deoxypyridinoline as a specific marker for measuring bone resorption. *Journal of Bone and Mineral Research,* **9**, 1643–9.
10 Hanson, D.A., Weis, M.A., Bollen, A.M., Maslan, S.L., Singer, F.R. and Eyre, D.R. (1992). A specific immunoassay for monitoring human bone resorption: quantitation of type I collagen cross-linked N-telopeptides in urine. *Journal of Bone and Mineral Research,* **7**, 1251–8.
11 Robins, S.P. (1995). Collagen crosslinks in metabolic bone disease. *Acta Orthopaedica Scandinavica Supplement,* **266**, 171–5.
12 Clemens, J.D., Herrick, M.V., Singer, F.R. and Eyre, D.R. (1997). Evidence that serum NTx (collagen-type I N-telopeptides) can act as an immunochemical marker of bone resorption. *Clinical Chemistry,* **43**, 2058–63.
13 Risteli, J., Elomaa, I., Niemi, S., Novamo, A. and Risteli, L. (1993). Radioimmunoassay for the pyridinoline crosslinked carboxy-terminal peptide of type I collagen: a new serum marker of bone collagen degradation. *Clinical Chemistry,* **39**, 635–40.
14 Sassi, M., Eriksen, H., Risteli, L. et al. (2000). Immunochemical characterization of assay for carboxyterminal telopeptide of human type I collagen: loss of antigenicity by treatment with cathepsin K. *Bone,* **26**, 367–73.
15 Bonde, M., Qvist, P., Fidelius, C., Riis, B. J. and Christiansen, C. (1994). Immunoassay for quantifying type I degradation products in urine evaluated. *Clinical Chemistry,* **40**, 2022–5.
16 Fledelius, C., Johnsen, A. H., Cloos, P. A. C., Bonde, M. and Qvist, P. (1997). Characterization of urinary degradation products derived from type I collagen. Identification of a beta-isomerized Asp-Gly sequence within the C-terminal telopeptide ($\alpha$1) region. *Journal of Biological Chemistry,* **272**, 9755–63.
17 Cloos, P.A.C. and Fledelius, C. (2000). Collagen fragments in urine derived from bone resorption are highly racemized and isomerized: a biological clock of protein aging with clinical potential. *Biochemical Journal,* **345**, 473–80.
18 Brady, J.D., Ju, J. and Robins, S.P. (1999). Isoaspartyl bond formation within N-terminal sequences of collagen type I: implications for their use as markers of collagen degradation. *Clinical Science,* **96**, 209–15.
19 Rosenquist, C., Fledelius, C., Christgau, S. et al. (1998). Serum CrossLaps One Step ELISA. First application of monoclonal antibodies for measurement in serum of bone-related degradation products from C-terminal telopeptides of type I collagen. *Clinical Chemistry,* **44**, 2281–9.
20 Tobias, J., Laversuch, C., Wilson, N. and Robins, S.P. (1996). Neridronate preferentially suppresses the urinary excretion of peptide-bound deoxypyridinoline in postmenopausal women. *Calcified Tissue International,* **59**, 407–9.
21 Garnero, P., Gineyts, E., Arbault, P., Christiansen, C. and Delmas, P.D. (1995). Different effects of bisphosphonate and estrogen therapy on free and peptide-bound bone cross-links excretion. *Journal of Bone and Mineral Research,* **10**, 641–9.
22 Robins, S.P., Duncan, A., Wilson, N. and Evans, B.J. (1996). Standardization of pyridinium

crosslinks, pyridinoline and deoxypyridinoline, for use as biochemical markers of collagen degradation. *Clinical Chemistry,* **42**, 1621–6.

23 Allevi, P., Longo, A. and Anastasia, M. (1999). Total synthesis of deoxypyridinoline, a biochemical marker of collagen turnover. *Chemical Communications,* **6**, 559–60.

24 Adamczyk, M., Johnson, D.D. and Reddy, R.E. (2000). Collagen cross-links. Synthesis of immunoreagents for development of assays for deoxypyridinoline, a marker for diagnosis of osteoporosis. *Bioconjugate Chemistry,* **11**, 124–30.

25 Adamczyk, M., Johnson, D.D. and Reddy, R.E. (1999). An efficient one-pot synthesis of (+)-deoxypyridinoline. *Tetrahedron Letters,* **40**, 8993–4.

26 Eyre, D.R., Dickson, I.R. and VanNess, K.P. (1988). Collagen crosslinking in human bone and cartilage: age-related changes in the content of mature hydroxypyridinium residues. *Biochemical Journal,* **252**, 495–500.

27 Lau, K.H., Onishi, T., Wergedal, J.E., Singer, F.R. and Baylink, D.J. (1987). Characterization and assay of tartrate-resistant acid phosphatase activity in serum: potential use to assess bone resorption. *Clinical Chemistry,* **33**, 458–62.

28 Halleen, J.M., Hentunen, T.A., Karp, M., Kakonen, S.M., Pettersson, K. and Vaananen, H.K. (1998). Characterization of serum tartrate-resistant acid phosphatase and development of a direct two-site immunoassay. *Journal of Bone and Mineral Research,* **13**, 683–7.

29 Saxne, T., Zunino, L. and Heinegard, D. (1995). Increased release of bone sialoprotein into synovial fluid reflects tissue destruction in rheumatoid arthritis. *Arthritis and Rheumatism,* **38**, 82–90.

30 Karmatschek, M., Maier, I., Seibel, M.J., Woitge, H.W., Ziegler, R. and Armbruster, F.P. (1997). Improved purification of human bone sialoprotein and development of a homologous radioimmunoassay. *Clinical Chemistry,* **43**, 2076–82.

31 Seibel, M.J., Woitge, H.W., Pecherstorfer, M. et al. (1996). Serum immunoreactive bone sialoprotein as a new marker of bone turnover in metabolic and malignant bone disease. *Journal of Clinical Endocrinology and Metabolism,* **81**, 3289–94.

12

# The clinical application of biomarkers in osteoporosis

Patrick Garnero
INSERM and SYNARC, Lyon, France

Osteoporosis is a disease characterized by low bone mass and by architectural deterioration of bone tissue. Both are related to abnormalities of bone turnover. Biochemical markers of bone turnover reflect the degree of increase in overall bone turnover, and, so far, there are no data indicating that levels of markers of bone formation and resorption can be combined to assess remodelling imbalance. The rate of bone formation or degradation can be assessed either by measuring an enzymatic activity of the osteoblastic or osteoclastic cells – such as alkaline and acid phosphatase activity – or by measuring components of the bone matrix released into the circulation during formation or resorption, such as osteocalcin and pyridinoline crosslinks (Table 12.1). In osteoporosis, bone turnover markers have been suggested to predict the rate of postmenopausal bone loss, to predict the occurrence of osteoporotic fractures and to monitor the efficacy of treatment, especially antiresorptive therapy (hormone replacement therapy [HRT], bisphosphonates and calcitonin). It has also been suggested that the measurement of bone turnover before treatment might be useful in selecting the type of therapy (antiresorptive or bone-stimulating agent) and in predicting the amplitude of the response to oestrogen and bisphosphonate treatment; however, there is little solid evidence for these two concepts. In this chapter, we will briefly review the clinical use of biochemical markers of bone turnover in predicting fracture risk and in monitoring treatment efficacy.

## Bone markers and fracture risk

With the emergence of effective – but rather expensive – treatments, it is essential to detect those women at higher risk of fracture. Several prospective studies have shown that a standard deviation (SD) decrease of bone mineral density (BMD) measured by dual X-ray absorptiometry (DXA) or heel ultrasound is associated with a 2–4-fold increase in relative fracture risk including of the hip, spine and forearm. In this context, the question arises as to what extent bone markers can add to bone mass measurements in order to improve the assessment of fracture risk.

**Table 12.1.** Biochemical markers for bone remodelling

| |
|---|
| FORMATION |
| Serum |
| *osteocalcin* (bone Gla-protein) |
| total and *bone alkaline phosphatase* |
| collagen type I C-terminal and N-terminal propeptides (PICP and *PINP*) |
| RESORPTION |
| Plasma/serum |
| tartrate-resistant acid phosphatase (TRAP) |
| free pyridinoline and deoxypyridinoline |
| N-terminal (NTx) and *C-terminal (CTx) telopeptide of type I collagen* |
| Bone sialoprotein (BSP) |
| Urine |
| *free pyridinoline and deoxypyridinoline* |
| *N-terminal (NTX) and C-terminal (CTX) telopeptide of type I collagen* |
| calcium |
| hydroxyproline |
| galactosylhydroxylysine |

*Note:*
The markers with the best performance characteristics in osteoporosis are in italics.

Several studies have shown that increased levels of bone turnover markers are associated with faster rates of bone loss in the subsequent years [1, 2]. For example, in a recent study of 305 untreated postmenopausal women [3], the author found that baseline values of serum osteocalcin, serum procollagen type I N-terminal propeptide (PINP), serum and urinary C-terminal crosslinking telopeptide of type I collagen (CTx) and urinary N-terminal crosslinking telopeptide of type I collagen (NTx) were negatively correlated with the rate of bone loss at the forearm over the next 4 years. Interestingly, women whose marker levels exceeded mean values in premenopausal women by more than two SDs lost bone two to six times faster than those whose marker levels were normal. However, the most important issue from a clinical standpoint is the ability of bone turnover markers to predict fracture risk.

Recent data obtained in three prospective studies (Epidemiologie des l'Osteoporose [EPIDOS], Rotterdam and Os des Femmes de Lyon [OFELY]) indicate that increased levels of bone turnover markers and, more specifically, markers of bone resorption are associated with an increased risk of hip, vertebral and nonhip and nonvertebral fractures over follow-up periods ranging from 1.8 to 5

**Table 12.2.** Combined use of bone mineral density (BMD) and of bone resorption marker (urinary CTx or free deoxypyridinoline [Dpd]) to improve hip fracture risk in elderly women: the EPIDOS study.

| | % of women at risk | Relative risk (95% CI) |
|---|---|---|
| Low BMD | 56% | 2.7 (1.5–5.0) |
| High CTx | 24% | 2.2 (1.3–3.6) |
| High free Dpd | 22% | 1.9 (1.1–3.2) |
| Low BMD + High CTx | 16% | 4.8 (2.4–9.5) |
| Low BMD + high free Dpd | 14% | 4.1 (2.0–8.2). |

*Note:*
Mean age: 83 years; follow-up: 22 months; 109 incident hip fractures; 292 age-matched controls. Low BMD was defined as a T score ≤ −2.5; high bone resorption marker as a T score >2.
*Source:* From Garnero et al. [5].

years [4–7]. This predictive value is consistently in the order of a 2-fold increase in the risk of fracture for levels above the upper limit of the premenopausal range. Both increased levels of serum and urine CTx and free deoxypyridinoline have been shown to be associated with a higher risk of hip, vertebral and other nonvertebral fractures. Increased bone resorption is associated with an increased risk of fracture only for values which exceed a threshold, suggesting that bone resorption becomes deleterious for bone strength only when it exceeds the normal physiological range. As bone resorption rates predict fracture independently of BMD, these data suggest that increased bone resorption can lead to increased skeletal fragility by two factors. Firstly, a prolonged increase in bone turnover will lead after several years to a lower BMD [3], which is a major determinant of reduced bone strength. Secondly, increased bone resorption above the upper limit of the normal range may induce microarchitectural deterioration of bone tissue such as perforation of the trabeculae, a major component of bone strength.

Thus, the combination of BMD and bone marker measurements could be useful in order to improve the identification of women at high risk for fracture. Using the database of the EPIDOS study, it was shown that combining a bone resorption marker and hip BMD measurement can detect women at very high risk of fracture. Indeed, women with both low hip BMD (according to the World Health Organization definition of osteoporosis) and high bone resorption had a 4–5-fold higher risk compared with the general population (Table 12.2). This has been confirmed for vertebral and nonvertebral and nonhip fractures. In addition, by using this combination, the specificity of hip fracture prediction is increased without a loss of sensitivity. In those women who are still asymptomatic, combining a bone marker measurement with a BMD measurement is likely to be helpful in order to predict fracture risk. The place of biochemical markers of bone resorption in the assessment of fracture risk is likely to be in combination with other important risk factors, including low BMD, personal and maternal history of fracture and low body weight.

## Bone markers to monitor antiresorptive therapy

Monitoring the efficacy of treatment of osteoporosis is a challenge. The goal of treatment is to reduce the occurrence of fragility fractures. Measurement of BMD by DXA is a surrogate marker of treatment efficacy that has been widely used in clinical trials. Its use in the monitoring of treatment efficacy in the individual patient, however, has not been validated. Given a short-term precision error of 1–1.5% of BMD measurement at the spine and hip, the individual change must be greater than 3–5% to be seen as significant. With bisphosphonates such as alendronate, repeating BMD 2 years after initiating therapy will determine if a patient is responding to therapy, i.e. shows a significant increase in BMD – at least at the lumbar spine which is the most responsive site. With treatments such as raloxifene or nasal calcitonin that induce much smaller increases in BMD, DXA is not appropriate to monitor therapy and, with any treatment, DXA does not allow the identification of all responders within the first year of therapy. Failure to respond may be due to noncompliance, to poor intestinal absorption (i.e. with the use of bisphosphonates) or to other unidentified factors.

Several randomized, placebo-controlled studies found that resorption-inhibiting therapy was associated with a prompt decrease of bone resorption markers that can be seen as early as 2 weeks with a plateau reached within 3–6 months. The decrease in bone formation markers is delayed – reflecting the physiological coupling of formation to resorption – and a plateau is usually achieved within 6–12 months. In addition, in these studies, the magnitude of the short-term bone marker level decrease was significantly correlated with the magnitude of the long-term BMD increase [8–10]. Although bone markers do not allow for the accurate prediction of bone gain in individual patients, measurement of a marker for resorption and/or formation can provide the same information on therapeutic efficacy as the measurement of BMD. Indeed, for the clinician, the primary concern is the identification of nonresponders, i.e. of those patients who will fail to demonstrate a significant increase in BMD after 2 years of treatment. Several methods have been suggested in order to identify responders/nonresponders according to bone marker response to therapy. One approach is to consider the least significant change of a bone marker (based on the short term or long term within subject variability), regardless of the BMD response. The percentage change of the marker under treat-

**Table 12.3.** Measurement of the early changes in bone remodelling markers predicts the efficacy of oestrogen replacement therapy with 90% specificity in individual patients. In this study, 569 postmenopausal women aged 40–60 years with a time since menopause shorter than 6 years were given either a placebo or transdermal oestrogen in a dosage of 25, 50 or 75 μg twice a week for 28 days (continuous treatment) or 50, 75 or 100 μg twice a week for 25 days per cycle (cyclic therapy). Bone mineral density (BMD) at the spine was measured at baseline and after 2 years using dual-energy X-ray absorptiometry (DXA). Women with a BMD increase versus baseline greater than 2.26% (i.e. twice the short-term coefficient of variation for DXA) were classified as treatment responders and women with a BMD decrease versus baseline of more than 2.26% as nonresponders. The table shows the sensitivity and the likelihood of a positive response obtained using a 3-month bone marker decrease cut-off associated with 90% specificity.

| Marker | Cut-off value for the bone marker decrease after 3 months | Sensitivity* | Likelihood of a positive response** |
|---|---|---|---|
| Serum CTx | −33% | 68% | 87% |
| Urinary CTx | −45% | 60% | 88% |

*Notes:*

* Proportion of women whose bone marker value decrease 3 months into therapy was equal to or greater than the cut-off, among the women with a greater than 2.26% BMD increase 2 years into therapy.

** Proportion of women with a greater than 2.26% BMD increase 2 years into therapy, among the women whose bone marker value decrease 3 months into therapy was equal to or greater than the cut-off.

*Source:* From Delmas et al. [11].

ment can be used, and cut-off values can be obtained with a prespecified sensitivity or specificity. Using this strategy, several recent studies have shown that bone markers are reliable indices of therapeutic efficacy in individual patients (Table 12.3) [10, 11]. The author also recently showed that combining an absolute bone marker value obtained 3–6 months into therapy with the percentage decrease in the same marker over the same period in a logistic regression model improved the ability of the marker to identify nonresponders to alendronate or oestrogen replacement therapy [11, 12].

The value of BMD changes in predicting the risk of fracture under treatment is debated, especially because some treatments – such as raloxifene – can induce a 30–50% reduction in vertebral fracture rate despite a small, 2–3%, increase in BMD at all skeletal sites. Thus, BMD changes may not be an adequate surrogate endpoint for analysing the ability of bone markers to predict fracture risk. In a recent report, it was shown that short-term changes in serum osteocalcin and bone alkaline phosphatase following raloxifene therapy were associated with a subsequent risk of vertebral fractures in a large subgroup of osteoporotic women enrolled in the MORE (Multiple Outcomes of Raloxifene Evaluation) study, while changes in BMD were not predictive [13].

In conclusion, long-term treatment of symptom-free patients raises special challenges because the benefits of the treatment are not perceived by the patients. In this situation, an improvement in a laboratory test result may allow the physician to convince the patient that the treatment is having beneficial effects. Several studies have shown that the percentage decrease of some bone markers after 3–6 months of HRT or alendronate can be used to predict the 2-year response in BMD with adequate sensitivity and specificity. However, prospective studies looking for a favourable effect of bone marker monitoring on treatment compliance and, ultimately, fracture risk reduction are needed.

## REFERENCES

1 Riis, S.B.J., Hansen, A.M., Jensen, K., Overgaard, K. and Christiansen, C. (1996). Low bone mass and fast rate of bone loss at menopause equal risk factors for future fracture. A 15 year follow-up study. *Bone*, **19**, 9–12.

2 Ross, P.D. and Knowlton, W. (1998). Rapid bone loss is associated with increased levels of biochemical markers. *Journal of Bone and Mineral Research*, **13**, 297–302.

3 Garnero, P., Sornay-Rendu, E., Duboeuf, F. and Delmas, P.D.(1999). Markers of bone turnover predict postmenopausal forearm bone loss over 4 years: the OFELY Study. *Journal of Bone and Mineral Research*, **14**, 1614–21.

4 van Daele, P.L.A., Seibel, M.J., Burger, H. et al. (1996). Case-control analysis of bone resorption markers, disability, and hip fracture risk: the Rotterdam study. *British Medical Journal*, **312**, 482–3.

5 Garnero, P., Hausher, E., Chapuy, M.C. et al. (1996) Markers of bone resorption predict hip fracture in elderly women: the EPIDOS prospective study. *Journal of Bone and Mineral Research*, **11**, 1531–8.

6 Garnero, P., Sornay-Rendu, E., Claustrat, B. and Delmas, P.D. (2000). Biochemical markers of bone turnover, endogenous hormones and the risk of fractures in postmenopausal women: the OFELY study. *Journal of Bone and Mineral Research*, **15**, 1526–36.

7 Ross, P.D., Kress, B.C., Parson, R.E., Wasnich, R.D., Armour, K.A. and Mizrahi, I.A. (2000). Serum bone alkaline phosphatase and calcaneus bone density predict fractures: a prospective study. *Osteoporosis International*, **11**, 76–82.

8 Garnero, P., Shih, W.J., Gineyts, E. et al. (1994). Comparison of new biochemical markers of bone turnover in late postmenopausal osteoporotic women in response to alendronate treatment. *Journal of Clinical Endocrinology and Metabolism*, **79**, 1693–700.

9 Rosen, C.J., Chesnut, C.H. and Mallinak, N.J.S. (1997). The predictive value of biochemical markers of bone turnover for bone mineral density in early postmenopausal women treated with hormone replacement therapy or calcium supplementation. *Journal of Clinical Endocrinology and Metabolism*, **82**, 1904–10.

10 Ravn, P., Hosking, D., Thompson, G.C. et al. (1999). Monitoring of alendronate treatment and prediction of effect on bone mass by biochemical markers in early postmenopausal intervention cohort of study. *Journal of Clinical Endocrinology and Metabolism*, **84**, 2363–8.

11 Delmas, P.D., Hardy, P., Garnero, P. and Dain, M.P. (2000). Monitoring individual response to hormone replacement therapy with bone markers. *Bone*, **26**, 553–60.

12 Garnero, P., Darte, C. and Delmas, P.D. (1999). A model to monitor the efficacy of alendronate treatment in women with osteporosis using a biochemical marker of bone turnover. *Bone*, **24**, 603–9.

13 Bjarnason, N.H., Christiansen, C., Sarkar, S. et al. for the MORE Study Group (1999). Six months' changes in biochemical markers predict 3-year response in vertebral fracture rate in postmenopausal, osteoporotic women : results from the MORE study. *Journal of Bone and Mineral Research*, **14**(Suppl 1), S157.

13

# Sources of preanalytical variability in the measurement of biochemical markers of bone turnover

Markus J Seibel

University of Sydney, Sydney, Australia

Nonspecific variability in the measurement of chemical analytes is a major issue in clinical chemistry. Consideration and, wherever possible, control of factors not related to the specific process in question are essential for the correct interpretation of laboratory results. At present, these caveats seem to be of particular relevance for the various molecular markers of bone turnover, as the degree of variability of some of these analytes has been shown to be rather substantial.

In general, sources of variability include preanalytical and analytical influences. Total variability (CVT) is therefore defined as the sum of preanalytical (CVPA) and analytical (CVA) variability and may be calculated as $CVT^2 = CVPA^2 + CVA^2$. Among the preanalytical factors affecting biochemical measurements are, firstly, preanalytical issues such as dietary influences, the type of specimen collected and the mode of sample processing and storage. Secondly, there are biological effects such as age, growth, gender, diurnal variation and other physiological or pathological influences. Finally, factors related to the clearance and metabolism of the marker component (e.g. renal or hepatic function) are considered to be biological variables.

The term 'analytical variability' mainly concerns issues of assay performance, i.e. measures of assay precision and accuracy (e.g. linearity, intra-assay and inter-assay variability, quality control and standardization). Recent studies have shown that results obtained from most bone marker assays (including commercial ones) vary greatly between laboratories, either due to the lack of appropriate standards or to suboptimal assay performance.

The ideal bone marker should have a high signal to noise ratio, i.e. a substantial signal in response to changes in bone turnover and as little 'background noise' as possible. While the signal is usually a function of a marker's sensitivity and specificity (which can be optimized by choosing the appropriate analyte), 'background noise' is generated by various factors which can only be partly controlled. Low background noise is characterized by minimal and predictable variability and can, at least in part, be achieved by standardizing (pre)analytical conditions. At present, however, there is no ideal marker and, in most assays used today, the signal to noise ratio is rather low. Therefore, changes in marker measurements (e.g. as a result of therapeutic interventions) should be interpreted against the background of the respective marker's total variability. As a rule of thumb, markers that change 'dramatically' in response to interventions usually show similar or even more pronounced degrees of nonspecific variability.

## Introduction

The meaningful interpretation of laboratory results has many aspects, but consideration of the potential sources of nonspecific variability is among the most important tasks. In general, nonspecific variability comprises both preanalytical (i.e. mostly subject related; CVPA) and analytical (i.e. mostly assay related; CVA) factors. Total variability is considered to be the sum of preanalytical and analytical variation and is defined as $CVT^2 = CVPA^2 + CVA^2$.

The ideal marker and assay is characterized by (i) good analytical performance (i.e. high precision and accuracy) and (ii) minimal and predictable preanalytical variability. Unfortunately, no method in clinical chemistry meets all of these criteria at the same time. However, most of the currently available assays for biochemical markers of bone turnover are characterized by substantial analytical and preanalytical variability.

This chapter deals with selected preanalytical sources of variability which confound measurements of biochemical markers and describes their impact on the interpretation of bone marker results. The more relevant preanalytical factors affecting marker variability are summarized in Table 13.1.

## Preanalytical variability

### Technical aspects of preanalytical variability

The choice of sample (i.e. serum versus urine), the mode of urine collection (i.e. 24-hour collection versus first or second morning void), the appropriate preparation of the patient (i.e. minimizing the effect of diet or exercise before phlebotomy) and the correct processing and storage of specimens all influence the final analytical result. Therefore, special care must be taken of these more technical issues, particularly as they are modifiable and controllable.

#### Handling, processing and storage of samples

Some, but not all, biochemical components used as markers of bone turnover are sensitive to ambient conditions such as temperature or ultraviolet (UV) radiation.

**Table 13.1.** Sources and categories of preanalytical variability

| |
|---|
| Preanalytical (sample and assay related) |
| type of specimen (serum versus urine) |
| type of urine sample (24-hour versus FMV versus SMV) |
| sample processing |
| sample storage (analyte stability) |
| diet |
| Biological (subject related) |
| Intraindividual |
| diurnal variability |
| day to day variability |
| menstrual rhythms |
| seasonal rhythms |
| exercise |
| Interindividual |
| age |
| gender |
| race |
| pregnancy and lactation |
| menopausal status |
| growth |
| nonskeletal diseases |
| skeletal disease (including fractures) |
| immobility |
| drugs (including oral contraceptives) |
| clearance-related factors |

In components susceptible to these influences, thermodegradation or photolysis will lead to a reduction in the measured signal and, if not noted, to a misinterpretation of the laboratory result.

While alkaline phosphatase and the procollagen propeptides have been shown to be rather stable at room temperature, serum osteocalcin (OC) is readily degraded even at temperatures as low as 4 °C [1, 2]. When using assays for intact OC (1–49), rapid enzymatic cleavage of the peptide into smaller fragments will lead to significant signal losses if the serum sample is kept at room temperature for more than 1–2 hours (Figure 13.1). Adding protease inhibitors will delay, but by no means prevent, this process [1]. Signal reduction is less pronounced when measuring smaller (already degraded) fragments of OC, e.g. OC(1–43) (Figure 13.1) [1, 3].

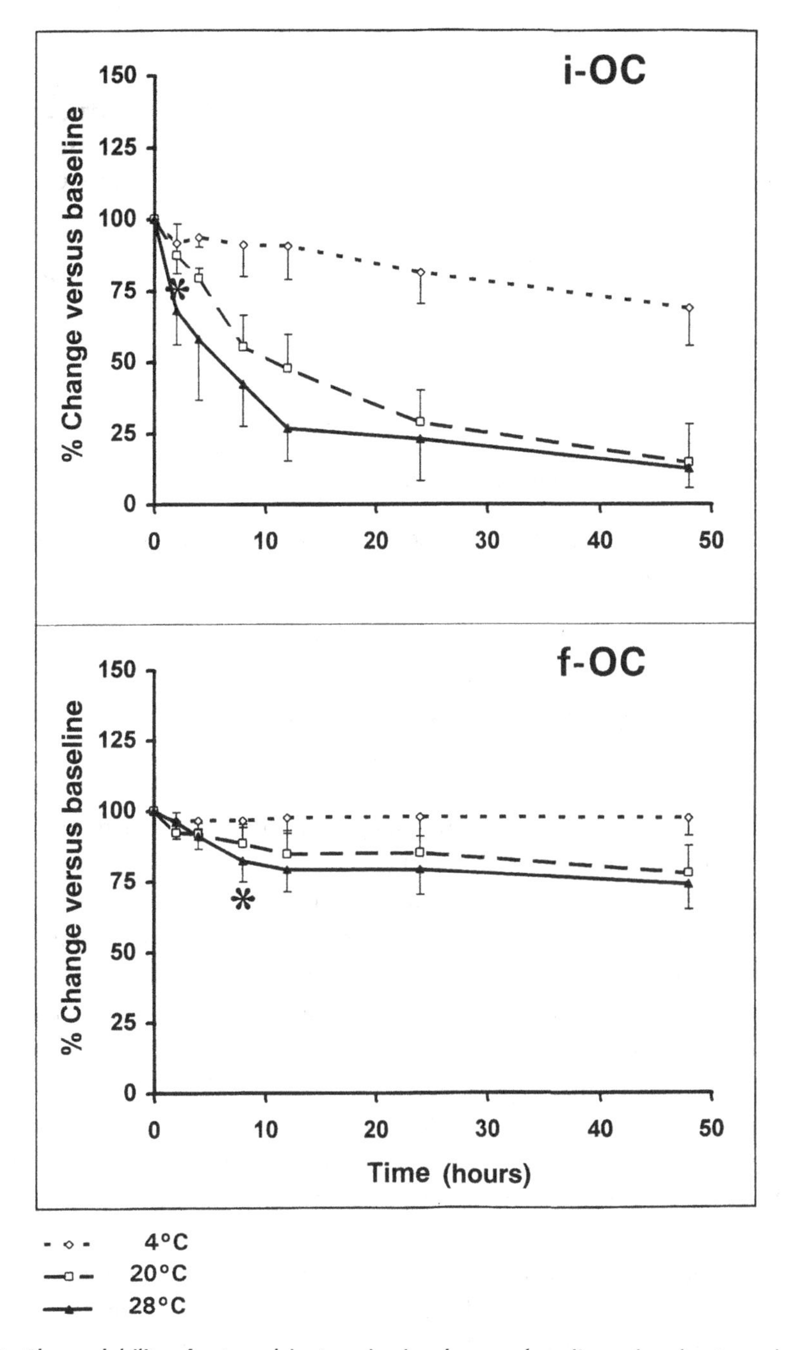


Figure 13.1 Thermolability of osteocalcin. Loss in signal versus baseline value due to prolonged storage at various temperatures. Assays for intact osteocalcin (1–49) (i-OC, upper panel) and for osteocalcin (1–43) fragments (f-OC, lower panel) are compared. * = denotes time point at and after which signal differs significantly ($p < 0.05$) from baseline. Values are means from five different samples and experiments. Data adapted from reference [1].

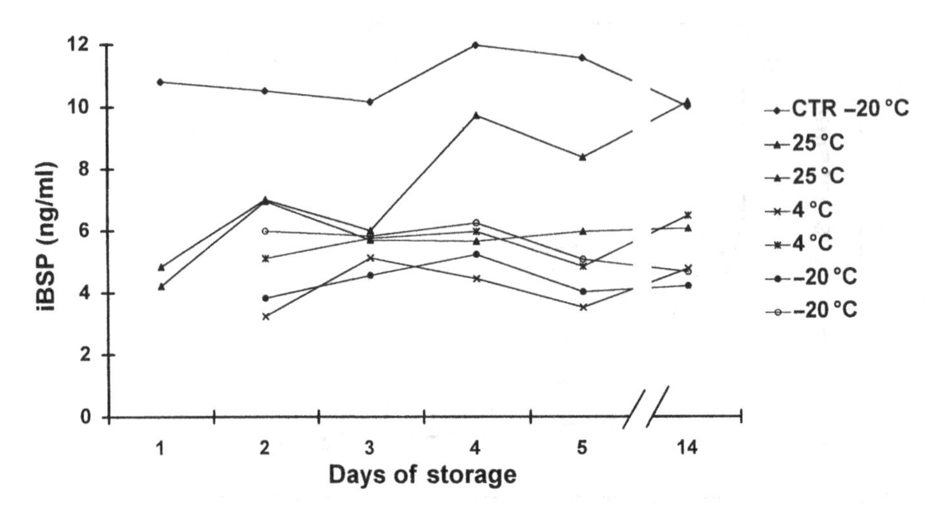


Figure 13.2 Effect of storage on apparent epitope levels of immunoreactive bone sialoprotein (iBSP) in serum samples. From reference [14].

In contrast, both the free and conjugated forms of pyridinoline (Pyr) and deoxypyridinoline (Dpd) have been shown to be stable in urine samples kept at room temperature for several weeks. Several reports show that pyridinium crosslinks can be stored at −20 °C for years and that repeated freeze–thaw cycles of urine samples have no effect on the concentrations of Pyr and Dpd [4–7]. Similar stability has been reported for the urinary N-terminal (NTx) and C-terminal (CTx) collagen type I telopeptides, while C-terminal telopeptide in serum (ICTP) loses up to 12% of the signal when stored at room temperature for 5 days [8–10]. The stability of glycosylated hydroxylysine residues has not been fully characterized yet, but it may be necessary to add boric acid to preserve the urine samples [11]. The activity of serum tartrate-resistant acid phosphatase (TRAP) declines rapidly during storage at room temperature or even at −20 °C, but it is stable when stored at −70 °C or lower [12]. Multiple freeze–thaw cycles usually have a deleterious effect on the serum TRAP activity [13]. In contrast, serum levels of bone sialoprotein (BSP) appear rather stable – at room temperature, 4 °C and −20 °C – and have been shown not to change significantly during repeated freeze–thaw cycles [14]. However, when samples are being exposed to temperatures above 30 °C, an increase in signal is usually seen with the radioimmunoassay [14]. This effect may be due to the generation of neoepitopes during the degradation of BSP–protein complexes (Figure 13.2).

Some assays and marker components are sensitive to haemolysis of the sample, resulting in values that are either too low or too high. This is usually the case for osteocalcin and BSP, but has also been described for TRAP and some other serum markers.

Pyridinium crosslinks in aqueous solutions are unstable when subjected to intensive UV irradiation [15–17]. The increased rate of photolysis in the aqueous standards compared with pyridinoline in urinary samples may be attributed to the fact that the intensity of UV light is higher in the aqueous solutions. Photolysis of the pyridinoline standard solution is prevented by storage in amber vials [16]. The rate of pyridinoline photolysis increases with rising pH [15]. It has also been shown that the sensitivity to UV light is greater for free pyridinoline than for total pyridinoline [17]. The collagen type I telopeptides in urine, NTx and CTx, are not affected by UV light exposure [17].

## Timing and mode of urine collection

In general, random samples can be used for the measurement of most urinary parameters (except urinary calcium, which always requires a 24-hour sample). For convenience, measurement of bone turnover markers is usually performed either in first or second morning voids, or in 2-hour collections. In each case, values need to be corrected for urinary creatinine which introduces additional preanalytical and analytical variability. The use of a urine creatinine correction is considered to be one of the reasons why urinary parameters show a higher degree of total variability than serum indices. Creatinine output has been reported to be fairly constant with time (variations within 10%) and to correlate with lean body mass [18], but there are also reports suggesting that the correction for creatinine in a urine spot sample could be misleading. As an alternative, a 24-hour urine collection can be ordered and the excretion rate of the marker determined. However, 24-hour urine collections are subject to inevitable inaccuracies due to collection errors. Accordingly, experienced clinical chemists consider the collection of a correctly timed 24-hour urine sample to be one of the more demanding tasks in laboratory medicine.

With most markers, similar results (i.e. high degree of correlation) are obtained from either 24-hour, 2-hour, or spot urine (first morning voids (FMV), second morning voids (SMV)) collections. It should be borne in mind, however, that for most marker compounds the slope of diurnal changes is steepest during the morning hours (see below), which is usually the time at which patients visit their doctor's office and urine samples are being collected. Therefore, if serial spot urine samples are used for the measurement of bone turnover markers, the time of collection should be controlled as closely as possible.

## Influence of diet or acute exercise

The only bone turnover marker that is markedly affected by dietary influences is urinary hydroxyproline, a degradation product of newly synthesized and mature collagens. Urinary levels of hydroxyproline will rise considerably after the ingestion

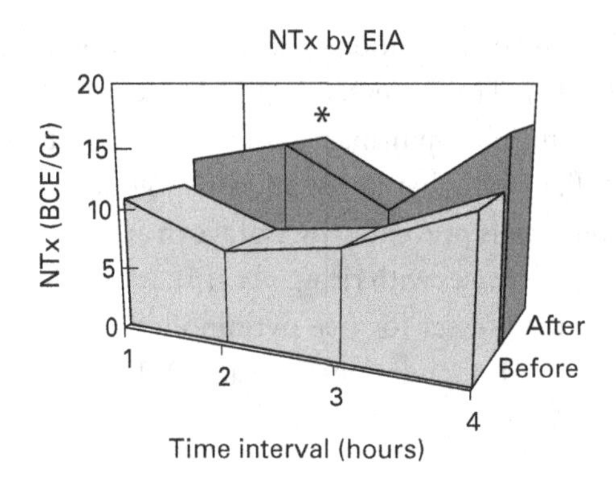


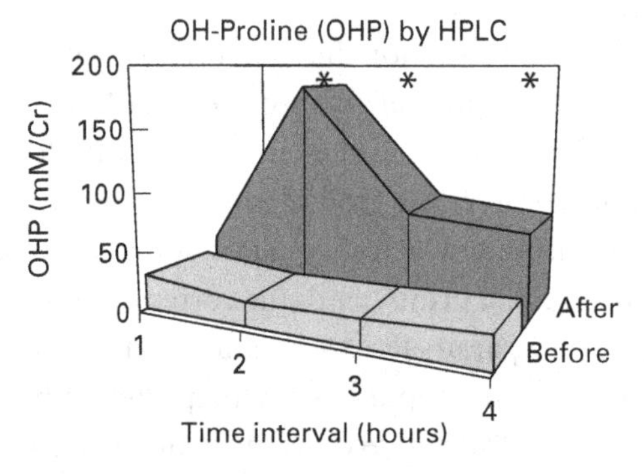


Figure 13.3 Effect of gelatine ingestion on urinary levels of immunoreactive NTx (upper panel) and of hydroxyproline (lower panel). Urine samples were collected after 48 hours of collagen-free diet at 08:00, 12:00, 16:00 and 20:00 (lighter bars; 'Before'), and at the same time points after subjects had ingested 100 g of porcine gelatine at 22:00 the day before collection (darker bars; 'After'). Note that the diurnal rhythm is only visible in the first series of collection. Representative plot from a series of study subjects. * denotes significant change versus collagen-free sampling. Data adapted from reference [3].

of hydroxyproline-rich foods, such as meat or gelatine (Figure 13.3) [3, 19]. Therefore, it is necessary to instruct patients to maintain a collagen-free diet for at least 24 hours before collecting their urine for hydroxyproline measurements. However, free and total Dpd, NTx and CTx are all affected by ingestion of large amounts (100 g) of pure gelatine (Figure 13.3) [3].

While overnight fasting has no significant influence on the levels of most serum and urine markers of bone turnover, recent results suggest that diurnal changes in serum CTx may be attenuated by an overnight fast [20]. Although no general recommendation has been put forward so far, samples for bone turnover markers should be taken in a standardized manner and, if possible, after an overnight fast.

The effect of acute exercise immediately or shortly before phlebotomy for bone turnover markers has been studied, but results are equivocal [21–23]. While some markers appear to rise by as much as 30–40% of their baseline value, others seem to be unaffected by these activities. These short-term effects need to be distinguished from the well-documented influences of long-term endurance exercise on bone turnover [24].

### A word about standardization

Although standardization is part of the quality control process and belongs to the discussion of analytical rather then preanalytical variability, the striking lack of standardization of most bone marker assays and its impact on the practical utility of these parameters deserve a few comments. Markers of bone turnover are now offered by a large number of commercial laboratories and are widely used among practising physicians in many countries. Nevertheless, there are presently no routine proficiency testing programmes for these parameters. A recent trial between laboratories in Europe showed marked variability of most commercialized test kits, with interlaboratory coefficients of variation of up to 40%. Results obtained from identical blood and urine samples using the same assay and the same method differed up to 7-fold between laboratories [25]! Therefore, results from one laboratory cannot be compared with those of a different laboratory, even if the same method and the same reference range has been used. It is obvious that immunoassays for bone turnover markers need to be standardized and should be included in routine proficiency testing programmes.

## Biological aspects of preanalytical variability

### Diurnal variation

Most biochemical markers show significant diurnal variations, with the highest values in the early morning hours and lowest values during the afternoon and at night. Thirty years ago, Mautalen [26] demonstrated that the excretion of urinary hydroxyproline (OHP) follows a diurnal rhythm. Diurnal variation in pre- and postmenopausal women, as well as in men, has now been reported by several investigators for all biochemical markers of bone turnover [14, 20, 27–39]. Although some authors have reported diurnal changes as high as 100% of the daily mean [29], most studies find amplitudes between 15% and 30%. Serum markers usually show less pronounced changes during the day than urine-based indices. This observation is also true for the new serum assays for the bone resorption markers such as serum NTx, serum CTx and BSP [20, 37; M.J. Seibel et al., unpublished data]. Wichers and colleagues [35], however, reported a daily amplitude of serum CTx of up to 66%, which exceeds the diurnal amplitude of most urinary markers. According to Schlemmer and Hassager [20], the diurnal variation in serum CTx can be reduced by acute fasting. In our hands, serum CTx tends to show a higher degree of diurnal variability than serum NTx or BSP, even when serum was obtained after an overnight fast [14, 74].

Markers of bone formation, such as serum alkaline phosphatase (AP), osteocalcin, undercarboxylated osteocalcin or the procollagen type I propeptides, are also affected by diurnal rhythms. The amplitude of change is usually between 10%–20% around the mean daily concentration, with a peak in the early morning hours and a nadir at noon and early afternoon.

Diurnal variation appears not to be affected by age, menopause, bed rest, physical activity or season [31, 40]. Although in postmenopausal women bone turnover is higher than in premenopausal women, the circadian variation is similar for both pre- and postmenopausal women and, thus, is not influenced by sex hormones [31, 33]. In osteoporotic women, the increase in bone resorption at night may persist longer into the morning, thus accounting in part for the greater bone loss in these individuals [28]. However, this could not be confirmed for elderly women with osteopenia [31]. The aetiology of diurnal variations is unknown. Several hormones, such as parathyroid hormone, growth hormone, or cortisol, show diurnal changes and may, therefore, be involved in the generation of diurnal changes in bone metabolism (Figure 13.4) [41–44]. However, so far, no conclusive evidence has been generated to support any of these associations. Independent of this, there is agreement that controlling the time of sampling is crucial in order to obtain clinically relevant information from bone markers. The concentration, for example, of serum CTx may differ by 60% or more between a serum sample collected at 08:00 and a second serum sample collected at 14:00 on the same day. This difference can also occur after 3–6 months of antiresorptive treatment with hormone replacement therapy or bisphosphonates.

### Day-to-day variability

When measured in the same individual, biochemical markers of bone turnover not only vary within a single day but, in most cases, also between consecutive days [4, 27, 34]. This phenomenon is called between day or day to day variability and is apparently due to genuine variations in marker levels and not to analytical imprecision. In general, serum markers show less day to day variability than markers of bone turnover measured in urine [14, 37–39]. In the latter, errors introduced by variations in urine collection, kidney function, urinary creatinine and other factors play a role. Thus, the day to day variation in the urinary excretion of Pyr and Dpd, measured by high performance liquid chromatography (HPLC) and corrected for creatinine, ranges between 16% and 26% [4, 45]. Similar results have been reported for free pyridinoline by enzyme immunoassay (7–25%), for NTx (13–35%), for CTx (12–35%) and for TRAP (10–12%) [35, 40, 46–48]. Long-term variability adds considerably to the total variation of

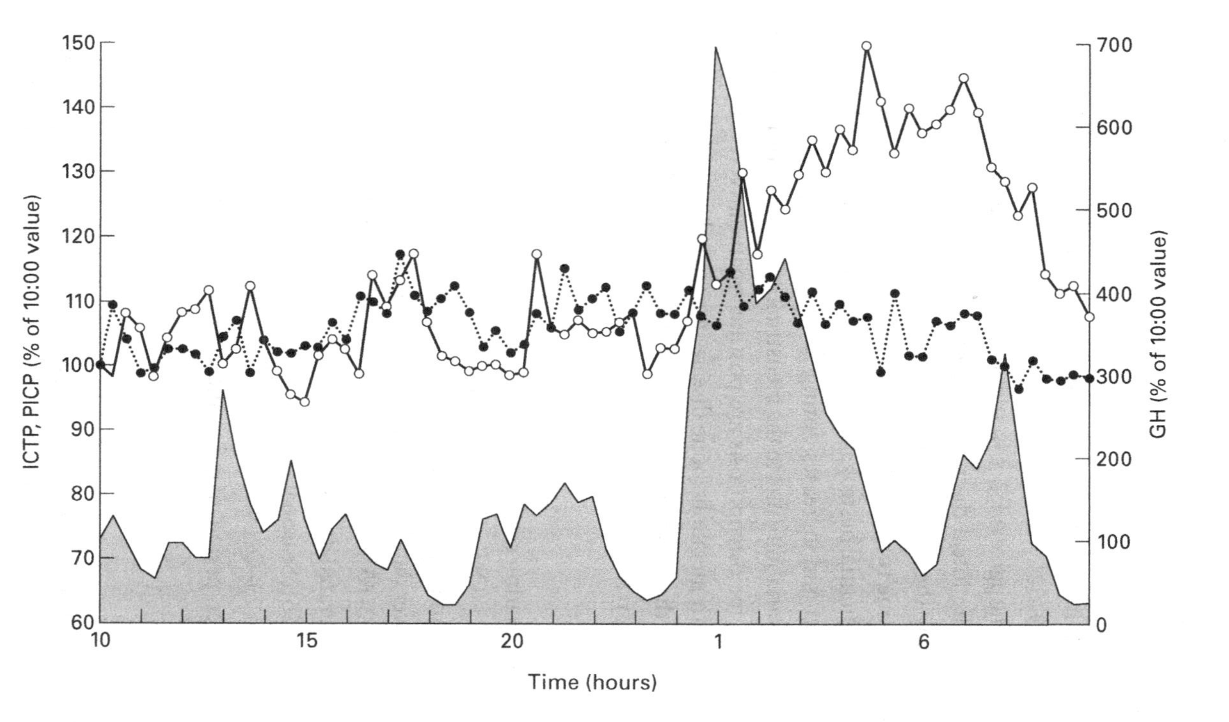


Figure 13.4 Diurnal variation of growth hormone (GH), serum ICTP and serum PICP in healthy subjects. Samples were taken every 20 min during a 24-hour period at the clinical research centre. Values are expressed as a percentage of the value measured at 10:00 hours. The X axis denotes clock time (Wüster C; Müller C and Seibel MJ; unpublished observations).

biochemical markers of bone turnover and is a major determinant when calculating the minimum (least) significant change needed to detect nonrandom changes in marker levels [47]. In contrast to diurnal variations, day to day variability cannot be controlled.

### Menstrual variability

Bone turnover varies with the menstrual cycle with an overall amplitude of approximately 10–20% [2, 14, 49–52]. Although not all available data show the same degree of change, there is enough evidence to support the suggestion that bone formation is higher during the luteal than the follicular phase [49], whereas bone resorption is higher during the midfollicular, late follicular and early luteal phase [51]. Cyclical changes in bone turnover have also been reported in postmenopausal women treated with sequential oestrogen/gestagen regimens, showing decreases during oestrogen treatment and increases during gestagen treatment [53, 54]. In clinical practice, however, variability due to the menstrual cycle is of little relevance, as most women evaluated for osteoporosis are postmenopausal. Nevertheless, in premenopausal women with metabolic bone disease, menstrual variability should be taken into account, and the timing for sampling is probably best during the first 3–7 days of the menstrual cycle.

### Seasonal variability

Bone turnover and its regulation seem to vary with seasonal changes. Many studies have shown that serum 25–OH vitamin D and urinary calcium are higher in summer than in winter, and that parathyroid hormone levels tend to increase during winter [55–61]. Fewer data are available as to whether the season influences bone turnover itself [57, 62–65]. The author and colleagues have recently shown that bone turnover is accelerated during winter, with more pronounced changes in females than in males [62, 66]. Seasonal high-turnover tends to coincide with a significant reduction in serum 25–OH vitamin D, suggesting that the increase in bone turnover during the winter period may be due, at least in part, to subclinical vitamin D deficiency [62].

### Growth, age and gender

During early childhood, and then again during the pubertal growth spurt, biochemical markers of bone turnover are significantly higher than during adulthood [67–70]. In girls, peak bone marker levels are observed approximately 2 years earlier than in boys, and oestradiol seems to be the major determinant of the increase in bone turnover. In men between 20 and 30 years of age, bone turnover markers are usually higher than in women of the same age bracket. After the age of 50, most bone turnover markers tend to increase with further ageing,

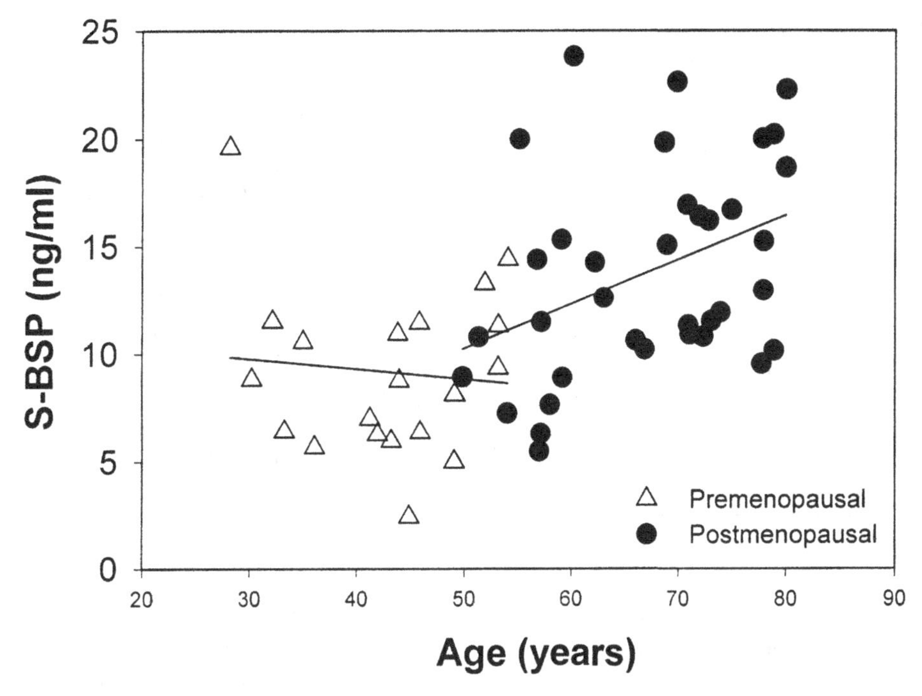


Figure 13.5 Serum levels of bone sialoprotein (BSP) in healthy pre- and postmenopausal women. Note the increase of serum BSP levels after the menopause, and the correlation with age in postmenopausal women.

but less in men than in women (Figure 13.5). In the latter, the age-related increase in bone turnover is more pronounced due to the menopause, when both bone resorption and formation markers increase by about 50–100% [71–75].

### Nonskeletal diseases and clearance-related factors

A number of nonskeletal diseases have been shown to affect markedly the bone turnover markers. These conditions mostly relate to impairments in the clearance and/or metabolism of the components measured. Thus, even moderate impairment of renal function (a glomerular filtration rate of 50 ml/min) has been shown to have significant effects on the serum levels of osteocalcin [76, 77], of BSP [71], and of the collagen type I telopeptides NTx and CTx (Figure 13.6) [78]. In contrast, the excretion of pyridinium crosslinks usually does not change until a creatinine clearance of <25 ml/min is reached [45, 79, 80]. Impairment of hepatic function usually leads to a nonspecific rise in the serum levels of the collagen type I propeptides (PINP, PICP), as these components are cleared by liver endothelial cells [81].

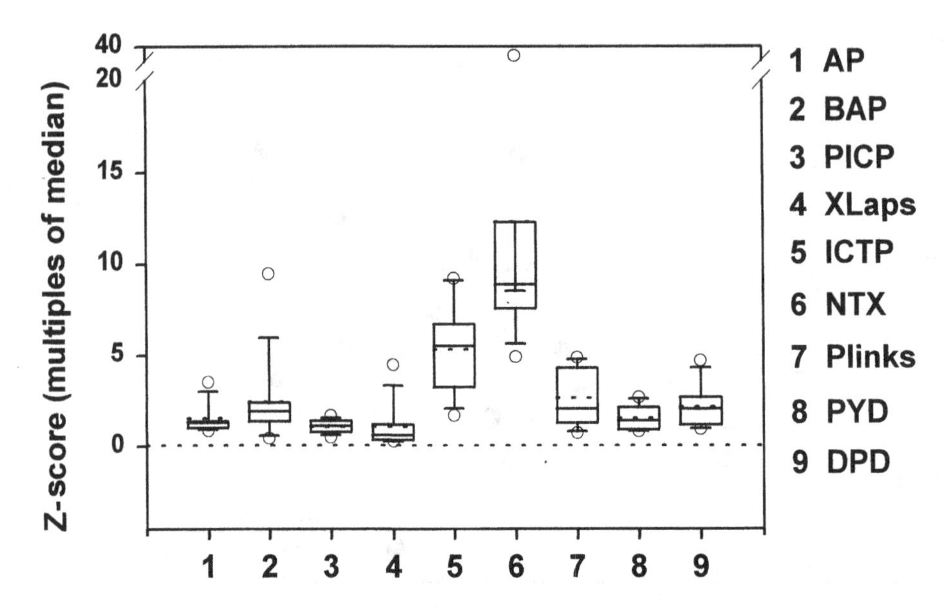


Figure 13.6 Effect of renal failure on serum and urine markers of bone turnover. Box and whisker plot of z-scores for various bone turnover markers measured in 25 patients with renal failure (creatinine clearance [cc] below 30 ml/ min; mean cc = 23 ml/min). Boxes represent the median 50% of values, whiskers the 10–90% range of values. Circles are values in the lower and upper 5% of the distribution. Solid lines inside the box are medians and dashed lines are means. The dashed line through 0 represents the median of healthy premenopausal women (Seibel, unpublished data).

## Summary

Numerous factors influence bone turnover, but there are even more sources of variability that need to be taken into account when measuring bone turnover with biochemical markers. To minimize preanalytical and analytical variability, the mode of sample collection, handling and storage needs to be standardized. Controllable factors such as diurnal and menstrual rhythms, presampling exercise and diet should be taken into account and eliminated whenever possible. Laboratories are encouraged to establish their own reference ranges and to use gender- and age-specific reference intervals. In order to reduce variability further, standardization of bone marker assays and the implementation of routine proficiency testing programmes are strongly recommended.

Even when all controllable influences are either eliminated or corrected for, a substantial degree of variability will still remain with most bone markers. Therefore, changes in marker measurements always need to be interpreted against the background of the respective marker's total variability. A true change in marker levels can only be assumed when it exceeds the long-term coefficient of variability by at least 2-fold. As a rule, markers showing 'dramatic' changes in response to disease processes or therapeutic interventions usually show equally 'dramatic' degrees of nonspecific variability.

## REFERENCES


1 Lang, M., Seibel, M.J., Zipf, A. and Ziegler, R. (1996). Influence of a new protease inhibitor on the stability of osteocalcin in serum. *Clinical Laboratory*, **42**, 5–10.

2 Blumsohn, A., Hannon, R.A. and Eastell, R. (1995). Apparent instability of osteocalcin in serum as measured with different commercially available immunoassays. *Clinical Chemistry*, **41**, 318–19.

3 Lang, M., Haag, P., Schmidt-Gayk, H., Seibel, M.J. and Ziegler, R. (1995). Influence of ambient storage conditions and of diet on the measurement of biochemical markers of bone turnover. (Abstract). *Calcified Tissue International*, **56**, 497.

4 Colwell, A., Russell, R.G.G. and Eastell, R. (1993). Factors affecting the assay of urinary 3-hydroxypyridinium crosslinks of collagen as markers of bone resorption. *European Journal of Clinical Investigation*, **23**, 341–9.

5 Gerrits, M.I., Thijssen, J.H. and van Rijn, H. (1995). Determination of pyridinoline and deoxypyridinoline in urine, with special attention to retaining their stability. *Clinical Chemistry*, **41**, 571–4.

6 Beardsworth, L.J., Eyre, D.R. and Dickson, I.R. (1990). Changes with age in the urinary excretion of lysyl- and hydroxylysylpyridinoline, two new markers of bone collagen turnover. *Journal of Bone and Mineral Research*, **5**, 671–6.

7 Seibel, M.J., Gartenberg, F., Silverberg, S.J. et al. (1992). Urinary hydroxypyridinium crosslinks of collagen in primary hyperparathyroidism. *Journal of Clinical Endocrinology and Metabolism*, **74**, 481–6.

8 Ju, H.-S.J., Leung, S., Brown, B. et al. (1997). Comparison of analytical performance and biological variability of three bone resorption assays. *Clinical Chemistry*, **43**: 1570–6.

9 Bonde, M., Qvist, P., Fledelius, C., Riis, B.J. and Christiansen, C. (1994). Immunoassay for quantifying type I collagen degradation products in urine evaluated. *Clinical Chemistry*, **40**, 2022–5.

10 Risteli, J., Elomaa, I., Niemi, S., Novamo, A. and Risteli, L. (1993). Radioimmunoassay for the pyridinoline cross-linked carboxyterminal telopeptide of type I collagen: a new serum marker of bone collagen degradation. *Clinical Chemistry*, **39**, 635–40.

11 Blumsohn, A., Hannon, R.A. and Eastell, R. (1995). Biochemical assessment of skeletal activity. *Physical Medicine and Rehabilitation Clinics of North America*, **6**, 483–505.

12 Lau, K.H.W., Onishi, T., Wergedal, J.E., Singer, F.R. and Baylink, D.J. (1987). Characterization and assay of tartrate-resistant acid phosphatase activity in serum: potential use to assess bone resorption. *Clinical Chemistry*, **33**, 458–62.

13 Schmeller, N.T. and Bauer, H.W. (1983). Circadian variation of different fractions of serum acid phosphatase. *Prostate*, 4, 391–5.

14 Li, Y., Woitge, W., Kissling, C. et al. (1998). Biological variability of serum immunoreactive bone sialoprotein. *Clinical Laboratory*, **44**, 553–5.

15 Colwell, A., Hamer, A., Blumsohn, A. and Eastell, R. (1996). To determine the effects of ultraviolet light, natural light and ionizing radiation on pyridinium cross-links in bone and urine using high-performance liquid chromatography. *European Journal of Clinical Investigation*, **26**, 1107–14.

16 Walne, A.J., James, I.T. and Perret, D. (1995). The stability of pyridinium crosslinks in urine and serum. *Clinica Chemica Acta*, **240**, 95–7.

17 Blumsohn, A., Colwell, A., Naylor, K.E. and Eastell, R. (1995). Effect of light and gamma-irradiation on pyridinolines and telopeptides of type I collagen in urine. *Clinical Chemistry*, **41**, 1195–7.

18 Heymsfield, S.B., Artega, C., McManus, B.S., Smith, J. and Moffitt, S. (1983). Measurements of muscle mass in humans: validity of the 24 hour urinary creatinine method. *American Journal of Clinical Nutrition*, **37**, 478–94.

19 Devgun, M.S., Paterson, C.R. and Martin, B.T. (1981). Seasonal changes in the activity of serum alkaline phosphatase. *Enzyme*, **26**, 301–5.

20 Schlemmer, A. and Hassager, C. (1999). Acute fasting diminishes the circadian rhythm of biochemical markers of bone resorption. *European Journal of Endocrinology*, **140**, 332–7.

21 Ashizawa, N., Ouchi, G., Fujimura, R., Yoshida, Y., Tokuyama, K. and Suzuki, M. (1998). Effects of a single bout of resistance exercise on calcium and bone metabolism in untrained young males. *Calcified Tissue International*, **62**, 104–8.

22 Wallace, J.D., Cuneo, R.C., Lundberg, P.A. et al. (2000). Responses of markers of bone and collagen turnover to exercise, growth hormone (GH) administration, and GH withdrawal in trained adult males. *Journal of Clinical Endocrinology and Metabolism*, **85**, 124–33.

23 Welsh, L., Rutherford, O.M., James, I., Crowley, C., Comer, M. and Wolman, R. (1997). The acute effects of exercise on bone turnover. *International Journal of Sports Medicine*, **18**, 247–51.

24 Woitge, H.W., Friedmann, B., Suttner, S. et al. (1998). Changes in bone turnover induced by aerobic and anaerobic exercise in young males. *Journal of Bone and Mineral Research*, **13**, 1797–804.

25 Seibel, M.J., Lang, M. and Geilenkeuser, W.J. (2001). Interlaboratory variation of biochemical markers of bone turnover. Results of a European trial. *Clinical Chemistry*, **47**, 1443–50.

26 Mautalen, C.A. (1970). Circadian rhythm of urinary total and free hydroxyproline excretion and its relation to creatinine excretion. *Journal of Laboratory and Clinical Medicine*, **75**, 11–18.

27 Jensen, J.B., Kollerup, G., Sorensen, A. and Sorensen, O.H. (1997). Intraindividual variability of bone markers in the urine. *Scandinavian Journal of Clinical and Laboratory Investigation*, **57**, 29–34.

28 Eastell, R., Calvo, M.S., Burritt, M.F., Offord, K.P., Russell, R.G.G. and Riggs, B.L. (1992). Abnormalities in circadian patterns of bone resorption and renal calcium conservation in type I osteoporosis. *Journal of Clinical Endocrinology and Metabolism*, **74**, 487–94.

29 Schlemmer, A., Hassager, C., Jensen, S.B. and Christiansen, C. (1992). Marked diurnal variation in urinary excretion of pyridinium cross-links in premenopausal women. *Journal of Clinical Endocrinology and Metabolism*, **74**, 476–80.

30 Pagani, F. and Panteghini, M. (1994). Diurnal rhythm in urinary excretion of pyridinium crosslinks. *Clinical Chemistry*, **40**, 952–3.

31 Schlemmer, A., Hassager, C., Pedersen, B. and Christiansen, C. (1994). Posture, age, menopause, and osteopenia do not influence the circadian variation in the urinary excretion of pyridinium crosslinks. *Journal of Bone and Mineral Research*, **9**, 1883–8.

32 Bollen, A.M., Martin, M.D., Leroux, B.G. and Eyre, D.R. (1995). Circadian variation in urinary excretion on bone collagen cross-links. *Journal of Bone and Mineral Research*, **10**, 1885–90.

33 Eastell, R., Simmons, P.S., Assiri, A.M., Burritt, M.F., Russel, R.G.G. and Riggs, B.L. (1992). Nyctohemeral changes in bone turnover assessed by serum bone gla-protein concentration and urinary deoxypyridinoline excretion: effect of growth and aging. *Clinical Science*, **83**, 375–82.

34 Ju, H.-S.J., Leung, S., Brown, B. et al. (1997). Comparison of analytical performance and biological variability of three bone resorption assays. *Clinical Chemistry*, **43**, 1570–6.

35 Wichers, M., Schmidt, E., Bidlingmaier, F. and Klingmüller, D. (1999). Diurnal rhythm of Crosslaps in human serum. *Clinical Chemistry*, **45**, 1858–60.

36 Sarno, M., Powell, H., Tjersland, G. et al. (1999). A collection method and high-sensitivity enzyme immunoassay for sweat pyridinoline and deoxypyridinoline crosslinks. *Clinical Chemistry*, **45**, 1501–9.

37 Gertz, B.J., Clemens, J.D., Holland, S.D., Yuan, W. and Greenspan, S. (1998). Application of a new serum assay for type I collagen crosslinked N-telopeptides: assessment of diurnal changes in bone turnover with and without alendronate treatment. *Calcified Tissue International*, **63**, 102–6.

38 Greenspan, S.L., Dresner-Pollak, R., Parker, R.A., London, D. and Ferguson, L. (1997). Diurnal variation of bone mineral turnover in elderly men and women. *Calcified Tissue International*, **60**, 419–23.

39 Rosen, H.N., Moses, A.C., Garber, J. et al. (2000). Serum CTX: a new marker of bone resorption that shows treatment effect more often than other markers because of low coefficient of variability and large changes with bisphosphonate therapy. *Calcified Tissue International*, **66**, 100–3.

40 Nielsen, H.K., Brixen, K. and Mosekilde, L. (1990). Diurnal rhythm and 24-hour integrated concentrations of serum osteocalcin in normals: Influence of age, sex, season, and smoking habits. *Calcified Tissue International*, **47**, 284–90.

41 Markowitz, M.E., Arnaud, S., Rosen, J.F., Thorpy, M. and Laximinarayan, S. (1988). Temporal interrelationships between the circadian rhythms of serum parathyroid hormone and calcium concentrations. *Journal of Clinical Endocrinology and Metabolism*, **67**, 1068–73.

42 Calvo, M.S., Eastell, R., Offord, K.P., Bergstralh, E.J. and Burritt, M.F. (1991). Circadian variation in ionized calcium and intact parathyroid hormone: evidence for sex differences in calcium homeostasis. *Journal of Clinical Endocrinology and Metabolism*, **72**, 69–76.

43 Nielsen, H.K., Laurberg, P., Brixen, K. and Mosekilde, L. (1991). Relations between diurnal

variations in serum osteocalcin, cortisol, parathyroid hormone, and ionized calcium in normal individuals. *Acta Endocrinologica (Copenhagen)*, **124**, 391–8.

44 Ebeling, P.R., Butler, P.C., Eastell, R., Rizza, R.A. and Riggs, B.L. (1991). The nocturnal increase in growth hormone is not the cause of the nocturnal increase in serum osteocalcin. *Acta Endocrinologica (Copenhagen)*, **73**, 368–72.

45 McLaren, A.M., Isdale, A.H., Whitings, P.H., Bird, H.A. and Robins, S.P. (1993). Physiological variations in the urinary excretion of pyridinium crosslinks of collagen. *British Journal of Rheumatology*, **32**, 307–12.

46 Popp-Snijders, C., Lips, P. and Netelenbos, J.C. (1996). Intra-individual variation in bone resorption markers in urine. *Annals of Clinical Biochemistry*, **33**, 347–8.

47 Hannon, R.A., Blumsohn, A., al-Dehaimi, A.W. and Eastell, R. (1998). Short-term intraindividual variability of markers of bone turnover in healthy adults. *Journal of Bone and Mineral Research*, **13**, 1124–33.

48 Panteghini, M. and Pagani, F. (1995). Biological variation in bone-derived biochemical markers in serum. *Scandinavian Journal of Clinical and Laboratory Investigation*, **55**, 609–16.

49 Nielsen, H.K., Brixen, K., Bouillon, R. and Mosekilde, L. (1990). Changes in biochemical markers of osteoblastic activity during the menstrual cycle. *Journal of Clinical Endocrinology and Metabolism*, **70**, 1431–7.

50 Schlemmer, A., Hassager, C., Risteli, J., Risteli, L., Jensen, S.B. and Christiansen, C. (1993). Possible variation in bone resorption during the normal menstrual cycle. *Acta Endocrinologica (Copenhagen)*, **129**, 388–92.

51 Gorai, I., Chaki, O., Nakayama, M. and Minaguchi, H. (1995). Urinary biochemical markers for bone resorption during the menstrual cycle. *Calcified Tissue International*, **57**, 100–4.

52 Zittermann, A., Schwarz, I., Scheld, K. et al. (2000). Physiologic fluctuations of serum estradiol levels influence biochemical markers of bone resorption in young women. *Journal of Clinical Endocrinology and Metabolism*, **85**, 95–101.

53 Johansen, J.S., Jensen, S.B., Riis, B.J. and Christiansen, C. (1990). Time-dependent variations in bone turnover parameters during 2 months' cyclic treatment with different doses of combined estrogen and progestagen in postmenopausal osteoporosis. *Metabolism: Clinical and Experimental*, **39**, 1122–6.

54 Christiansen, C., Riis, B.J., Nilas, L., Rodbro, P. and Deftos, L. (1985). Uncoupling of bone formation and resorption by combined oestrogen and progestagen therapy in postmenopausal osteoporosis. *Lancet*, **2**, 800–1.

55 Juttman, J.R., Visser, T.J., Buurman, C., De Kam, E. and Birkenhäger, J.C. (1981). Seasonal fluctuations in serum concentrations of vitamin D metabolites in normal subjects. *British Medical Journal*, **282**, 1349–52.

56 Overgaard, K., Nilas, L., Sidenius, J. and Christiansen, C. (1988). Lack of seasonal variation in bone mass and biochemical estimates of bone turnover. *Bone*, **9**, 285–8.

57 Scharla, S., Scheidt-Nave, C., Leidig, G., Seibel, M.J. and Ziegler, R. (1996). Lower serum 25-hydroxyvitamin D is associated with increased bone resorption markers and lower bone density at the proximal femur in normal females: a population-based study. *Experimental and Clinical Endocrinology and Diabetes*, **104**, 289–92.

58 Morgan, D.B., Rivlin, R.S. and Davis, R. (1972). Seasonal changes in the urinary excretion of calcium. *American Journal of Clinical Nutrition*, **25**, 652–4.
59 Lips, P., Hakeng, H.L., Jongen, M.J.M. and van Ginkel, F.C. (1983). Seasonal variation in serum concentrations of parathyroid hormone in elderly people. *Journal of Clinical Endocrinology and Metabolism*, **57**, 204–6.
60 Krall, E.A., Sahyoun, N., Tannenbaum, S., Dallal, G.E. and Dawson-Hughes, B. (1989). Effect of vitamin D intake on seasonal variations in parathyroid hormone secretion in postmenopausal women. *New England Journal of Medicine*, **321**, 1777–83.
61 Chapuy, M.-C., Schott, A.M., Garnero, P., Delmas, P.D. and Meunier, P.J. (1996) Healthy elderly French women living at home have secondary hyperparathyroidism and high bone turnover in winter. *Journal of Clinical Endocrinology and Metabolism*, **81**, 1129–33.
62 Woitge, H., Scheidt-Nave, C. Kissling, C. et al. (1998). Seasonal variation of biochemical indices of bone turnover: results of a population-based study. *Journal of Clinical Endocrinology and Metabolism*, **83**, 68–75.
63 Hyldstrup, L., McNair, P., Jensen, G.F. and Transbol, I. (1986). Seasonal variations in indices of bone formation precede appropriate bone mineral changes in normal men. *Bone*, **7**, 167–70.
64 Vanderschueren, D., Gevers, G., Dequeker, J. et al. (1991). Seasonal variation in bone metabolism in young healthy subjects. *Calcified Tissue International*, **49**, 84–9.
65 Thomsen, K., Eriksen, E.F., Jorgensen, J., Charles, P. and Mosekilde, L. (1989). Seasonal variation in serum bone gla protein. *Calcified Tissue International*, **49**, 605–11.
66 Woitge, H.W., Knothe, A., Witte, K. et al. (2000). Circannual rhythms and interactions of vitamin D metabolites, parathyroid hormone and biochemical markers of skeletal homeostasis. A prospective study. *Journal of Bone and Mineral Research*, **15**, 2443–50.
67 Hodgkinson, A. and Thompson, T. (1982). Measurement of the fasting urinary hydroxyproline: creatinine ratio in normal adults and its variation with age and sex. *Journal of Clinical Pathology*, **35**, 807–11.
68 Stepan, J.J., Tesarova, A., Havranek, T., Jodl, J., Normankova, J. and Pacovsky, V. (1985). Age and sex dependency of the biochemical indices of bone remodelling. *Clinica Chemica Acta*, **151**, 273–283.
69 Krabbe, S., Hummer, L. and Christiansen, C.K. (1984). Longitudinal study of calcium metabolism in male puberty. Bone mineral content; and serum levels of alkaline phosphatase; phosphate and calcium. *Acta Paediatrica Scandinavia*, **73**, 750–5.
70 Hyldstrup, L., McNair, P., Jensen, G.F., Nielsen, H.R. and Transbol, I. (1984). Bone mass as referent for urinary hydroxyprolin excretion: age and sex-related changes in 125 normals and in primary hyperparathyroidism. *Calcified Tissue International*, **36**, 639–44.
71 Seibel, M.J., Woitge, H.W., Pecherstorfer, M. et al. (1996). Serum immunoreactive bone sialoprotein as a new marker of bone turnover in metabolic and malignant bone disease. *Journal of Clinical Endocrinology and Metabolism*, **81**, 3289–94.
72 Eastell, R., Delmas, P.D., Hodgson, S.F., Eriksen, E.F., Mann, K.G. and Riggs, B.L. (1988). Bone formation rate in older normal women: concurrent assessment with bone histomorphometry, calcium kinetics, and biochemical markers. *Journal of Clinical Endocrinology and Metabolism*, **67**, 741–8.

73 Kelly, P.J., Pocock, N.A., Sambrook, P.N. and Eisman, J.A. (1989). Age and menopause-related changes in indices of bone turnover. *Journal of Clinical Endocrinology and Metabolism*, **69**, 1160–5.

74 Hassager, C., Colwell, A., Assiri, A.M. and Christiansen, C. (1992). Effect of menopause and hormone replacement therapy on urinary excretion of pyridinium crosslinks: a longitudinal and cross-sectional study. *Clinical Endocrinology*, **37**, 45–50.

75 Seibel, M.J., Woitge, H.W., Scheidt-Nave, C. et al. (1994). Urinary hydroxypyridinium crosslinks of collagen in population-based screening for overt vertebral osteoporosis: results of a pilot study. *Journal of Bone and Mineral Research*, **9**, 1433–40.

76 Gundberg, C.M., Hanning, R., Liu, A., Zlotkin, S., Balfe, J. and Cole, D. (1987). Clearance of osteocalcin by peritoneal dialysis in children with end-stage renal disease. *Pediatrics Research*, **21**, 296–300.

77 Gundberg, C.M. and Weinstein, R.S. (1986). Multiple immunoreactive forms of osteocalcin in uremic serum. *Journal of Clinical Investigation*, **77**, 1762–7.

78 Woitge, H.W., Oberwittler, H., Farahmand, I., Lang, M., Ziegler, R. and Seibel, M.J. (1999). New serum assays for bone resorption. Results of a cross-sectional study. *Journal of Bone and Mineral Research*, **14**, 792–801.

79 Seibel, M.J. (1997). Clinical use of pyridinium crosslinks. In *Calcium Regulating Hormones and Markers of Bone Metabolism: Measurement and Interpretation*, Ed. by H. Schmidt-Gayk et al., pp. 157–69. Heidelberg: Clin Lab Publications.

80 Robins, S.P. (1990). Collagen markers in urine in human arthritis. In *Methods in Cartilage Research*, Ed. by A. Maroudas et al., pp. 348–52. London: Academic Press.

81 Lang, M. (2000). Influence of pre-analytical factors on the measurement of biochemical markers of bone metabolism. MD Thesis, University of Heidelberg.

14

# Genetic approaches to the study of complex diseases: osteoporosis

Andre G Uitterlinden

Erasmus University Medical School, Rotterdam, The Netherlands

## Genetics of osteoporosis

Certain aspects of osteoporosis have strong genetic influences. This finding is derived from genetic epidemiological analyses which show that, in women, a maternal family history of fracture is positively related to fracture risk [1]. Most evidence, however, has come from twin studies on bone mineral density (BMD) [2–4]. For BMD, heritability has been estimated to be high: 50–80% [2–4]. Thus, although twin studies can overestimate heritability, a considerable part of the variance in BMD values might be explained by genetic factors, while the remaining part may be due to environmental factors. This also suggests that there are 'bone density' genes, variants of which will result in BMD levels that are different between individuals. These interindividual differences in BMD levels can become apparent in different ways – as differences in either peak BMD or rate of bone loss at an advanced age.

The heritability estimates of osteoporosis suggest considerable influences from environmental factors which can modify the effect of genetic predisposition. The gene–environment interactions could include diet, exercise and exposure to sunlight, for example. While genetic predisposition will be constant during life, environmental factors tend to change during the different periods of life resulting in different 'expression levels' of the genetic susceptibility. Ageing is associated with a general functional decline in overall activity resulting in less exercise, less time spent outdoors, changes in diet, etc. This can result in specific genetic susceptibilities being exposed only later in life, after a period in life when they went unnoticed due to sufficient exposure to a positive environmental factor.

Taking all this into account, it becomes evident that osteoporosis is, not surprisingly, a 'complex' genetic trait. This complexity is shared with other common, and often age-related, traits with genetic influences such as diabetes, schizophrenia, osteoarthritis, cancer, etc. 'Complex' means that a trait is multifactorial, as well as being multigenic. Thus, genetic risk factors (i.e. certain alleles or gene variants) will be transmitted from one generation to the next, but the expression of these factors in the phenotype will be dependent on their interaction with other gene variants and with environmental factors.

## Genome scans and candidate genes

A first step in the dissection of the genetic factors in osteoporosis is the 'genomics' of osteoporosis, i.e. the identification, mapping and characterization of the set of genes responsible for contributing to the genetic susceptibility to different aspects of osteoporosis. Finding the gene responsible for monogenic disorders has now become almost a routine exercise for specialized laboratories. However, the complex character of osteoporosis makes it quite resistant to standard methods of analysis which, in the past, have worked so well for the monogenic diseases. Therefore, different and often more cumbersome approaches have to be applied (see, for example, reference [5]).

In a *top-down* approach, large-scale genome searches are initially performed to identify which chromosomal areas might contain osteoporosis genes. In an optimal setting, such searches are performed in hundreds of relatives (sibs, pedigrees, etc.) with hundreds of DNA markers (mostly microsatellites) evenly spread over the genome. Genome searches are based on the assumption that relatives who share a certain phenotype will also share one or more chromosomal areas identical by descent containing one or more gene variants eliciting (to a certain extent) the phenotype of interest (e.g. low BMD). The gene is then considered to be linked with the DNA marker used to 'flag' a certain chromosomal region. Upon positive linkage, subsequent research will focus on the identification of the dozens of genes in the chromosomal area involved in bone metabolism and then on identification of the particular sequence variant that gives rise to (aspects of) osteoporosis. This approach is illustrated in Figure 14.1.

In contrast, the *bottom-up* approach builds on the known involvement of a particular gene in certain types of osteoporosis, e.g. bone metabolism, as established by biological and/or animal experiments. In this gene, sequence variants have to be identified which are associated with differences in the function of the encoded protein. Such variants can then be tested in association or linkage analyses to evaluate their contribution to the phenotype of interest. Naturally, top-down and bottom-up approaches will meet each other somewhere down the line, leading to the mapping of the candidate osteoporosis genes and the generation of maps of genome areas containing putative osteoporosis genes which will either completely or partially overlap. Sequence analysis of a 'candidate' osteoporosis gene in a number of different individuals will identify sequence variants, some of which will be polymorphic but without functional consequences, while others will have an

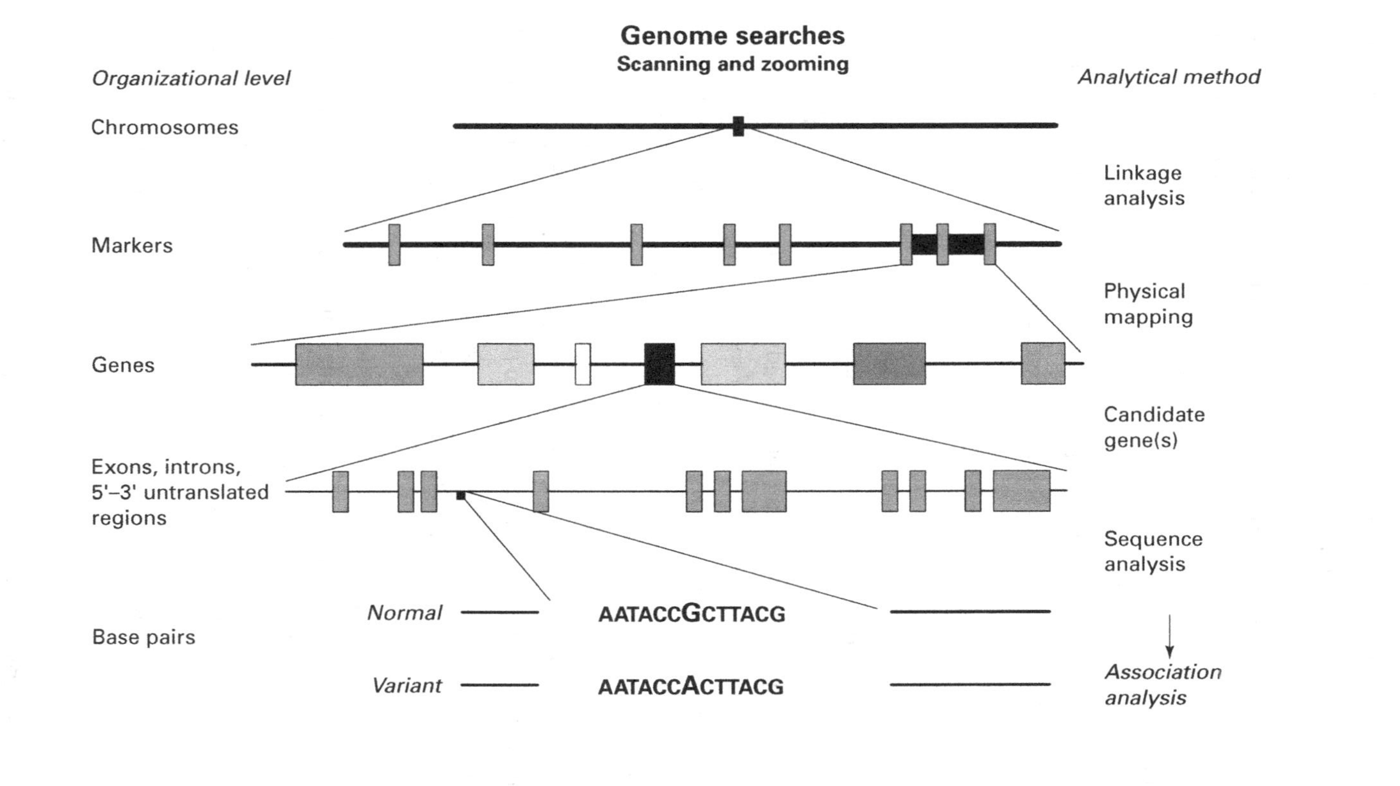


Figure 14.1 Schematic flow diagram depicting the different steps in a genome search. On the left, the level of organization of the DNA molecules is presented, whereas, on the right, the different analytical steps in the process are shown.

effect on function. This can include, for example, sequence variations leading to alterations in the amino acid composition of the protein and changes in the 5′ promotor region, leading to differences in expression, and/or polymorphism in the 3′ region leading to differences in mRNA degradation. The functional polymorphisms are of prime interest for testing in association analyses whether the candidate gene is a true osteoporosis gene or not. Because functional polymorphisms lead to meaningful biological differences in the function of the encoded 'osteoporosis' protein, this also makes the interpretation of association analyses using these variants quite straightforward. For functional polymorphisms, it is expected that the same allele will be associated with the same phenotype in different populations.

## Complicating factors

From several studies, it has become evident that, on average, one out of every 500 base pairs is a variant in the population. This is highlighted by the recent comprehensive sequence analysis of 9.7 kilobases of genomic DNA encoding part of the human lipoprotein lipase gene in 71 different individuals [6]. In this relatively small stretch of genomic sequence, 88 different sites of sequence variation were found, with a mean heterozygosity of 20%. Of these, 79 were single nucleotide polymorphisms (SNPs) and nine involved an insertion–deletion type of variation. Thus, candidate gene analysis will have to focus on which of the many variant nucleotides are the ones that actually matter – that is, which sequence variation is functionally relevant, by changing expression levels, changing codons, etc. Given the average size of a gene and the relatively young age of human populations, it can be predicted that several sequence variations 'that matter' will coexist in a given sample from a study population. A major challenge of fundamental research will be to unravel the functionality of these variations.

Yet, in spite of such complicating factors, genetic research will contribute to a further understanding of complex diseases. The identification of new genes, or of new roles of already known genes, will allow insights into mechanistic pathways which might help in designing targeted therapeutic protocols. The description of a genetic variant underlying phenotypic variation can be used, in concert with existing easy-to-assess risk factors, to predict the risk of osteoporosis. In this respect, insights into gene–environment interactions may enable novel therapies to be offered to patients.

## Osteoporosis candidate genes: collagen type Ia1 and the vitamin D receptor

The vitamin D receptor (VDR) gene is the candidate gene that initiated the molecular genetic studies on osteoporosis published by Morrison, Eisman and colleagues

[7]. They initially reported that anonymous 3′ polymorphisms, detected as *Bsm* I, Apa l, and *Taq* restriction fragment length polymorphisms, were associated with differences in BMD and they claimed that up to 80% of the population variance in BMD could be explained by this single gene. Although some of their results were later retracted [8], this study paper left behind the notion that there may be a single osteoporosis gene. For a complex disease such as osteoporosis, however, this is very unlikely and, indeed, several large population-based association analyses [9] and meta-analyses have indicated that – at most – these variants explain only a few per cent of the population variance of BMD. Thus, for the prediction of BMD, the gene for VDR does not seem to be the answer. However, in view of its pleiotropic involvement in a number of biological pathways, VDR polymorphisms have been analysed in relation to a number of other diseases. For example, the author and colleagues have recently shown that VDR is associated with (bony) aspects of osteoarthritis, i.e. with osteophytosis [10]. Yet, a major problem with the VDR gene polymorphisms is still the lack of insight into the functional consequences of the polymorphisms used to date. Therefore, a major effort is being focused on finding additional polymorphic sites – but in functionally relevant areas of the gene, such as the 3′ untranslated region and the 5′ promotor area of this gene.

An example of a functional sequence variant in an osteoporosis candidate gene that has shown promising associations is the G to T substitution in the Sp1 binding site in the first intron of the collagen type I$\alpha$1 gene (COLIA1). After it discovery and initial association analysis by Grant et al. [11], a large-scale population analysis in the Rotterdam Study showed the T allele to be associated with low BMD and increased fracture risk [12]. An interesting observation in the author's cohort of 1782 elderly women was of the age-dependent, genotype-dependent differences in BMD and the fact that the genotype-dependent fracture risk was independent of the differences in BMD. Meanwhile, several other studies have confirmed that the T allele is a promising osteoporosis candidate gene polymorphism that could affect BMD and fracture risk. Studies by Ralston et al. (personal communication) have shown that the T allele is associated with an increased affinity for the Sp1 binding factor, increased mRNA and protein production and, in biomechanic experiments on bone biopsies, has been shown to be associated with decreased bone strength. It has been suggested that the overrepresentation of COLIA1 homotrimers might explain the decreased bone strength, although this still has to be proven.

The findings with the COLIA1 Sp1 binding site polymorphism clearly illustrate how and when epidemiological association analyses should be performed and used. Firstly, a functional polymorphism is found, then its molecular mode of action is established and, finally, association analysis is performed to see what relevant (disease) end-points will result from this genotype difference. This approach is much more useful than simply taking a random anonymous polymorphism and seeing what end-point is associated with it. Establishing functionality of polymorphisms is, therefore, a major requirement in putting the application and outcome of the human genome project into perspective.

## REFERENCES


1 Cummings, S.R., Nevitt, M.C., Browner, W.S. et al. (1995). Risk factors for hip fracture in white women. *New England Journal of Medicine*, **332**, 767–73.

2 Smith, D.M., Nance, W.E., Kang, K.W., Christian, J.D. and Johnston, C.C. (1973). Genetic factors in determining bone mass. *Journal of Clinical Investigation*, **52**, 2800–8.

3 Pocock, N.A., Eisman, J.A., Hopper, J.L., Yeates, G.M., Sambrook, P.N. and Ebert, S. (1987). Genetic determinants of bone mass in adults: a twin study. *Journal of Clinical Investigation*, **8**, 706–10.

4 Flicker, L., Hopper, J.L., Rodgers, L., Kaymakci, B., Green, R.M. and Wark, J.D. (1995). Bone density determinants in elderly women: a twin study. *Journal of Bone and Mineral Research*, **10**, 1607–13.

5 Lander, E.S. and Schork, N.J. (1994). Genetic dissection of complex traits. *Science*, **265**, 2037–48.

6 Nickerson, D.A., Taylor, S.L., Weiss, K.M. et al. (1998). DNA sequence diversity in a 9.7 kb region of the human lipoprotein lipase gene. *Nature Genetics*, **19**, 233–40.

7 Morrison, N.A., Qi, J.C., Tokita, A. et al. (1994). Prediction of bone density from vitamin D receptor alleles. *Nature*, **367**, 284–7.

8 Morrison, N.A., Qi, J.C., Tokita, A. et al. (1997). Prediction of bone density from vitamin D receptor alleles (Correction). *Nature*, **387**, 106.

9 Uitterlinden, A.G., Pols, H.A.P., Burger, H. et al. (1996). A large scale population based study of the association of vitamin D receptor gene polymorphisms with bone mineral density. *Journal of Bone and Mineral Research*, **11**, 1242–8.

10 Uitterlinden, A.G., Burger, H., Huang, Q. et al. (1997). Vitamin D receptor genotype is associated with osteoarthritis. *Journal of Clinical Investigation*, **100**, 259–63.

11 Grant, S.F.A., Reid, D.M., Blake, G., Herd, R., Fogelman, I. and Ralston, S.H. (1996). Reduced bone density and osteoporotic vertebral fracture associated with a polymorphic Sp1 binding site in the collagen type Ia1 gene. *Nature Genetics*, **14**, 203–5.

12 Uitterlinden, A.G., Burger, H., Huang, Q. et al. (1998). Relation of alleles at the collagen type Ia1 gene to bone density and risk of osteoporotic fractures in postmenopausal women. *New England Journal of Medicine*, **338**, 1016–21.

# Part 4

# Biomarkers of liver disease and dysfunction

15

# Biomarkers of hepatic disease

Michael Oellerich

Georg-August-Universität, Göttingen, Germany

## Introduction

Appropriate biomarkers are needed to detect hepatic disease, to direct diagnostic work-up, to estimate disease severity, to assess prognosis and to evaluate therapy. The available tests can be divided into essential and special static tests and dynamic tests. On the one hand the traditional static tests are only an indirect measure of hepatic function or damage, and these tests involve the measurement of endogenous substances at a single point in time. Dynamic tests, on the other hand, reflect real-time hepatic function. In these tests, the dimension of time is also considered. The clearance of a test substance or the formation rate of a metabolite reflects the actual performance of the liver.

## Static tests

The essential static tests for hepatobiliary disease are summarized in Table 15.1. The pattern of these conventional tests indicates which more specialist tests are likely to be valuable. There are several limitations associated with these conventional liver function tests. Test results of liver enzymes and clotting factors can be affected by the substitution of blood components. Falling aminotransferase values are only reassuring when accompanied by a restoration of metabolic function. Aminotransferases, alkaline phosphatase and bilirubin lack organ specificity. The long plasma half-life of cholinesterase does not allow the detection of rapid changes in hepatic function. Most importantly, the conventional tests are only of limited prognostic value. More specialist static tests include bile acids indicating excretory function and portosystemic shunting, ammonia as a marker of reduced urea synthesis and parameters reflecting fibrotic activity such as aminoterminal procollagen type III peptide and various other well-known tests [1]. Further diagnostic criteria are the immunoglobulins, indicating humoral immunoresponse, autoantibodies for the assessment of autoimmune liver diseases and viral hepatitis markers [1].

In addition to the well-established conventional biomarkers of hepatic disease, there are various static tests with special applications. Hyaluronic acid is basically

**Table 15.1.** Essential static tests in hepatobiliary disease

| Test | Application |
|---|---|
| Aspartate aminotransferase (AST)<br>Alanine aminotransferase (ALT)<br>Glutamate dehydrogenase (GLDH) | Hepatocellular damage |
| Alkaline phosphatase (AP)<br>$\gamma$-glutamyltransferase ($\gamma$-GT) | Cholestasis, hepatic infiltrations<br>Cholestasis, alcohol abuse |
| Bilirubin | Conjugation, excretory function, to assess severity |
| Prothrombin time (post vitamin K)<br>Cholinesterase<br>Albumin | Synthetic function, to assess severity |
| $\gamma$-globulin | Chronic hepatitis, cirrhosis, follow course of chronic disease |

the only direct marker of sinusoidal endothelial cell function and is a marker of liver perfusion. Serial determinations are useful in monitoring patients at risk of progressive fibrosis [2, 3]. Serum hyaluronic acid has been proposed as a noninvasive index of the severity of fibrosis in chronic viral hepatitis and as a measure of response to antiviral therapy. In one study [3], hyaluronic acid showed a sensitivity and specificity for stage four and five fibrosis of ≥85% – exceeding those of alanine aminotransferase and glutathione S-transferase. In primary biliary cirrhosis and cirrhotic alcoholic liver disease, serum hyaluronic acid discriminates between early and advanced liver disease. In alcoholic liver disease, serum hyaluronic acid can be applied for the assessment of haemodynamic changes. In liver transplantation, early graft function can be predicted by this test [2]. Furthermore, hyaluronic acid has been proposed as a marker of acute liver rejection. An increase of hyaluronic acid in the early post-transplant phase, however, is not specific for acute rejection as it also occurs in patients with septicaemia [4]. So far, this test has not found widespread application.

The ketone body ratio in arterial blood reflects mitochondrial function and is used to assess primary graft function. In transplant recipients with an uncomplicated postoperative course, there is a rapid increase in the ketone body ratio within the first 6 hours after reperfusion [5]. In liver recipients with a prognostically unfavourable course, an increase in the ketone body ratio was not observed within 24 hours of transplantation. In the interpretation of the ketone body ratio, however, it has to be considered that acetoacetate and 3-hydroxybutyrate may be metabolized to a different extent in extrahepatic tissues. This effect may influence the ketone body ratio in arterial blood. During routine application, false positive and

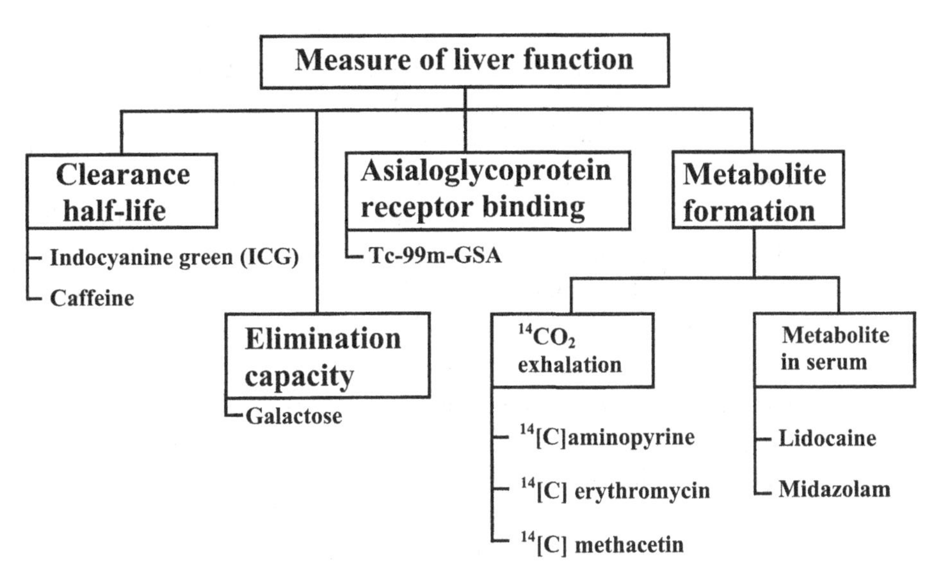


Figure 15.1 Dynamic liver function tests.

false negative test results have both been observed at the author's transplant centre. Within these limitations, the test can be used to assess primary graft function.

Serum alpha-glutathione *S*-transferase (GST) is an emerging static test indicating hepatocellular damage. As the in vivo plasma half-life of this enzyme is less than 60 min, this hepatic enzyme offers advantages in post-transplant monitoring as an aid to the diagnosis of acute allograft rejection. Recent investigations have demonstrated that the monitoring of GST reduced both mortality and morbidity in liver transplant recipients. GST showed a distinctly higher sensitivity (89%) for acute rejection than conventional liver function tests; the specificity, however, was also relatively low (45%). In combination with a blood eosinophil count, GST has facilitated the early diagnosis of acute rejection [6]. The rational for this approach is the observation that portal tract eosinophil infiltration is a particularly specific histological marker of acute rejection. Recent investigations suggest that serum fructose-1,6 bisphosphatase may offer a comparable sensitivity to that of GST for the detection of acute rejection. This test could provide a less expensive, rapid and automated alternative to GST.

## Dynamic tests

The best-studied dynamic liver function tests are outlined in Figure 15.1. In the first group of tests, the clearance or half-life of indocyanine green (ICG) or caffeine has been used as a measure of liver function. A further related approach involves the measurement of galactose elimination capacity. In addition, a technetium-99-m galactosyl serum albumin (GSA) test has been developed in which asialoglycoprotein receptor binding is used as a measure of synthetic liver function. Tests based on the rate of metabolite formation are of particular interest. In the group of breath tests, $^{14}CO_2$ exhalation gives a measure of hepatic oxidative function. Such breath tests are now being adapted to stable isotope technology, thus avoiding the need of radioactivity. Midazolam has also been proposed as a probe of human cytochrome P450 IIIA (CYP3A4). All of these tests, however, have not been widely adopted in clinical practice. Due to its ease of use and rapid turnaround time, monoethylglycinexylidide (MEGX) formation from lidocaine has found widespread application as a dynamic liver function test in recent years [7]. Lidocaine is metabolized primarily by liver CYP3A4 through sequential oxidative *N*-dealkylation, the major initial metabolite in humans being MEGX. On account of the relatively high extraction ratio of lidocaine, this liver function test depends not only on hepatic metabolic capacity but also on hepatic blood flow. MEGX can be determined in serum either by a highly sensitive and specific HPLC (high performance liquid chromatography) method with fluorescence detection or by a liquid chromatography tandem mass spectrometry method.

The integrity of the hepatic cytochrome P450 (CYP) system is essential for the metabolism of lidocaine. The structure and function of the hepatic CYP system are closely linked. Severe hypoxia or lipid peroxidation can damage the endoplasmic CYP environment, resulting in a rapid breakdown of this enzyme. Using an isolated perfused pig liver model, Mets et al. [8] have demonstrated that, in livers subjected to hypoxia, MEGX formation from lidocaine showed an early significant decrease which was paralleled by an impaired hepatic oxygen consumption and ATP production. This seems to be consistent with the fact that CYP is predominantly located in acinar zone 3, which is particularly prone to anoxic liver injury. Furthermore, certain cytokines like interleukin-6 (IL-6) may influence the expression of CYP isoforms.

The clinical usefulness of the MEGX test has been primarily studied in the field of liver transplantation and intensive care. Reference intervals have been established in normal subjects. Women aged <45 years not taking oral contraceptives showed significantly lower MEGX concentrations (*c.* 27%) than men [9]. The lowest MEGX values were observed in women taking contraceptives.

In cadaveric liver donors, MEGX test results were 50–60% higher at 15 min compared with normal subjects [9] and there was no significant gender-specific difference in MEGX formation. Conflicting results have been reported regarding the usefulness of MEGX for donor rating [9]. Reviewing all the available donor data, it appears that the MEGX test is an interesting research tool that allows real-time assessment of functional properties of the donor liver that may be related to early outcome.

An important area for the clinical application of MEGX is the assessment of pretransplant prognosis in patients with terminal cirrhosis. Shiffman et al. demon-

**Table 15.2.** Mortality risk in patients with chronic liver disease related to MEGX test results

| Patients | MEGX test | Predictive for |
|---|---|---|
| Transplant candidates | | |
| adults (cirrhosis) | <10 μg/l | 1-year spontaneous survival 46% versus 82%* |
| | <25 μg/l | 1-year spontaneous survival 53% versus 97%* |
| paediatric cirrhosis | <10 μg/l | 1-year spontaneous survival 38% versus 97%* |
| Chronic liver disease | <30 μg/l | 10 months' survival 63% versus 92%* |
| Paediatric liver disease | 11 ± 4 μg/l | high } mortality risk |
| | 42 ± 6 μg/l | low } mortality risk |

*Notes:*
* Probability of survival (%) at MEGX test results below or above the indicated cut-off point.
*Source:* Data are from reference [9].

strated that MEGX declines in a stepwise fashion with advancing histology in patients with chronic liver disease [10]. Due to a wide interindividual variability, the MEGX test cannot be used to diagnose the initial stages of chronic hepatitis. MEGX test results, however, do relate to the histology activity index in patients with advanced liver disease. In patients with cirrhosis, MEGX decreases with worsening Child–Pugh score. Most patients with MEGX test results below 20 μg/l had cirrhosis of Child–Pugh class C confirmed on histological evaluation. There is general agreement that MEGX concentrations of less than 20 μg/l at 15 min reflect poor liver function.

The problems of patient selection have been discussed in detail [9]. Patients awaiting transplantation may be stable for long periods of time, but may suddenly decompensate. The degree of abnormality of conventional liver tests does not necessarily predict those patients at risk of rapid clinical deterioration. Prospective studies have evaluated the usefulness of the MEGX test for predicting pre-transplant complications and survival. It was found that life-threatening complications of cirrhosis were observed only in patients with MEGX concentrations <30 μg/l [9]. Furthermore, death as a direct result of these complications was observed only in those patients with MEGX <10 μg/l.

In various independent, prospective studies, MEGX test results were related to survival in patients with chronic liver disease. Low MEGX test results were associated with 1-year spontaneous survival of about 50% or less (Table 15.2). On the other hand, MEGX test results above the indicated cut-off points at 10 or 25 μg/l were associated with a survival of 82% and 97%, respectively. All these findings suggest that the MEGX test is a useful tool for the improvement of the decision-making process with respect to the selection of transplant candidates. Recent investigations in patients with primary biliary cirrhosis (PBC) have demonstrated an independent prognostic value only for the MEGX test and the Mayo score [11]. It was found that the asymptomatic progressive functional deterioration occurring during the natural history of PBC could be monitored by the MEGX test because it appeared to be able to identify abnormalities prior to the onset of alterations in conventional laboratory and/or clinical parameters which are likely to affect the Mayo score.

Early post-transplant liver function is a further area where serial measurements of MEGX have proven to be useful. After liver transplantation, primary graft nonfunction or dysfunction may be related to pre-existing or ischaemia/reperfusion-induced liver damage. Primary graft nonfunction is associated with very low MEGX test results (range of 2–14 μg/l) within the first 3 days after transplantation [4]. The concentration of hyaluronic acid, however, was highly elevated in these patients, indicating impaired endothelial cell function. Haemodynamic changes resulting from endothelial injury may explain why flow-dependent tests like MEGX, and also ICG [12], are substantially decreased in primary graft dysfunction. A combination of MEGX and hyaluronic acid has been found to predict liver graft survival. The probability of 120-day graft survival was significantly higher for those patients in whom MEGX tests results on post-operative day 1 were $>22$ μg/l and hyaluronic acid concentrations were $\leq 730$ μg/l. These findings suggest that the inclusion of MEGX and hyaluronic acid in postoperative monitoring of liver transplant recipients may be helpful in the early prediction of graft survival. There is further evidence from experiments with the isolated perfused pig liver that vascular changes may be reflected by MEGX test results. The impairment of MEGX formation after 20 hours of cold ischaemia could be significantly attenuated by the pre-treatment of animals with the antioxidant idebenone that is known to decrease sinusoidal endothelial damage. MEGX formation in liver microsomes isolated from these perfused organs after 20 hours of cold ischaemia was not decreased, indicating that the CYP system was intact. These results after 20 hours of cold ischaemia suggest that impaired MEGX formation during perfusion was due to diminished sinusoidal blood flow associated with endothelial injury (E. Wieland et al., unpublished data).

Long-term hepatic regeneration and function in infants and children following liver resection has been studied by Shamberger et al. [13]. A retrospective evaluation was made in 10 children following hepatic resection for benign or malignant tumours. Evaluations were performed a median of 4.3 years after hepatic resection. Tests of synthetic function were essentially normal. Aminotransferases and ammonia were mildly elevated, and hepatic volumes were below but close to the normal curve in most cases. MEGX test results were normal in all of the children except one, who was hepatitis C virus positive. The authors suggest that the MEGX test may also be used to assess regeneration and function following liver resection. However, a major shortcoming of this study is the fact that baseline MEGX test results were not obtained shortly after resection.

There is emerging evidence that the MEGX test may be of prognostic value in assessing the liver function of critically ill patients at risk for developing multiple organ failure (MOF). Purcell et al. [14] studied MEGX production in an animal model of acute lung injury managed with positive end-expiratory pressure (PEEP) ventilation. Hepatic blood flow and hepatic oxygen delivery were significantly decreased after lung injury and PEEP, and this was accompanied by a significant impairment in the formation of MEGX. The authors reasoned that decreased MEGX production may be a useful clinical indicator of reduced hepatic blood flow and oxygen supply in critical illness.

In a clinical study conducted by Lehmann et al., the MEGX test was serially performed in 28 critically ill patients admitted to an intensive care unit (ICU) after multiple trauma [15]. Nine of the 10 patients who subsequently developed MOF without signs of bacterial sepsis displayed a sharp decrease in their median MEGX values between days 1 and 3 after trauma, from 67 to 15 μg/l. In the patients who did not develop MOF, MEGX test results remained stable. When compared with conventional liver function tests, receiver operating characteristic (ROC) curve analysis revealed that the results of the MEGX test on day 3 provided the greatest discriminating power between patients with and without MOF. In one MOF patient the syndrome was induced by bacterial sepsis that became clinically evident on the sixth day after trauma. This was paralleled by a marked decrease in MEGX values from days 3 to 7. After removal of the septic focus, the patient showed gradual clinical improvement, which was accompanied by a return of MEGX test results to within the reference range.

In another study [16], involving 27 consecutive critically ill patients with evidence of inadequate tissue perfusion requiring pulmonary artery catheterization and mechanical ventilation, MEGX formation and ICG clearance were compared with standard liver function tests and related to gastric intramucosal pH (pHim); the latter is considered as an alternative measure of adequacy of splanchnic oxygenation. The 12 nonsurvivors displayed a dramatic decrease in their MEGX test results from days 1 (20 μg/l) to 3 (2.4 μg/l). In survivors, MEGX test results remained relatively constant over this period. None of the other liver function tests displayed any significant difference between survivors and nonsurvivors. ROC

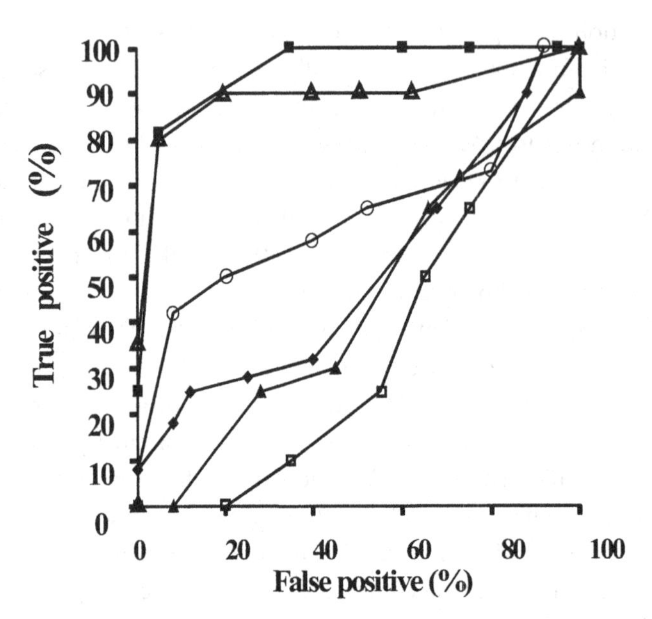


Figure 15.2 ROC curves for liver function tests (MEGX —■—, ICG clearance —⊟—, alkaline phosphatase —◆—, aspartate aminotransferase —⊖—, bilirubin —▲—) and gastric pHim (—△—). Data are from reference [16].

curve analysis showed that the MEGX test and pHim on day 3 were the most discriminatory with regard to death and survival (Figure 15.2). The authors of this study concluded that critically ill patients develop hepatic dysfunction presumably due to a mismatch between hepatic metabolic demand and hepatic blood flow. The MEGX test appears to be extremely effective in assessing liver function and hepatic blood flow in this group of patients.

Further investigations from the author's group are in progress in order to clarify the pathogenesis of MOF/multiple organ dysfunction syndrome (MODS) with regard to early hepatic dysfunction in ICU patients after polytrauma or with sepsis [17]. In patients with sepsis there is a highly significant negative correlation between MEGX test results on days 1–5 with the SAPS II mortality risk score ($r=-0.60$; $p<0.0001$), IL-6 ($r=-0.71$; $p<0.0001$), C reactive protein ($r=-0.62$; $p<0.0001$) and hyaluronic acid ($r=-0.38$; $p=0.0025$). The correlation of MEGX with IL-10 was less pronounced but still significant. No correlation was found between MEGX and conventional liver function tests such as transaminases, albumin and bilirubin (Igonin et al., unpublished data). The data suggest that hepatic dysfunction as assessed by MEGX is related to the systemic inflammatory response.

## Future aspects

The emerging static and dynamic tests discussed in this chapter offer new strategies to assess prognosis and real-time hepatic function in patients with liver disease. The author concurs with Sheila Sherlock [1] who stated that there is no magic liver function test and that it is unnecessary to use a large number of methods – this type of shotgun investigation only adds to the confusion. Rather, a few single tests of established value should be used. Dynamic liver function tests such as MEGX can complement established and emerging static tests in areas where prognostic information is of interest. This is the case, for instance, in the area of transplant candidate selection, the assessment of progressive functional deterioration in PBC and in the prediction of MOF/MODS in ICU patients after polytrauma or with sepsis.

## REFERENCES

1 Sherlock, S. and Dooley, J. (1993). Assessment of liver function. In *Diseases of the Liver and Biliary System*, 9th edition, eds S. Sherlock and J. Dooley. Oxford: Blackwell Scientific Publications, pp. 17–32, 353.
2 Lindqvist, U. (1997). Is serum hyaluronic acid a helpful tool in the management of patients with liver diseases? *Journal of Internal Medicine*, **242**, 67–71.
3 Wong, V.S., Hughes, V., Trull, A., Wight, D.G.D., Petrik, J. and Alexander, G.J.M. (1999). Serum hyaluronic acid is a useful marker of liver fibrosis in chronic hepatitis C virus infection. *Journal of Viral Hepatitis*, **5**, 187–92.
4 Schütz, E., Luy-Kaltefleiter, M., Kaltefleiter, M. et al. (1996). The value of serial determination of MEGX and hyaluronic acid early after orthotopic liver transplantation. *European Journal of Clinical Investigation*, **26**, 907–16.
5 Taki, Y., Gubernatis, G., Yamaoka, Y. et al. (1990). Significance of arterial ketone body ratio measurement in human transplantation. *Transplantation*, **49**, 535–9.
6 Hughes, V., Trull, A.K., Joshi, O. and Alexander, G.J.M. (1998). Monitoring eosinophil activation and liver function after liver transplantation. *Transplantation*, **65**, 1334–9.
7 Oellerich, M., Raude, E., Burdelski, M. et al. (1987). Monoethylglycinexylidide formation kinetics: a novel approach to assessment of liver function. *Journal of Clinical Chemistry and Clinical Biochemistry*, **25**, 845–53.
8 Mets, B., Hickman, R., Allin, R., Van Dyk, J. and Lotz, Z. (1993). Effect of hypoxia on the hepatic metabolism of lidocaine in the isolated perfused pig liver. *Hepatology*, **17**, 668–76.
9 Potter, J.M. and Oellerich, M. (1996). The use of lidocaine as a test of liver function in liver transplantation. *Liver Transplantation and Surgery*, **2**, 211–24.
10 Shiffman, M.L., Luketic, V.A., Sanyal, A.J. et al. (1994). Hepatic lidocaine metabolism and liver histology in patients with chronic hepatitis and cirrhosis. *Hepatology*, **19**, 933–40.

11 Gindro, T., Arrigoni, A., Martinesso, G. et al. (1997). Monoethylglycinexylidide (MEGX) test evaluation in primary biliary cirrhosis: comparison with Mayo Score. *European Journal of Gastroenterology and Hepatology*, **9**, 1155–9.
12 Jalan, R., Plevris, J.N., Jalan, A.R., Finlayson, N.D.C. and Hayes, P.C. (1994). A pilot study of indocyanine green clearance as an early predictor of graft function. *Transplantation*, **58**, 196–200.
13 Shamberger, R.C., Leichtner, A.M., Jonas, M.M. and LaQuaglia, M.P. (1996). Long-term hepatic regeneration and function in infants and children following liver resection. *Journal of the American College of Surgeons*, **182**, 515–19.
14 Purcell, P.N., Branson, R.D., Schroeder, T.J., Davis, K. and Johnson, D.J. (1992). Monoethylglycinexylidide production parallels changes in hepatic blood flow and oxygen delivery in lung injury managed with positive end-expiratory pressure. *Journal of Trauma*, **33**, 482–6.
15 Lehman, U., Armstrong, V.W., Schütz, E., Regel, G., Pape, D. and Oellerich, M. (1995). Monoethylglycinexylidide as an early predictor of posttraumatic multiple organ failure. *Therapeutic Drug Monitoring*, **17**, 125–32.
16 Maynard, N.D., Bihari, D.J., Dalton, R.N., Beale, R., Smithies, M.N. and Mason, R. (1997). Liver function and splanchnic ischemia in critically ill patients. *Chest*, **111**, 180–7.
17 Oellerich, M. and Armstrong, V.W. (2001). The MEGX test: a tool for the real-time assessment of hepatic function. *Therapeutic Drug Monitoring*, **23**, 81–92.

16

# The immunogenetics of metabolic liver disease

Peter T Donaldson

Centre for Liver Research, University of Newcastle, Newcastle-upon-Tyne, UK

## Introduction

While most metabolic diseases result from single gene defects or the direct action of a toxic agent, there are circumstances where disordered metabolism or exposure to toxic agents gives rise to diseases which have a distinct immune basis. Such diseases are unlikely to arise from a single gene and fall into the category of 'complex' diseases, involving one or more genes as well as additional factors. The classical examples of immune-mediated metabolic disease are immune–allergic drug reactions (e.g. co-amoxiclav-induced hepatic cholestasis or minocyclin-induced autoimmune hepatitis) and metabolic diseases where repeated exposures induce chronic inflammation (e.g. alcoholic cirrhosis). The clinical outcome following acute toxic injury may also be influenced by the immune response (e.g. the outcome of acute liver failure following paracetamol overdose). Immune response genes may also play a role in determining the extent and severity of liver involvement in a number of single gene disorders including cystic fibrosis and $\alpha$-1-antitrypsin deficiency.

The immune response genes do not themselves cause disease, but specific genes may be considered as risk factors with a potential to increase (i.e. susceptibility genes) or reduce (i.e. resistance genes) the risk of disease. Genes may also determine the rate of disease progression and/or the severity of symptoms (i.e. the disease phenotype). It must be remembered that these genes are often present in a large number of nonaffected individuals in the population and most are not defective under normal circumstances.

The greatest progress in understanding the genetics of complex liver disease has come from studies of the human leucocyte antigen (HLA) genes [1]. However, the human genome mapping project (HGMP) has also identified huge numbers of polymorphic immunoregulatory genes as potential candidates for association studies. Much recent interest has been directed towards polymorphism in cytokine genes (see http//www.pam.bris.ac.uk/services/GAI/cytokine-4.htm) [2]. The role of immunoregulatory genes in the T- and B-cell immune response can be considered on a temporal basis. The first event, antigen recognition, is a process which is governed mostly by polymorphism of HLA, T-cell receptor (TCR) and immunoglobulin (Ig) genes. Secondly, there is a phase of cosignalling governed by the interaction of cytotoxic T lymphocyte antigen-4 (CTLA-4) with either CD28 or B7 and various other accessory molecules. Thirdly, there is a phase of inflammation regulated by a network of interacting cytokines, particularly the interleukins (IL), but also tumour necrosis factor (TNF) and the interferons (IFN). In this third phase, clonal expansion of both T and B cells occurs, regulated by cytokines. The T-cell expansion may be directed towards either a predominantly Th1 or Th2 response under strict regulation of the cytokine cascade (see below). Inflammation results in tissue damage and causes further inflammation and repair leading in the liver to fibrosis and ultimately cirrhosis. Nearly all the genes within this interacting series of cascades are known to be polymorphic and, consequently, most appear as potential candidate risk factors in disease development.

To understand which candidate(s) may be worthy of investigation in metabolic liver disease, we need to consider the possible mechanisms that give rise to immune–metabolic interactions. These diseases may occur when a drug or toxic metabolite directly neutralizes or enhances the action of a specific element of the immune response. An example of such a direct interaction is hydralazine-induced lupus, where hydralazine may interact with the C4 complement protein causing a temporal complement deficiency. Susceptibility to this adverse drug reaction is enhanced by the presence of specific complement isoforms with particularly high affinity for hydralazine or its metabolites [3].

The most common pathway for immune–metabolic interaction is through induction of autoimmunity. Autoimmunity occurs when the immune system recognizes self as nonself. This may occur when: (i) self(auto)-antigens are modified (altered self); (ii) self-antigens develop enhanced immunogenicity due to modified presentation by a carrier or hapten; (iii) immune tolerance to self breaks down; or (iv) immune tolerance for the self-antigen does not exist. The latter intolerance may arise because the antigen is expressed in such a way that it is not seen during the thymic maturation of T cells (i.e. 'cryptic epitopes' are usually unseen by the immune system) and becomes a target as a consequence of abnormal expression at the cell surface [4]. In halothane-induced hepatitis, halothane metabolites cause trifluroacetylation of lysine-rich proteins and antibodies are formed against these trifluroacetylated lysine peptides. This appears to be sufficient to induce damage through antibody-dependent cell-mediated cytotoxicity during which antibody–trifluroacetyl–lysine complexes on the hepatocyte surface are recognized by killer cells and induce hepatocyte lysis. This gives rise to an acute cytolytic hepatitis which may present with symptoms ranging from mild jaundice to fulminant liver failure [5].

The hapten effect may underlie tienilic acid-induced hepatitis where patients present with jaundice and antibodies to antinuclear antigens (ANA), a major component of which is anticytochrome P450 (CYP) 2C9. The 2C9 isoform of CYP is particularly important in tienilic acid metabolism when hapten formation may occur. Alternatively, and because CYP 2C9 is an intracellular protein normally unseen by T cells (cryptic), it may be one for which T-cell tolerance is not established in the thymus during development. There are many other examples involving immune responses to cryptic epitopes in autoimmune liver disease, including the pryruvate dehydrogenase (PDC) 2E epitope which is the principle target for both T- and B-cell recognition in primary biliary cirrhosis (PBC) [6]. In PBC, PDC-E2 is expressed on the surface of hepatocytes, but, in healthy liver, the antigen is only expressed in the mitochondria within the cell.

The delicate balance which exists between predominantly humoral (B cell/antibody) immunity and T-cell immune responses is maintained by a host of interacting regulatory cytokines. The cytokines may favour either a Th1 response (predominantly, IL-2, IFN$\gamma$-producing T cells) which enhances macrophage IL-1 production, TNF and clonal expansion of T cells or a Th2 response (predominantly IL-4, IL-5 and IL-10 producing T cells) which enhances expansion of B cells and the antibody response. Within these two subsets, there is also thought to be a population of cells which is responsible for downregulation of the inflammatory response (the so-called suppressor T cells). The switch from Th1 to Th2, and vice versa, is known as 'immune deviation' and is viewed as a mechanism of suppression of one limb of the immune system in favour of the other. Genetic polymorphism in these key regulatory cytokines, or any event which favours one side of this balance over the other, may be particularly important in the response to a metabolic injury or insult and may determine prognosis for individuals with specific immune-mediated diseases.

There is also evidence that long-term or chronic tissue injury may give rise to autoimmune processes, although the precise mechanism of this is unclear. Cryptic autoimmunity may occur where there is damage to other organs not usually associated with the primary disease – for example, liver damage in the single gene disorders cystic fibrosis or $\alpha$-1-antitrypsin deficiency (see below). This damage may occur more commonly than is suspected, accounting for much of the heterogeneity in disease phenotype. Whether this is a consequence of the specific metabolic defect in these patients, or whether it is due to other complications such as repeated infection or a particular therapy, is unknown.

To illustrate these theories, the current data on the immunogenetics of metabolic liver disease are reviewed under four headings below. Some important practical considerations applicable to experimental interpretation have been considered by Tredger (Chapter 17) and will not be repeated here.

## Immune allergic drug reactions

Drug-induced hepatitis is rare during therapeutic dosage regimens. Although of unknown aetiology, several features suggest an immunological basis including: (i) frequent symptoms of hypersensitivity, with fever, rash and eosinophilia; (ii) the presence of specific autoantibodies; (iii) liver parenchymal lesions, usually including a mixed cellular infiltrate; and (iv) a more rapid and more severe response following rechallenge. Several hundred commonly prescribed drugs are associated with liver toxicity. Berson et al. [7], reviewing 71 patients treated with 41 different agents, identified no single unifying HLA association. Although none of the individual study groups was of sufficient size for HLA association studies (the maximum was seven patients), a number of interesting observations were reported for individual drugs (Table 16.1), including associations of reactions to tricyclic antidepressants with HLA A11, diclofenac with A11 and chlorpromazine with DR6. In addition, patients with B17 seemed to be more likely to develop severe and prolonged cholestasis. Earlier studies also suffered from small numbers, but associations have been reported for HLA A11 with halothane hepatitis [8], DR6 and DR2 with nitrofurantoin [9], and B8 with clometacin toxicity [10] (Table 16.2). Collectively, these data represent interesting preliminary observations on which to base future hypotheses.

There have been two recent independent studies of HLA associations in co-amoxiclav-associated cholestasis [11, 12]. Co-amoxiclav (amoxicillin–clavulanate potassium) extended the antibacterial spectrum of amoxicillin when introduced in the UK in 1981 and has become one of the most frequently prescribed antibiotics. Hepatotoxicity is uncommon, is mostly cholestatic and rarely fatal. It is associated with infiltration of the biliary epithelium by various immunocytes including neutrophils, eosinophils and macrophages and with focal injury to interlobular bile ducts. Both reports cite an elevated frequency of the *DRB1*1501–DQB1*0602* haplotype. In the first study from Belgium [11], 57% of the 35 patients tested had this haplotype and, in the second study from Scotland [12], 70% of 20 compared with 11% and 20% (respectively) of matched 'healthy' controls. The findings suggest a highly significant, 6.5-fold increased likelihood of cholestasis in patients with *DRB1*1501–DQB1*0602* treated with co-amoxiclav. The second study also reported an increased frequency of patients (35%) who were homozygous for this haplotype (odds ratio [OR] = 8.7).

The same extended *DRB1*1501–DQB1*0602* haplotype has been associated with halothane hepatitis in Japan [15], with liver disease in cystic fibrosis [13] (see below), with primary sclerosing cholangitis, an autoimmune cholestatic liver disease [16] and with a high incidence of allergies, particularly ragweed pollen hypersensitivity [17]. Interestingly, this haplotype may encode resistance to type 1

**Table 16.1.** Summary of the principal HLA associations with drug-induced idiosyncratic hepatitis

| Drug | n | Antigen | % patients | % controls | Reference |
|---|---|---|---|---|---|
| Tricyclic antidepressants | 12 | A11 | 50 | 12 | 7 |
| Diclofenac | 4 | A11 | 75 | 12 | 7 |
| | 29 | Class II negative | – | – | |
| Halothane | 17 | A11 | 29 | 11 | 8 |
| | | DR2 | n/a | n/a | |
| Chloropromazine | 5 | DR6 | 80 | 22 | 7 |
| Nitrofurantoin | 9 | DR6 | 56 | 29 | 9 |
| | | DR2 | 56 | 29 | |
| Clometacin | 10 | B8 | 70 | 16 | 10 |
| Co-amoxiclav | 35 | DRB1*1501 | 57 | 10 | 11 |
| | 20 | DRB1*1501 | 70 | 20 | 12 |

**Table 16.2.** Role of HLA in determining susceptibility to liver disease in the single gene disorders cystic fibrosis (F) and $\alpha$-1-antitrypsin deficiency ($\alpha$-1-AT↓)

| | | Percentage | | |
|---|---|---|---|---|
| CF | n | B7 | DR15 | DQ6 |
| With liver disease | 175 | 33 | 44 | 67[a,b] |
| No liver disease | 76 | 21 | 26 | 33 |
| Controls | 141 | 23 | 29 | 29 |
| $\alpha$-1-AT↓ | n | DR3 | DR4 | |
| With liver disease | 75 | 47[c] | 39 | |
| No liver disease | 28 | 18 | 61[d] | |
| Controls | 100 | 23 | 36 | |

*Note:*

In CF, [a] $p<0.001$, OR = 5.0 for liver disease versus controls and [b] $p<0.005$, OR = 4.1 for liver disease versus no liver disease.

In $\alpha$-1-AT↓, [c] $p<0.001$, OR = 2.92 for liver disease versus controls and [d] $p<0.02$, OR = 2.75 for no liver disease versus controls.

*Source:* Based on Duthie et al. [13] and Doherty et al. [14].

autoimmune hepatitis [18]. The debate continues as to whether the susceptibility allele is *DRB1*1501* itself or a linked gene such as *HLA A3, HLA B7, MICA*008, TNFA*2* or the distantly linked gene mutation *C282Y* (responsible for the iron storage disease, haemochromatosis [19]). Finally, there is some evidence that T-cell recognition of amoxicillin is DR restricted but may not require antigen processing [20].

## Immunogenetic markers as determinants of outcome following paracetamol overdose

The nature and magnitude of the inflammatory immune response to tissue injury is an important element in the development of multiorgan failure (MOF) in the critically ill. In some cases, the immune response may be far greater than that required to deal with the original insult and promote tissue injury rather than resolution and healing. Patients respond very differently to paracetamol overdose once the critical threshold of 15 g has been exceeded. Outcome may vary from self-limiting coagulopathy to rapidly progressing MOF and the syndrome of hyperacute (fulminant) liver failure (POD-ALF). Evidence that immune activation occurs after paracetamol overdose is mainly based on the observation that high circulating levels of the cytokines IL-1, IL-6 and TNF-$\alpha$ can be detected at this time (reviewed in [21]). Not only do these cytokines profoundly affect events close and distant to their release, but experimental evidence suggests that they may also impair liver regeneration [22].

In a series of studies performed at King's College Hospital, London [21, 23], the author and colleagues have begun to investigate the relationship between genetic polymorphisms in specific proinflammatory cytokines and outcome following paracetamol overdose. So far, the results indicate a complex relationship between TNF and IL-1 genes and prognosis, but no relationship with either HLA class II (included because of linkage with TNF) or IL-6.

The TNF gene variants examined were two A to G single nucleotide polymorphisms (SNPs) in the TNF-$\alpha$ promoter region at −238 and −308, as well as an A to G SNP in the lymphotoxin-$\alpha$ (or TNF-$\beta$) gene sequence. These polymorphisms have been linked with outcome in other critical illnesses including cerebral malaria and severe sepsis [24, 25] and are thought to be functionally significant [26, 27]. Earlier investigations [21, 23] indicated no overall association with TNF-$\alpha$ polymorphism. However, the TNF-$\beta$ gene was associated with the severe encephalopathy (grade 3 or 4 on a scale of 0–4, where grade 4 equals coma) present in 46 of 97 patients. The *TNFB*1,TNFB*1* genotype was found in 18% of patients with grade 0–2 and 2.2% of those with grade 3–4 encephalopathy, whereas the *TNFB*1,TNFB*2* genotype was found in 39% and 54%, respectively ($p<0.03$) [23]. Further investigations have suggested an association independent of HLA and perhaps due to the interrelation between encephalopathy and sepsis.

Polymorphism in the IL-1 genes has been widely investigated in inflammatory conditions and autoimmune liver disease [28–30] and was recently linked to an increased risk of gastric cancer from *Helicobacter pylori* infection [31]. The IL-1 gene family on chromosome 2q13–14 encodes three proteins: IL-1$\alpha$, IL-1$\beta$ and the IL-1 receptor antagonist (IL-1Ra). IL-1Ra competes with IL-1$\beta$ (and IL-1$\alpha$) for binding to the IL-1 receptors and is a potent inhibitor of IL-1 activity. Four nucleotide substitutions of the IL-1$\beta$ gene (*IL-1B*) have been described, but only one (at position +3953 in exon 5) has been shown to affect protein production. The presence of the less common allele 2 (*IL-1B*2*) at this site has been associated with increased IL-1$\beta$ secretion in vitro [32]. Studies of the gene encoding IL-1Ra (*IL-1RN*) have also indicated that genetic polymorphism (in the variable number of tandem repeat sequences in intron 2) may influence biological activity [33]. Five alleles are described, expressing two to six repeats, but only two variants are commonly found: the 4 (*IL-1RN*1*) and 2 (*IL-1RN*2*) repeat variants [34]. In vitro studies have shown that the *IL-1RN*2* allele is associated with high IL-1Ra and IL-1$\beta$ production in healthy volunteers and linkage has been recognized between *IL-1RN*2* and *IL-1B*1* [35].

Analysis of the distribution of these IL-1 gene polymorphisms in 140 patients with POD-ALF has shown a reduced frequency of the *IL-1B*1,2* and an increased frequency of the *IL-1B*2,2* genotype ($p<0.02$) *versus* 106 controls. The frequencies of both the *IL-1RN*1,2* and *IL-1RN*2,2* genotypes ($p<0.009$) were also increased in the POD-ALF patients. Furthermore, the *IL-1B* association was independent of *IL-1RN*, but the *IL-1RN* association was due to linkage with *IL-1B*. Patients with the *IL-1B*2,2* genotype were also found to have significantly higher peak INR values (a measure of liver function and coagulation status) than their counterparts with other *IL-1B* genotypes. Overall, these results were not unexpected, since the study was designed to detect clinical correlates and prognostic markers. However, the associations of genotype with susceptibility to POD-ALF were not anticipated, although confounding factors not taken into account in the original study design include referral patterns. As a tertiary referral centre, the Liver Unit at King's College Hospital accepts only a minority of very ill POD patients out of the total numbers presenting at regional casualty departments throughout the country. These patients are more likely to have a poor prognosis and will include a high percentage with pre-existing sepsis and other complications.

## Immunogenetic markers as determinants of outcome and fibrogenesis in alcoholic liver disease

Alcoholic liver disease (ALD) progresses to liver cirrhosis in less than 10% of heavy drinkers. This clinical variability, and disappointing results from studies of genes involved in alcohol metabolism, has led to considerations that cirrhosis, fatty liver and alcoholic hepatitis are determined to some degree by genetic factors which regulate the immune system. An immune hypothesis for alcoholic cirrhosis has been suggested, although there is some evidence for a dose response. Twin studies suggest that genes do play a role in determining progression beyond simple alcoholic steatohepatitis, and candidate associations have been sought either with HLA or with other immunoregulatory genes.

Early investigations of the relationship between HLA and alcoholic liver disease were mixed and, for the most part, disappointing. However, most of those studies were performed long before the introduction of molecular genotyping. Nevertheless, recent studies investigating HLA class II in alcoholic liver disease have failed to find any relationship between alcoholic liver disease and individual *HLA DRB1*, *DQA1* and *DQB1* alleles (Donaldson et al., unpublished data).

More recently, attention has focused on the cytokine genes. Ongoing studies at the Centre for Liver Research in Newcastle, UK, have linked TNF-238 and *IL-10-627* with alcoholic cirrhosis but not with alcoholic hepatitis [36, 37]. The *IL-10* gene promoter encodes three SNPs which have been linked with susceptibility to autoimmune disease [38] but not with autoimmune liver disease [28, 29]. These are A to G at position −1082, C to T at position −819 and A to C at position −592 (sometimes termed −627 [2]). In 285 patients with ALD, 50% of those with advanced ALD expressed at least one A allele at position −592 (−627) compared with 33% of controls ($n=227$) and 34% of heavy drinkers without ALD ($n=107$) [37]. The authors also reported a strong association with the A allele at position −1082 (−1117) which was assigned to linkage between *A −627* and *A −1117* [37].

The same group from Newcastle [36] has also reported an association between *TNFA −238* and alcoholic steatohepatitis. In that study, there was no relationship with *TNFA −308* polymorphism, but an excess of the *TNF −238 A* allele was recorded in 150 patients with biopsy-proven alcoholic liver disease (58 with acute steatohepatitis) versus 145 controls. Even more recently, the same group has described links between *CTLA-4 Ig* and *IL-4R* gene polymorphisms and alcoholic cirrhosis (CP Day, unpublished data). Studies at King's College Hospital have failed to link any of the *IL-1* gene family or *IL-6* with either alcoholic hepatitis or alcoholic cirrhosis (M Phillips, unpublished data). Overall, a picture of alcoholic liver disease as a polygenic disorder involving a degree of autoimmunity is beginning to emerge. Many of these latter results are unpublished and independent confirmation of the *TNF* and *IL-10* data would strengthen the Newcastle hypothesis.

## Immunogenetic markers as determinants of liver disease complicating the single gene disorders, cystic fibrosis and $\alpha$-1-antrypsin deficiency

### Cystic fibrosis

Approximately 4% of children and up to 25% of adults with cystic fibrosis (CF) develop features of severe chronic liver disease including cirrhosis [39, 40] and this carries a poor prognosis. There is no association with any of the major CF gene mutations, *DF508, G551D* or *R553X* [41, 42] but familial clustering of liver disease and inappropriate immune responses may be responsible [43]. HLA A, B, DR and DQB typing performed at King's College Hospital in 247 CF children and adults, 82 of whom had chronic liver disease (Table 16.2), showed a greater prevalence of DQ6 in those with liver disease (66% versus 33% without). The study concluded that *B7–DR15–DQ6* was associated with an increased risk of liver disease in CF patients, particularly males [13].

### $\alpha$-1-antitrypsin deficiency

In $\alpha$-1 antitrypsin deficiency, up to 20% of patients with the primary PiZZ defect may develop chronic liver disease [44]. Familial histories and immune responses to liver autoantigens have suggested that an autoimmune phenomenon may be responsible [45]. Doherty et al. [14] investigated the HLA A, B DR and DQ serotypes of 140 PiZZ individuals (92 with liver disease) and a further 206 first-degree relatives (Table 16.2). A positive correlation was identified between the presence of liver disease and DR3 (particularly the DR3–Dw25 genotype, equivalent to *DRB1*0301–DRB3*0201/0202*). This haplotype is distinct from that associated with primary sclerosing cholangitis and type 1 autoimmune hepatitis [1], encoding a different *DRB3* gene product – Dw25 (*DRB3*0201/0202*) rather than Dw24 (*DRB3*0101*).

The technology used in these two studies is slightly dated and the major preposition that liver disease develops from autoimmune reactions following chronic injury to other primary organs in both diseases has not been retested since the early 1980s. However, studies have not identified a liver disease-specific mutation in the genes responsible for CF or $\alpha$-1-antitrypsin deficiency. Therefore, the possibility remains that immunoregulatory genes may influence the development of liver disease and that these genes may be useful biomarkers for this subgroup of patients.

## Conclusion

The evidence reviewed in this chapter suggests that there is a role for immunogenetics in metabolic disease. Because there have been only a limited number of reports, many of which do not meet the rigorous criteria now demanded for such studies, it is important that future investigators continue to consider immunogenetic markers in metabolic disease, especially where there is marked clinical heterogeneity. At this stage, the existing data should be considered preliminary and should be used to focus future candidate gene studies. Investigators should particularly note the errors in the design of reported studies when embarking on a similar voyage of discovery.

## REFERENCES

1 Donaldson, P.T. and Manns, M.P. (1999). Immunogenetics of liver disease. In *Oxford Textbook of Clinical Hepatology*, eds. J. Bircher, J.-P. Benhamou, N. McIntyre, M. Rizetto and J. Rodes. Oxford: Oxford University Press, pp. 173–88.

2 Bidwell, J., Keen, L., Gallagher, G. et al. (1999). Cytokine gene polymorphisms in human disease: on-line database. *Genes and Immunity*, **1**, pp. 3–19.

3 Spiers, C., Fielder, A.H.L., Chapel, H., Davey, N.J. and Batchelor, J.R. (1989). Complement system protein C4 and susceptibility to hydralazine-induced systemic lupus erythematosus. *Lancet*, **1**, 922–4.

4 Rose, N.R. (1998). The concept of autoimmune liver disease. In *Autoimmune Liver Diseases*, eds. E. Krawitt, M. Nishioka and R. Wiesner. Oxford: Elsevier, pp. 1–20.

5 Pessayre, D., Larrey, D. and Biour, M. (1999). Drug-induced liver injury. In *Oxford Textbook of Clinical Hepatology*, eds. J. Bircher, J.-P. Benhamou, N. McIntyre, M. Rizetto and J. Rodes. Oxford: Oxford University Press, pp. 1261–315.

6 Yeaman, S.J., Kirny, J.A. and Jones, D.E.J. (2000). Autoreactive responses to pyruvate dehydrogenase complex in the pathogenesis of primary biliary cirrhosis. *Immunology Reviews*, **174**, 238–49.

7 Berson, A., Fréneaux, E., Larrey, D. et al. (1994). Possible role of HLA in hepatotoxicity: an exploratory study in 71 patients with idiosyncratic drug-induced hepatitis. *Journal of Hepatology*, **20**, 336–42.

8 Eade, O.E., Grice, D., Krawitt, E.L. et al. (1981). HLA A and B locus antigens in patients with unexplained hepatitis following halothane anaesthesia. *Tissue Antigens*, **17**, 428–32.

9 Stricker, BHCh, Block, R., Claas, F.H.J., Parys, G.E.V. and Desmet, V.J. (1988). Hepatic injury associated with the use of nitrofurans: a clinico-pathological study of 52 reported cases. *Hepatology*, **8**, 599–606.

10 Pariente, A., Hamoud, A., Goldfain, D. et al. (1989). Hépatites à la clométacine (Duperan®). Etude rétrospective de 30 cas. Un modèle d'hépatite immunoallergique. *Gastroenterologie Clinique de Biologie*, **13**, 769–74.

11 Hautkeete, M.L., Horsmans, Y., Van Waeyenberge, C. et al. (1999). HLA association of amoxycillin-clavulanate-induced hepatitis: implications for the pathogenesis of drug-induced immunoallergic hepatitis. *Gastroenterology*, **117**, 1181–6.

12 O'Donohue, J., Oien, K., Donaldson, P. et al. (2000). Co-amoxiclav cholestasis: clinical and histological features and HLA class II association. *Gut*, **47**, 717–20.

13 Duthie, A., Doherty, D.G., Donaldson, P.T. et al. (1995). The major histocompatibility complex influences the development of chronic liver disease in male children and young adults with cystic fibrosis. *Journal of Hepatology*, **23**, 532–7.

14 Doherty, D.G., Donaldson, P.T., Whitehouse, D.B. et al. (1990). HLA phenotypes and gene polymorphisms in juvenile liver disease associated with α-1-antitrypsin deficiency. *Hepatology*, **12**, 218–23.

15 Otsuka, S., Yamamoto, M., Kasuya, S., Ohtomo, H., Yamamoto, Y., Akaza, T. (1985). HLA antigens in patients with unexplained hepatitis following halothane anaesthesia. *Acta Anaesthesiologica Scandinavia*, **25**, 497–501.

16 Donaldson, P.T., Farrant, J.M., Wilkinson, M.L., Hayllar, K., Portmann, B.C. and Williams, R. (1991). Dual association of HLA DR2 and DR3 with primary sclerosing cholangitis. *Hepatology*, **13**, 129–33.

17 Blumenthal, M., Marcus-Bagley, D., Awdeh, Z., Johnson, B., Yunis, E.J. and Alper, C.A. (1992). HLA-DR2, [HLA-B7, SC31, DR2], and [HLA-B8–SC01, DR3] haplotypes distinguish subjects with asthma from those with rhinitis only in ragweed pollen allergy. *Journal of Immunology*, **148**, 411–16.

18 Strettell, M.J.D., Donaldson, P.T., Thompson, L.J. et al. (1997). Allelic basis for HLA-encoded susceptibility to type 1 autoimmune hepatitis. *Gastroenterology*, **112**, 2028–36.

19 Feder, J.N., Gnirke, A., Thomas, W. et al. (1996). A novel MHC class I-like gene is mutated in patients with hereditary haemochromatosis. *Nature Genetics*, **13**, 399–408.

20 Horton, H., Weston, S.D. and Hewitt, C.R. (1998). Allergy to antibiotics: T cell recognition of amoxicillin is HLA-DR restricted and does not require antigen processing. *Allergy*, **53**, 83–8.

21 Bernal, W., Donaldson, P.T. and Wendon, J. (1999). Pro-inflammatory cytokine genomic polymorphism in critical illness. In *1999 Yearbook of Intensive Care and Emergency Medicine*, ed. J.L. Vincent. New York: Springer-Verlag, pp. 110–18.

22 Boulton, R., Woodman, A., Calnan, D., Selden, C., Tam, F. and Hodgson, H. (1997). Nonparenchymal cells from regenerating rat liver generate interleukin-1a and 1b: a mechanism of negative regulation of hepatocyte proliferation. *Hepatology*, **26**, 50–8.

23 Bernal, W., Donaldson, P., Wendon, J., Underhill, J. and Williams, R. (1998). Association of tumour necrosis factor polymorphisms and encephalopathy in acute acetominophen induced toxicity. *Journal of Hepatology*, **29**, 53–9.

24 Stuber, F., Udalova, I.A., Book, M. et al. (1996). −308 tumor necrosis factor (TNF) polymorphism is not associated with survival in severe sepsis and is unrelated to lipopolysaccharide inducibility of the human TNF promoter. *Journal of Inflammation*, **46**, 42–50.

25 Westendorp, R.G.J., Langermans, J.A.M., Huizinga, T.W.J. et al. (1997). Genetic influence on cytokine production and fatal meningococcal disease. *Lancet*, **349**, 170–3.

26 Wilson, A.G., Symons, J.A., McDowell, T.L., McDevitt, H.O. and Duff, G.W. (1997). Effects

of a polymorphism in human tumor necrosis factor α promoter on transcriptional activation. *Proceedings of the National Academy of Sciences USA*, **94**, 3195–9.

27 Louis, E., Franchimont, D., Piron, A. et al. (1998). Tumour necrosis factor (TNF) gene polymorphism influences TNF-α production in lipopolysaccharide (LPS)-stimulated whole blood cell culture in healthy humans. *Clinical and Experimental Immunology*, **113**, 401–6.

28 Donaldson, P. T., Norris, S., Constantini, P.K., Harrison, P. and Williams, R. (2000). The interleukin-1 and interleukin-10 gene polymorphisms in primary sclerosing cholangitis: no associations with disease susceptibility/resistence. *Journal of Hepatology*, **32**, 882–6.

29 Cookson, S., Constantini, P.K., Clare, M. et al. (1999). The frequency and nature of cytokine gene polymorphisms in type 1 autoimmune hepatitis. *Hepatology*, **30**, 851–6.

30 Donaldson, P. T., Agarwal, K., Craggs, A., Craig, W., James, O.F.W. and Jones, D.E.J. (2001). HLA and interleukin-1 gene polymorphisms in primary biliary cirrhosis: associations with disease progression and disease susceptibility. *Gut*, **48**, 397–402.

31 El-Omar, E.M., Carrington, M., Chow, W.-H. et al. (2000). Interleukin-1 polymorphisms associated with increased risk of gastric cancer. *Nature*, **404**, 398–402.

32 Pociot, F., Molvig, J., Wogensen, L., Worsaae, H. and Nerup, J. (1992). A *Taq1* polymorphism in the human interleukin-1 beta (IL-1 beta) gene correlates with IL-1 beta secretion in vitro. *European Journal of Clinical Investigation*, **22**, 396–402.

33 Danis, V.A., Millington, M., Hyland, V.J. and Grennani, D. (1995). Cytokine production by normal human macrophages: inter-subject variation and relationship to an IL-1 receptor antagonist (IL-1Ra) gene polymorphism. *Clinical and Experimental Immunology*, **99**, 303–10.

34 Tarlow J.K., Blakemore, A.I., Lennard, A. et al. (1993). Polymorphism in the human IL-1 receptor antagonist gene intron 2 is caused by variable numbers of an 86-bp tandem repeat. *Human Genetics*, **91**, 403–4.

35 Hurme, M. and Santtila, S. (1998). IL-1 receptor antagonist (IL-1Ra) plasma levels are coordinately regulated by both IL-1Ra and IL-1B genes. *European Journal of Immunology*, **28**, 2598–602.

36 Grove, J., Daly, A.K., Bassendine, M.F. and Day, C.P. (1997). Association of a tumour necrosis factor promoter polymorphism with susceptibility to alcoholic steatohepatitis. *Hepatology*, **26**, 143–6.

37 Grove, J., Daly, A.K., Bassendine, M.F., Gilvarry, E. and Day, C.P. (2000). Interleukin-10 promoter region polymorphisms and susceptibility to advanced alcoholic liver disease. *Gut*, **46**, 540–5.

38 Turner, D.M., Williams, D.M., Sankaran, D., Lazarus, M., Sinnott, P.J. and Hutchinson, I.V. (1997). An investigation of polymorphism in the interleukin-10 promoter. *European Journal of Immunogenetics*, **24**, 1–8.

39 Mowat, A.P. (1987). Hepatobiliary lesions in cystic fibrosis. In *Liver Disorders in Childhood*, 2nd edition, ed. A.P. Mowat. London: Butterworths, pp. 277–86.

40 Scott-Jupp, R., Lama, M. and Tanner, M.S. (1991). Prevalence of liver disease in cystic fibrosis. *Archives of Diseases in Childhood*, **66**, 698–701.

41 Duthie, A., Doherty, D.G., Williams, C. et al. (1992). Genotype analysis for DF508, G551D and R553X mutations in children and young adults with cystic fibrosis with and without chronic liver disease. *Hepatology*, **15**, 660–4.

42 Cystic Fibrosis Genotype–Phenotype Consortium (1993). Correlation between genotype and phenotype in patients with cystic fibrosis. *New England Journal of Medicine*, **329**, 1308–13.

43 Mieli-Vergani, G., Psacharopoulos, H.T., Nicholson, A.M., Eddleston, A.W.L.F., Mowat, A.P. and Williams, R. (1980). Immune responses to liver membrane antigens in patients with cystic fibrosis and liver disease. *Archives of Diseases in Childhood*, **55**, 696–701.

44 Mowat, A.P. (1984). Alpha-1-antitrypsin deficiency in liver disease. In *Butterworth's International Medical Reviews. Gastroenterology 4*, eds. R. Williams and W.C. Maddrey. London: Butterworths, pp. 52–75.

45 Mondelli, M., Mieli-Vergani, G., Eddleston, A.W.L.F., Williams, R. and Mowat, A.P. (1984). Lymphocytotoxicity to autologous hepatocytes in α-1-antitrypsin deficiency. *Gut*, **25**, 1044–9.

17

# Toxicogenetic markers of liver dysfunction

J Michael Tredger

Institute of Liver Studies, Guy's, King's and St Thomas' School of Medicine, London, UK

## Introduction

Toxicogenetics is the study of the genetic basis for variability in toxic injury – in this instance, toxic *liver* injury. It differs in three ways from pharmacogenetics – the genetic basis for variability in drug disposition – through its consideration of: (i) toxic (liver) injury rather than drug disposition; (ii) genetic variability in pathways that both induce and prevent toxic liver injury; and (iii) causes of damage wider than iatrogenic (drug-induced) injury. These additional causes of toxic liver injury may include other xenobiotics (e.g. environmental toxins), liver trauma and pathology (e.g. ischaemia/reperfusion injury), pathogens (e.g. hepatitis viruses) and extremes of physiological function (e.g. oxidative stress secondary to iron accumulation in haemochromatosis). Characteristic sequelae of all these primary mediators of toxic liver injury are a progression to oxidative stress and the production of secondary mediators of liver damage. The former reflects a shift of the cell's redox status away from its customary reduced status and the latter includes production of tumour necrosis factor (TNF) and cytokines by nonparenchymal cells.

The quantitative study of the consequences of toxicogenetic gene expression – toxicogenomics – is considered by Tugwood and Beckett in Chapter 28.

### How can genetic diversity affect toxic liver injury?

Toxicogenetic diversity may determine the extent of expression of both the original toxic insult and the secondary response it elicits, as well as the function of endogenous cytoprotection. Variable responses to the primary insult may be the result of genetically determined diversity in the xenobiotic metabolizing enzymes such as cytochrome P450 (CYP), in receptor structure and drug transport (e.g. the *MDR*-1 gene and p-glycoprotein), and in the immune response (considered by Donaldson in Chapter 16). Variability in the secondary response (e.g. TNF production) and in endogenous cytoprotection (e.g. glutathione synthesis and conjugation) depends equally on genetic diversity and contributes just as significantly to the variable final outcome. It is important to stress that multiple variants in several of these pathways are the most likely explanation for the diverse response to toxic

**Table 17.1.** Contributions of genetic variants and environmental modulators to induced toxicity

| Toxic outcome | Phenotypic association (underlying genetic dependency) | Additional predisposing factors (Reference) |
| --- | --- | --- |
| Familial unconjugated hyperbilirubinaemia | Defective bilirubin glucuronidation (*UGT1A*) | None noted (Burchell and Coughtrie [1]) |
| Chlorpromazine-induced hepatotoxicity | Fast debrisoquine hydroxylation (*CYP2D6*) | None noted |
| | Slow sulphoxidation (?) | (Watson et al. [2]) |
| Hepatocellular carcinoma | High activity precarcinogen metabolism (*CYP1A*)<br>Low activity carcinogen GSH-detoxication (*GSTM1*) | Increased predisposition in heavy smokers, habitual alcohol consumers and subjects with low $\beta$-carotene E.g. odds ratios increase to:<br>3.54 in heavy smokers with *GSTM1* null and low $\beta$-carotene<br>8.24 in habitual drinkers with *GSTM1* null and low $\beta$-carotene<br>3.15 in heavy smokers with *GSTM1* null and *CYP1A1* msp1 promoter variant<br>(Yu et al. [3]) |

injury observed within the human population, with severe defects in a single pathway rarely responsible exclusively for an adverse outcome. Examples illustrating an increasing complexity of the components contributing towards toxicity are shown in Table 17.1. In considering these processes in greater detail, the remainder of this chapter will illustrate how genetic variability may influence the expression of toxic liver injury and how candidate biomarkers may be selected, drawing particularly on the example of paracetamol-induced hepatotoxicity.

## Metabolism: the first major determinant of toxic potential

Multiple enzymic transformations occur during the metabolism of most xenobiotics and the same enzymes transform different drugs to inert or, occasionally, toxic products. A typical example (Figure 17.1) shows how the production of a toxic metabolite (M1), catalysed by enzyme (*a*), may compete with a detoxifying pathway (*b*) to an inert metabolite (M2). However, under normal circumstances, M1 will be detoxified by a further enzymic step (*c*) to another inert metabolite (M3). No toxicity will be observed when M2 production (via *b*) dominates, or when M3 formation (via *c*) occurs faster than M1 formation (via a). However, excessive M1 production (and potential toxicity) may result if: (i) defective or low activity

Types of drug-induced liver damage

- *Augmented*
  - excessive 'pharmacology'
- *Idiosyncratic*
  - metabolic
  - immunological
- *Chemical*
  - predictable reactivity
- *Delayed*
  - late developing tumours
- *Withdrawal*

Classification from Park *et al.* [4]

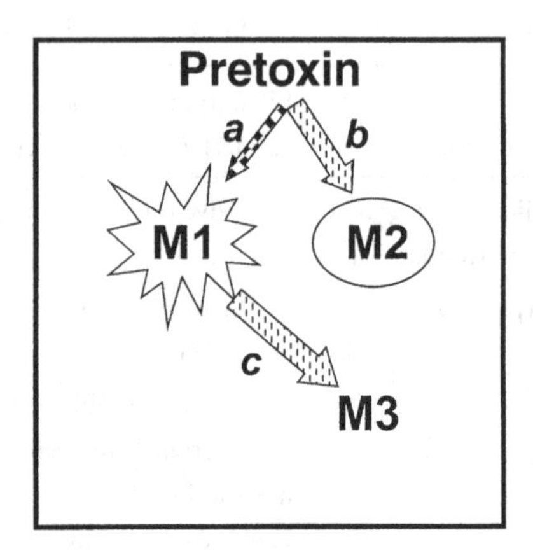


Figure 17.1 The metabolic basis for drug-induced liver injury.

variant enzymes of *b* exist; (ii) defective or low activity variant enzymes of *c* exist; and (iii) if a highly active variant enzyme of *a* prevails. These types of variant enzyme isoforms, and their corresponding genetic variants, are already being identified (and increasingly so) for the xenobiotic metabolizing enzymes catalysing both phase I (e.g. CYP-dependent oxidations) and phase II pathways (e.g. conjugations with glucuronic acid) [1, 5]. There is evidence that they may be responsible for a range of toxic reactions [4]:

- *idiosyncratic hepatotoxicity* – where, in the example above, M1 mediates either a metabolic or an immune cytotoxic response only in the individuals affected;
- *chemical hepatotoxicity* – where only the extent of the toxic response varies from one individual to another; and
- *augmented toxicity* – where defective clearance of the parent compound, by defects in steps *a* and *b* above, elicits an excessive pharmacological response.

Toxicity may be expressed both in the short term (e.g. via extensive necrosis or apoptosis) or over longer periods (e.g. via neoplasia or cirrhosis). Respective examples are shown in Table 17.1. The first is that of a defect in one enzyme, UDP-glucuronosyltransferase (UDPGT), associated with variants in the *UGT1* gene. Toxicity results when defective UDPGT activity in the liver is untreated, causing cerebral impairment due to an accumulation of unconjugated bilirubin in Crigler–Najjar syndrome [1]. In the second example, chlorpromazine hepatotoxicity, two coexisting genetic variants appear to be responsible for the idiosyncratic cholestasis that may result from both immune- and reactive-metabolite-mediated events. The variants are those associated with impaired sulphoxidation and high activity isoforms catalysing CYP2D6 7-hydroxylation [2]. In the third, most complex, example, the development of hepatocellular carcinoma is associated with multiple determinants. These not only include the coexistence of high activity variants of the CYPs responsible for precarcinogen activation and low activity variants of the glutathione *S*-transferase enzymes detoxifying the electrophilic products, but also alcohol consumption, cigarette smoking and dietary intake of $\beta$-carotene [3].

## Cytoprotection: the second major determinant of toxic potential

The progression to cell damage after the production of primary or secondary cytotoxic mediators is not inevitable, nor is it irreversible. A battery of chemical and enzymic cytoprotectants prevails within the liver. These cytoprotectants combat the endogenous production of potential cytotoxins (such as reactive oxygen species [ROS] and cytokines) and reverse their adverse effects (by the repair of oxidized proteins, damaged membranes and methylated DNA). The cytoprotectants include the antioxidant vitamins A, C and D, reduced glutathione (GSH) and numerous chemicals, proteins and enzymes with complementary functions in maintaining redox status with its normal reductive bias. Particularly important are the enzymes detoxifying ROS, e.g. superoxide dismutase, catalase and the glutathione-dependent enzymes. The latter include glutathione peroxidase, which reduces hydrogen peroxide in mitochondria where catalase is absent, and the glutathione transferases which catalyse the detoxication of electrophiles with GSH. Within normal liver, the activity of these multiple, interacting, cytoprotective pathways retains a bias towards continued cellular homeostasis. This prevails despite the deleterious influence of dietary restrictions and minor bacterial/viral infections, and the customary presence of toxic insults of short-term duration and/or mild severity (e.g. ROS produced in mitochondria, peroxisomes and the endoplasmic reticulum, and by activated Kupffer cells). However, the balance will tip towards liver damage if the impairment of cytoprotection is severe and the load of intoxicants exceeds cytoprotective capacity in either the short or long term. Clearly, multiple defects in the cytoprotective armoury coupled with a tendency to an excessive toxin production (or secondary response) will predispose individuals to damage. Genetic variants recognized in detoxication pathways are being increasingly associated with such predisposition.

### Genetic variants may encode inactive or superactive gene products

It should be stressed that the presence of a genetic variant may be associated with extremes of phenotypic expression; in one case, this may result in inactive gene products, while, in another, superactivity results. This can equally apply to the opposing pathways producing cytotoxins as well as to those maintaining cytoprotection. For example, single nucleotide polymorphisms (SNPs) are often associated with gene products of reduced activity or function when expressed as truncated or aberrant proteins/enzymes, but may result in higher activity if they affect gene promoters or prolong the half-life of the mRNA or protein encoded. Likewise, entire genes may be absent or replicated, with contrasting consequences to phenotype and a further divergence in effect when affecting intoxication rather than detoxication.

## Selection of toxicogenetic biomarkers: lessons from paracetamol toxicity

Paracetamol (acetaminophen)-induced hepatotoxicity provides a useful illustration of the potential for toxicogenetic variants to determine outcome following overdose with a drug and these are currently under investigation in the author and colleagues' laboratories. Paracetamol is a safe and effective analgesic when used in the recommended doses (four × 1 g daily) but is considered the archetypal, predictable, iatrogenic hepatotoxin if excessive amounts are ingested [5]. Approximately 150 deaths from acute liver failure occur annually in the UK following paracetamol overdose. Considerable differences exist in the severity of liver damage, as well as in the outcome. For example, there are contrasting case reports of fatality after 6 g [6], but an absence of toxicity with repeated daily doses of 20 g sustained for 5 years [7]. While liver transplantation offers a final treatment modality, the early and accurate prediction of patients requiring this expensive and limited resource are essential. Therefore, prognostic biomarkers that can be used to identify those at risk of an adverse outcome are needed to improve current models. Unlike the current criteria that are based predominantly on initial liver damage (see Chapter 18), these should take into account cytoprotective capacity or secondary responses to the initial toxic event. This is particularly so in severe cases where multiorgan failure ensues and death frequently results for extrahepatic events such as cerebral oedema.

### Practical issues in selecting toxicogenetic biomarkers

The discussion above suggests that the selection of toxicogenetic biomarkers of outcome after paracetamol overdose should be based on key metabolic steps in the intoxication and detoxication of paracetamol and on events perceived to be critical in mediating secondary responses to paracetamol-induced hepatic necrosis. This candidate gene approach contrasts with the alternative technique of whole genome scanning of chromosomal DNA, which is often used to identify areas linked to diseases of interest. While the candidate gene approach is frequently applied to comparisons between selected populations including families, genome scanning almost exclusively considers families with the inherited disorder. Because of the relative rarity of hepatic drug reactions, only the candidate gene approach is usefully applied in toxicogenetics. Usually, case-control studies examine genetic associations in the group of interest and an 'unaffected' or healthy control cohort. In such cases, genetic polymorphisms are analysed and a comparison made between the distribution/frequency of the different alleles in adequate numbers of affected cases and this appropriately matched control group.

Gene frequencies are usually compared by means of nonparametric chi-square or Fisher's exact probability tests. An adjustment of *p* values (e.g. Bonferroni's correction) must be applied where multiple alleles are tested. Frequencies may be compared as *relative risks* (the ratio of the proportion of cases with the gene of interest in the test and control groups) or as *odds ratios* (the ratio in each of the two groups of the affected and unaffected cases). Odds ratios approach relative risks at low gene frequencies and their interpretation may vary with sample size and the frequency at which variants normally occur [8, 9]. These same discussions emphasize the value of corroborating proposed associations in a second population of patients where possible and of confirmatory phenotypic data. Searches for secondary associations may also be made after correction for contributions from the dominant primary association and may be valuable for identifying subgroups.

## Biomarkers of paracetamol intoxication and detoxication

The metabolic basis of paracetamol hepatotoxicity was elucidated over 25 years ago [5]. Figure 17.2 illustrates the kinetic changes in paracetamol disposition that underlie the development of liver damage. These principally involve the rate and extent of production of a reactive metabolite, *N*-acetyl-*p*-benzoquinoneimine (NAPQI), in amounts that exceed the capacity for detoxication by conjugation with GSH. The formation of paracetamol sulphate and glucuronide conjugates represent pathways that compete with paracetamol intoxication to NAPQI. GSH availability limits the arylation of protein sulphydryl groups with NAPQI and the metabolic sequelae which culminate in hepatic necrosis that also include secondary toxicity e.g. via Kupffer cell activation and TNF release. Toxicogenetic biomarkers of paracetamol-induced liver damage might include: (i) those which accentuate intoxication (e.g. production of NAPQI via high activity CYP1A2, 2E1 and 3A4 variants, and mediators of an exaggerated secondary response), or (ii) those that impair detoxication (e.g. defective glucuronide, sulphate or GSH conjugation and reduced GSH synthesis). There is already phenotypic evidence that confirms these biomarkers to be relevant determinants of variability in the response to paracetamol hepatotoxicity.

*Glucuronide and sulphate conjugation* of a therapeutic dose of paracetamol vary complementarily and by 4–5-fold in normal individuals, with urinary recoveries ranging from 13% to 60% and 14% to 52%, respectively [10]. Extensive conjugation was observed in 15.1% and 24.5%, respectively [11]. Paracetamol glucuronidation is catalysed principally by the UGT1A6 isoenzyme of UDPGT, but evidence that the UGT1A1 isoform is also important was presented by De Morais et al. [12].

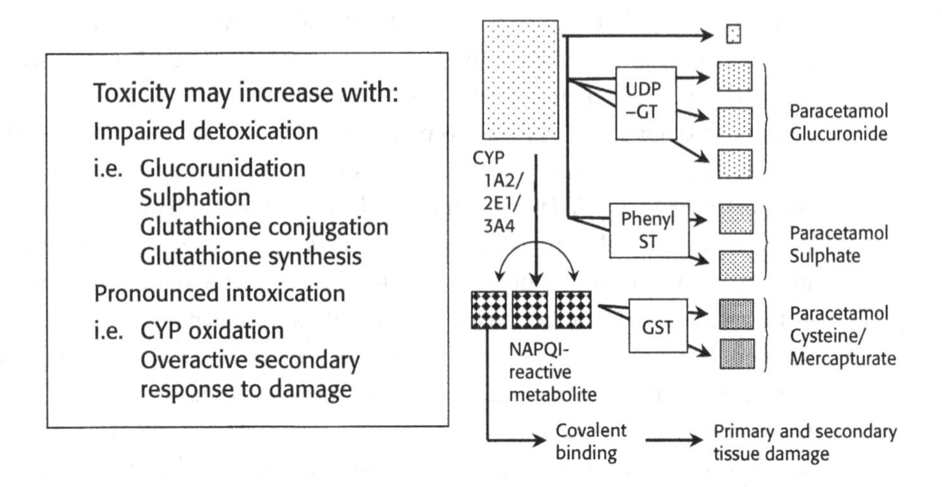


Figure 17.2 Paracetamol hepatotoxicity and candidate pathways for biomarker studies. Abbreviations: CYP, cytochrome P450; GST, glutathione *S*-transferase; NAPQI, *N*-acetyl-*p*-benzoquinoneimine; Phenyl-ST, phenylsulphotransferase; UDP–GT, UDP–glucuronosyltransferase.

These workers showed that subjects with Gilbert's syndrome, an inherited defect of UGT1A1 causing mild unconjugated hyperbilirubinaemia, excreted 31% less paracetamol glucuronide than controls. A potential for greater toxicity was suggested by the 70% increase in excretion of NAPQI-derived GSH conjugates [12]. So far, there is no corresponding evidence for an association between increased susceptibility to paracetamol-induced hepatotoxicity and the recently discovered variants of the phenylsulphotransferases (PSTs) that catalyse conjugation of paracetamol with inorganic sulphate. Some of these PST variants have been shown to exhibit reduced enzymatic activity.

*CYP1A2, 2E1 and 3A4* are the major isoforms of cytochrome P450 which catalyse NAPQI formation and variants that may lead to high enzyme activity (due to promoter polymorphisms) have been detected for all three isoforms (http://www.imm.ki.se/CYPalleles/). A doubling in paracetamol clearance was reported when therapeutic doses were ingested by individuals expressing the corresponding *CYP2E*5B* variant. Paracetamol hepatotoxicity was also strongly associated with the expression of high activity CYP1A variants expressed over 18 generations in inbred mice [13].

The GSTs are, like the CYPs, a superfamily of multiple enzymes in which diverse genetic variants exist within the $\alpha$, $\mu$, $\pi$, $\theta$ and $\zeta$ families. Notable among these are the null phenotypes of GSTM1 and GSTT1, expressed in up to 50% of individuals from different ethnic origins and associated with an increased risk of neoplasms including skin and lung tumours. In a preliminary study, the author and colleagues examined the association between GST $\pi$ variants and outcome in a small cohort of paracetamol overdose patients and detected an odds ratio of 2.4 for an adverse outcome in subjects with the *GSTP1*A* variant. An analogous, unexpected, observation of decreased toxicity has been obtained by Park and colleagues in a GST $\pi$ knockout mouse (B.K. Park, personal communication).

A reduced capacity for conjugation with GSH may also be the result of impaired GSH synthesis. Spielberg and Gordon [14] investigated such a possibility using lymphocytes from individuals with an inherited deficiency in glutathione synthetase, the terminal enzyme involved in GSH formation. When these lymphocytes were incubated with paracetamol, they exhibited more rapid GSH depletion than control lymphocytes from healthy individuals without this defect. This suggested that defective variants in the enzymic steps catalysing GSH synthesis may represent potential biomarkers of susceptibility to paracetamol intoxication yet to be investigated.

An indication of the role of secondary mediators of toxicity in influencing outcome after paracetamol overdose is evident from the work of Bernal et al. [15]. They showed that individuals who produce low amounts of TNF (the *TNF B1B1* variant) were underrepresented among subjects developing severe hepatic encephalopathy following paracetamol overdose. This phenomenon is considered in greater detail in Chapter 16.

### Phenotypic variability

Although a close correlation is usually observed between genotype and the expressed phenotype, modulation of phenotype by environmental factors (e.g. enzyme-inducing agents and physiological regulators) is an important source of diversity that may also need to be considered. Such factors may have profound implications for xenobiotic metabolism. Equally influential may be physiological regulation, with a particularly relevant example to paracetamol hepatotoxicity being the increase in UDPGT activity (via enhanced UDP–glucose turnover) when GSH depletion occurs. Already, genes and their associated variants are being identified that regulate such responses. This development may enhance the potential value of toxicogenetic biomarkers even further.

## REFERENCES

1 Burchell, B. and Coughtrie, M.W.H. (1997). Genetic and environmental factors associated with variation of human xenobiotic glucuronidation and sulfation. *Environmental Health Perspectives*, **105**(Suppl 4), 739–47.

2 Watson, R.G.P., Olomu, A., Clements, D., Waring, R.H., Mitchell, S. and Elias, E. (1988). A proposed mechanism for chlorpromazine jaundice – defective hepatic sulphoxidation combined with rapid hydroxylation. *Journal of Hepatology*, **7**, 72–8.
3 Yu, M.W., Chiu, Y.-H., Chiang, Y.-C. et al. (1999). Plasma carotenoids, glutathione S-transferase M1 and T1 genetic polymorphisms and risk of hepatocellular carcinoma: independent and interactive effects. *American Journal of Epidemiology*, **149**, 621–9.
4 Park, B.K., Pirmohamed, M. and Kitteringham, N.R. (1998). Role of drug disposition in drug hypersensitivity: a chemical, molecular and clinical perspective. *Chemical Research in Toxicology*, **11**, 969–88.
5 Zimmerman, H.J. (1999). *Hepatotoxicity: The Adverse Effect of Drugs and Other Chemicals on the Liver*. Philadelphia: Williams and Wilkins, Lippincott, pp. 30–1.
6 Prescott, L.F. (1996). *Paracetamol (Acetaminophen): A Critical Bibliographic Review*. London: Taylor and Francis.
7 Tredger, J.M., Thuluvath, P., Williams, R. and Murray-Lyon, I.M. (1994). Metabolic basis for high paracetamol dosage without hepatic injury. *Human and Experimental Toxicology*, **14**, 8–12.
8 Anon (1999). Freely associating. *Nature Genetics*, 22, 1–2.
9 Todd, J.A. (1999). Interpretation of results from genetic studies of multifactorial disease. *Lancet*, 354(Suppl 1), 15–16.
10 Caldwell, J., Davies, S. and Smith, R.L. (1980). Inter-individual differences in the conjugation of paracetamol with glucuronic acid and sulphate. *British Journal of Pharmacology*, **70**, 112p–13p.
11 Rona, K., Ary, K., Szuts, I., Kovacs, L. and Gachalyi, B. (1994). Polymorphic paracetamol conjugation: phenotyping in a Hungarian population. *Acta Medica Hungarica*, **50**, 65–74.
12 De Morais, S.M.F., Uetrecht, J.P. and Wells, P.G. (1992). Decreased glucuronidation and increased bioactivation of acetaminophen in Gilbert's syndrome. *Gastroenterology*, **102**, 577–86.
13 Casley, W.L., Menzies, J.A., Mousseau, N., Girard, M., Moon, T.W. and Whitehouse, L.W. (1997). Increased basal expression of CYP1A1 and CYP1A2 genes in inbred mice selected for susceptibility to acetaminophen-induced hepatotoxicity. *Pharmacogenetics*, **7**, 283–93.
14 Spielberg, S.P. and Gordon, J.B. (1981). Glutathione synthetase-deficient lymphocytes and acetaminophen toxicity. *Clinical Pharmacology and Therapeutics*, **29**, 51–5.
15 Bernal, W., Donaldson, P., Underhill, J., Wendon, J. and Williams, R. (1998). Tumor necrosis factor genomic polymorphism and outcome of acetaminophen (paracetamol)-induced acute liver failure. *Journal of Hepatology*, **29**, 53–9.

18

# Prognosis and management of patients with acute liver failure

Fin Stolze Larsen

Rigshospitalet, Copenhagen University Hospital, Denmark

## Introduction

Acute liver failure (ALF) is a devastating disease with a high fatality rate. The clinical picture closely resembles that of septic shock, with arterial hypotension, increased cardiac output and progressive multiorgan dysfunction. Cerebral oedema and intracranial hypertension are the most common causes of death. The pathophysiology of circulatory instability, multiorgan failure (MOF) and cerebral oedema is not fully understood but seems to be related to a decrease in vascular resistance and oxygen extraction, i.e. tissue metabolism. This chapter first defines ALF and then describes the most important aetiological causes of ALF. The features of ALF are then considered in relation to various prognostic markers and disease management.

### Incidence and aetiology

ALF is a rare but dramatic disease that often affects previously healthy and young people. The incidence is ~8 per million in Denmark and is probably similar in other western countries. Recently, it was estimated that there are 2000 cases of ALF each year in the USA.

The single most common cause of ALF is paracetamol intoxication, accounting for ~50% of the ALF patients in UK and Denmark and for ~20% of ALF patients in the USA. Acute viral hepatitis B is the predominant cause of ALF in central and southern Europe, with acute viral hepatitis E a frequent cause of subacute liver failure in India. Other causes of ALF include mushroom intoxication, drug-induced hepatotoxicity (ecstasy, halothane, valproate and disulfiram), autoimmune hepatitis, cardiac failure, and inherited metabolic diseases.

### Prognosis and complications

The prognosis of patients with ALF depends on the time from first symptoms to the development of hepatic encephalopathy (HE), the aetiology and clinical complications of which have been associated with various biomarkers of outcome.

## Time from jaundice to HE

Fulminant hepatic liver failure was defined as a form of ALF with an onset of HE within 8 weeks of the first clinical symptoms of liver disease, e.g. malaise, nausea, right abdominal discomfort or jaundice [1]. For a subgroup of patients with a longer interval to the development of HE, the terms subfulminant hepatic failure [2] or late-onset fulminant hepatic failure were adopted. Although this subgroup of patients may not develop severe HE, the prognosis is as poor as in patients with fulminant hepatic failure where a fatality rate of 82% has been reported. As the time from jaundice to development of HE is of prognostic value, a redefinition of patients with ALF into subgroups with hyperacute, acute and subacute liver failure has been proposed [3].

## Aetiology, sex and age

In a multivariate statistical analysis of more than 500 patients with ALF, aetiology was found to be the single most important independent predictor of outcome [4]. Fatality rates are invariably 100% in ALF due to Wilson's disease and may exceed 70% in cases due to hepatitis nonA–E and drug-induced liver failure without liver transplantation. In contrast, patients with ALF due to acute viral hepatitis A and B have survival rates of 40–70%. Females with acute fatty liver of pregnancy and patients with paracetamol-induced ALF are those with the best prognosis [3, 4].

Patients aged $<10$ years or $>40$–50 years show increased mortality from ALF due to hepatitis [1–4], but gender is of no prognostic value.

## Clinical features

The stage of HE, cerebral oedema, cerebral perfusion pressure and haemorrhage and the presence of renal failure have been considered as important indicators of outcome. However, such events often evolve 'late' in the clinical course of the disease, and may be less useful as prognostic indicators for planning management strategies in individual patients.

### Severe HE

Severe HE is a poor prognostic sign. In two studies, survival fell from 66% in stage II, to 42% in stage III and 18% in stage IV [1] or from 57% in stages I–III to 10% in stage 4 [5]. Data from the author's liver failure unit have shown that 21% of patients in stages I–III survived, but only 11% if there was progression to stage IV [6]. Such progression has a poor prognosis as cerebral oedema and haemodynamic instability become more prominent. However, this prognosis is influenced by the time from the development of jaundice to clinically overt HE. Patients with subacute liver failure and HE stage II may have a worse prognosis than patients with paracetamol intoxication in HE stage IV.

## Convulsions

Convulsions due to evident or subclinical epileptic activity are frequent in patients with liver failure. Development of convulsions is a poor prognostic indicator, with survival figures of 0–6% reported versus ~25% in those without convulsions [5]. Prophylactic administration of phenytoin was recently reported to increase survival, probably by preventing subclinical fitting and cerebral oedema [7].

## Cerebral oedema

Cerebral oedema is one of the most common causes of death in ALF and develops in ~60–80% of patients. Simple elevation of intracranial pressure contains little prognostic information in the author's experience, i.e. a predictive value for death of 0.69 using a cut-off of 40 mm Hg. An alternative is the calculation of the cerebral perfusion pressure (mean arterial pressure minus intracranial pressure) as a possible marker of cerebral blood flow (CBF). In the author's experience, a positive predictive value (for death) of 0.5 prevails using a cerebral perfusion pressure below 40 mm Hg for more than 2 hours, while measurement of cerebral oxygenation status predicts cerebral herniation with a positive predictive value of 0.85 [8]. While supported by data from a large series of patients with ALF, the favourable prognostic value of a high CBF has not been confirmed and is still under evaluation.

## Arterial hypotension, sepsis and lactic acidosis

Despite an increased cardiac output and heart rate in ALF, arterial pressure is reduced. Systolic pressures <80 mm Hg for >1 hour were noted in 82 of 94 patients with ALF [9] and low systemic vascular resistance and cardiac filling pressures are frequent. Systolic arterial pressures are now maintained above 80 mm Hg with volume replacement in most patients with ALF, and this supports the concept that 'peripheral' arteriolar dilatation contributes to the development of arterial hypotension.

## Endotoxins

Endotoxins leak from the gut to portal blood as a result of developing portal hypertension in patients with ALF [10]. Systemic endotoxin levels may rise because portal blood bypasses the liver via intrahepatic shunting, thus decreasing clearance by Kupffer cells. Endotoxins induce the release of a variety of cytokines, including tumour necrosis factor (TNF)-$\alpha$, and the interleukins IL-1 and IL-6, which are potent stimulators of the inducible isoform of nitric oxide (NO) synthetase. In ALF, plasma concentrations of TNF-$\alpha$ and IL-1$\beta$ and IL-6 in the inflammatory host defence system are increased. There is accumulating evidence that the hyperdynamic circulation in ALF may result from cytokine activation of the endothelium, leading to release of excessive amounts of endothelium-derived NO [10]. However, plasma levels of cytokines have no established prognostic value.

### Lactate

Patients who develop an increased mixed venous blood lactate level after adequate volume replacement have a poor prognosis [7]. Mixed venous lactate concentration is inversely correlated with systemic vascular resistance. Arterial hypotension and a low systemic vascular resistance have been suggested to result from the development of arteriovenous shunts [7]. These arteriovenous shunts increase perfusion of non-nutritive capillaries and are considered to be a major reason for the development of tissue hypoxia, lactic acidosis and MOF. It has been suggested that systemic lactic acidosis results from insufficient blood flow especially in the splanchnic bed and brain. However, recent studies of systemic and regional circulation have failed to support this concept of 'pathological supply dependency'. This apparent discrepancy may result from methodological problems because the same technique was used to calculate both systemic oxygen delivery and consumption in the initial studies. Using two independent techniques, Walsh et al. [11] recently demonstrated no profound changes in systemic oxygen consumption and oxygen delivery in patients with ALF during *N*-acetylcysteine infusions. Additional results indicate that both the splanchnic organs and the brain may receive sufficient perfusion for the maintenance of oxidative metabolism. For example, splanchnic blood flow may increase without causing tissue hypoxia [12] and cerebral oxygen consumption remains constant despite increases in CBF after noradrenaline [13]. Thus, it remains questionable if tissue hypoxia is of importance for outcome in patients with ALF.

### Haemorrhage

Following early conflicting data on the adverse effects of gastrointestinal bleeding in ALF, bleeding is presently considered to be of no prognostic value. However, most liver failure units now administer $H_2$ blockers to patients with ALF as a prophylactic step.

### Renal failure

Renal failure is a common complication of ALF and is particularly prevalent after paracetamol overdose, where 75% of patients develop renal failure compared with only 30% in HE stage IV nonparacetamol-induced ALF. A fatal outcome is no longer inevitable if renal failure develops, e.g. 50% survival in patients with paracetamol overdose and 30% in patients with viral hepatitis A and B versus a 100% fatality rate in older studies [10, 14].

## Biomarkers

### Single factors

Any proposed laboratory method for assessing prognosis in patients with ALF should be simple and highly reproducible. The method should also have high predictive accuracy, both for death (and in helping to select patients for liver transplantation) and survival (identifying where a transplant is not required). Of clinical importance is a method that can predict death/survival early in the course of ALF in order to delineate a valuable management plan.

The extent of liver injury is the single most important determinant of outcome for patients with ALF. No single biochemical liver function test can reliably predict outcome, probably because of the diverse homeostatic metabolic functions the liver performs and the similarly diverse features of ALF. However, some of the tests described below provide valuable prognostic information.

## Coagulation factors

Severely compromised synthesis of coagulation factors is both characteristic of ALF and closely related to the severity of the liver damage. Absolute and serial measurements of factor VII were an early candidate for prognostic outcome, either alone or in conjunction with coma grade. More recently, factor V measurements have been favoured, e.g. in cases with ALF due to hepatitis B [2]. However, there was a significant overlap of factor V levels in survivors and nonsurvivors of a paracetamol overdose [15, 16].

## Prothrombin time

The prothrombin time (PT) (either absolute or normalized as the international normalized ratio [INR]) is widely available as a test for coagulation. Multivariate analysis has demonstrated its value as an independent predictor of outcome in all aetiological groups. In patients with ALF due to paracetamol overdose, a poor prognosis was obtained when PT>100 s was considered or when serial changes on days 3/4 after presentation were considered [17]. A rise in PT (or INR) from day 3 to 4 was associated with only 7% survival, while survival was 79% in those patients with no rise in INR.

## $\alpha$-l-fetoprotein

$\alpha$-1-fetoprotein (AFP) was identified as a possible prognostic marker in ALF patients in several smaller studies, but Nusinovici et al. [5] found no predictive value for survival. Mitotic index data demonstrate that regeneration occurs in all patients with ALF, including the subsequently fatal cases, so the rationale for use of AFP is weak and it is not used in the selection of patients for transplantation [6].

## Bilirubin

An elevated serum bilirubin is the single most important indicator of outcome in patients with viral- or drug-induced ALF [18]. Its value is evident irrespective of aetiological group, although death may occur before bilirubin has increased significantly in those patients with hyperacute liver failure, e.g. paracetamol overdose.

**Table 18.1.** Positive (PPV) and negative (NPV) predictive values for some reported prognostic markers in acute liver failure

| Criteria | PPV | NPV | Reference |
|---|---|---|---|
| King's College criteria | 0.98 | 0.82 | [4] |
| | 0.96 | 0.5 | [16] |
| | 0.69 | 0.57 | [20] |
| Clichy criteria | 0.90 | 0.94 | [2] |
| | 0.90 | 0.28 | [16] |
| GEC (<10 μmol/min/kg) | 0.95 | 0.68 | [19] |
| GC (<34 μg/l) | 1.00 | 0.68 | [21] |
| (<100 μg/l) | 0.79 | 0.60 | [20] |
| $NH_3$ (>150 μmol/l) | 1.00 | 0.70 | [22] |

*Notes:*
Abbreviations: GEC, galactose elimination capacity; GC, actin scavenger GC globulin; $NH_3$, plasma ammonia.

### Galactose elimination capacity

Low galactose elimination capacity has been shown to be a good predictor of outcome in patients with ALF [6, 19]. A suggested discrimination limit of 15 μmol/kg/min differentiated groups with 57% or 19% survival (at lower values) in a prospective study. Using a discrimination limit of 10 μmol/kg/min, current findings are of a positive predictive value (for death/need for transplantation) of 0.95 and a negative predictive value (for survival) of 0.68 (Table 18.1).

### Actin scavenger GC globulin

Levels of the actin scavenger GC globulin decrease in patients with ALF in all aetiological groups [20]. Its predictive value for selecting patients for transplantation is nearly 100%, although NPV is less accurate, as with most other biomarkers.

### Nitrogen metabolism

Colombi [23] reported that a plasma ammonia concentration above 200 μg/ml was associated with a 9% survival while survival was 71% at lower ammonia levels. Caroli et al. [24] reported that concentrations >400 μg/ml were inconsistent with survival (19% survived at <400 μg/ml). Others [5, 6] found no difference between survivors and nonsurvivors. However, ammonia content differs between venous and arterial blood because ammonia is converted to glutamine in the muscles and brain and this may account for the apparent discrepancies reported above. Arterial

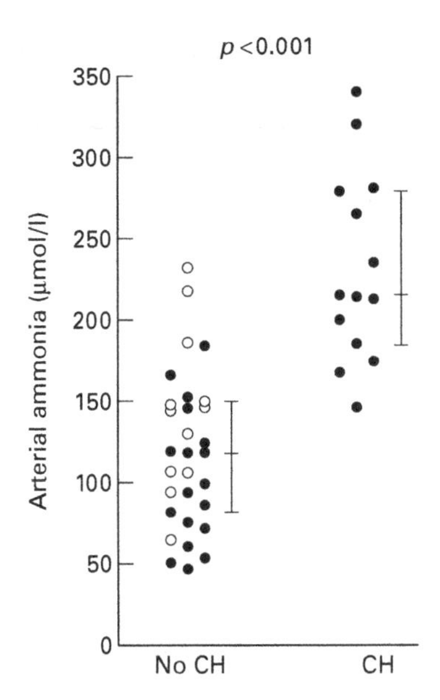


Figure 18.1 Arterial plasma ammonia concentration in 30 patients who did not develop cerebral herniation (no CH) and 14 patients who died from cerebral herniation (CH). Medians and error bars (25th and 75th percentiles) are shown to the right of each group. Closed symbols – patients who underwent liver transplantation; open symbols – patients who died from other causes. Reproduced with permission [22].

blood ammonia >150 μmol/l was recently reported to predict reliably cerebral herniation in patients with ALF (Figure 18.1) [22]. Lower values were associated with a 60% survival. Although needing confirmation, arterial ammonia concentrations >150 μmol/l may be valuable in selecting patients who need intracranial pressure monitoring.

## Combination of factors

Multivariate analysis has been used to identify combinations of markers that enhance the predictive accuracy of those used in isolation. One of the first attempts, based on 21 variables in 33 patients, identified a discriminant score based on age, sex, duration of illness, aetiology, PT, albumin, potassium, glucose, leucocyte count and blood type [18]. Scores of >0.5 were associated with an 86% survival and those of <0.5 with a survival of only 5%. This score was never prospectively evaluated.

The Clichy group [2] demonstrated that the presence of HE and a factor V level below 10% of normal was highly predictive for death. Based on a retrospective analysis of 588 patients with ALF treated at King's College Hospital, the combination of a high PT, pH <7.3, stage of HE and serum creatinine was found to predict the death of patients with paracetamol intoxication (Table 18.1). In patients with

**Table 18.2.** King's College criteria for liver transplantation in patients with acute liver failure

| | |
|---|---|
| *Paracetamol intoxication* | |
| Arterial pH <7.30 after adequate volume expansion (irrespective of stage of hepatic encephalopathy) | |
| *or* | |
| • INR >7 | *and* |
| • Serum creatinine >300 μmol/l | *and* |
| • Hepatic encephalopathy stage 3 and 4 | |
| *All other aetiological causes* | |
| INR >7 (irrespectively of stage of HE) | |
| *or* | |
| Any *three* of the following variables: | |
| • Age <10 or age >40 years | |
| • Aetiology of nonA–nonB hepatitis (nonviral) or idiosyncratic drug reactions | |
| • Duration of jaundice before onset of hepatic encephalopathy >7 days | |
| • INR>3.5 | |
| • Serum bilirubin >300 μmol/l | |

*Notes:*
From O'Grady et al. [4].

other causes of ALF, PT, age, time from jaundice to HE and serum bilirubin – but not serum creatinine – were of prognostic value. For the entire group, irrespective of aetiology, the positive predictive value (for death) was found to be 0.95 (Table 18.1) and this has been largely confirmed subsequently. However, both the King's College and Clichy criteria (Table 18.2) are less valuable for selecting survivors (who do not need a liver transplant), i.e. negative predictive values are around 0.60. Confirmed using other tests, this implies that up to 40% of patients with ALF fulfilling the selection criteria for transplantation may be transplanted, although they may have otherwise survived.

## Management

Transfer of the patient to an experienced intensive care unit at a liver transplantation centre familiar with handling ALF patients should be considered. Patients with hepatic encephalopathy stage 3–4 should be switched to mechanical ventilation with $PaCO_2$ of ~4.0–4.5 kPa. Metabolic acidosis and arterial hypotension should be corrected by aggressive volume expansion. Patients not responding to such volume therapy may benefit from insertion of a pulmonary artery catheter to regulate the infusion rate of noradrenalin. Rarely, a low cardiac output syndrome may develop which is associated with a poor prognosis. Sudden deterioration in arterial pressure and/or systemic vascular resistance often results from the development of sepsis.

In order to preserve normal intracranial compliance, treatment of patients with ALF has traditionally been based on the monitoring of intracranial pressure (ICP) for the calculation of cerebral perfusion pressure (mean arterial pressure minus intracranial pressure) [25]. Cerebral monitoring should focus on cerebral oxygenation in addition to intracranial pressure. Changes in CBF (assuming a constant metabolism) can be monitored by internal jugular vein oxygen saturation ($SvjO_2$). A $SvjO_2$ below 55% is associated with cerebral symptoms both in healthy individuals and patients with liver failure. A prolonged (measured in minutes) decrease in $SvjO_2$ below 55% may result in cerebral hypoxia and oedema. Accordingly, mean arterial pressure should be increased instantaneously by volume expansion and/or noradrenalin. In some patients, CBF increases during the course of ALF possibly due to gradual cerebral arteriolar vasodilatation, and $SvjO_2$ may increase to above 75% [8].

Mannitol infusion is the main treatment for a raised ICP and not only reduces ICP but also increases CBF and metabolism [26]. This effect is probably the result of an increase in the colloid osmotic pressure in the cerebral capillaries and a reduction in the interstitial water content.

Acute hyperventilation decreases CBF, $SvjO_2$ and ICP. Although hyperventilation has been considered inappropriate in intensive care units, hyperventilation is an indispensable and powerful part of the available treatment modalities for intracranial hypertension and re-establishes normal regulation of CBF [25]. Short-term hyperventilation should be instituted to terminate intracranial hypertensive episodes as long as the $SvjO_2$ can be maintained above 55% [8].

Recently, mild hypothermia has been demonstrated to be a new powerful treatment modality to reduce and control intracranial pressure in ALF [27, 28]. Mild hypothermia (33–35 °C) reduces CBF and cerebral blood volume. It probably also restores the normal balance between the Starling forces, so that oedema is prevented. While the rapid and dramatic effect of mild hypothermia on intracranial pressure seems promising, it awaits further study in clinical controlled trials.

In patients with ALF and severe intracranial hypertension, the intravenous injection of indomethacin decreases the intracranial pressure within a few minutes by reducing CBF [29]. Further experimental and clinical trials are underway to establish its indications and side effects.

## Conclusion and perspective

The continuing definition of the distinct pathophysiological mechanisms responsible for cardiovascular instability and cerebral oedema will allow for the evaluation of specific treatments to prevent these devastating complications within the near future. However, predicting the prognosis will continue to be of pivotal importance for optimizing the chance of survival.

A wide range of prognostic factors has been proposed to predict prognosis in patients with ALF and these markers are now adopted to select patients for emergency liver transplantation. Although the King's College criteria are the most widely used prognostic marker for selecting patients for liver transplantation, other proposed criteria seem equally sensitive. It is remarkable, however, that all of these criteria are less valuable for forecasting survival accurately. Thus, new, or at least additional, prognostic markers are needed to improve the negative predictive value of the present set of prognostic criteria.

## REFERENCES

1 Trey, C. and Davidson, L.S. (1970). The management of fulminant hepatic failure. In *Progress in Liver Disease*, eds. H. Popper and F. Schaffner. New York: Grune and Stratton, pp. 282–98.

2 Bernuau, J., Rueff, B. and Benhamou, J. (1986). Fulminant and sub-fulminant hepatic failure. Definition and causes. *Seminars in Liver Diseases*, **6**, 97–106.

3 O'Grady, J.G., Schalm, S.W. and Williams, R. (1993). Acute liver failure: redefining the syndromes. *Lancet*, **342**, 273–5.

4 O'Grady, J.G., Alexander, G.J.M., Hayllar, K. and Williams, R. (1989). Early indicators of prognosis in fulminant hepatic failure. *Gastroenterology*, **97**, 439–45.

5 Nusinovici, V., Crubille, C., Opolon, P., Touboul, J.P., Darnis, F. and Caroli, J. (1977). [Fulminant hepatitis: an experience based on 137 cases. II. Course and prognosis]. *Gastroenterologie Cliniques Biologie*, **1**, 875–86.

6 Tygstrup, N. and Ranek, L. (1986). Assessment of prognostic factors in fulminant hepatic failure. *Seminars in Liver Diseases*, **6**, 129–37.

7 Wendon, J.A. and Ellis, A.J. (1997). Circulatory derangements, monitoring, and management: heart, kidney, and brain. In *Acute Liver Failure*, eds. W.M. Lee and R. Williams. Cambridge: Cambridge University Press, pp. 132–43.

8 Larsen, F.S., Knudsen, G.M. and Hansen, B.A. (1997). Pathophysiological changes in cerebral circulation and oxidative metabolism in patients with acute liver failure. Tailored cerebral oxygen utilization. *Journal of Hepatology*, **27**, 231–8.

9 Trewby, P.N., Hanid, M.A., Mackenzie, R.L., Mellon, P.J. and Williams, R. (1978). Effect of cerebral oedema and arterial hypotension on cerebral blood flow in an animal model of hepatic failure. *Gut*, **19**, 999–1005.

10 Wilkinson, S.P., Arroyo, V., Gazzard, B.G., Moodie, H. and Williams, R. (1978). Relation of

renal impairment and haemorrhagic diethesis to endotoxaemia in fulminant hepatic failure. *Gastroenterology*, **74**, 859.

11 Walsh, T.S., Hopton, P. and Lee, A. (1998). A comparison between the Fick method and indirect calorimetry for determining oxygen consumption in patients with fulminant hepatic failure. *Critical Care Medicine*, **26**, 1200–7.

12 Clemmesen, J.O., Gerbes, A., Hansen, B.A. et al. (1999). Hepatic blood flow and metabolism in patients with fulminant hepatic failure before and after high-volume plasmapheresis. *Hepatology*, **29**, 347–55.

13 Larsen, F.S. (1996). Cerebral circulation in liver failure. *Seminars in Liver Diseases*, **16**, 281–93.

14 Ring-Larsen, H. and Palazzo, U. (1981). Renal failure in fulminant hepatic failure and terminal cirrhosis: a comparison between incidence, types, and prognosis. *Gut*, **22**, 585–91.

15 Pereira, L.M., Langley, P.G., Hayllar, K.M., Tredger, J.M. and Williams, R. (1992). Coagulation factor V and VIII/V ratio as predictors of outcome in paracetamol induced fulminant hepatic failure: relation to other prognostic indicators. *Gut*, **33**, 98–102.

16 Pauwels, A., Mostefa-Kara, N., Florent, C. and Levy, V.G. (1993). Emergency liver transplantation for acute liver failure. Evaluation of London and Clichy criteria. *Journal of Hepatology*, **17**, 124–7.

17 Harrison, P.M., O'Grady, J.G., Keays, R.T., Alexander, G.J. and Williams, R. (1990). Serial prothrombin time as prognostic indicator in paracetamol induced fulminant hepatic failure. *British Medical Journal*, **301**, 964–6.

18 Christensen, E., Bremmelgaard, A., Bahnsen, M., Andreasen, P.B. and Tygstrup, N. (1984). Prediction of fatality in fulminant hepatic failure. *Scandinavian Journal of Gastroenterology*, **19**, 90–6

19 Tygstrup, N., Larsen, F.S. and Hansen, B.A. (1997). Treatment of acute liver failure by high volume plasmapheresis. In *Acute Liver Failure*, eds. W.M. Lee and R. Williams. Cambridge: Cambridge University Press, pp. 267–77.

20 Schiodt, F.V., Bondesen, S., Petersen, I., Dalhoff, K., Ott, P. and Tygstrup, N. (1996). Admission levels of serum Gc-globulin: predictive value in fulminant hepatic failure. *Hepatology*, **23**, 713–18

21 Lee, W.M., Galbraith, R.M., Watt, G.H. et al. (1995). Predicting survival in fulminant hepatic failure using serum Gc protein concentrations. *Hepatology*, **21**, 101–5.

22 Clemmesen, J.O., Larsen, F.S., Kondrup, J., Hansen, B.A. and Ott, P. (1999). Cerebral herniation in patients with acute liver failure is correlated with arterial ammonia concentration. *Hepatology*, **29**, 648–53.

23 Colombi, A. (1970). Early diagnosis of fatal hepatitis. *Digestion*, **3**, 129–45.

24 Caroli, J., Opolon, P., Scotto, J., Hadchouel, P., Thomas, M. and Lageron, A. (1971). [Prognostic factors during coma due to acute hepatic atrophy]. *La Presse Medicale*, **79**, 463–6.

25 Blei, A.T. and Larsen, F.S. (1999). Pathophysiology of cerebral oedema in fulminant hepatic failure. *Journal of Hepatology*, **31**, 771–7.

26 Wendon, J.A., Harrison, P.M., Keays, R. and Williams, R. (1994). Cerebral blood flow and metabolism in fulminant liver failure. *Hepatology*, **19**, 1407–13.

27 Cordoba, J., Crespin, J., Gottstein, J. and Blei, A.T. (1999). Mild hypothermia modifies

ammonia-induced brain edema in rats after portacaval anastomosis. *Gastroenterology*, **116**, 686–93.

28 Jalan, R., Damink, S.W., Deutz, N.E., Lee, A. and Hayes, P.C. (1999). Moderate hypothermia for uncontrolled intracranial hypertension in acute liver failure. *Lancet*, **354**, 1164–8.

29 Clemmesen, J.O., Hansen, B.A. and Larsen, F.S. (1997). Indomethacin normalizes intracranial pressure in acute liver failure: a 23 year-old woman treated with indomethacin. *Hepatology*, **25**, 1423–6.

19

# Biomarkers in artificial and bioartificial liver support

Robin D Hughes
Institute of Liver Studies, Guy's, King's and St Thomas' School of Medicine, London, UK

## Introduction

The original concept of artificial liver support was based on the removal of toxic substances from the systemic circulation in acute liver failure (ALF). A large number of potentially toxic substances have been identified which are either produced in the gut, released from the necrotic liver or result from failure of metabolism in the damaged liver, and these substances lead to coma and the development of multiorgan failure. The early systems were based on adsorbents or dialysis membranes to remove substances directly from blood [1]. Substances such as ammonia, aromatic amino acids and related molecules, fatty acids, mercaptans, bile acids and bilirubin were used as biomarkers to assess the effects of a device on the patient. This type of device only replaced the excretory function of the failed liver and there was thus considerable interest in the development of devices which incorporated biological function, i.e. liver cells with the capability of replacing deficient metabolic and synthetic function. It should be emphasized that it is still not clear what exact missing function is required to promote recovery in the patient. Stimulation of liver regeneration is often suggested to be most important, but progression of the complications of liver failure is often more life threatening, particularly if a raised intracranial pressure develops due to brain oedema. Much effort has been made to develop bioartificial liver support systems to replace all functions of the liver. There are now quite a large number of devices reported in the literature, which in general consist of hepatocytes of different types retained in hollow fibre (bioreactor) membranes [2]. A limited number of systems have been evaluated in patients with liver failure, but there is a paucity of properly controlled data from these studies. The inclusion of a biological component in liver support systems has increased the range of biomarkers that have been used to assess 'liver' synthetic and metabolic function during extracorporeal liver support.

One key question related to bioartificial liver devices is: what quantity of liver cells is required to supply significant functional benefit? The figure of 20% of normal (1500 g) is often considered to be a reasonable target minimum value for survival. This is supported by results from both animal experiments and partial auxiliary liver transplantation, where similar amounts of tissue are transplanted with good clinical effect. Some of the current bioartifical liver devices (see below) contain this amount of hepatocytes, although these will not function as efficiently in the bioreactor as in the highly organized structure of the liver.

## The need for biomarkers

There is a clear need for specific biomarkers that can be used in research studies for the testing of both novel artificial and bioartificial devices. In acute liver failure, there is considerable interest in prognostic markers (see Chapter 18) to predict outcome in patients – particularly in relation to selection for liver transplantation. Transplant criteria currently used at King's College Hospital are based on the presence of encephalopathy, the prothrombin time (alone or standardized as the international normalized ratio [INR]), blood bilirubin and creatinine levels. These and other prognostic markers can be used to evaluate the effect of the device on the patient. When developing a device, in vitro experiments are performed to evaluate the activity of the device, usually with the measurement of basic liver functions such as production of a specific protein, e.g. albumin, metabolism of ammonia to urea or drug metabolizing cytochrome P450 (CYP) activity for lignocaine. Determination of device function in vitro is straightforward, but this becomes more difficult in vivo, whether in animal experiments or in humans due to the competing function of the native liver, even when severely damaged. With artificial systems, it is often possible to look at extraction (clearance) across the device. Otherwise, highly sensitive tests are required to determine function across the device in the extracorporeal circuit. At present, these are not readily available. This may, to some extent, reflect the limited function of the currently available bioartificial devices. The most likely approach is infusion of an isotopically labelled compound into the inlet line of the device with determination of its metabolites in the outlet line. This is less of a problem when xenogeneic hepatocytes are involved, e.g. porcine hepatocytes in humans, where it is possible to determine production of porcine albumin. The converse situation is how to determine the effect of the device on native liver function. The device could remove/metabolize substances with inhibitory effects on liver metabolic function, and produce stimulators or inhibitors of liver regeneration.

Another objective is to produce tests that can be easily used to standardize the performance of different devices and their use at different centres. Finally, the clinical use of a device with its associated extracorporeal circuit can have direct, adverse effects on the patient. These could include haemodynamic changes related to the extracorporeal volume and blood dilution. Activation of blood cells, clotting factors and complement due to contact with polymer membrane surfaces can also have detrimental effects. Patients with acute liver failure have pre-existing abnormalities in many of these parameters, which make them more susceptible to adverse reactions than might be expected.

## Biomarkers in clinical studies

Over the years, a number of liver support systems have been evaluated at King's College Hospital, London. Charcoal haemoperfusion was used extensively, and pilot studies of resin haemoperfusion and different forms of haemodialysis were also undertaken [3]. More recently, two forms of liver support were used in small controlled studies and these will be used to illustrate the use of biomarkers in the treatment of ALF. The BioLogic-DT is a computerized dialysis machine in which the dialysis fluid contains a suspension of powdered charcoal to enhance the removal of toxic substances. Five patients with ALF were treated daily with this machine for up to 5 days with five control patients followed for a similar period [4]. One of the most important symptoms (a clinical biomarker) is encephalopathy, which reflects the overall changes in brain neurochemistry. There is no circulating marker which closely correlates with encephalopathy, although arterial blood ammonia is of considerable relevance, since it features in the brain glutamine theories of the pathogenesis of encephalopathy and oedema [5]. The full range of haemodynamic parameters (mean arterial pressure, cardiac output, systemic vascular resistance index and oxygen delivery and consumption) were monitored in the study with the BioLogic-DT. Blood ammonia, plasma lactate and plasma bile acids were measured to determine the effects of treatment on the patient, but any reductions in levels observed were not significant due to the small numbers of patients in the study. There was a fall in blood fibrinogen indicative of binding of this protein to the dialysis membrane. Another dialysis system which has shown quite promising results for function, particularly in removing protein-bound toxins, is the Molecular Adsorbents Recirculating System (MARS) which has an albumin-enriched dialysis fluid which is regenerated through adsorbents. This machine has marked effects on bilirubin levels, with 20–25% removal [6]. It may be that these types of artificial system are of most use in acute exacerbations of chronic liver disease, particularly in terms of improvement in grade of coma.

The bioartificial liver system which has had greatest clinical use is the bioartificial liver (BAL) developed by Demetriou's group [7] in the USA and which is now produced by Circe BioMedical. Normal pig hepatocytes (50 g) attached to microcarriers are contained in the extracapillary compartment of a hollow fibre membrane. Plasma produced by a centrifugal separator is first perfused through a charcoal column to remove toxic substances, then an oxygenator, followed by the cell bioreactor. Twenty-three patients were treated for 6 hours each day (1–5 times), in most cases as a bridge to liver transplantation [7]. There was a significant fall in the intracranial pressure, an important cerebral parameter as mentioned previously, in association with a fall in blood ammonia. There were significant changes in blood bilirubin as well as other markers of liver function. With devices based on pig hepatocytes, there are a number of concerns – firstly, the functional compatibility of pig and human proteins and, secondly, the possibility of infection with porcine endogenous retroviruses (PERV). However, studies in the BAL-treated patients do not show evidence of PERV infection using polymerase chain reaction (PCR) techniques, which may reflect an immunoisolating effect of the hollow fibre membrane [8].

Another interesting bioartificial system has been developed by Gerlach in Berlin which is based on plasma perfusion of pig hepatocytes (up to 400 g) contained in a complex three-dimensional capillary structure, allowing oxygenation, nutrient supply or inclusion of different types of cells. Eight patients have been bridged to liver transplant with treatment for up to 46 hours continuously [2]. Production of porcine albumin and a reduction in bilirubin were observed.

Controlled trials will determine whether these bioartificial systems have significant clinical benefit.

## The use of the ELAD at King's College Hospital

At King's College Hospital, the author and colleagues have evaluated the ELAD bioartificial liver system, based on a human liver cell line (C3A) which is considered to retain good hepatocyte function and which is contained in the extracapillary compartment of a hollow fibre dialyser [9]. Two cartridges (400 g cells?) in series were perfused continuously with blood from patients with ALF for up to 1 week. Twelve patients were treated with the ELAD and 12 were followed similarly as controls [10]. A series of clinical and biochemical tests was performed to assess the effects of this treatment on the patients. Changes in encephalopathy and haemodynamic variables were used as clinical biomarkers. The intracranial pressure was only measured in a small number of the cases. For metabolic function, arterial blood ammonia resulting from the failure of urea synthesis in the liver was measured using the Ammonia Checker II (Biomen Ltd.) and showed some decrease with ELAD treatment initially, but this was not statistically significant. Blood lactate, which is normally cleared by the liver, was measured using an analyser (YSI Ltd.) and increased slightly in both treatment and control groups. The galactose elimination capacity is a measure of dynamic liver function and is used as a prognostic marker in ALF [11]. Galactose is infused into the patient and the plasma clearance (GEC) determined by enzyme assay (Boehringer Mannheim). A small but significant increase in GEC was seen in the ELAD-treated group in the first 6 hours. The

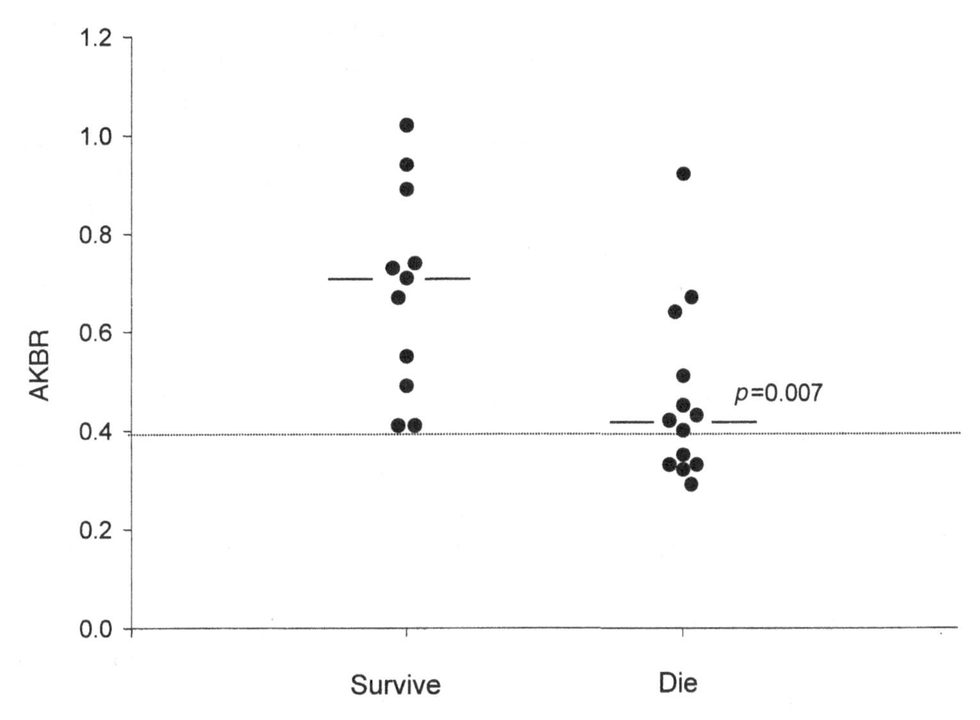


Figure 19.1 The arterial ketone body ratio (AKBR) on admission in acute liver failure patients entered into the ELAD study. Patients who underwent liver transplantation are included as nonsurvivors. Horizontal bars show median values.

arterial ketone body ratio of acetoacetate: $\beta$-hydroxybutyrate (AKBR) developed by Ozawa et al. [12] is considered to reflect hepatic mitochondrial redox state – which is a measure of hepatic energy charge (ATP production). Normal values of AKBR are around 1, with values $<0.4$ denoting patients at risk and those of $<0.25$ indicating fatality. Initial AKBR values assayed using an enzymatic assay (Ketorex kit, Sanawa Co.) in the ELAD patients are shown in Figure 19.1 related to outcome, AKBR being significantly lower in those that died. The difficulty of measuring the function of the device in use was mentioned above, but the oxygen extraction/consumption across it can give a measure of overall cellular metabolic activity. With the ELAD, maximum oxygen extraction was observed after 48 hours of perfusion with a value of about 2.5 ml/min, which, if related to normal human liver values, approximates to a maximum of about 80–90 g of functioning cells. Good oxygenation is an essential component to ensure optimal cell function is maintained, but diffusion of oxygen across the polymer fibres of the bioreactor cartridge may be limiting. Another key measure of liver function is the metabolism of bile acids but measurement of total bile acids in the study showed little change over the first 24 hours in both ELAD-treated and control patients. There is no equivalent of the biliary system in a bioartificial liver device unless suitable adsorbents are incorporated into the circuit, so excretion is dependent on bile flow in the native liver. Similarly, determination of conjugation of bilirubin should be a useful metabolic parameter, which is also dependent on final biliary elimination, but no additional conjugation was observed with the ELAD. In this study, any assays of CYP-dependent drug-metabolizing activity were not performed, which might have been helpful. There are limited substrates for such tests, with the clearance of lignocaine (monoethylglycinexylidide [MEGX]) test often being used, but there were concerns about its possible cardiac effects with repeated use in patients with ALF. This type of function is an area for the development of future tests that can be used in vivo.

Protein synthetic function of the device was monitored mainly with blood clotting tests. Factor V is a recognized and widely used prognostic marker in ALF. No significant effects were observed with the ELAD compared with the controls, but there was a progressive increase in factor V in both treated and untreated patients reflecting the overall survival of around 60% in both groups. The balance of clotting factors to inhibitors is also important, with the tendency to disseminated intravascular coagulation in ALF. Measurements of antithrombin III showed a tendency to decrease in both patient groups, indicating a less rapid recovery in this inhibitory activity. For fibrinogen, a significant reduction was seen in the first 6 hours of treatment, probably reflecting the binding of fibrinogen to components of the perfusion circuit, but this did not persist subsequently. In terms of other markers of the effects of the extracorporeal circuit, plasma levels of the terminal complement component (C5b-9) were determined. There were small increases in the first 6 hours with the ELAD, indicating activation of complement, but these increases were not greatly outside the normal range.

## Cytokines

Cytokine production is increased in ALF and has been associated with the development of the multiorgan complications of the disease. Ideally, a liver support system should reduce increased cytokine levels. Plasma levels of tumour necrosis factor-$\alpha$ (TNF-$\alpha$) and interleukin-6 (IL-6) were both determined by enzyme-linked immunosorbent assay (ELISA) in the patients treated with the ELAD. There were small but significant increases in both cytokines during the first 6 hours of ELAD perfusion, which decreased subsequently towards baseline levels after 48 hours [13]. Release of cytokines has been shown in other extracorporeal circuits including dialysis and cardiopulmonary bypass as a response to blood contact with foreign surfaces and the effect of pumping. It is likely that this explains the increased levels in the patients during ELAD treatment. The situation with regard to the effects of

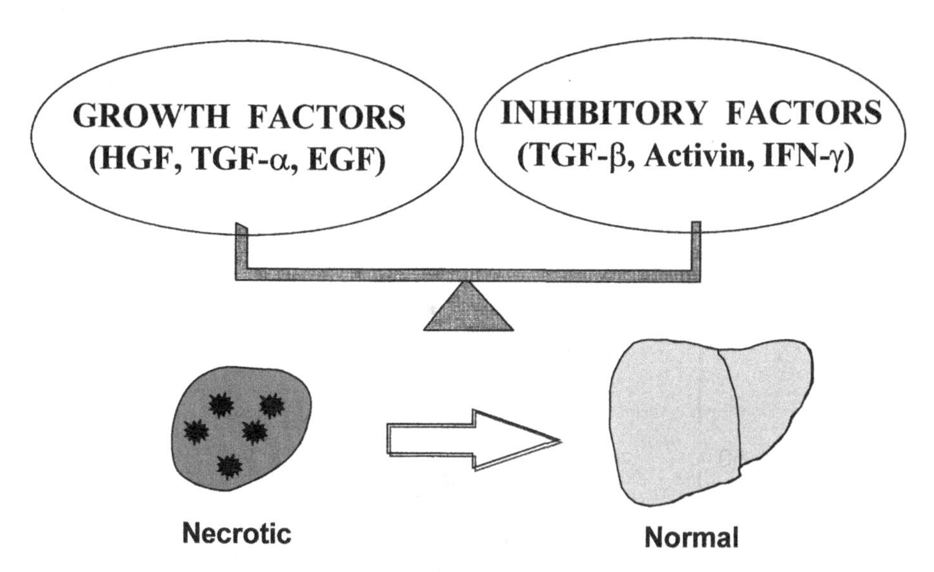


Figure 19.2 The balance of endogenous factors involved in liver regeneration in acute liver failure. HGF is hepatocyte growth factor; TGF-$\alpha$ is transforming growth factor-$\alpha$; EGF is epidermal growth factor; TGF-$\beta$ is transforming growth factor-$\beta$; and IFN-$\gamma$ is interferon-$\gamma$.

cytokines has become more complex with the recent findings that both TNF-$\alpha$ and IL-6 are involved in the early stages of liver regeneration (see Chapter 22). However, excessive production of cytokines as part of an inflammatory response is still likely to be deleterious.

## Regeneration

Recovery from ALF has always been considered to be dependent on adequate liver regeneration with functional recovery being more important than morphological recovery. This may be the result of the balance between endogenous factors promoting, and factors inhibitory to, regeneration (Figure 19.2), although toxic metabolites accumulating in ALF probably have negative effects. Thus, the influence of bioartificial liver support systems on this balance may be of considerable clinical importance. Hepatocyte growth factor (HGF) is one of the most potent mitogens for hepatocyte DNA synthesis. Plasma levels of HGF determined by ELISA are greatly increased (over 50-fold) in ALF, but this is considered to be due more to a lack of clearance rather than the presence of a powerful regenerative stimulus, particularly as it correlates with the degree of liver damage. When plasma levels of HGF were determined in the ELAD study, there was a further 3-fold increase in the first 6 hours of treatment compared with the

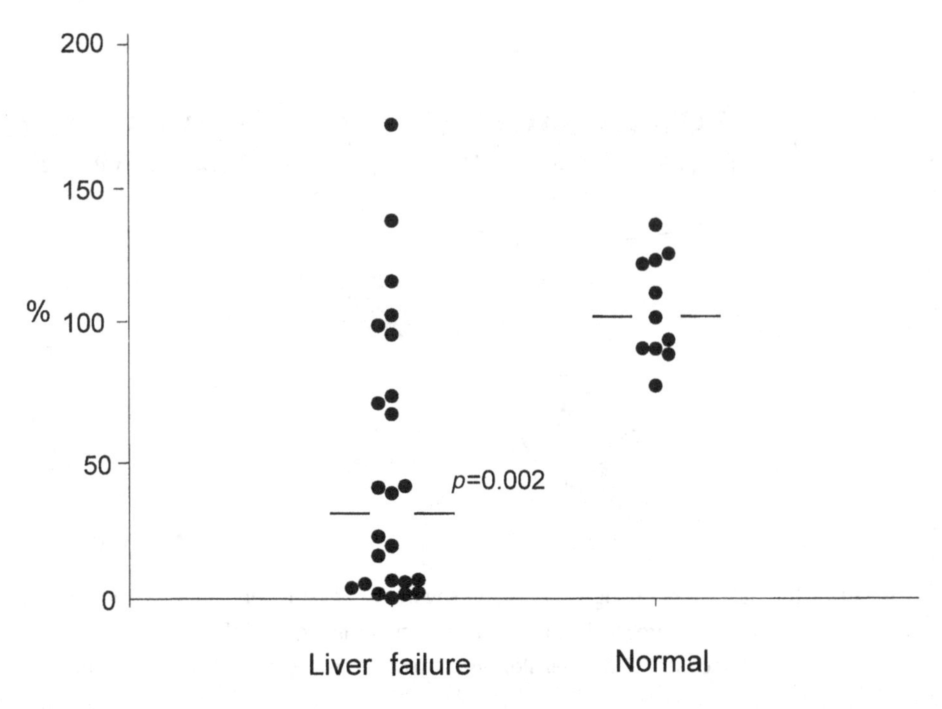


Figure 19.3 Assay to detect inhibitory substances in acute liver failure serum based on incorporation of $^{3}$H-thymidine into HepG2 cells. Horizontal bars show median values [15].

control nonELAD-treated patients [14]. This surprising result was found to be due to the administration of heparin as anticoagulant for extracorporeal circulation, which displaces HGF from endothelial binding sites. It is unclear whether there is any biological significance in further increasing HGF in ALF. Transforming growth factor $\beta_1$ (TGF-$\beta_1$) is an endogenous inhibitor of DNA synthesis which may control liver size. Plasma TGF-$\beta_1$, which is only slightly increased in acute liver failure, remained at similar levels in both ELAD and control patients.

As mentioned above, ALF serum has toxic effects on hepatocytes to which regenerating hepatocytes may be more susceptible. An assay to detect such inhibitory substances was developed based on the incorporation of $^{3}$H-thymidine into human-derived HepG2 cells, which are quite similar to the C3A cells in the ELAD cartridge. Marked inhibition of $^{3}$H-thymidine incorporation (30% of normal sera value) was observed with ALF serum (Figure 19.3), strongly supporting the presence of inhibitory/toxic substances in the circulation [15]. When samples from the ELAD patients were tested in the assay, the inhibitory activity was found to increase rapidly, rather than decrease as was hoped. These changes appeared to mirror the changes in HGF and it is thus possible that they may also be related to the heparinization of the patient. Further experiments are required to understand these effects and determine whether the assay is useful in monitoring treatment with liver support systems.

The data from the use of the ELAD in acute liver failure were obtained with the original design. A second generation version of the ELAD is now undergoing clinical trials in ALF. It incorporates four cartridges instead of two, and the devices are perfused with a plasma ultrafiltrate in a secondary circuit in which there is an oxygenator. The results of these trials should show greater metabolic function than that obtained with the original system and it remains to be seen whether there will be a significant benefit to the survival of patients with ALF.

## REFERENCES


1 Hughes, R.D. and Williams, R. (1986). Clinical experience with charcoal and resin haemoperfusion. *Seminars in Liver Disease*, **6**, 164–73.

2 Busse, B., Smith, M.D. and Gerlach, J.C. (1999). Treatment of acute liver failure: hybrid liver support. A critical overview. *Langenbeck's Archives of Surgery*, **383**, 588–99.

3 Hughes, R.D. and Williams, R. (1996). Use of bioartificial and artificial liver support devices. *Seminars in Liver Disease*, **16**, 435–44.

4 Hughes, R.D., Pucknell, A., Routley, D., Langley, P.G., Wendon, J.A. and Williams, R. (1994). Evaluation of the BioLogic-DT sorbent-suspension dialyser in patients with fulminant hepatic failure. *International Journal of Artificial Organs*, **17**, 657–62.

5 Butterworth, R.F. (1997). Hepatic encephalopathy and brain edema in acute hepatic failure: does glutamate play a role. *Hepatology*, **25**, 1032–4.

6 Stange, J., Mitzner, S.R., Risler, T. et al. (1999). Molecular adsorbent recycling system (MARS): clinical results of a new membrane-based blood purification system for bioartificial liver support. *Artificial Organs*, **23**, 319–30.

7 Arkadopoulos, N., Detry, O., Rozga, J. and Demetriou, A.A. (1998). Liver assist systems: state of the art. *International Journal of Artificial Organs*, **21**, 781–7.

8 Pitkin, Z. and Mullon, C. (1999). Evidence of absence of porcine endogenous retrovirus (PERV) infection in patients treated with a bioartifical liver support system. *Artificial Organs*, **23**, 828–33.

9 Gislason, G.T., Lobdell, D.D., Kelly, J.H. and Sussman, N.L. (1994). A treatment system for implementing an extracorporeal liver assist device. *Artificial Organs*, **18**, 385–9.

10 Ellis, A.J., Wendon, J.A., Hughes, R.D. et al. (1996). Pilot controlled trial of the extracorporeal liver assist device in acute liver failure. *Hepatology*, **24**, 1446–51.

11 Ranek, L., Buch Andreasen, P. and Tygstrup. N. (1976). Galactose elimination capacity as a prognostic index in patients with fulminant hepatic failure. *Gut*, **17**, 959–64.

12 Ozawa, K., Aoyama, H., Yasuda, K. et al. (1983). Metabolic abnormalities associated with postoperative organ failure: a redox theory. *Archives of Surgery*, **118**, 1245–51.

13 Hughes, R.D., Nicolaou, N., Langley, P.G. et al. (1998). Plasma cytokine levels and coagulation and complement activation during treatment with the ELAD in acute liver failure. *Artificial Organs*, **10**, 854–8.

14 Miwa, Y., Ellis, A.J., Hughes, R.D. et al. (1996). Effect of ELAD liver support on plasma HGF and TGF-$\beta_1$ in acute liver failure. *International Journal of Artificial Organs*, **19**, 240–4

15 Anderson, C., Thabrew, M.I. and Hughes, R.D. (1999). Assay to detect inhibitory substances in serum in acute liver failure. *International Journal of Artificial Organs*, **22**, 113–17.

20

# Prognostic markers in liver disease

Martin Burdelski

University Hospital Hamburg Eppendorf, Hamburg, Germany

## Introduction

Acute and chronic end-stage liver disease may be treated successfully by liver transplantation. With organ and patient survival now approaching 80–90% [1], these results represent a considerable improvement compared with the early days of transplantation. As a natural consequence, increasing numbers of disorders have been accepted for transplantation. In order to achieve optimal patient survival and efficient use of limited organ resources, objective criteria are needed for transplant candidate selection. Thus, accurate estimates of prognosis became an essential element in hepatology.

Initially, prognosis was derived from the natural course of a given disorder. In paediatric liver disorders, for instance, there is a rapid progressive course in cholestatic syndromes. The time between diagnosis and death from chronic end-stage liver cirrhosis ranges from months in extrahepatic biliary atresia to a few years in progressive familial intrahepatic cholestasis [2, 3]. This rapid progressive course is not observed in adult liver disorders such as primary biliary cirrhosis (PBC), sclerosing cholangitis (PSC) or hepatitis B and C [4–6] where liver cirrhosis is observed between 10 and 20 years after the initial symptoms.

The option to perform liver transplantation in these disorders changed requirements for the quality of prognostic information. Clinical scores such as the Child–Pugh score in adults and the Malatack score in paediatric patients were developed in order to identify those patients most at risk of dying after being placed on the transplant waiting list [7, 8]. Later, dynamic liver function tests were included in the evaluation of patients in order to improve the quality of the prognostic information available in the pre-transplant setting [9].

After liver transplantation was shown to be effective in acute and chronic end-stage liver disease, a third phase of prognostic evaluation of patients began: the optimal timing of transplantation to avoid wasting organs. It was no longer enough simply to identify a patient unlikely to survive the wait until transplantation and identification was required of those patients who would benefit most with regard to post-transplant survival. These three periods of the development of prognostic markers in clinical hepatology are discussed below.

## Prognosis by diagnosis

The natural course of hepatic disorders is determined by age, gender, the type of the disorder and its biological activity. The resulting prognosis is greatly variable, as shown in Table 20.1. The worst prognosis is observed in paediatric liver diseases such as extrahepatic biliary atresia [2], and some transporter defects [3], and hepatic metabolic disorders [1]. The later the manifestation of a hepatic disorder, the longer its natural course will be. In general, chronic end-stage liver disease in PBC, PSC or hepatitis B and C is only observed after 10–20 years [4, 5]. In acute liver failure, both children and adults have a poor prognosis of only a few days. In both age groups, the absolute prothrombin time (PT) or its international normalized equivalent (INR) have been shown to be the best prognostic predictors: a PT result below 20% of normal indicating almost no chance of spontaneous survival (Table 20.1) [11].

Diagnostic markers have been detected in recent years for various paediatric cholestatic disorders collectively termed progressive familial intrahepatic cholestasis (PFIC) types [1–3]. Underlying defects have been detected in the *FIC1* gene [17], the bile salt export pump (BSEP, formerly SPGP) [18], the multiple drug resistance 3 gene (*MDR3*) [19] and in Alagille's syndrome with a defective jagged 1 protein [20]. In clinical practice, the gene and protein markers identified cannot be used widely since only single laboratories can identify variants in these transporters. However, the transport defects identified have contributed greatly to our understanding of the pathophysiology of cholestasis. The essential serological marker for differentiating cholestatic syndromes – at least in paediatrics – is serum $\gamma$-glutamyltranspeptidase ($\gamma$GT) activity (Table 20.1) [10]. Serum alkaline phosphatase does not play an essential role in paediatrics because there is interference from the bone isoenzyme due to the rapid growth of children. Cholestatic disorders represent two-thirds of the indications for liver transplantation in children. In adults, however, serum alkaline phosphatase activity plays an important role with regard to the diagnosis of PBC (Table 20.1) [4].

## Pre-transplant survival

In the pre-transplant phase, the key issue is whether a patient with chronic end-stage liver disease will survive the wait until a donor organ becomes available. In paediatric patients, the Malatack score has been used to identify the next transplant recipient [8]. However, this score was only suitable for nonPFIC patients since it

**Table 20.1.** Prognostic and diagnostic markers in acute and chronic end-stage liver disease

| Marker | Quality | Disorder | Pre liver Transplant | Post liver transplant | Reference |
|---|---|---|---|---|---|
| $\gamma$GT | Diagnostic | Paediatric cholestasis | Natural course | | 10 |
| ALP | Diagnostic | Adult cholestasis | Natural course | | 4 |
| PT | Prognostic | Acute liver failure | Survival | | 11 |
| Hanover score | Prognostic | Paediatric chronic liver failure | Pre-transplant survival | | 12 |
| Malatack score | Prognostic | Paediatric chronic liver failure | Pre-transplant survival | | 8 |
| Pugh score | Prognostic | Adult chronic liver failure | Pre-transplant survival | | 7 |
| MEGX test | Prognostic | Paediatric and adult chronic liver failure | Pre-transplant survival | | 13,14 |
| Aminopyrine breath test | Prognostic | Adult chronic liver failure | Pre-transplant survival | | 15 |
| Bilirubin, albumin | Prognostic | Paediatric chronic liver failure | | Post-transplant survival | 16 |

considers low cholesterol as a significant covariate in the multivariate analysis. While low cholesterol is a characteristic finding in PFIC, it bears no relationship to liver function. This is a major disadvantage of the Malatack score, since PFIC patients represent the second largest group for liver transplantation.

Another risk score for this age group was the Hanover score, which incorporated serum bilirubin concentrations, PT test results, the activity of serum pseudocholinesterase (CHE) and the body weight. Cut-off values were 300 μmol/l (bilirubin), 50% (PT), 1.5 kU/l (CHE) and the third percentile (body weight) (Table 20.1) [12]. Using this risk score, patients could be stratified as showing a high or low risk of pre-transplant survival. Although there was a marked difference between cholestatic and noncholestatic disorders, the score showed significant differences between survivors and nonsurvivors in both groups after 3 and 18 months, respectively.

In adult patients, the Child–Pugh score still yields useful prognostic information [7]. For specific disorders such as PBC and PSC, disease-specific scores have been developed [4].

Finally, dynamic liver function tests have been used in order to improve the prognostic capacity of liver function tests, sometimes in combination with clinical scores for pre-transplant survival assessment. Such dynamic tests may use compounds with a high to intermediate extraction ratio where their metabolism is dependent on the hepatic blood flow. These include indocyanine green (ICG) (extraction ratio 0.5–0.8), galactose (0.3–0.7) and monoethylglycinexylidide (MEGX) formation from lidocaine (0.5–0.8). Alternative tests using drugs with a low extraction rate (<0.3), such as caffeine, aminopyrine and antipyrine, reflect the metabolic capacity of the liver, being related to the activity of different cytochrome P450 (CYP) isoenzymes [15]. Several prospective trials have been undertaken to evaluate the prognostic value of these tests (Table 20.1) [13].

In adults, the combination of the ICG half-life or MEGX formation 30 min after lidocaine injection with the Child–Pugh risk score suggested that the appropriate combination of a flow-dependent dynamic liver function test result with a risk score could be useful in improving transplant candidate selection and the timing of tranplantation [13]. In paediatric patients, prospective studies have demonstrated that MEGX formation after lidocaine injection (1 mg/kg body weight over 2 min) provided particularly useful prognostic information with regard to pre-transplant survival [14]. The combination of MEGX test results with a clinical risk score did not provide better prognostic information. The serial determination of dynamic (and static) liver function tests has provided useful prognostic information in that a consistently low rate of MEGX formation (<10 μg/l after 30 min) is an unfavourable prognostic sign in these transplant candidates [14].

Since liver blood flow measurements have become possible using noninvasive methods such as Doppler ultrasound, their influence on MEGX formation has been studied in more detail. Despite shortcomings in the accuracy of blood flow measurements using the Doppler technique, there is a disappointing lack of correlation between liver blood flow and MEGX formation in vivo or in microsomal preparations. There is also no correlation of blood flow with the mRNA expression of the CYP responsible for MEGX formation, CYP3A4, or its protein concentration in patients with compensated and decompensated paediatric liver cirrhosis [21]. There is, however, a significant difference between in vivo MEGX fomation, microsomal MEGX formation, mRNA expression of CYP3A4 and its protein concentration in the compensated and decompensated patients. These data suggest that CYP3A4 activity is the major determinant of MEGX formation and that liver blood flow seems to be of lesser importance, despite its role in determinations of the extraction rate (which equals blood flow $\times$ extraction ratio).

## Post-transplant survival

A further change in the demand for prognostic information has followed recent advances in liver transplantation relating to improvements in intensive care, immunosuppression, the diagnosis and treatment of bacterial and virological infections and, not least, innovative surgical techniques. Now it has become necessary not simply to identify the patient with the greatest need for liver transplantation but to predict which patient would be most likely to survive transplantation. This novel prognostic consideration has been studied in several retrospective studies.

In one such study, static liver function tests such as bilirubin and albumin showed significant and independent prognostic value as covariates influencing the hazard function of paediatric post liver transplant survival in the Cox model [16]. A further significant independent predictor of post-transplant survival was the standard deviation score (SDS) for weight. The cut-off values for bilirubin, albumin and SDS weight were calculated at 340 μmol/l, 33 g/l and $-2.2$, respectively. None of the 16 remaining variables used in the model was a significant predictor of post-transplant survival. This included the MEGX test result used as a dynamic liver function test, suggesting that CYP3A4 activity had less influence on post-transplant survival than the excretory function represented by bilirubin.

In conclusion, the relevant markers of liver disorders may be differentiated with regard to diagnostic and prognostic information. Prognostic information needs to be separated further for assessing pre- and post-transplant survival since different qualities of liver function are addressed.

## REFERENCES


1 Burdelski, M., Nolkemper, D., Ganschow, R. et al. (1999). Liver transplantation in children: long-term outcome and quality of life. *European Journal of Pediatrics*, **158**(Suppl), S34–42.

2 Davenport, M., Kerkar, N., Mieli-Vergani, G., Mowat, A.P. and Howard, E.R. (1997). Biliary atresia: the King's College Hospital experience (1974–1995). *Journal of Pediatric Surgery*, **32**, 479–85.

3 Whitington, P.F., Freese, D.K., Alonso, E.M., Schwarzenberg, S.J. and Sharp, H.L. (1994). Clinical and biochemical findings in progressive familial intrahepatic cholestasis. *Journal of Pediatric Gastronterology and Nutrition*, **18**, 134–41.

4 Wiesner, R.H., Porayko, M.K., Dickson, E.R. et al. (1992). Selection and timing of liver transplantation in primary biliary cirrhosis and primary sclerosing cholangitis. *Hepatology*, **16**, 1290–9.

5 Weissberg, J.I., Andres, L.L., Smith, C.I. et al. (1984). Survival in chronic hepatitis B. *Annals of Internal Medicine*, **101**, 613–16.

6 Alter, M.J. and Mast, E.E. (1994). The epidemiology of viral hepatitis in the United States. *Gastroenterology Clinics of North America*, **23**, 437–55.

7 Pugh, R.N.H., Murray-Lyon, I.M., Dawson, J.L., Pietroni, M.C. and Williams, R. (1973). Transection of the oesophagus for bleeding oesophageal varices. *British Journal of Surgery*, **60**, 646–9.

8 Malatack, J.J., Schaid, D.J., Urbach, A.H. et al. (1987). Choosing a pediatric recipient for orthotopic liver transplantation. *Journal of Pediatrics*, **111**, 479–89.

9 Oellerich, M., Raude, E., Burdelski, M. et al. (1987). Monoethylglycinexylidide formation kinetics: a novel approach to assessment of liver function. *Journal of Clinical Chemistry and Clinical Biochemistry*, **25**, 845–53.

10 Maggiore, G., Bernard, O., Hadchouel, M., Lemmonier, A. and Alagille, D. (1991). Diagnostic value of serum γGT activity in liver disease. *Journal of Pediatric Gastroenterology and Nutrition*, **12**, 21–6.

11 O'Grady, J.G., Alexander, G.J.M., Hayllar, K.M. and Williams, R. (1989). Early indicators of prognosis in fulminant hepatic failure. *Gastroenterology*, **97**, 439–45.

12 Burdelski, M., Ringe, B., Rodeck, B., Hoyer, P.F., Brodehl, J. and Pichlmayer, R. (1988). Indications and results of liver transplantation in children. *Kindesalter Monatsschrift fur Kinderheilkunde*, **36**, 317–22.

13 Oellerich, M., Burdelski, M., Lautz, H.-U. et al. (1991). Assessment of pre-transplant prognosis in patients with cirrhosis. *Transplantation*, **51**, 801–6.

14 Burdelski, M., Schutz, E., Nolte-Buchholtz, S., Armstrong, V.W. and Oellerich, M. (1996). Prognostic value of the monoethylglycinexylidide test in pediatric liver transplantation. *Therapeutic Drug Monitoring*, **18**, 378–82.

15 Brockmöller, J. and Roots, J. (1994). Assessment of liver metabolic function. Clinical implications. *Clinical Pharmacokinetics*, **27**, 216–48.

16 Rodeck, B., Melter, M., Kardoff, R. et al. (1996). Liver transplantation in children with chronic end stage liver disease. *Transplantation*, **62**, 1071–6.

17 Bull, L.N., Carlton, V.E.H., Stricker, N.L. et al. (1997). Genetic and morphological findings in

progressive familial intrahepatic cholestasis (Byler disease [PFIC-1] and Byler syndrome): evidence for heterogeneity. *Hepatology*, **26**,155–64.

18 Jansen, P.L.M., Strautnieks, S.S., Jacquemin, E. et al. (1999). Hepatocanalicular bile salt export pump deficiency in patients with progressive familial intrahepatic cholestasis. *Gastroenterology*, **117**, 1370–9.

19 DeVree, J.M.L., Jacquemin, E., Sturm, E. et al. (1998). Mutations in the MDR3 gene cause progressive familial intrahepatic cholestasis. *Proceedings of the National Academy of Sciences USA*, **95**, 282–7.

20 Oda, T., Elkahloun, A.G., Pike, B.L. et al. (1997). Mutations in the human Jagged 1 gene are responsible for Alagille syndrome. *Nature Genetics*, **16**, 235–42.

21 Badur, N., Burdelski, M., Schutz, E. and Oellerich, M. (1999). Hepatic hecNOS and CYP3A4 mRNA expression in chronic pediatric liver failure. *Therapeutic Drug Monitoring*, **21**, 215. (Abstract).

21

# Apoptosis: biomarkers and the key role of mitochondria

Kelvin Cain

MRC Toxicology Unit, University of Leicester, Leicester, UK

## Introduction

It is now widely recognized that apoptosis is a major pathway for the deletion of aberrant or unwanted cells. For many years, research efforts were targeted at identifying morphological and biochemical features that could be used for distinguishing apoptosis from necrosis. This is not a trivial exercise as secondary necrosis can be induced quite often in the surrounding cells as the result of apoptotic cell death. Nevertheless, distinct phenotypic differences between apoptosis and necrosis can be defined, as shown in Table 21.1. Many of these differences now form the basis of selective bioassay markers for apoptosis.

Significantly, it was originally believed that mitochondria were not involved in apoptosis. This is a view that has changed radically in the last few years as new research has demonstrated that mitochondria play a key role in the execution of stress-induced cell death. Another key development in our understanding has been the recognition that the cell death pathway is genetically conserved from the nematode worm *Caenorhabditis elegans* to mammals. The *C. elegans* death gene *ced-3* was discovered to be homologous to the mammalian enzyme, interleukin-1$\beta$-converting enzyme (ICE), that is now known as caspase-1. Since then, a further 13 caspases have been identified which have different and varying roles in apoptosis and inflammation (see [1] for review). It is now widely accepted that the activation of the caspases is central to the apoptotic process and that mitochondrial factors play an important role in caspase activation. While it is beyond the scope of this chapter to review all the relevant literature which support this conclusion, it is helpful to summarize the current status as it highlights the potential for the use of specific biomarkers for apoptosis.

## Caspase activation as the central execution event

Caspases (cysteinyl aspartate-specific proteases) are synthesized as proenzymes/zymogens ($M_r$ of ~30–50 kD), which have a characteristic domain structure

**Table 21.1.** Characteristics of cell death

| Apoptosis | Necrosis |
|---|---|
| Cell shrinkage/plasma | Cell swelling/disruption of membrane/cell blebbing |
| Characteristic nuclear changes | Undefined nuclear/chromatin changes |
| Mitochondria intact | Mitochondria disrupted |
| No release of intracellular contents | Release of cellular contents |
| No inflammation | Inflammation |
| Specific common biochemical changes | Variable biochemical changes |

containing an *N*-terminal prodomain of varying length, followed by ~20 kD and ~10 kD domains which, after cleavage, form the large and small subunits of the active caspase, respectively [2]. X-ray crystallography studies show that the active enzymes are heterotetrameric with two small and two large subunits arranged in a p20:p10:p10:p20 sandwich-like structure. The active site is situated at one end of each p10:p20 interface and the tetramer thus has two active sites in a head-to-tail configuration. The active site is a unique feature of the caspases, and confers cleavage specificity for short tetrapeptide motifs with an absolute requirement for an aspartate in the $P_1$ position. The caspases tolerate large variations in the $P_2$ position while generally preferring a glutamate in the $P_3$ position. The $P_4$ position is the key determinant of specificity and is defined by its corresponding binding site $S_4$ on the enzyme which varies considerably between the caspases [2]. It is on the basis of this specificity that a number of caspase-'specific' substrates and inhibitors have been developed and marketed commercially as tools for apoptosis research.

The tetrapeptide motif specificity defines the pivotal role of the caspases – specific cleavage sites are found in so-called death substrates, e.g. poly-ADP-ribose polymerse (PARP), DNA-protein kinase Cs, Gelsolin and sterol regulatory element-1 binding protein (SREBP) (see [2] for a more comprehensive list). In some cases, such as PARP, the particular protein or enzyme is disabled by cleavage. In others, e.g. caspase-activated deoxyribonuclease and its inhibitor complex (CAD/ICAD) or the Bcl-2 family protein Bid, the cleavage of the protein leads to the activation of a particular apoptotic function (Figure 21.1). The caspases also contain specific tetrapeptide sites, which link the prodomain, large and small subunits and require cleavage autocatalytically or by another caspase. On the basis of their substrate specificity, the caspases can be divided into three groups. Group I caspases (1, 4, 5 and 13) are primarily involved in inflammation. Group II caspases (2, 3 and 7) have short prodomains and prefer the DExD motif (aspartate–glutamate–x–aspartate) which is present in so many caspase targets. Group III caspases

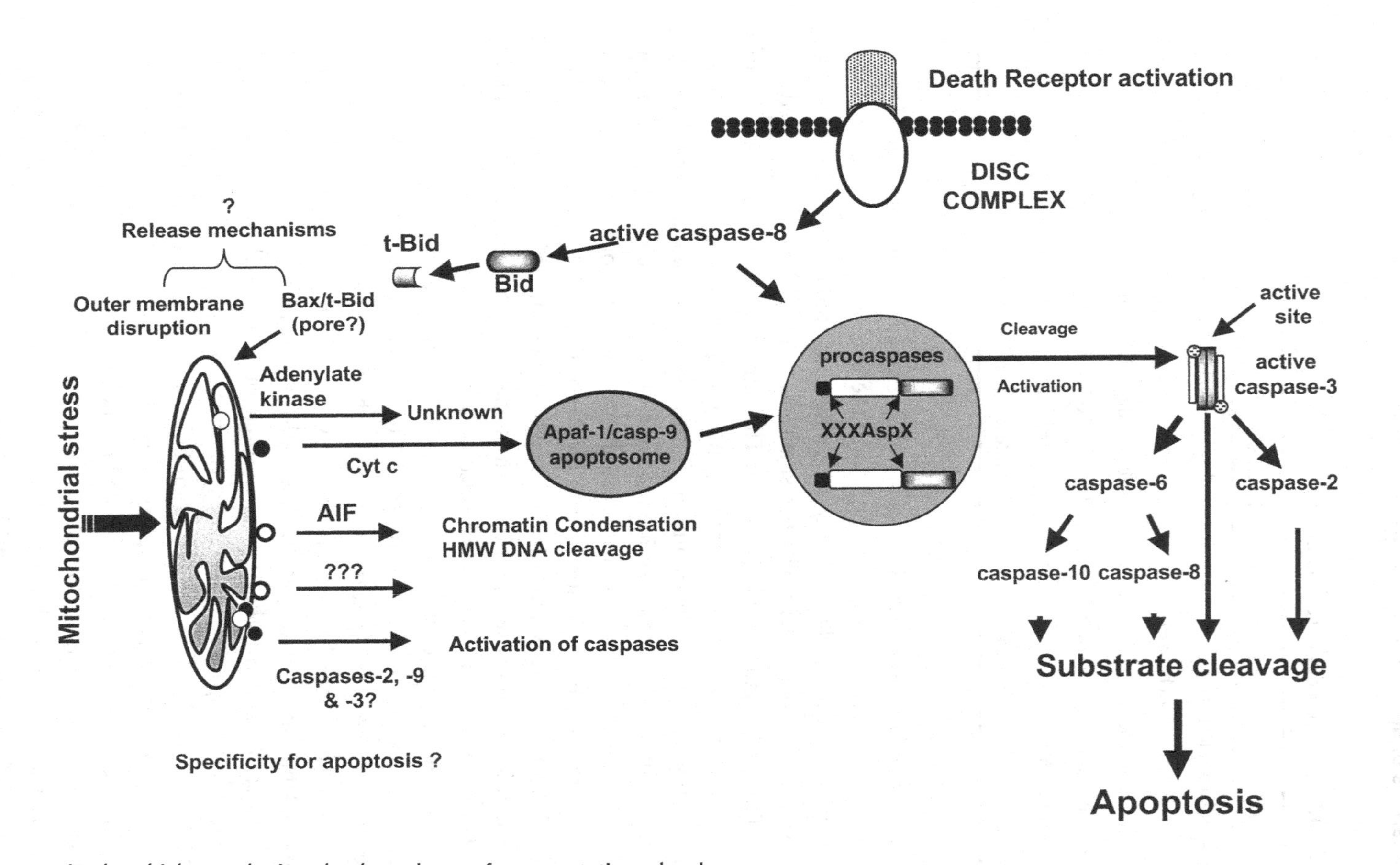


Figure 21.1 Mitochondrial perturbations lead to release of proapoptotic molecules.
Abbreviations: Apaf-1, apoptotic protease-activating factor 1; AIF, apoptosis-inducing factor; Bax and Bid are Bcl-2 family proteins; Cyt c, cytochrome c; DISC, death-inducing signalling complex; HMW, high molecular weight.

(6, 8, 9 and 10) have long prodomains and prefer branched chain aliphatic amino acids such as isoleucine in the $P_4$ position. The latter sequence (IxxD) is the key cleavage motif in the effector (group II) caspases, which results in caspase activation. Thus, caspase activation is essentially a caspase cascade in which initiator caspases are activated and, in turn, process and activate downstream caspases which cleave the essential proteins necessary to dismantle and kill the cell (Figure 21.1) [3]. Caspase activation appears to be specific to apoptosis and thus provides a unique bioassay signature for measuring and quantifying apoptotic cell death.

## Caspase activation and the role of mitochondria

The exact mechanisms of caspase activation are only now being elucidated. As shown in Figure 21.1, the caspases can be activated primarily by two mechanisms: (i) receptor-mediated activation or (ii) mitochondrially dependent activation. In the case of receptor-mediated activation, the plasma membrane contains a number of receptors that are part of the tumour necrosis factor (TNF) superfamily. These receptors, which include CD95/Fas/Apo1, TNFR1, TNFR2, DR3/Wsl-1/Tramp, DR4/TRAIL-R1, DR5/TRAIL-R2/TRICK2/Killer and DR6, have a common structure, with a cysteine-rich extracellular domain and an intracellular cytoplasmic domain known as the death domain (DD). The ligated/activated receptors assume a trimerized configuration. They recruit adaptor proteins such as Fas-associated DD (FADD) to bind to the receptor and a *C*-terminal DED (death effector domain) which binds to a similar motif on caspase-8 (see [3, 4] for review). The formation of this receptor/adaptor/caspase-8 death-inducing signalling complex (DISC) results in the autoactivation of caspase-8, which then cleaves and activates downstream caspases.

In the case of the mitochondrially dependent caspase pathway, the key event appears to be the release of cytochrome c that binds to apoptotic protease-activating factor-1 (Apaf-1), a mammalian homologue of another *C. elegans* death protein, Ced-4. In the presence of dATP/ATP, Apaf-1 oligomerizes to form a large ~700 kD apoptosome complex [5, 6] which recruits and processes procaspase-9. The active caspase-9 now recruits and activates caspase-3. The apoptosome complex is analogous to the DISC, which is formed during receptor-mediated cell death, in that it appears to act as a scaffold complex which can recruit the initiator caspase, thereby allowing contact-dependent autocatalysis and activation. This model also allows for the development of the concept in which the mitochondria are the intracellular receptors of cell damage and initiators of the caspase-activating apoptosome complex. The exact mechanism by which cytochrome c is released is still controversial, but evidence is growing that apoptogenic cytochrome c release can be triggered by proapoptotic members of the Bcl-2 family such as Bid and Bim.

These family members contain only one Bcl-2 (BH) domain that is known as the BH3 domain. Recent evidence suggests that they interact with other proapoptotic Bcl-2 members such as Bax and Bak which then oligomerize in the mitochondrial membrane [7], resulting in outer membrane permeabilization and cytochrome c release.

There is also increasing evidence that other apoptogenic proteins are released during apoptosis. These include apoptosis-inducing factor (AIF), adenylate kinase, caspases -2, -9 and -3 [8], and, more recently, second mitochondria-derived activator of caspase/direct IAP binding protein with low *pI* (Smac/DIABLO) [9]. AIF and caspases 2, 9 and 3 are believed to cause chromatin condensation and high molecular weight DNA cleavage, while Smac/DIABLO may bind to XIAP (inhibitor of apoptosis protein), which inhibits apoptosome function, so minimizing XIAP's inhibitory effects. These findings support the concept that the mitochondria can release proteins that cause apoptotic death. It is possible that the intracellular appearance of these proteins can also be used as a biomarker of apoptosis. However, caution is required because the disruption of mitochondria observed during necrosis will also lead to the release of many mitochondrial proteins without inducing caspase activation and apoptotic cell death.

## Biomarkers of apoptosis

Good assays for biomarkers of apoptosis should be specific, noninvasive, quantitative, speedy and relatively cost-effective. It is probably true that there is no single method that satisfies all these criteria. While all the available methods will not be discussed here, this brief summary will highlight the principles of the various methods and the stage of the apoptotic process they are measuring. In this respect, it is worth emphasizing that an apoptotic cell takes about 2–4 hours to die and to be completely phagocytosed by scavenging cells. The paradigm in Figure 21.2 shows the typical time-scale of the various biochemical and morphological changes which occur in apoptosis. On the basis of this, it is possible to identify potential markers and there are a number of methods now available for assaying these apoptotic biomarkers (Table 21.2).

The very earliest change appears to be cell shrinkage, which can be very easily measured in suspension cultures using fluorescence-activated cell sorting (FACS) technology or instruments based on the Coulter principle. However, shrinkage is much harder, even impossible, to detect easily when using adherent cell cultures or tissues. However, changes in plasma membrane integrity are the key feature of the apoptotic process and can be exploited to assay apoptosis. Phosphatidyl serine (PS) is normally situated on the inner leaflet of the plasma membrane, but very early on in the apoptotic programme this translocates (flips) to the outer leaflet. Annexin V,

**Table 21.2.** Cytoplasmic and membrane bioassay markers for apoptosis

| Apoptotic change | Method of analysis |
|---|---|
| Cell shrinkage | FACS: Coulter |
| Plasma membrane changes (PS exposure) | Annexin V, FACS, fluorescence microscopy |
| Membrane permeability | FACS nuclear dyes, e.g. Hoechst 33342 |
| Nuclear changes | PI, DAPi etc., ISEL (TUNEL), FACS, microscopy<br>DNA cleavage into large (FIGE) and internucleosomal (CAGE) fragments |
| Caspase activation | Enzymic assays, Western blotting<br>Specific protein cleavage, e.g. PARP |
| Ultrastructure | Electron microscopy |

*Notes:*
Abbreviations: CAGE, conventional agarose gel electrophoresis; DAPi, 4′, 6-diamidino-2-phenylindole; FACS, fluorescence-activated cell sorter; FIGE, field inversion gel electrophoresis; ISEL, in situ end-labelling; PI, propidium iodide; TUNEL, terminal deoxynucleotidyl transferase-mediated dUTP nick end-labelling.

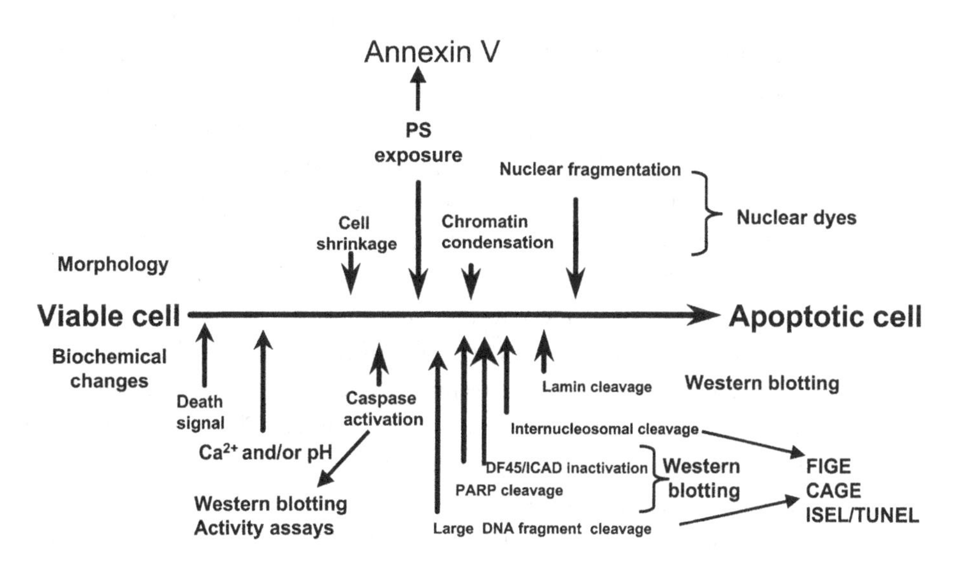


Figure 21.2 Biomarkers of apoptotic cell death: time course of morphological and biochemical events. Abbreviations: CAGE, conventional agarose gel electrophoresis; FIGE, field inversion gel electrophoresis; ICAD, inhibitor of caspase-activated deoxyribonuclease; ISEL, in situ end-labelling; PARP, poly-ADP-ribose polymerase; PS, phosphatidylserine; TUNEL, terminal deoxynucleotidyl transferase-mediated dUTP nick end-labelling.

a phospholipid-binding protein, can then be used to detect the externalized PS, which in normal cells is inaccessible to Annexin V. Necrotic cells, however, are leaky and Annexin V can penetrate the cells and bind to the internal PS. This potential drawback is overcome in most methods by using a fluorescein-tagged Annexin V in conjunction with the DNA-binding dye, propidium iodide, which can only penetrate necrotic cells. This enables differentiation between low green/low red fluorescent (control), high green/low red fluorescent (apoptotic) and high green/high red fluorescent (necrotic) cells. Standard FACS technology can be used to quantify the apoptotic and necrotic cells. The method is quick, reliable, sensitive and can also be used on trypsinized cells. It can be also used on adherent cells where fluorescence microscopy can be used to visualize and count the apoptotic and necrotic cells. The Annexin V method is relatively costly in terms of reagents and hardware, but its many advantages have meant that it has become a very popular and standard method for studying apoptosis in cell culture. In this respect, it has tended to supersede the standard FACS method of assaying for a subdiploid peak as a measure of apoptosis. Other methods have also been employed to exploit changes in the plasma membrane permeability. In some cells, for example, the dye Hoechst 33342 is taken up by apoptotic cells before the uptake of propidium iodide and this property can then be used to differentiate between apoptotic and necrotic cells.

Although increasing evidence suggests that the PS changes are caspase independent, it is clear that the activation of the caspases is an early and critical step in the execution phase of both receptor-mediated and mitochondrial stress-induced apoptosis. In the case of stress-induced apoptosis, an early permeabilization of mitochondrial membranes causes the release of apoptogenic factors such as cytochrome c that can be observed using immunofluorescence- and green fluorescent protein (GFP)-tagged proteins and confocal microscopy [9]. While this methodology is useful for mechanistic studies, it is not an easy bioassay for apoptosis. However, subsequent to the release of mitochondrial apoptogenic factors, there are a series of downstream events that are easily assayed. The first of these, the formation of the apoptosome, has been identified in apoptotic lysates [10] using gel filtration and Western blotting techniques. As yet, this is a time-consuming exercise, which requires significant amounts of material. The activation of the caspases, however, can easily be measured in a variety of ways and provides a distinctive biomarker for apoptotic cells.

The activated caspase activity is best measured using appropriately *C*-terminally tagged tetrapeptide substrates. Both colorimetric (p-nitroaniline) and fluorimetric (7-amino-methylcoumarin [AMC] or 7-amino-4-trifluormethylcoumarin [AFC]) substrates have been developed to measure caspase activity. The product can be measured and quantitated easily, using both conventional and microtitre plate spectrophotometers and fluorimeters. Microtitre plate technology is particularly useful as it allows for the minimum usage of expensive tetrapeptide substrates coupled with high throughput and kinetic measurements. Both whole cell/tissue homogenates or lysates can be assayed. Normal cells have very low caspase activity and it is only when apoptosis is induced that significant increases in caspase activity occur that can be correlated with other markers of apoptosis (e.g. Annexin V binding). The use of specific tetrapeptide motifs for the assay of individual caspases has often been proposed as a method for determining the order of activation of the caspases. However, this approach must be treated with caution because several of the caspases, e.g. caspase-3, can also hydrolyse nonideal peptide motifs. While the rates of hydrolysis may be much slower than for the optimal motif, it must be stressed that there will be widely different amounts of the individual caspases in cell lysates, each contributing to hydrolysis of the substrates to different extents. Also in this context, single end-point analysis of caspase activity should be avoided unless the linearity of the reaction has already been predetermined. Alternative substrates, such as recombinant radiolabelled substrates (e.g. PARP), can also be used, but problems of quantitation, preparation and the widespread prevalence of tetrapeptide substrates argue against their use as the method of choice.

The other methods for demonstrating caspase activation are sodium dodecyl sulphate polyacrylamide gel electrophoresis (SDS-PAGE) and Western blotting, followed by enhanced chemiluminescence (ECL). There are many antibodies currently available which can be used in these assays. Ideally, a caspase antibody should recognize both the proform and at least one of its cleaved products. With an appropriate panel of antibodies, it is possible to detect how many caspases are being activated and this is usually a very convincing demonstration of apoptotic cell death. Quantitation is, at best, poor, however, due to the well-documented problems of nonlinearity of exposed film. The alternative approach with SDS-PAGE and Western blotting is to probe for the cleavage of death substrates, such as PARP and lamins. This approach is more often used to confirm unequivocally that it is apoptotic cell death that is being observed.

Subsequent to caspase activation, the late-stage events of apoptosis are usually assayed by examining the nuclear changes. There are, basically, two distinctive features to be observed: (i) chromatin condensation and (ii) DNA cleavage. Chromatin condensation can be visualized in whole cells and tissue slices, and basically involves fixation of tissue and staining of the nucleus with either fluorimetric or colorimetric dyes. The apoptotic nuclei are easily identified by microscopy as condensed structures with marginated or fragmented chromatin. The method is simple and cheap, but requires skilled operators to count a sufficient number of cells to give good statistical accuracy. DNA cleavage offers two methods of analysis. In one, the DNA is simply analysed on conventional agarose gel electrophoresis (CAGE) or field inversion gel electrophoresis (FIGE) for internucleosomal (laddering) and large fragment DNA cleavage, respectively. These techniques are relatively easily established in the laboratory using published protocols, but a number of companies market kits for this purpose. Quantitation is usually subjective and many studies have used the gel assays to show that the DNA cleavage is proof of apoptotic cell death. Other methods can also be used, e.g. enzyme-linked immunosorbent assays (ELISAs) to measure nucleosome release after DNA cleavage, but these are not totally specific for apoptotic cell death. An alternative and popular method of assaying for DNA cleavage is in situ end-labelling (ISEL) or terminal deoxynucleotidyl transferase-mediated dUTP nick-3-OH end-labelling (TUNEL) of double-strand breaks occurring during apoptosis. This is relatively (not totally) specific and modified X-dUTP nucleotides are incorporated (where X is biotin, digoxigenin or fluorescein) onto the cleaved DNA. The labelled ends are detected/amplified with fluorescent antibodies or with antibodies conjugated with suitable marker enzymes. The labelled cells can then be detected by microscopy or with FACS (for isolated cells). This method is quantitative and, in the case of tissue slices, can be automated with appropriate slide scanning technology. This offers advantages over the conventional condensed nuclei counting method.

## Future and emerging biomarkers

All the methods based on the nuclear changes suffer from the fact that these are end-stage events and it is possible to underestimate the true number of apoptotic cells. Other techniques, such as caspase activity assays and SDS-PAGE analysis, are based on cell lysates or homogenates and, therefore, are usually considering mixtures of apoptotic and normal cells. The emerging technology is very much focused on individual cell analysis. In one such technique, cell-permeable fluorogenic caspase substrates have been developed. These contain 18 amino acids with tetrapeptide recognition sites and rhodamine derivatives covalently attached near their termini [11]. The rhodamine derivatives undergo intramolecular complex formation and cause noncovalent cyclization of the peptide that is nonfluorescent and cell permeant. Within an apoptotic cell, the tetrapeptide motif is cleaved and the liberated rhodamine peptides are highly fluorescent. With appropriate red and green fluorescent rhodamine derivatives, it is possible to visualize the activities of two separate caspases simultaneously (e.g. initiator versus effector caspase). This could be particularly useful for detecting receptor-mediated apoptosis as caspase-8 is the first caspase to be activated.

An alternative approach to screening the activation of the caspases in individual cells is to use antibodies that are specific for the active caspase. This is done by raising an antibody that only recognizes the active tetrameric caspase. The cells/tissues can then be fixed, and analysed by confocal microscopy [12]. This technique also allows analysis of the subcellular distribution of the caspase during apoptosis and can be used to identify potential substrates.

## Conclusions

There are now several reliable and quantitative methods for assaying apoptosis. In cell cultures, the Annexin V method of measuring PS externalization is probably the quickest and easiest method. However, it is always wise to use another method to confirm that apoptosis is indeed occurring. Caspase activation can be demonstrated enzymatically or by immunoblotting and can be used in isolated cells, lysates and homogenates. Caspase activation is in many ways the hallmark of apoptosis and is usually conclusive proof that apoptotic cell death is occurring, either by a receptor-mediated or mitochondrially induced activation pathway. However, when it comes to tissue slices, the most common methods rely on dye or TUNEL staining of the nucleus.

## REFERENCES

1 Thornberry, N.A. and Lazebnik, Y. (1998). Caspases: enemies within. *Science*, **281**, 1312–16.
2 Nicholson, D.W. (1999). Caspase structure, proteolytic substrates, and function during apoptotic cell death. *Cell Death and Differentiation*, **6**, 1028–42.
3 Bratton, S.B., MacFarlane, M., Cain, K. and Cohen, G.M. (2000). Protein complexes activate distinct caspase cascades in death receptor and stress-induced apoptosis. *Experimental Cell Research*, **256**, 27–33.
4 Walczak, H. and Krammer, P.H. (2000). The CD95 (APO-1/Fas) and the TRAIL (APO-2L) apoptosis systems. *Experimental Cell Research*, **256**, 58–66.
5 Cain, K., Brown, D.G., Langlais, C. and Cohen, G.M. (1999). Caspase activation involves the formation of the aposome, a large (similar to 700 kDa) caspase-activating complex. *Journal of Biological Chemistry*, **274**, 22686–92.
6 Green, D.R. (2000). Apoptotic pathways: paper wraps stone blunts scissors. *Cell*, **102**, 1–4.
7 Eskes, R., Desagher, S., Antonsson, B. and Martinou, J.C. (2000). Bid induces the oligomerization and insertion of Bax into the outer mitochondrial membrane. *Molecular and Cellular Biology*, **20**, 929–35.
8 Loeffler, M. and Kroemer, G. (2000). The mitochondrion in cell death control: certainties and incognita. *Experimental Cell Research*, **256**, 19–26.
9 Kroemer, G. and Reed, J.C. (2000). Mitochondrial control of cell death. *Nature Medicine*, **6**, 513–19.
10 Cain, K., Bratton, S.B., Langlais, C. et al. (2000). Apaf-1 oligomerizes into biologically active ~700-kDa and inactive ~1.4–MDa apoptosome complexes. *Journal of Biological Chemistry*, **275**, 6067–70.

11 Komoriya, A., Packard, B.Z., Brown, M.J., Wu, M.L. and Henkart, P.A. (2000). Assessment of caspase activities in intact apoptotic thymocytes using cell-permeable fluorogenic caspase substrates. *Journal of Experimental Medicine*, **191**, 1819–28.

12 MacFarlane, M., Merrison, W., Dinsdale, D. and Cohen, G.M. (2000). Active caspases and cleaved cytokeratins are sequestered into cytoplasmic inclusions in TRAIL-induced apoptosis. *Journal of Cell Biology*, **148**, 1239–54.

22

# Liver regeneration: mechanisms and markers

Nelson Fausto, Jean Campbell

University of Washington School of Medicine, Seattle, Washington, USA

## Introduction

During liver regeneration, quiescent differentiated hepatocytes replicate to restore hepatic tissue. Regeneration can be triggered by the surgical removal of liver tissue or by hepatocyte loss caused by chemical or viral injury. Regardless of the cause, hepatocytes proliferate in a relatively synchronous way to restore the functional capacity of the liver. The most extensively studied model of liver regeneration is that which occurs after removal of two-thirds of the liver (partial hepatectomy). Remarkably, this process, which is referred to as 'regeneration', does not involve true regenerative growth [1]; the hepatic lobes removed by the operation do not grow back. Instead, hepatic mass increases by compensatory hyperplasia of the remaining lobes. These lobes increase in size as a consequence of hepatocyte proliferation and the process terminates when the mass of the enlarged lobes reaches that of the original liver. This chapter will highlight the precise regulatory controls which are activated during this remarkable growth process and identify those key components that may be considered as candidate biomarkers of regeneration. In rodents, 90–95% of hepatocytes replicate within 2 days after partial hepatectomy. DNA replication is preceded by a prereplicative phase in which a large number of genes are activated [2].

## Experimental and clinical features

Liver regeneration is important both from a scientific perspective and clinically. The same growth factors that regulate liver regeneration in rodents also appear to be active in humans. Moreover, the precise regulation of liver growth that occurs in liver regeneration in laboratory animals is also present in human liver transplantation. This is clearly illustrated by the regulated growth of 'small for size' transplants and the lack of growth of 'large for size' transplants. In both cases, the transplanted liver adapts to the functional demands of the new host and reaches a state of quiescence in which both proliferation and apoptosis are negligible. Transplantation from living donors is now frequently used in medical practice. In this procedure, a partial hepatectomy is performed on the donor to remove 50–70% of the liver which is then transplanted to the recipient. The liver remnant in the donor as well as the transplanted organ grow until they reach the optimal liver:body weight ratio for each individual [2].

## Initiating responses

A very large number of genes are either newly expressed or increase their expression after partial hepatectomy [3–5]. Many of these genes, if not the majority, code for proteins that are necessary for liver function and may not be directly related to cell proliferation or to specific biomarkers of proliferation. Nevertheless, it is possible to distinguish separate phases of gene activation in the regenerating liver. Almost immediately after partial hepatectomy, there is an increase in transcription of about 70 genes. This phase, which is referred to as the immediate early gene response, includes the activation of the proto-oncogenes, c-fos, c-jun and c-myc, as well as transcription factors, binding proteins and protein phosphatases. These genes have little in common functionally, but they can all be activated in the absence of protein synthesis. As the process progresses, new sets of genes are progressively expressed in an immediate delayed phase and in a cell cycle progression phase. This latter phase corresponds to the progression of hepatocytes from GI to S in the cell cycle and includes the activation of cyclin D1, mdm2, p53 and p21 among others.

## Growth factors and priming

It would be logical to expect that all of the events of liver regeneration would be orchestrated by growth factors which could act on hepatocytes to bring these cells from quiescence into DNA replication. Indeed, at least two growth factors – hepatocyte growth factor (HGF) and transforming growth factor alpha (TGF$\alpha$) – play important roles in liver regeneration [6]. Surprisingly, however, hepatocytes in the normal liver in situ are relatively insensitive to these growth factors. Infusions of relatively high doses of these factors over a 24 hour period induce proliferation in less than 10% of hepatocytes in rats and mice [7]. Large doses of HGF can induce proliferation of hepatocytes in situ. However, this effect requires supraphysiological amounts of the factor and either prolonged infusion or repeated injections. These observations made it clear that quiescent hepatocytes need to become 'competent' to proliferate. By acquiring proliferative competence, hepatocytes become capable of fully responding to growth factors. The authors and colleagues have referred to the competence process as priming.

To understand the priming process in molecular terms, it is necessary to analyze the patterns of gene expression that occur during the first 2–4 hours after partial hepatectomy. Of great importance in this analysis was the discovery that at least four transcription factors, NF$\kappa$B, STAT3, AP-1 and CEBP/$\beta$, are rapidly activated after partial hepatectomy. NF$\kappa$B and STAT3 activation have been extensively studied and their regulation by the cytokines tumour necrosis factor (TNF) and interleukin-6 (IL-6) has been demonstrated. Some of the salient findings [8–10] describing the role of these cytokines in liver regeneration are: (i) TNF and IL-6 increased both in liver and plasma very shortly after partial hepatectomy; (ii) antibodies to TNF partially block liver regeneration; (iii) knockout mice deficient in TNF receptor type 1 (TNFR-1) or IL-6 have deficient regeneration; (iv) NF$\kappa$B and STAT3 activation is inhibited in TNFR-1 knockouts; and (v) in both TNFR-1 and IL-6 knockouts, IL-6 injection corrects the defect in liver regeneration and restores STAT3 activation. Taken together, the results suggest that liver regeneration has two main restriction points: (i) a priming phase regulated by cytokines in which quiescent hepatocytes move into the cell cycle and (ii) a cell cycle progression phase in which primed hepatocytes respond to growth factors and progress towards DNA replication. It appears that an important stage in cell cycle progression corresponds to the expression of cyclin D1. Activation of this cyclin may correspond to a stage in which hepatocyte proliferation proceeds in an autonomous way and may no longer depend on growth factor activity.

The two most important factors in liver regeneration, HGF and TGF$\alpha$, differ in their mode of production. HGF is produced by liver nonparenchymal cells, as well as by most mesenchymal cells in other tissues. This growth factor is found in an inactive form bound to extracellular matrix components, from which it is released, and circulates in the blood. Both the precursor and active HGF may be present in the blood, but the precursor does not bind to the c-met receptor. Because of these features, HGF can act on hepatocytes both by endocrine or paracrine mechanisms. After partial hepatectomy, HGF concentrations in the blood increase during the first few hours. However, there is no complete agreement about the proportion of active HGF in plasma after partial hepatectomy. HGF mRNA also increases in liver nonparenchymal cells (Kupffer cells, stellate cells and endothelial cells) starting at approximately 6 hours after partial hepatectomy. In contrast to HGF, TGF$\alpha$ is an autocrine factor for hepatocytes – that is, it is produced by hepatocytes which also have the up receptors for this growth (the epidermal growth factor (EGF) receptor) in the plasma membrane. TGF$\alpha$ production is a good marker for hepatocyte proliferation, as demonstrated in studies of liver development, regeneration and carcinogenesis, as well as in hepatocyte cultures. Similarly, the levels of TGF$\alpha$ in the blood show good correlation with regenerative activity in the human liver. Because HGF is produced by many different cells of the body, establishing a correlation between plasma HGF and hepatocyte replication is often difficult. The highest levels of HGF are detected in patients with acute liver failure. In these cases, plasma HGF levels have an inverse relationship with patient survival. Under these circumstances, it is not possible to determine whether high HGF levels in these patients constitute a marker for proliferative activity in the liver or whether they are a consequence of liver damage.

## TNF$\alpha$ and IL-6

The realization that cytokines participate in and may be required for the initiation of liver regeneration provides an interesting dilemma. Both TNF and IL-6 are mediators of the hepatic acute phase response. Moreover, TNF can have a strong cytopathic effect on the liver. How different then are the TNF-activated pathways that may signal such different biological responses? Interestingly, signalling for these responses is through the TNFR-1. Regarding the relationship between an apoptotic or a proliferative response, the key elements that decide the outcome are the amount of reactive oxygen species (ROS) generated by TNF and the activity of antioxidant defences by the hepatocyte. During liver regeneration, ROS do not accumulate and are channelled through NF$\kappa$B activation. In contrast, whenever ROS are produced in excess or the antioxidant defences of the hepatic cells are compromised, ROS accumulate and produce mitochondrial damage and subsequent apoptosis. The major antioxidant defence of hepatocytes is reduced glutathione. TNF is not by itself an apoptotic agent for hepatocytes. Its apoptotic effect requires blockage of RNA transcription or protein synthesis. Yet, TNF by itself causes apoptosis in hepatocytes with decreased glutathione levels.

## Termination

A major puzzle in studies of liver regeneration is understanding how the process terminates at a predictable point. TGF$\beta$ and activins which are expressed in the regenerating liver have been proposed as potential candidates for 'shut-off' agents. Nevertheless, it is equally possible that cessation of proliferation is determined very early in the process by an interplay between cell cycle stimulation and inhibitory genes which are activated during liver regeneration.

## Conclusions

Progress is being made on identifying the major protagonists initiating, sustaining and terminating the process of regeneration. These include the proto-oncogenes c-fos, c-jun and c-myc, cell cycle regulators such as cyclin D1, mdm2, p53 and p21, growth factors (especially HGF and TGF$\alpha$), components of the priming response

(NF$\kappa$B, STAT3, AP-1 and CEBP/$\beta$), cytokine regulators (especially IL-6 and TNF$\alpha$) and potential mediators terminating the process (TGF$\beta$ and activins). In terms of the value of these regulators as biomarkers of the regenerative process, no single component so far demonstrates the specificity required. Combinations of, and sequential changes in, these potential candidates currently offer the greatest value as surrogate markers of regeneration.

## REFERENCES

1 Fausto, N. and E.M. Webber. (1994). Liver regeneration. In *The Liver: Biology and Pathobiology*, eds. I. Arias, J. Boyer, N. Fausto, W. Jakoby, D. Schachter and D. Shafritz. New York: Raven Press Ltd, pp. 1059–84.
2 Fausto, N. (2000). Liver regeneration. *Journal of Hepatology*, **32**, 19–31.
3 Diehl, A.M. and Rai, R.M. (1996). Regulation of signal transduction during liver regeneration. *FASEB Journal*, **10**, 215–27.
4 Fausto, N. (1996). Hepatic regeneration. In *Hepatology: A Textbook of Liver Disease*, eds. D. Zakim and T.D. Boyer. Philadelphia: W.B. Saunders, pp. 32–58.
5 Taub, R. (1996). Transcriptional control of liver regeneration. *FASEB Journal*, **10**, 413–27.
6 Michalopoulos, G.K. and DeFrances, M.C. (1997). Liver regeneration. *Science*, **276**, 60–6.
7 Webber, E.M., Bruix, J., Pierce, R.H. and Fausto, N. (1998). Tumor necrosis factor primes hepatocytes for DNA replication in the rat. *Hepatology*, **28**, 1226–34.
8 Akerman, P.P., Cote, S.Q., Yang, C. et al. (1992). Antibodies to tumor necrosis factor-alpha inhibit liver regeneration after partial hepatectomy. *American Journal of Physiology*, **263**(4 Part l): G579–85.
9 Cressman, D.E., Greenbaum, L.E., DeAngelis, R.A. et al. (1996). Liver failure and defective hepatocyte regeneration in interleukin-6-deficient mice. *Science*, **274**, 1379–83.
10 Yamada, Y., Kirillova, I., Peschon, J.J. and Fausto, N. (1997). Initiation of liver growth by tumor necrosis factor: deficient liver regeneration in mice lacking type 1 tumor necrosis factor receptor. *Proceedings of the National Academy of Sciences USA*, **94**, 1441–6.

23

# Determinants of responses to viruses and self in liver disease

Michael P Manns, Petra Obermayer-Straub

Medical School of Hannover, Hannover, Germany

## Induction of autoimmune responses by viral infections

### Current concepts

Infections by hepatotropic viruses are suspected as triggers for autoimmune liver diseases. Evidence for this hypothesis is provided by numerous observations of the manifestation of autoimmune hepatitis (AIH) after infections with hepatotropic viruses. The list of viruses which have been suspected to have triggered AIH is long and includes hepatitis A virus, hepatitis B virus, hepatitis C virus (HCV), Epstein–Barr virus, herpes simplex virus (HSV), human herpes virus 6 and the measles virus [1]. These viruses show profound differences in replication, tissue distribution and clinical outcome of infection, yet all may cause inflammation in hepatic tissues and may activate humoral and cellular defence mechanisms. This chapter considers potential markers of the autoimmune response induced by viral infection, the detection of which employs conventional serological and genomic analysis. Several postulates have considered how infections with hepatotropic viruses may trigger autoimmune reactions and three are considered below.

#### Molecular mimicry

In order to minimize the risk of recognition as nonself by the immune system, it is believed that the sequences of viral proteins evolved to resemble closely domains on host proteins which are protected by tolerance. During an immune response, B cells may be generated which show cross-reactivity towards self-proteins. Also, naive T cells with a potential to bind to peripheral self-peptides presented by major histocompatibility complex (MHC) class I or MHC class II receptors may leave the thymus. Usually, these cells will become anergic by peripheral tolerance mechanisms. However, during a viral infection, these naive T cells may also get activated by cross-reactivity with viral components. In the presence of inflammatory cytokines, these reactions may overshoot a critical threshold level beyond which they can be considered autoimmune reactions. An example of a domain on a viral protein showing a sequence homology of six amino acids with a hepatic host

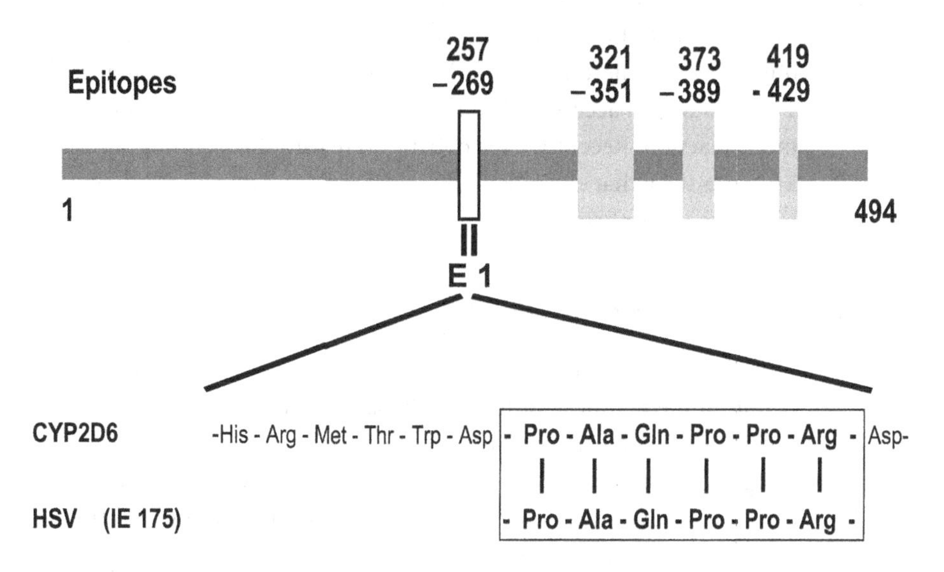


Figure 23.1 The major epitopes of LKM1 autoantibodies in AIH.
A sequence of six amino acids is homologous with the HSV transcription factor, IE 175.

protein is the PAQPPR sequence (Figure 23.1) [2]. This sequence is found on the immediate early protein 175 (IE 175), a transcription factor of HSV, and also on CYP2D6, an enzyme expressed in hepatic microsomes. Interestingly, this sequence is the major autoepitope of liver–kidney microsome autoantibodies (LKM1) autoantibodies, which are detected in patients with autoimmune hepatitis type 2 (AIH-2) [2].

## Bystander activation

In the course of inflammatory reactions, interferon-$\alpha$ (IFN-$\alpha$), proinflammatory cytokines and chemokines are released. Somatic cells, which are not actively involved in these processes, are exposed to these immune modulators and may express MHC class II molecules and activation markers. Such a bystander activation may drive autoreactive T cells towards an (auto)immune response instead of anergy. Clinical experiences with the role of immune modulators are available for IFN-$\alpha$ in humans. IFN-$\alpha$ therapy has been tested in clinical trials in patients with different forms of cancer and is now in use in therapy for chronic hepatitis C. Interestingly, patients receiving IFN-$\alpha$ treatment show an increased prevalence and increased titres of hepatic and thyroid autoantibodies, with manifestations of autoimmune hepatitis noted repeatedly after IFN-$\alpha$ therapy in patients with chronic HCV infection.

### Induction of cross-reactive T cells by superantigens

The third hypothesis on how viruses might induce autoimmune diseases is via unspecific stimulation of large T-cell subsets by viral superantigens. The activated T cells may gain access to the target tissue and initiate a self-perpetuating autoimmune reaction. Recently, a retrovirus was reported which was postulated to express a superantigen that expands a subset of T cells carrying a V$\beta$7 chain [3]. This report has yet to be confirmed.

## Autoimmunity in chronic hepatitis C

Hepatic infection with HCV is known to induce several hepatic and extrahepatic autoimmune manifestations. Extrahepatic manifestations include mixed cryoglobulinaemia, membranoproliferative glomerulonephritis, porphyria cutanea tarda, Sjörgen syndrome and autoimmune thyroid disease [4]. Not surprisingly, numerous autoantibodies are found to be associated with chronic hepatitis C, i.e. antinuclear antibody (ANA), smooth muscle antibody (SMA), LKM-1 and antithyroid autoantibodies [5]. However, in the majority of cases, these antibodies are not markers of real autoimmune diseases. LKM1 autoantibodies, for example, are detected in patients with AIH-2 and in 0–7% of patients with chronic hepatitis C [6]. Patients with LKM1 autoantibodies and chronic HCV infection show clinical features which are typical for an HCV infection and pathogenesis is clearly different from that in patients with AIH. They are usually older than 40 years, inflammatory activity is low and response to corticosteroids is not convincing. HCV patients with antiLKM1 autoantibodies may be treated with IFN-$\alpha$. However, they should be closely followed, since about 10% of those patients may actually have developed a real AIH-2 and hepatitis may be exacerbated under IFN-$\alpha$ treatment [7]. After exacerbation, LKM1-positive patients may be treated successfully with immunosuppressive therapy in spite of concurrent HCV infection [7, 8]. This finding corroborates the coexistence of AIH-2 and HCV infection in these patients.

The role of HCV in the development of AIH remains unclear, but case studies exist which demonstrate the development of LKM1 autoantibodies directly after a primary infection with HCV virus.

## Autoimmunity in chronic hepatitis D

Two RNA viruses causing hepatitis are known: HCV and hepatitis D virus (HDV). In contrast to infections with DNA viruses, both RNA viruses are strongly associated with autoimmune manifestations. While HCV infection is associated with LKM1 autoantibodies, HDV infection is associated with LKM3. These LKM3 autoantibodies are detected in 13% of all patients with chronic hepatitis D and are found also in AIH-2, but there is no overlap of HCV infection with LKM3 autoantibodies. The molecular target of the LKM3 autoantibody was identified as uridine diphosphate (UDP)–glucuronosyltransferases of family 1 (UGT1) [9, 10]. Interestingly, LKM3 autoantibodies recognize a conformation-dependent epitope of amino acids 264–373.

The presence of LKM1 autoantibodies in HCV infection and of LKM3 autoantibodies in HDV infection demonstrates that different hepatotropic viruses are able to induce specific autoantibodies, which are not detected in other virus infections. This specificity cannot be explained by tissue damage and exposure of autoantigens from necrotic tissue to activated macrophages. Specific processes must be postulated, which result in the formation of autoantibodies which are restricted to specific virus infections.

## The role of genetic predisposition in autoimmune diseases

AIH is detected with a mean annual incidence of 1.9 in 100 000 patients [11]. When patients and their first-degree relatives were investigated, 30–40% of them were found to be affected by other autoimmune diseases. The most prevalent concurrent autoimmune diseases were autoimmune thyroiditis, Graves' disease, ulcerative colitis, insulin-dependent diabetes mellitus and vitiligo [12]. This finding indicates that patients with AIH and their first-degree relatives share a genetic predisposition for autoimmune diseases. To date, we know of several factors, such as specific human leucocyte antigen (HLA) alleles, which increase the risk of developing AIH. However, these risk factors cannot account for all patients with AIH and other pathogenic causes must be postulated. Most likely to be involved are other susceptibility genes regulating the immune responses and/or the establishment and maintenance of tolerance. A model gene for such factors is the autoimmune regulator gene (AIRE). Defects in AIRE cause a syndrome characterized by multiple autoimmune diseases in a single patient which is called the autoimmune polyglandular syndrome type 1 (APS1) [13] (see below).

## Known risk factors associated with AIH

### Female gender

Typical for many autoimmune diseases is a strong female preponderance. For AIH, the female:male ratio is about 4:1. Interestingly, in type 1 AIH (AIH-1), there are also two peaks of disease onset, the first peak around puberty and the second around the menopause, indicating that changes in the hormone balance of the patients may increase the risk of developing AIH-1.

### The HLA A1-B8-DRB1*0301 and the HLA DRB1*0401 haplotypes

In addition to female gender, organ-specific autoimmune diseases often associate with specific HLA alleles. A susceptibility marker for AIH-1 in Caucasians is the HLA A1-B8-DRB1*0301 haplotype. These HLA alleles are located in close proximity to each other on the MHC complex and are subject to a strong linkage disequilibrium, forming a tightly linked HLA A1-B8-DRB1*0301 haplotype. Since linkage is strongest with the DR alleles, it is assumed that increased risk associates with the DRB1*0301 alleles. The HLA A1-B8-DRB1*0301 haplotype not only confers an increased risk for developing AIH-1, but it is also associated with a different clinical profile [14]. Patients with the HLA A1-B8-DRB1*0301 haplotype develop AIH at a significantly younger age, they relapse more frequently under immunosuppressive treatment and they are more frequently referred to liver transplantation [14]. In Caucasian AIH-1 patients, the HLA A1-B8-DRB1*0301 haplotype is strongly overrepresented. When all HLA DRB1*0301-positive patients are eliminated from the evaluation, a secondary association of HLA DRB1*04 with AIH-1 is noted. HLA DRB1*0401 is associated with a more favourable clinical course and with the presence of concurrent autoimmune diseases. Those associated autoimmune diseases with the highest prevalences were autoimmune thyroiditis, Graves' disease, ulcerative colitis and pernicious anaemia [15]. The effect of homozygosity was different for HLA DRB3*0101 and HLA DRB1*0401: while homozygosity of DRB1*0401 had no additional effect on the risk of developing AIH-1, homozygosity for DRB1*0301 increased the relative risk from 4.2 to 14.7 and was associated with a reduced 10-year survival.

The genetic basis for the association of certain HLA alleles with an increased susceptibility to AIH is not fully understood, but the HLA alleles DRB1*0301 and DRB1*0401 share a common six amino acid motif, LLEQKR, at positions 67–72 of the DR$\beta$ polypeptide (Figure 23.2). Within this motif, the critical amino acid seems to be a lysine residue located at position 71 on the lip of the antigen-binding groove and possibly influencing the conformation of the antigen-presenting complex [14].

### Polymorphisms in the human complement C4 genes

Early studies showed that AIH-1 was frequently associated with persistently low serum complement levels. C4 phenotyping revealed an association of null allotypes at the *C4A* or the *C4B* loci with AIH-1 and lower serum C4 levels in patients with *C4A* [16]. Furthermore, *C4A* deletions are associated with increased mortality and a higher tendency to relapse during immunosuppressive treatment. Homozygosity for *C4A* deletion correlated with a much stronger increase in the relative risk (RR) for hepatitis (RR = 18.1) than did a single deletion (RR = 3.3) [16]. While the increased risk for AIH-1 in patients with a single gene deletion may be the result of a strong linkage disequilibrium with HLA B3*0101, the strong increase in risk in homozygous patients suggests an additional role for C4 in disease susceptibility.

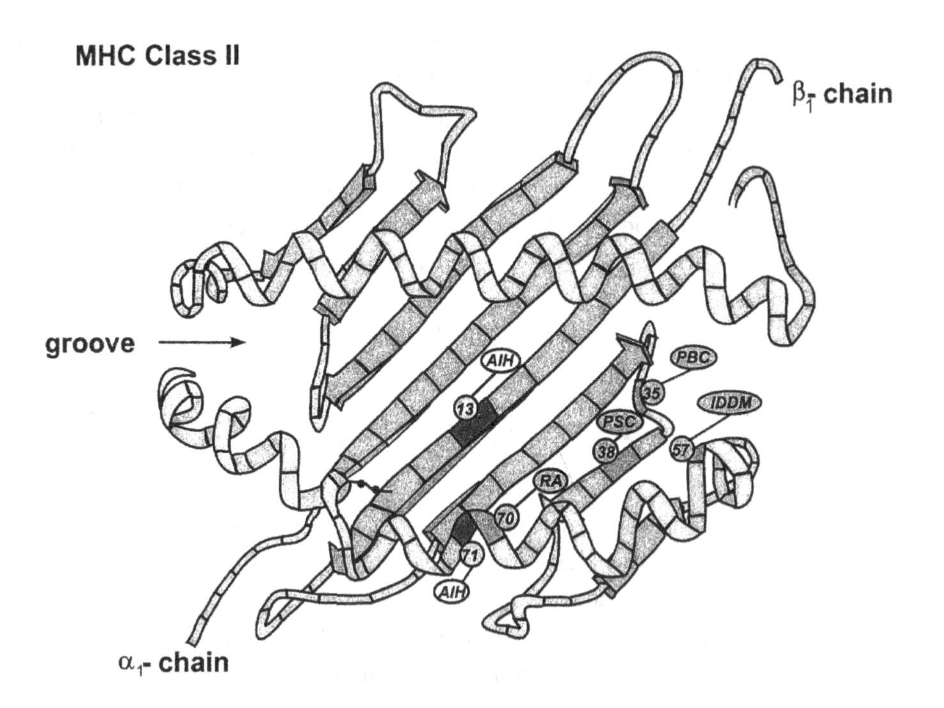


Figure 23.2 Structure of the peptide-binding groove of HLA-DR.
The position of residues that confer increased risk for different autoimmune diseases is shown. Abbreviations: autoimmune hepatitis (AIH): insulin-dependent diabetes mellitus (IDDM); major histocompatibility complex (MHC); primary biliary cirrhosis (PBC): primary sclerosing cholangitis (PSC); rheumatoid arthritis (RA).

## The −308A allele of tumour necrosis factor $\alpha$

Since immune responses are regulated by multiple cytokines, polymorphisms in the induction of cytokine responses will exert profound effects on immune and autoimmune responses. Interestingly, a polymorphism in the promoter region of tumour necrosis factor $\alpha$ (TNF$\alpha$), a −380A allele, was associated with a risk of developing AIH and with an unfavourable clinical course [17, 18]. This polymorphism associates with an increased transcription of TNF$\alpha$. TNF$\alpha$ is encoded in the class III region of the MHC locus and a linkage disequilibrium between the −380A TNF$\alpha$ allele and the A1-B8-DRB1*0301 haplotype exists. It is still unclear whether this linkage equilibrium causes the association of the −308A TNF$\alpha$ allele with AIH or whether this allele represents a separate risk factor for AIH.

## Cytotoxic T lymphocyte antigen 4

Cytotoxic T lymphocyte antigen 4 (CTLA-4) is believed to play an important role in the maintenance of peripheral tolerance by the modulation of T-cell immune responses. Therefore, Agarwal and colleagues [19] investigated the role of CTLA4 polymorphisms in AIH and were able to associate an A to G substitution at position 49 with an increased susceptibility for AIH. However, the relevance of this finding remains unclear, since this substitution results in a threonine to alanine substitution in the leader sequence of this protein, which theoretically should have little effect on post-translational processing or protein localization. This Thr/Ala polymorphism is also, however, in strong linkage with a downstream $(AT)_n$ repeat, which may affect RNA stability. Alternatively, this polymorphism may serve as a linkage marker for an unidentified susceptibility gene for AIH-1.

## AIRE, a model gene for the development of organ-specific autoimmune diseases

A model gene for a general predisposition for organ-specific autoimmune diseases is the autoimmune regulator (AIRE), which was recently cloned. Severe reduction of AIRE function causes a genetic disease, called autoimmune polyendocrine syndrome type 1 (APS1) [13]. Patients with APS1 usually develop mucocutaneous candidiasis at least once in their lives and a broad spectrum of different autoimmune diseases [20]. The most frequent autoimmune diseases are hypoparathyroidism (80%) and adrenal insufficiency (70%). In addition to endocrine disorders, ectodermal dystrophies are frequently noted as well as hepatogastrointestinal dysfunction (Table 23.1). Ten to twenty per cent of patients with APS1 develop AIH.

AIRE is the first gene known outside of the MHC locus which is strongly associated with the development of autoimmune diseases in humans. AIRE encodes a nuclear transcription factor [22], characterized by two plant homeodomain-type (PHD-type) zinc finger motifs, a newly described SAND domain, a putative nuclear targeting signal, a proline-rich region and four LXXLL nuclear receptor-binding motifs [13] (Figure 23.3). The AIRE protein shows colocalization with cytoskeletal filaments and is also found in spotlike domains of the cell [23, 24]. The AIRE protein is not detected in the target cells of autoimmune destruction, but in cells involved in the induction and maintenance of tolerance, i.e. dendritic and epithelial cells of the thymus medulla and dendritic cells of the spleen and lymph nodes [23, 24]. Thirty different mutations have been described to date in patients with APS1, many of which cause frameshift mutations and destruction of at least one of the two PHD-finger motives. For several mutations, it has been shown that mutated AIRE proteins show an altered subcellular localization and that transcriptional activity is altered. While defects in the AIRE coding sequence do not play a major role for AIH [25], AIRE is a model gene for demonstrating that genes involved in T-cell selection and antigen presentation may play a pivotal

**Table 23.1.** Autoimmune diseases caused by the AIRE gene defect

| Disease components | Prevalence (%) |
| --- | --- |
| Mucocutaneous candidiasis | 100 |
| *Endocrine components* | |
| Hypoparathyroidism | 79 |
| Adrenal failure | 72 |
| Insulin-dependent diabetes mellitus | 12 |
| Autoimmune thyroid disease | 2 |
| Gonadal failure | 50 |
| *Hepatogastrointestinal components* | |
| Chronic hepatitis | 12 |
| Parietal cell atrophy | 13 |
| Intestinal malabsorption | 18 |
| *Others* | |
| Alopecia | 29 |
| Vitiligo | 13 |
| Keratoconjunctivitis | 35 |
| Enamel hypoplasia | 77 |
| Nail dystrophy | 52 |

*Notes:*
Data in this table are from Ahonen et al. [21]

role in the development of autoimmune diseases. The AIRE gene defect also demonstrates that it is susceptibility and not the specific autoimmune disease that is inherited with such defects. In spite of a homogeneous gene defect in Finns, in whom 90% share the same disease allele, the clinical spectrum of disease components in AIRE is extremely heterogeneous. Therefore, for a specific basic gene defect, outcome of the clinical disease is modified by several other genes. The genetic basis of these 'risk factors' contrasts with others that may be environmental – e.g. immune reactions induced by virus infections – which may act as triggers for specific autoimmune diseases.

## Summary

AIH is a rare disease of unknown aetiology. Since AIH tends to associate with other autoimmune diseases, a genetic predisposition for autoimmune diseases has to be postulated. On the one hand, this genetic predisposition will consist of known risk

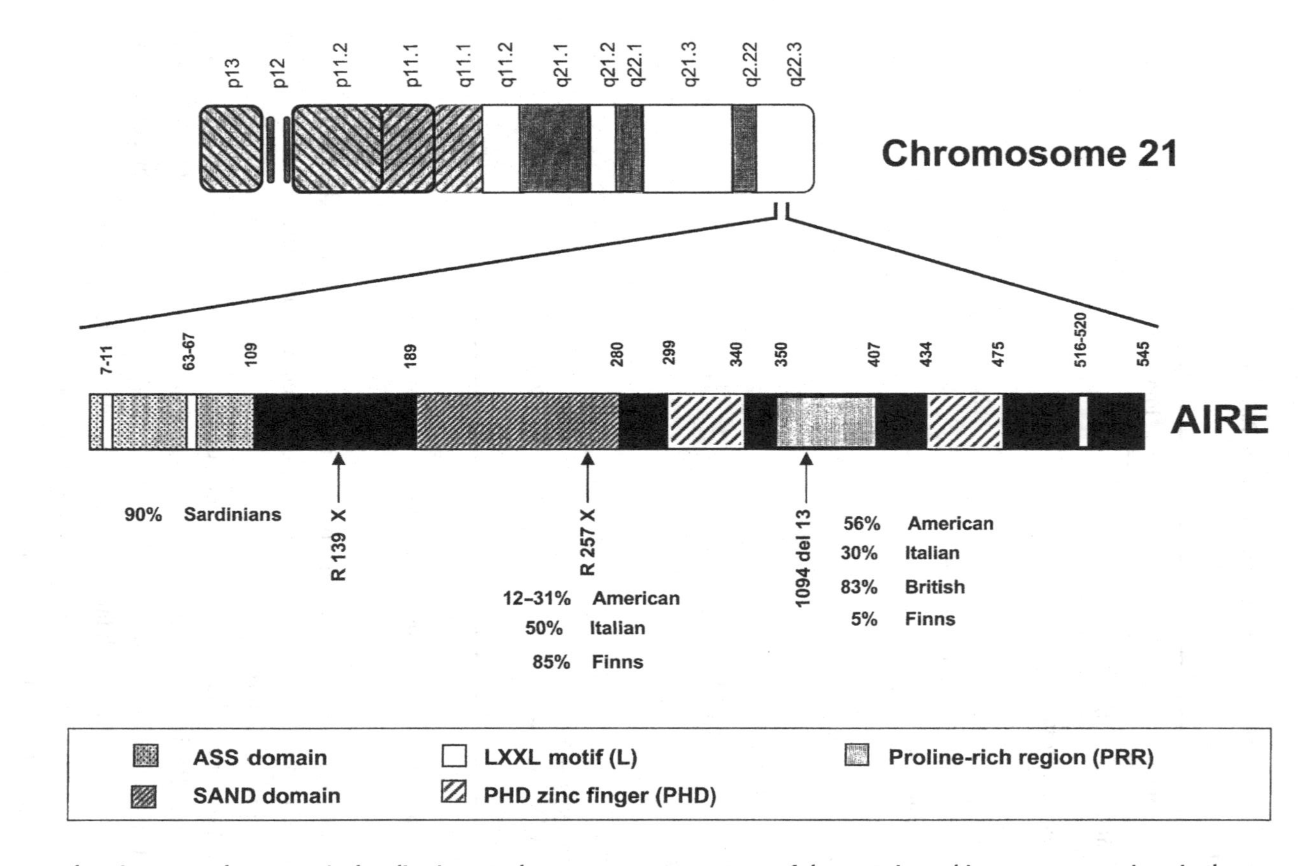


Figure 23.3 The Aire gene: shown are its localization on chromosome 21, structure of the protein and important mutations in the AIRE coding sequence.

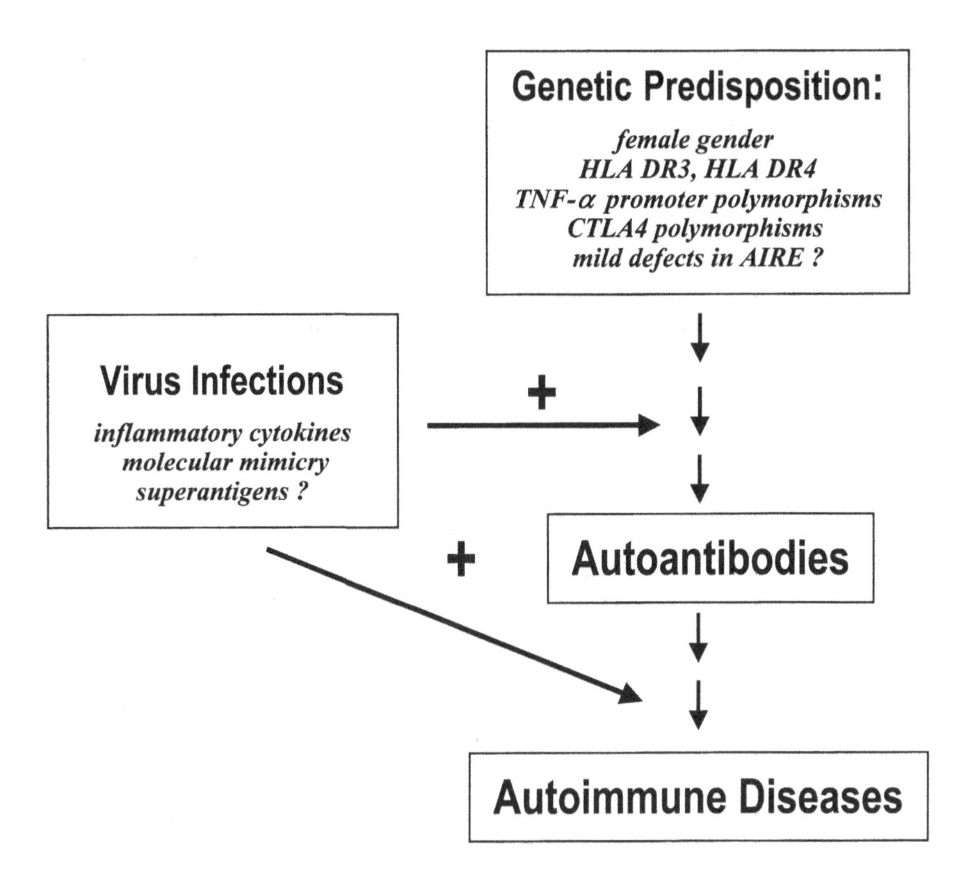


Figure 23.4 Genetic predisposition and the role of virus infections in the development of autoimmune diseases.

factors but, on the other hand, additional genes active in the regulation of tolerance and immune mechanisms may play an important role. Even in identical twins, both siblings are rarely affected. Therefore, in addition to a multigenic genetic predisposition, other triggers of autoimmune reactions have to exist (Figure 23.4). Because infections by hepatotropic viruses are repeatedly reported to have preceded AIH, such virus infections may be the triggers. Consistent with this hypothesis is evidence that infections with the hepatotropic RNA viruses, HCV and HDV, are well known to induce autoimmune manifestations. Additional mechanisms involved in the induction of autoimmune phenomena by viruses may include molecular mimicry, activation of bystander cells and the expression of superantigens by viruses.

## REFERENCES

1 Manns, M.P. and Obermayer-Straub, P. (1997). Viral induction of autoimmunity: mechanisms and examples in hepatology. *Journal of Viral Hepatology*, **4**, 42–7.

2 Manns, M.P., Griffin, K.J., Sullivan, K.F. and Johnson, E.F. (1991). LKM-1 autoantibodies recognize a short linear sequence in P450IID6, a cytochrome P-450 monooxygenase. *Journal of Clinical Investigation*, **88**, 1370–8.

3 Conrad, B., Weissmahr, R.N., Boni, J., Arcari, R., Schupbach, J. and Mach B. (1997). A human endogenous retroviral superantigen as candidate autoimmune gene in type 1 diabetes. *Cell*, **90**, 303–13.

4 Pawlotsky, J.M., Ben Yahia, M., Andre, C. et al. (1994). Immunological disorders in C virus chronic active hepatitis: a prospective case-control study. *Hepatology*, **19**, 841–8.

5 Clifford, B.D., Donahue, D., Smith, S. et al. (1995). High prevalence of serological markers of autoimmunity in patients with chronic hepatitis C. *Hepatology*, **231**, 613–19.

6 Obermayer-Straub, P. and Manns, M.P. (1996). Cytochrome P450 enzymes and UDP-glucuronosyltransferases as hepatocellular autoantigens. *Molecular Biology Reports*, **23**, 235–42.

7 Todros, L., Saracco, G., Durazzo, M. et al. (1995). Efficacy and safety of interferon alpha therapy in chronic hepatitis C with autoantibodies to liver kidney microsomes. *Hepatology*, **22**, 1374–8.

8 Dalekos, G.N., Obermayer-Straub, P., Maeda, T., Tsianos, E.V. and Manns, M.P. (1998). Antibodies against cytochrome P4502A6 (CYP2A6) in patients with chronic viral hepatitis are mainly linked to hepatitis C virus infection. *Digestion*, **59**, S36.

9 Philipp, T., Durazzo, M., Trautwein, C. et al. (1994). Recognition of uridine diphosphate glucuronosyl transferases by LKM-3 antibodies in chronic hepatitis D. *Lancet*, **344**, 578–81.

10 Strassburg, C., Obermayer-Straub, P., Alex, B. et al. (1996). Autoantibodies against glucuronosyltransferases differ between viral hepatitis and autoimmune hepatitis. *Gastroenterology*, **11**, 1582–92.

11 Boberg, K.M., Aadland, E., Jahnsen, J., Raknerud, N., Stiris, M. and Bell, H. (1998). Incidence and prevalence of primary biliary cirrhosis, primary sclerosing cholangitis, and autoimmune hepatitis in a Norwegian population. *Scandinavian Journal of Gastroenterology*, **33**, 99–103.

12 Gregorio, G.V., Portmann, B., Reid, F. et al. (1997). Autoimmune hepatitis in childhood: a 20-year experience. *Hepatology*, **25**, 541–7.

13 Ahonen, J. and Björses, P. (1999). Cloning of the APECED gene provides new insight into human autoimmunity. *Annals of Medicine*, **31**, 111–16.

14 Donaldson, P., Doherty, D., Underhill, J. and Williams, R. (1994). The molecular genetics of autoimmune liver disease. *Hepatology*, **20**, 225–9.

15 Czaja, A.J., Carpenter, H.A., Santrach, P.J. and Moore, S.B. (1993). Significance of HLA DR4 in type 1 autoimmune hepatitis. *Gastroenterology*, **105**, 1502–7.

16 Scully, L.J., Toze, C., Sengar, D.P.S. and Goldstein, R. (1993). Early-onset autoimmune hepatitis is associated with a C4A gene deletion. *Gastroenterology*, **104**, 1478–84.

17 Cookson, S., Constantini, P.K., Clare, M. et al. (1999). Frequency and nature of cytokine gene polymorphisms in type 1 autoimmune hepatitis. *Hepatology*, **30**, 851–6.

18 Czaja, A.J., Cookson, S., Constantini, P.K., Clare, M., Underhill, J.A. and Donaldson, P.T.

(1999). Cytokine polymorphisms associated wth clinical features and treatment outcome in type 1 autoimmune hepatitis. *Gastroenterology*, **117**, 645–52.

19 Agarwal, K., Czaja, A.J., Jones, D.E. and Donaldson, P.T. (2000). Cytotoxic T lymphocyte antigen-4 (CTLA-4) gene polymorphisms and susceptibility to type 1 autoimmune hepatitis. *Hepatology*, **31**, 49–53.

20 Perheentupa, J. and Miettinen, A. (1999). Type 1 autoimmune polyglandular disease. *Annales de Medicine Interne*, **150**, 313–25.

21 Ahonen, P., Myllärniemi, S., Sipilä, I. and Perheentupa, J. (1990). Clinical variation of autoimmune polyendocrinopathy-candidiasis-ectodermal dystrophy (APECED) in a series of 68 patients. *New England Journal of Medicine*, **332**, 1829–36.

22 Pitkanen, J., Doucas, V., Sternsdorf, T. et al. (2000). The autoimmune regulator protein has transcriptional transactivating properties and interacts with common co-activator CBP. *Journal of Biological Chemistry*, **275**, 16802–9.

23 Heino, M., Peterson, P., Kudoh, J. et al. (1999) Autoimmune regulator is expressed in the cells regulating immune tolerance in thymus medulla. *Biochemical and Biophysical Research Communications*, **257**, 821–5.

24 Rinderle, C., Christensen, H.M., Schweiger, S., Lehrach, H. and Yaspo, M.L. (1999). AIRE encodes a nuclear protein co-localizing with cytoskeletal filaments: altered sub-cellular distribution of mutants lacking the PHD zinc fingers. *Human and Molecular Genetics*, **8**, 277–90.

25 Vogel, A., Liermann, H., Harms, A., Strassburg, C.P., Manns, M.P. and Obermayer-Straub, P. (2001). Autoimmune regulator AIRE: evidence for genetic differences between autoimmune hepatitis and hepatitis as part of the autoimmune polyglandular syndrome type I. *Hepatology*, **33**, 1047–52.

24

# IL-6-type cytokines and signalling in inflammation

Peter C Heinrich, Johannes G Bode, Lutz Graeve, Serge Haan, Astrid Martens, Gerhard Müller-Newen, Ariane Nimmesgern, Fred Schaper, Jochen Schmitz, Elmar Siewert

Institute of Biochemistry, Aachen, Germany

Interleukin (IL)-6 was identified as a hepatocyte-stimulating factor more than 10 years ago [1]. Subsequently, the authors have proposed IL-6 to be the major mediator of acute phase protein synthesis in liver cells. IL-6 belongs to the so-called IL-6-type cytokine family comprising, additionally, IL-11, leukaemia inhibitory factor (LIF), oncostatin M (OSM), ciliary neurotrophic factor (CNTF) and cardiotrophin-1 (CT-1). The members of this cytokine family exert pro- as well as anti-inflammatory activities via surface receptors. For IL-6 and IL-11, these surface receptors comprise specific $\alpha$-receptors and two identical gp130 signal transducer molecules, while, for other IL-6-type cytokines such as CNTF, OSM, LIF and CT-1, one gp130 and one LIF-receptor are involved. Resolution of the three-dimensional structures of IL-6, LIF and CNTF shows that the three cytokines form a bundle of four antiparallel long-chain alpha-helices of an up-up-down-down topology that are connected by loops.

The amino acid residues involved in the interaction between IL-6, its $\alpha$-receptor and the signal transducer gp130 have been identified by several groups including that of the authors. Recently, tyrosine 190, phenylalanine 191 and valine 252 of gp130 were identified as important residues for the binding and signalling of IL-6 and IL-11 [2].

## IL-6 signalling: within the cytoplasm

Early in 1994, the major steps in IL-6 signalling were elucidated [3, 4], clarifying how signals reach the cell nucleus and regulate acute phase protein genes such as C-reactive protein, fibrinogen, haptoglobin or hemopexin. A schematic representation is shown in Figure 24.1. IL-6 first binds to its $\alpha$-receptor (gp80). This complex cannot signal directly, but interacts with two molecules of the signal transducer gp130,

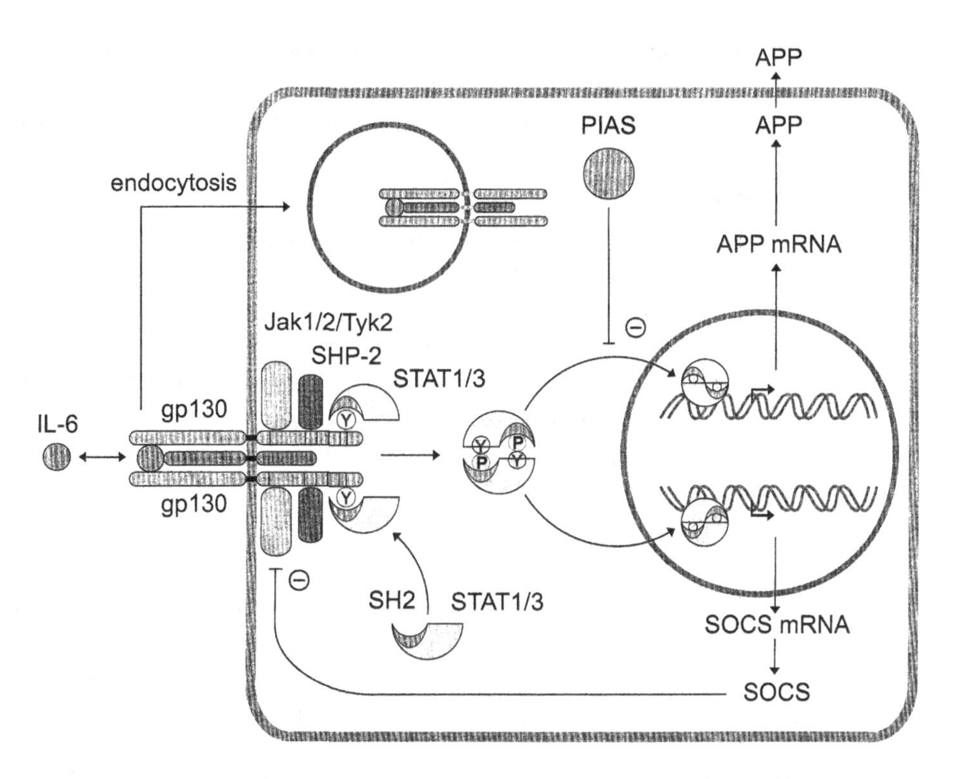


Figure 24.1 Negative regulation of the IL-6-type cytokine signal transduction pathway. Abbreviations: Jak, Janus kinase; STAT, signal transducer and activator of transcription; SH2, src homology domain; APP, acute phase protein; SOCS, suppressor of cytokine signalling; PIAS, protein inhibitor of activated STAT.

forming a ternary complex. Constitutively associated tyrosine kinases of the Janus family (Jak1, Jak2 and Tyk2 in the case of IL-6) phosphorylate each other upon ligand binding and, in turn, phosphorylate tyrosine residues in the cytoplasmic tail of gp130. These residues in gp130 are recruitment sites for transcription factors of the STAT (signal transducers and activators of transcription) family and, in the case of IL-6, STAT3 and STAT1 are recruited via their SH2-domains. STAT3 – originally named 'acute phase response factor' – was first characterized in the authors' laboratory as the major transcription factor activated by IL-6 stimulation [5]. After tyrosine phosphorylation by Jak kinases – Jak1 being particularly important – the STATs are released from the receptor complex, homo- or heterodimerize, are serine phosphorylated and translocate to the nucleus where they bind to enhancer elements of IL-6-type cytokine target genes [6].

Jak kinases not only contain a tyrosine kinase and a kinase-like domain but also five homology domains. The STAT factors have an *N*-terminal tetramerization domain, a DNA-binding domain, an Src homology (SH2) domain and a transactivation domain at the *C*-terminal part of the molecule. A tyrosine residue near the *C*-terminal end must be phosphorylated to allow binding of the STAT-dimers to enhancer elements.

Since there are six tyrosine residues in the cytoplasmic tail of gp130, the authors asked the question: which of the tyrosine residues activates which STAT factor? To answer this question, gp130 'add back' mutants were generated, i.e. all the six tyrosine residues in the cytoplasmic part were exchanged for phenylalanine and the individual tyrosine residues were added back. Transfection of Ba/F3-cells and stimulation with IL-6/soluble IL-6 receptor complexes showed that the four distal tyrosine residues of gp130 were able to activate STAT3.

## IL-6 nuclear signalling

Western blot analysis of immunoprecipitates of STAT3 from a cytosolic extract of HepG2 hepatoma cells showed coprecipatation with STAT1 in both naive cells and those stimulated with IL-6. Therefore, STAT1 and STAT3 appear to be preassociated in the cytoplasm. To investigate nuclear translocation, a fusion protein of STAT3 and green fluorescent protein was expressed in COS or HeLa cells. Although fluorescence was initially expressed all over the cells, stimulation with IL-6/soluble IL-6-receptor complexes for 10–15 minutes caused a rapid shift of fluorescence to the nucleus. While clearly rapid, the mechanism underlying nuclear translocation remains unknown since the STATs have no nuclear localization sequences.

Christoph Müller and coworkers succeeded in crystallizing the STAT3 dimer bound to its enhancer DNA sequence [7]. The complex has a nutcracker-like structure. Interestingly, the phosphotyrosine of one STAT3 interacts with the SH2-domain of the second STAT3 and vice versa.

To establish which STAT is the major player in the activation of acute phase protein gene transcription, the authors cotransfected HepG2 cells with cDNAs coding for various acute phase protein promoter/reporter genes. Constructs were produced with promoters of the $\gamma$-fibrinogen, CRP, haptoglobin or hemopexin genes, together with chimeric receptors consisting of the extracellular domain of the erythropoietin (epo) receptor, the transmembrane region of gp130 and 60 membrane-proximal amino acids of the cytoplasmic part of gp130 (linked to tyrosine motifs which specifically activate STAT1, STAT3 or STAT5). After epo stimulation of the cotransfected HepG2 cells, there was an unambiguous answer: only activated STAT3 led to the activation of acute phase protein gene transcription in liver cells.

## Features of Jak/STAT signalling

Additional key questions with respect to the Jak/STAT signal transduction pathway include: how is signalling terminated and how is it modulated? The following termination mechanisms have been proposed (Figure 24.1):

- tyrosine phosphatases in the nucleus;
- STAT degradation via the nuclear proteasome;
- tyrosine phosphatases dephosphorylate receptor-associated proteins (i.e. Jaks, STATs);
- specific inhibition of STAT proteins by protein inhibitors of activated STATs; and
- synthesis of Jak/STAT feedback inhibitors (suppressors of cytokine signalling, SOCS).

The following studies emphasize the respective importance of two of these – dephosphorylation of Jak/STATs and the synthesis of SOCS. In studies of the former with the gp130 mutant YFYYYY, it is evident that the tyrosine phosphatase SHP2, which is activated through Y759 of gp130, is counteracting Jak/STAT signalling. In Ba/F3 cells stably transfected with the gp130 Y759F mutant, STAT1/3 activation was shown to be prolonged compared with that in the gp130 wild type using both Western blotting and electrophoretic mobility shift assays. Furthermore, acute phase protein gene induction in HepG2 cells after IL-6 stimulation measured by a reporter assay clearly showed that exchange of tyrosine 759 of gp130 with phenylalanine gave rise to higher reporter gene activity than the wild type. In a further series of experiments, using a heterodimeric receptor system in which tyrosine 759 was exchanged by phenylalanine on the second gp130, the authors showed that tyrosine 759 on a single gp130 receptor chain is sufficient to attenuate signalling. The authors next asked the question of whether activation of SHP2 on one gp130 receptor chain was capable of affecting the tyrosine residues responsible for STAT recruitment on the other gp130 molecule. It was found that SHP2 and STAT recruitment sites do not have to be located on the same receptor chain. When two different receptor systems were transfected into HepG2 cells – one which recruits and activates SHP-2 but not STATs, the other unable to recruit SHP2 but able to activate STAT3 – no cross-talk was observed.

SOCS are molecules which have a so-called SOCS box near the *C*-terminal end and an SH2 domain important for function. Presently, there are seven known members of the SOCS protein family. After IL-6 stimulation of liver cells, SOCS1 and SOCS3 mRNAs are induced and the corresponding proteins act as very efficient feedback inhibitors by interacting with the catalytic domains of Jak1 or Jak2. Recently, the authors discovered that SOCS3 specifically binds to the phosphotyrosine peptide containing the tyrosine 759 motif of gp130. The other gp130 phosphotyrosine motifs do not bind SOCS3. As already mentioned, SHP2 also binds to the tyrosine 759 motif in gp130, i.e. SHP2 and SOCS3, but not SOCS1, can be coimmunoprecipitated together with the phosphotyrosine 759 peptide of gp130. The question then arose of whether SOCS3 was bound directly to phosphotyrosine 759 or whether it was recruited via SHP2. Using cell lysates containing SOCS3, but depleted of SHP2, it was shown that binding of SOCS3 to the phosphotyrosine motif around Y759 of gp130 occurred directly.

In collaboration with Nelson Fausto of the University of Washington, the authors have examined whether SOCS3 induction is a possible mechanism underlying the transient activation of STAT3 observed around 2 hours after partial hepatectomy of mice. A dramatic increase in SOCS3 mRNA has been observed. Consistent with these findings are reports from Fausto's group that TNF$\alpha$ injection into TNF$\alpha$ receptor 1-deficient or IL-6-deficient mice suppressed liver regeneration and the induction of SOCS3 mRNA. Interestingly, IL-6 injection overcame the defect, i.e. injection of IL-6 into TNF$\alpha$ receptor 1- or IL-6-deficient mice led to a strong SOCS3 induction.

## Modulation of Jak/STAT signalling

Besides the mechanisms of termination of IL-6-type cytokine signalling discussed above, signal transduction through the Jak/STAT pathway can be modulated by a number of mechanisms:

- differential expression of $\alpha$-receptors (IL-6, IL-11, CNTF);
- internalization of cytokine/receptor complexes;
- receptor shedding;
- different half-lives of the components of signalling cascades;
- induction of SOCS proteins by other cytokines;
- competition of STAT factors with other transcription factors for binding to overlapping response elements; and
- cytokine degradation.

To escape from being overstimulated, most cells internalize their surface receptors after ligand binding. The authors have shown that the IL-6-type cytokine receptor subunits gp130 and LIFR are internalized constitutively, i.e. ligand independently. A dileucine motif in the cytoplasmic tail of gp130 (but a leucine–isoleucine motif in the case of LIFR) has been found to be responsible for endocytosis [8]. This has been shown by confocal laser scanning microscopy and the use of agonistic and neutralizing monoclonal antibodies to gp130.

No internalization signal has been found in the cytoplasmic tail of the IL-6 receptor $\alpha$, gp80. Gp80/IL-6 complexes are taken up by liver cells as gp80/IL-6/gp130 ternary complexes, i.e. the internalization kinetics of IL-6 and gp80 are identical to those of gp130.

Although many functions of the cytoplasmic tail of gp130 have been described

– including binding of Jak kinases, activation of SHP-2, SOCS proteins and STAT factors as well as constitutive internalization – nothing is known so far about the role of the intracellular part of the IL-6 receptor $\alpha$-chain. Recently, work from the authors' laboratory has shown an important function of the cytoplasmic tail: the basolateral expression of the gp80. The experiments were carried out in stably transfected Madin–Darby canine kidney cells. Stable expression of wild-type gp80 resulted in basolateral expression, and deletion of the whole cytoplasmic tail in apical expression. Successive truncation of the cytoplasmic tail of gp80 led to the identification of two discontinuous sorting sequences for basolateral expression, one with a leucine–isoleucine motif and the other with a tyrosine motif.

Since the availability of the components involved in the Jak/STAT signal transduction pathway may be important for the modulation of signalling, pulse-chase experiments were performed and the half-lives of the IL-6 signalling components determined. The turnover rates were found to differ substantially [9], with three groups of signalling proteins readily discriminated. Thus, the feedback inhibitors SOCS1, SOCS2 and SOCS3 are very short lived (1–1.5 hours) while STAT1, STAT3 and SHP2 have an extremely slow turnover (8.5–20 hours) and the Janus kinases Jak1, Jak2, Tyk2 and gp130 show intermediate half-lives (2–3 hours).

Finally, there is recent evidence of the cross-talk between the Jak/STAT pathway and the MAP kinase cascade [10]. In this study [10], it was found that there was a decrease or abolition of STAT3 activation after IL-6 stimulation when either rat Kupffer cells or human macrophages were pretreated with the proinflammatory mediators TNF$\alpha$ or lipopolysaccharide (LPS). The inhibition closely correlates with the increase of SOCS3 mRNA by LPS or TNF$\alpha$, both of which are recognized activators of the p38 MAP kinase, implying that inhibition of this enzyme underlies the TNF$\alpha$-mediated effect on IL-6-induced STAT3 activation (Figure 24.2).

## REFERENCES

1 Andus, T., Geiger, T., Hirano, T. et al. (1987). Recombinant human B cell stimulatory factor 2 (BSF-2/IFN beta-2) regulates $\beta$-fibrinogen and albumin mRNA levels in Fao-9 cells. *FEBS Letters*, **221**, 18–22.

2 Horsten, U., Müller-Newen, G., Gerhartz, C. et al. (1997). Molecular modeling guided mutagenesis of the extracellular part of gp130 leads to the identification of contact sites in the IL-6/IL-6 receptor/gp130 complex. *Journal of Biological Chemistry*, **272**, 23748–57.

3 Lütticken, C., Wegenka, U.M., Yuan, J. et al. (1994). Association of transcription factor APRF and protein kinase JAK1 with the interleukin-6 signal transducer gp130. *Science*, **263**, 89–92.

4 Stahl, N., Boulton, T.G., Farruggella, T. et al. (1993). Association and activation of Jak-Tyk kinases by CNTF-LIF-OSM-IL-6$\beta$ receptor components. *Science*, **263**, 92–5.

5 Wegenka, U.M., Lütticken, C., Buschmann, J. et al. (1994). The interleukin-6-activated

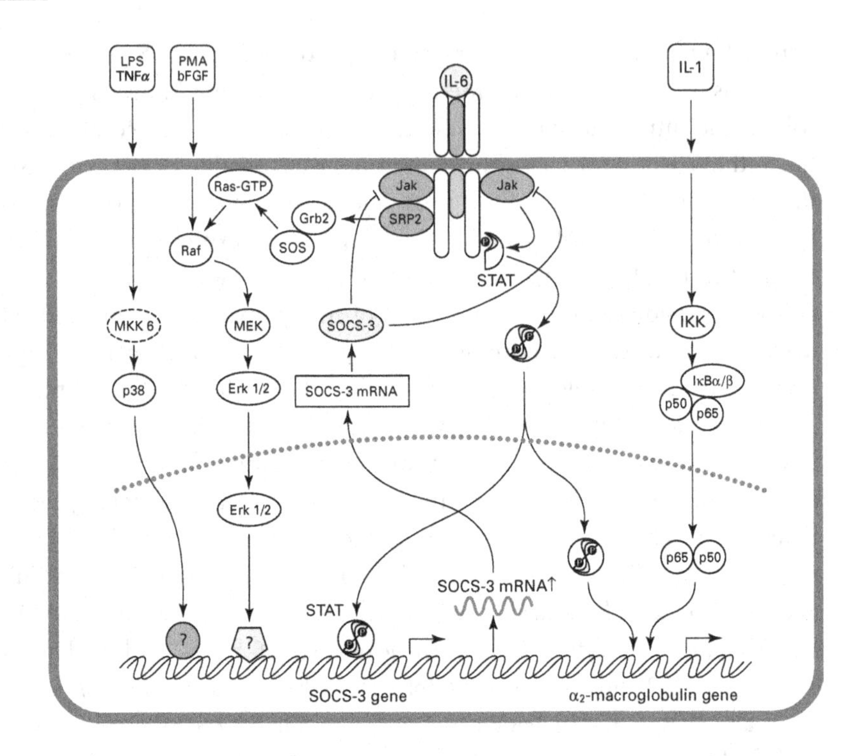


Figure 24.2 Modulation of IL-6 signalling through the Jak/STAT pathway *via* cross-talk with other signalling cascades. Abbreviations: Erk, extracellular signal-regulated kinase; FGF, fibroblast growth factor; Grb, growth factor receptor-bound protein; LPS, lipopolysaccharide; MKK/MEK, MAP kinase kinase; NF, nuclear factor; PMA, phorbol 12-myristate 13-acetate; TNF$\alpha$, tumour necrosis factor $\alpha$; IKK, inhibitor of KappaB Kinase; Ras, rat sarcoma.

acute-phase response factor is antigenically and functionally related to members of the signal transduction and transcription (Stat) factor family. *Molecular and Cellular Biology* **14**, 3186–96.

6 Heinrich, P.C., Behrmann, I., Müller-Newen, G., Schaper, F. and Graeve, L. (1998). IL-6-type cytokine signalling through the gp130/Jak/STAT pathway. *Biochemical Journal*, **334**, 297–314.

7 Becker, S., Groner, B. and Müller, C.W. (1998). Three-dimensional structure of the Stat3beta homodimer bound to DNA. *Nature* **394**, 145–51.

8 Dittrich, E., Renfrew-Haft, C., Muys, L., Heinrich, P.C. and Graeve, L. (1996). A di-leucine motif and an upstream serine in the interleukin-6 signal transducer gp130 mediate ligand-induced endocytosis and down-regulation of the IL-6 receptor. *Journal of Biological Chemistry*, **271**, 5487–94.

9 Siewert, E., Müller-Esterl, W., Starr, R., Heinrich, P.C. and Schaper, F. (1999). Different protein turnover of interleukin-6-type cytokine signalling components. *European Journal of Biochemistry* **265**, 251–7.

10 Bode, J.G., Nimmesgern, A., Schmitz, J. et al. (1999). LPS and TNF$\alpha$ induce SOCS3 mRNA and inhibit IL-6-induced activation of STAT3 in macrophages. *FEBS Letters* **463**, 365–70.

# Part 5

# Biomarkers of gastrointestinal disease and dysfunction

25

# Biomarkers in gastrointestinal disease

Humphrey JF Hodgson
Royal Free Hospital, London, UK

Gastroenterology remains the most general of specialties, encompassing psychological, functional, inflammatory and infectious and neoplastic disorders. The last 30 years have seen a major revolution in the way that gastroenterological clinicians think and work, brought about largely by the ready ability to inspect visually, and to biopsy, the upper gastrointestinal tract, colon and terminal ileum. Nevertheless, the relative nonspecificity of gastroenterological symptomatology, the relative inaccessibility of the small intestine (5–6 m in length) and the desire to define simple serological tests continues to provide a major role for laboratory assessments to detect and to assess gastrointestinal disease.

It is convenient, therefore, to assess what the straightforward first-line techniques of clinical gastroenterology can achieve, before considering the role of laboratory assessments. With respect to oesophageal disease, clinical history taking can localize disease in most cases, or at least to the oesophago-gastro-duodenal complex. With suggestive symptoms (reflux symptoms of acid heartburn, dysphagia), most clinicians will rapidly proceed to one of two approaches, both anatomical – endoscopy or radiology – to define the presence or absence of ulceration, inflammation, neoplasia or fibrotic stricture. Similarly, if the patient complains of clearly acid-related gastroduodenal symptoms, there will be rapid recourse to endoscopy.

This approach to 'acid-related dyspepsia' will be increasingly complemented with, or replaced by, seeking to identify the presence or absence of infection with *Helicobacter pylori*, the association of which with duodenal and gastric ulceration, but not with reflux oesophagitis, has become increasingly stressed in recent years. Testing for the presence of the organism, which predominantly colonizes the gastric antrum, can be done by a variety of tests (histology, culture or functional testing for the specific enzyme urease in biopsy specimens) at the same time as endoscopy, but can also be done indirectly [1]. In the future, there will be an increasing pressure to use indirect techniques – either serology or faecal testing for the organism, or breath tests based on $^{13}C$-labelled urea which rely on bacterial urease to release labelled $^{13}CO_2$. Pressure to use indirect tests for *H. pylori*, rather than identifying the bacteria in the gastric antrum at endoscopy, will grow. Most upper gastrointestinal symptomatology is benign. The most significant benign diseases (gastric and duodenal ulceration) are associated with *H. pylori* infection, and, most importantly, eradication of *H. pylori* dramatically reduces the recurrence rate for these diseases. There is thus an increasing tendency when considering the young patient with gastrointestinal symptomatology, in whom there are no worrying warning signals – (i.e. no weight loss, no anaemia), to screen for *H. pylori* infection first and then treat to eradicate the bacteria if it is shown to be present. This approach may be cheaper than resorting to endoscopy in all cases and is one well suited to primary care.

The most straightforward indirect test is for serum antibodies for *H. pylori* which are present while infection is present and for up to 6 months after eradication of the bacillus. It is most practical to test for efficacy of eradication by breath testing using $^{13}C$-labelled urea, which is split by bacterial urease to yield labelled $^{13}CO_2$ detectable by mass spectrometry. In the future, testing for *H. pylori* infection may take on a wider role – notably because of epidemiological associations of *H. pylori* infection with gastric cancer, probably mediated via the induction of gastric atrophy and intestinal metaplasia in the stomach. However, this whole area remains somewhat problematic because it is also clear that many people can carry *H. pylori* for decades without any apparent upper gastrointestinal disease.

Endoscopic approaches dominate the investigation of the key symptomatology of colorectal disease. Investigations of a history of rectal bleeding, or small volume diarrhoea with blood or pus, demand rapid direct inspection of the colonic mucosa by sigmoidoscopy (an instant outpatient procedure) or colonoscopy (which, of course, requires patient preparation with bowel cleansing). These investigations permit rapid diagnosis of the presence of neoplasia (polyps and cancer) and inflammation (ulcerative colitis, Crohn's disease, diverticulitis or ischaemic colitis). This pragmatic approach reduces the requirement for the use of biomarkers for disease detection. Tumour markers for colorectal cancer are of course available, but, for certainty, endoscopy and biopsy are required.

Nevertheless, the clinician remains highly tied to serological and haematological investigations for the detection and management of gastrointestinal disease. One of the imperatives for this is the sheer ubiquity of gastrointestinal symptomatology. It has been estimated that up to 10% of complaints leading to hospital referral relate to the gastrointestinal tract. However, most of the symptoms are reflections of 'functional' disease where the gastrointestinal tract itself is morphologically and histologically normal, but symptoms reflect either the presence of motility disorders (spasm, intestinal hurry) or, often, an oversensitivity to the normal workings of the gastrointestinal tract. Thus, a convenient starting point for this chapter is the assessment of a history which – at first glance – is likely to reflect functional disease.

Consider a 23-year-old female patient, with a 2–3 year history of intermittent diarrhoea, some abdominal pain, some distension and tiredness. Provided the initial physical examination, including proctosigmoidoscopy to rule out distal ulcerative proctitis, is normal, the following screening blood tests are indicated.

- *A full blood count* to identify or exclude:
  - anaemia;
  - haematinic deficiency;
  - leucocytosis – suggesting inflammation;
  - lymphopaenia – suggesting lymphocyte loss or immunodeficiency;
  - hyposplenism on blood film – suggesting coeliac or inflammatory bowel disease; and
  - thrombocytosis – suggesting inflammation
- *Routine biochemistry* to detect:
  - hypokalaemia – suggesting substantial loss of small intestinal fluid, prominent, for example, in surreptitious anthraquinone abuse;
  - hypoalbuminaemia – suggesting protein-losing enteropathy or reduced hepatic protein synthesis in the face of a prominent acute phase response;
  - calcium, phosphate and alkaline phosphatase disturbances – indicating potential malabsorption and osteomalacia
- *Haematinic levels* may be particularly helpful:
  - low iron levels may reflect either:
    - low intake – including malabsorption, blood loss, particularly in inflammatory conditions or neoplasia; or
    - low circulating iron binding proteins (notably transferrin) in the presence of chronic disease.
  - folate should be determined since levels are
    - low in malabsorption and conditions of increased cell turnover and also fall rapidly if intake is acutely diminished; or
    - high in the presence of small intestinal bacterial overgrowth, due to uptake of bacterially synthesized folate.
    - Contrary to previous dogma, there is little practical advantage in the more complex estimations of red cell folate, and serum folate estimations are adequate.
  - serum $B_{12}$ abnormalities may reflect disease in various intestinal sites due to the complex physiology of $B_{12}$ absorption [2]. Although uptake is restricted to the terminal ileum, intestinal processing involves:
    - the stomach – as a source of intrinsic factor to link with $B_{12}$ and provide the $B_{12}$–IF complex absorbed in the ileum;
    - the pancreas – required to break down competing $B_{12}$ binding proteins (R proteins) synthesized by the stomach;

  - the terminal ileum; and
  - small intestinal bacterial overgrowth – where competition for the bacteria in the bowel may lead to lack of available B12 for uptake.
- *Markers of inflammation:*
  - Erythrocyte sedimentation rate (ESR) – require investigation, but are of less practical use at the stage of screening for disease than the prototype acute phase reactant, C-reactive protein (CRP).
  - CRP – synthesis is rapidly induced in response to pro-inflammatory cytokines; its level rises many thousand-fold in severe disease, its half-life is short (18 h) and the cut off between normality and abnormality is clear. These features are all distinct advantages over the ESR, which is a resultant of the changes in many distinct serum proteins, some of which rise and some fall in the presence of inflammation, and the rate of change in response to changes in disease state are slow.

To return to our patient – in whom the differential diagnosis includes irritable bowel syndrome, inflammatory bowel disease and coeliac disease – normal or abnormal screening blood tests would be taken as an indication as to whether or not to proceed to further investigation. Normal tests would support the diagnosis of irritable bowel syndrome and, in most clinicians' minds, would obviate the need for further investigations.

Until recently, the possibility of coeliac disease would lead to small intestinal biopsy, now readily and routinely achieved at gastrointestinal endoscopy by duodenal biopsy. A key identifying feature is the flattened villi of gluten-sensitive enteropathy, and indeed such is the ready availability of the technique that this route to diagnosis remains probably the most common in adult patients. However, there is now available both specific and sensitive serology to make the diagnosis noninvasively. This is particularly valuable in screening the relatives of patients with coeliac disease and in the investigation of children. The many attempts to utilize circulating antibodies to wheat proteins (antigluten and antigliadin peptides) all suffered from a degree of insensitivity and nonspecificity. However, identification of circulating antiendomysial antibodies of IgA class in individuals on a normal diet has a very high specificity and sensitivity for the disease [3]. Recent work has identified the target antigen as tissue transpeptidase. The initial immunofluorescent technique utilizing monkey oesophageal muscle as substrate has been replaced by the use of human umbilical cord. The major pitfall in complete reliance on the serological approach to the diagnosis of coeliac disease is IgA deficiency in one in 50 patients with the condition. On a strict gluten-free diet, the antibody disappears, although how effective this is in monitoring dietary compliance has not yet emerged.

It is finally worth emphasizing that the physiological approach to documenting absorption, with, for example, urine-based absorption tests [4], has become much less widely used because of the ready availability of endoscopic approaches to identify the presence of abnormalities in the small intestinal mucosa.

In pursuing a suspected diagnosis of inflammatory bowel disease, the suspicion raised by one or more of the combination of anaemia, haematinic deficiency, low albumin or an elevated CRP would be pursued anatomically – by a combination of endoscopy and biopsy, contrast radiology or more specialized investigations such as leucocyte scanning. Biomarkers, however, have been extensively investigated as tools to manage patients and to assess disease activity in inflammatory bowel disease.

Despite the marked similarity of the inflammatory processes in Crohn's disease and ulcerative colitis, the latter has relied little on serological measurements. As ulcerative colitis virtually always involves the rectum, with involvement to a varying extent of the colon proximally, simple sigmoidoscopic examination and histology can judge the presence of disease. Extent and severity correlate well with symptomatology. Although many abnormalities can be documented in active disease – high CRP, low albumin, elevated faecal proteins (such as IgG and neutrophil-derived proteins) – their use in the average patient is limited [5]. There is some evidence that low albumin levels may define those least likely to respond to medical treatment in the major clinical event of ulcerative colitis – acute severe colitis. This medical emergency requires emergency admission and full supportive treatment including the administration of high-dose immunosuppression. A failure to lower CRP levels to $<45$ mg/l (normal $<8$ mg/l) within 3 days of high-dose corticosteroid treatment may define patients who will not respond to medical treatment and require colectomy [6]. These useful indicators, however, are also considered with the straightforward clinical indicators of disease severity – frequency of diarrhoea, macroscopic bleeding, fever and tachycardia – and many clinicians rely exclusively on the latter.

In Crohn's disease, there is a much more striking requirement for biomarkers. This reflects the fact that, unlike ulcerative colitis, the disease is very heterogeneous. In ulcerative colitis, the major variable is the extent to which the colon is affected, and the severity of the inflammation in the mucosal layer. In contrast, in Crohn's disease, any part of the bowel may be affected, so symptomatology may vary widely. Symptoms may reflect active inflammation either in the mucosa of the bowel or spreading through and outside the bowel, leading to fistula and abscess formation; alternatively, they may reflect chronic fibrotic consequences of previous inflammation, with subacute obstruction due to stenosis of the bowel. Biomarkers thus have a vital role in determining whether or not the disease is in an active inflammatory stage, and, to a lesser extent, in helping identify disease site.

A common example of this dilemma is in the patient in whom anatomic studies have identified terminal ileal disease, and who presents with obstructive symptoms. In this case, there may be varying contributions from fibrous obstruction and active inflammation – and the management is different. Inflammation may be expected to respond to medical management, while fibrous stenosis requires surgical resection. Elevations of acute phase proteins, particularly of CRP, are of immense value in making this distinction [7].

It is interesting to note the extent to which biomarkers have *not* been used in Crohn's disease. Reflecting the difficulties in identifying effective side effect-free treatments in this chronic disease, many thousands of patients have been admitted into controlled trials of medical treatment. A wide variety of assessments of disease activity have been investigated to produce a robust tool effective in the assessment of this heterogeneous disease in different centres in the clinical trial setting. There has been widespread agreement that biomarkers ought to have a major role to play. However, disappointingly, all the major clinical trials to date have relied almost exclusively on the use of clinical indices that either have no laboratory component at all, or only the most basic (e.g. haemoglobin, albumin). It is striking that trials of treatments as advanced as humanized chimeric antibodies to proinflammatory cytokines or recombinantly produced anti-inflammatory cytokines are assessed by a clinical scoring system. This routinely incorporates stool frequency, the presence or absence of pain and abdominal masses, and extraintestinal manifestations such as arthritis, the utilization of antidiarrhoeals, the patients' sense of well-being and their deviation from ideal weight [8]. Incorporating various weightings, these comprise the CDAI – the Crohn's Disease Clinical Activity Index. Its continuing use should be regarded as a challenge to laboratory scientists, because it is clear that such clinical indices have significant limitations. The first is that the end-point of such clinical indices – clinical improvement – is not the same as absence of disease. Indeed, in a very striking paper, it was demonstrated that, in a group of patients entering remission after corticosteroid treatment, the endoscopic appearances of the bowel were essentially unchanged [9]. In contrast, one of the striking advances that antiTNF (tumour necrosis factor) therapy has brought appears to be the rapid healing of the mucosa that this induces. Both treatments result in improvement using clinical indices [10]. Biomarkers that reflect such differences are certainly required but are not currently available in a valuable format. Desirable markers might reflect, for example, cell turnover, mucosal permeability and lack of activation of adhesion molecules. Many of these functions can be assessed only in complex procedures such as the assessment of mucosal permeability after the ingestion of probe molecules, followed by urinary analysis, adhesion molecule activation by monoclonal antibody scanning techniques and cell turnover by tissue studies.

Providing specific markers of active intestinal inflammation, in convenient form, and proving their validity and robustness, is likely to provide a worthwhile challenge to clinical investigators for some years to come.

## REFERENCES


1 Snyder, J.D. and Veldhuyzen Van Zanten, S. (1999). Novel diagnostic tests to detect *Helicobacter pylori* infection. *Canadian Journal of Gastroenterology*, **13**, 585–9.

2 Nickoloff, T.E. (1988). Schilling test: physiologic basis for and use as a diagnostic test. *Critical Reviews in Clinical Laboratory Sciences*, **26**, 263–76.

3 Bird, G., Fisher, P., Mahy, N. and Jewell, D. (1999). Coeliac disease in primary care: case finding study. *British Medical Journal*, **318**, 164–7.

4 Craig, R.M. and Ehrenpreis, E.D. (1999). D-xylose testing. *Journal of Clinical Gastroenterology*, **29**, 143–50.

5 Adeyemi, E.O. and Hodgson, H.J. (1992). Faecal elastase reflects disease activity in active ulcerative colitis. *Scandinavian Journal of Gastroenterology*, **27**, 139–42.

6 Travis, S.P., Farran, J.M., Ricketts, C. et al. (1996). Predicting outcome in severe ulcerative colitis. *Gut*, **38**, 905–10.

7 Saverymuttu, S.H., Hodgson, H.J., Chadwick, V.S. and Pepys, M.B. (1986). Differing acute phase responses in Crohn's disease and ulcerative colitis. *Gut*, **27**, 809–13.

8 Rutgeerts, P., D'Haens, G., Targan, S. et al. (1999). Efficacy and safety of retreatment with anti-tumor necrosis factor antibody (infliximab) to maintain remission in Crohn's disease. *Gastroenterology*, **117**, 761–9.

9 Mary, J.Y. and Modigliani, R. (1989). Development and validation of an endoscopic index of the severity of Crohn's disease: a prospective multicentre study. Groupe d'Etudes Therapeutiques des Affections Inflammatoires du Tube Digestif (GETAID). *Gut*, **30**, 983–9.

10 van Deventer, S.J. (1999). Targeting TNF alpha as a key cytokine in the inflammatory processes of Crohn's disease – the mechanisms of action of infliximab. *Alimentary Pharmacology and Therapeutics*, **13**(Suppl 4), 3–8.

26

# Tumour markers in gastrointestinal cancer

Anthony J FitzGerald, Nicholas A Wright

Hammersmith Hospital, London, UK

## Introduction

Biomarkers may be valuable in diverse aspects of both the investigation and treatment of gastrointestinal tumours. Useful in diagnosis, in screening and surveillance and in the study of the molecular epidemiology of tumours, biomarkers may have additional value in the treatment of such tumours by acting as surrogate end-points in clinical trials. They can also be used as predictors of survival after surgical treatment.

In diagnosis for example, some have advocated the use of serum markers – a typical example being $\alpha$-fetoprotein (AFP) for the detection of hepatocellular carcinoma in cases of cirrhosis associated with chronic hepatitis B or C viral infections. Perhaps the main role of the serum biomarkers used currently in clinical practice is for monitoring responses to chemotherapy and for the detection of tumour recurrence at an early stage. Good examples of the use of biomarkers in screening and surveillance include the identification of adenomas in the colon during colonoscopy or sigmoidoscopy and their removal to prevent the progression to carcinoma. Examining the extent of intestinal metaplasia in different populations at risk from gastric carcinoma, and its molecular pathology, gives insight into the size of such population risks. Surrogate end-points in the chemoprevention of colorectal carcinoma include the detection of adenomas at endoscopy. Similarly, aberrant crypt foci can be detected and used as surrogate end-points, now that magnifying endoscopy is becoming a reality. Biomarkers such as the extent of tubule formation, lymphocytic infiltration and the nature of the margin of the tumour – whether infiltrating or pushing – can be used to predict survival in colorectal carcinoma [1] and characterize the time-honoured Duke's method of classification, based on the extent of invasion of the bowel wall.

## Oesophagus and stomach

Epithelial cell proliferation is commonly increased in the oesophagus, stomach and colon of individuals at increased risk of gastrointestinal cancers, often long before the appearance of tumours. As such, the rate of cell proliferation and surrogate indicators of this rate have frequently been proposed as being useful.

Within the oesophagus, an increase in the size of the proliferative compartment leads to an increase in the numbers of proliferating cells in the basal layer, including immature epithelial cells which can reach the surface of the oesophageal lining. If these cells then undergo metaplasia, this can in turn lead to Barrett's oesophagus, and to an increased incidence of oesophageal cancer. Biomarkers of developing oesophageal cancer in tissues obtained at biopsy include histopathological/morphological changes associated with hyperplasia and dysplasia, and expression of cytokeratins not normally expressed by the normal tissue [2]. A feature which Barrett's oesophagus shares with the gastric mucosa is a high frequency of intestinal metaplasia, manifest when the phenotype of the gastric mucosa changes to that of the intestine. Early studies indicated that the intestinal type of gastric carcinoma arose from such metaplasia [3]. Several different forms of intestinal metaplasia have been distinguished: principally type 1, where the epithelium resembles that of the small intestinal crypt, complete with Paneth cells, goblet cells and enterocytic differentiation, and type II, where all cells contain mucus and there are no enterocytes. A further form, classed as type IIa or III by proponents, is where the usual acid sialomucin in the goblet cells is replaced with sulphomucin, detectable with the high iron diamine stain. This sulphomucin-containing type of intestinal metaplasia has been suggested as a biomarker for the future development of gastric carcinoma (see [4] for review), but this is hotly disputed.

No longer in real dispute is the observation that intestinal metaplasia, most often of the type II variety, can show important molecular changes which are replicated in gastric or Barrett's carcinoma, in the absence of changes in morphology such as dysplasia. For example, mutations in p53 and APC are seen, and intestinal metaplasia in Barrett's oesophagus shows clonal loss of heterozygosity (LOH) for APC. Other changes include mutations in c-*met* and changes in telomerase activity (see [4] for review).

In the stomach, there is increasingly compelling evidence that the carcinogenic process begins as a *Helicobacter pylori*-associated chronic gastritis which is thought to lead, via atrophic gastritis, intestinal metaplasia and dysplasia, to gastric carcinoma [5]. Consequently, *H. pylori* has been designated as a gastric carcinogen, and infection with the organism is now considered to be a possible biomarker for the development of gastric carcinoma. This has led to calls for the eradication of *H. pylori* as an intervention in the process. The main problem with this approach is the conflict over whether the precancerous biomarkers (or mucosal changes) are indeed reversible (see [4] for review), but the current consensus is that such changes are not. If confirmed, then the value of such markers and intervention by *H. pylori* eradication would be limited.

A number of other biomarkers for gastric carcinomas have been suggested. These range from realistic candidates to markers shown to be altered in only a few studies. Examples are: (i) brain-type glycogen phosphorylase (BGP) which is overexpressed in the intestinal type of gastric carcinoma – and indeed in both of the pathways of the adenoma–carcinoma sequence (ACS) as well as the 'de novo' colorectal carcinoma pathway [6]; (ii) mucosal ornithine decarboxylase (ODC), which reflects an increased rate of cell proliferation and is significantly elevated in areas of gastric atrophy and/or in cases of intestinal metaplasia [7]; and (iii) altered levels of apoptosis (possibly secondary to loss of function of the tumour suppressor gene, *p53*). Of potential for the future is an increase in gastric cell proliferation due to loss of growth inhibition by transforming growth factor-beta (TGF-$\beta$) as a result of mutation of the type-II TGF-$\beta$ receptor [5].

## Colorectal cancer

Colorectal cancer is the second highest cause of cancer death in the UK and 31 000 new cases of bowel cancer were diagnosed in 1997, with 17 349 deaths (Imperial Cancer Research Fund cancer statistics, 1998). There is a 20-fold variation in the incidence of colon cancer worldwide, emphasizing the great importance of environmental factors and the need for screening, surveillance and the prediction of long-term survival after diagnosis of the disease. Future new biomarkers of gastrointestinal tumours will have to be able to detect the early genetic changes within cells and tissues as well as the associated small biochemical changes. Genetic biomarkers will have to accommodate screening of individuals with a higher than normal susceptibility to cancer. This includes smokers, those with increased occupational and environmental exposure to carcinogens and members of high-risk families with familial adenomatous polyposis (FAP) or hereditary nonpolyposis colorectal cancer (HNPCC). New cellular biomarkers might assess nuclear morphological changes, alteration in levels of proliferation and mitotic index or DNA ploidy status. Genotypic markers might target LOH, inhibition or inactivation of tumour suppressor genes, such as *p53*, activation of oncogenes such as *k-ras* and *c-erb* B2 and gene mutations such as those for APC. Phenotypic and genotypic changes may identify new molecular and biochemical biomarkers, such as increased levels of cellular proliferation, cellular antigens Ki-67 and MIB1 and overexpression of growth factors – epidermal growth factor (EGF), transforming growth factor-alpha (TGF-$\alpha$) and insulin-like growth factor (IGF-1).

The process of carcinogenesis in the colorectum follows the well-known adenoma–carcinoma sequence and is marked by several discrete morphological correlates with potential as markers. It has been proposed that one of the earliest changes observed in this sequence is a stage of hyperproliferation in the colonic mucosa before the initiation of neoplastic changes [8]. Now a widely quoted first phase in the molecular process of carcinogenesis as envisaged by Vogelstein, this in turn leads to the hypothesis that increased rates of cell proliferation could be an important biomarker for predicting the risk of colorectal cancer. Confirmation of both elevated labelling indices and an increase in the size of the proliferative compartment within the colonic crypts was made initially using ex vivo labelling with $^{3}$H-thymidine or BrdU. Later, immunohistochemical markers of proliferation, such as PCNA or Ki67, were used. The hypothesis would continue with an increased rate of cell proliferation leading, in turn, to an increased risk of somatic mutation, with the potential for selection and clonal expansion.

Several studies apparently show increased labelling indices and proliferative compartment sizes in conditions associated with increased rates of colorectal carcinoma. Some human studies using proliferative indices as a marker of neoplasia have merely examined *rectal* epithelial proliferation without considering the site from which the biopsy was taken. Most studies reporting such changes have been in the flat mucosa of patients with familial polyposis coli or in the 'normal' mucosa from patients with adenomas or carcinomas [9–17]. Nevertheless, many reports now tacitly assume that elevated labelling indices and increased proliferative compartment size are established facets of the premalignant process and use these biomarkers in intervention and epidemiological studies.

In contrast, several comparable studies dispute these findings [18–22]. While a number of these used the classical methods with tissue sections referred to above, Wasan et al., who advocated the use of whole microdissected crypts and the counting of mitotic figures, suggested that section artifact might be responsible for the disparity between these results [21]. Moreover, Green et al., looking at crypt cell proliferation as a phenotypic biomarker in subjects with HNPCC, showed no significant change in the total labelling index, indicating that cell proliferation could not be used as a marker of gene carriage in HNPCC subjects [23]. This was confirmed by Jass et al. [24] using immunohistochemical staining of the cell cycle-associated nuclear proteins, proliferating cell nuclear antigen (PCNA) and Ki-67. They observed no difference between labelling indices of mucosal proliferation in biopsies obtained from the ascending colon from four mutation-positive members of an HNPCC family and in subjects with some positive family history of colorectal cancer but lacking the clinical and pathological features of HNPCC. Importantly, Jass and coworkers concluded from these and other studies that it was unlikely that the commoner types of colorectal carcinomas were associated with a prodromal phase of increased cell proliferation.

In experimental studies with carcinogens, particularly the rat dimethylhydrazine or azoxymethane models, tiny lesions called aberrant crypt foci (ACF) appear. These are best detected as raised lesions on whole mounts of colonic mucosa stained with methylene blue, initially one crypt in diameter, but growing by crypt fission to occupy several crypts. It has been claimed that these lesions show the same distribution in the colon as the adenomas and carcinomas which arise in these models [25] and that these are the lesions from which adenomas arise. Later, in their evolution to carcinomas, dysplasia occurs, with associated changes in oncogenes such as *ras.* Although the progression of ACF to tumours was disputed [26], Park et al. confirmed that the relationship was maintained when ACF were correlated with the histological tumour type [25]. The number of crypts involved is also important – aberrant crypt multiplicity (ACM) being proposed as an end-point or biomarker for colorectal cancer risk. With ACM, it is not simply the number of ACF that can act as the biomarker but rather the number of crypts contained within each ACF, i.e. the more crypts per ACF, the greater the predictor of tumour outcome.

ACFs were identified initially in colons excised from patients with FAP, but later in those from patients with sporadic colorectal cancer [27]. Until recently, the excised colon was essential for the identification of these lesions, but, with the advent of magnifying colonoscopy, it is possible to see these lesions in the colonic mucosa in vivo. An interesting development is the use of colonic ACFs as a biomarker in a pilot study on the effect of cyclooxygenase inhibitors as chemopreventative agents, with the lesions shown to regress over 12 months' therapy [28]. Possibly in response to these studies, Kinzler and Vogelstein have replaced the earliest stage of hyperproliferation with one describing 'dysplastic aberrant foci' in a newer model of the colorectal tumourigenesis sequence [29].

Screening for colorectal cancer has been based mainly on the identification of occult blood in stools but such blood may be due more commonly to large polyps (>2.0 cm in diameter). Current surveillance requires regular testing and follow up. This can in turn lead to further tests such as contrast studies with barium enemas and colonoscopy. More recently, methods depending on direct endoscopic examination of the colon, either by full colonoscopy or sigmoidoscopy, have proved popular. The rationale of both procedures is to identify the biomarker, here the adenoma, and to remove it. Removal not only removes the source of any further tumours, but also appears to reduce the risk of other lesions developing [30]. A further rationale of flexible sigmoidoscopy is that most adenomas and carcinomas occur within the reach of this instrument. With the peak incidence of such lesions at 55–60 years, a useful hypothesis is that a 'one-off' sigmoidoscopy at this age, with removal of any adenomas, will result in a large reduction in morbidity and indeed mortality from this disease. Confirmatory trials are ongoing, but the methods are relatively expensive, not always acceptable to the participants and there is only limited evidence of their effectiveness in reducing the mortality from gastrointestinal cancer.

## Current serum biomarkers of cancer

Serum biological markers are used currently in clinical practice for detecting the presence of disease or the recurrence of tumour after a course of treatment and for monitoring responses to chemotherapy. Most markers are overexpressed by the gastrointestinal tumour itself and thus are detectable in the plasma of the patient.

Those in clinical use at present include: carcinoembryonic antigen (CEA) for colorectal carcinomas, AFP for primary liver carcinomas, CA125 for colon carcinomas (but predominantly used for ovarian carcinomas) and CA 19–9 for pancreatic and gastric carcinomas. Both serum CA 19–9 and CA125 tests are predictive for diagnosing advanced stage carcinomas of the gastrointestinal tract. A frequent problem with these serum biomarkers is their lack of organ specificity (as in the case of CA 19–9), insensitivity and a poor correlation with disease of measurable changes in the blood/plasma. It is also impossible to diagnose early stages of cancers by elevated levels of these biomarkers.

CEA, for example, is a human tumour marker typically associated with malignant epithelial neoplasms. The CEA gene belongs to a family of more than 20 members, clustered on the long arm of chromosome 19 [31] and belonging to the immunoglobulin supergene family. Its main use is to monitor patients with colorectal carcinoma, but the antigen is not specific and elevated levels may be observed with some nonmalignant diseases including inflammatory bowel disease, liver disease [32], pancreatitis and renal failure [33]. In healthy individuals, plasma levels are in a range of 0–5 ng/ml and 3–10 ng/ml in smokers, whereas patients with colorectal cancer exhibit levels >20 ng/ml.

## Future possibilities

Other potential tumour biomarkers include nuclear lamin expression within the gastrointestinal tract, cytokeratins and Gal-GalNAc. Nuclear lamins are a group of filaments located on the inner nuclear membrane of cells. In a recent study, Moss et al. showed that expression of the lamins A/C and lamin B1 was reduced in patients with gastric and colonic adenocarcinomas [34]. The disaccharide tumour marker Gal-GalNAc is commonly present in cancer cells and in the rectal mucous of patients with colon cancer, whereas no expression is seen in the normal human large intestine. It is expressed throughout the entire colon of patients with precancer and cancer as well in experimental rodent models of carcinogenesis (e.g. after induction by azoxymethane). Detected by the simple technique of the galactose oxidase–Schiff sequence, it is a biomarker that appears during the very early stages of cancer progression [35]. Cytokeratins (CK) are constituent parts of cellular intermediate filaments of epithelial cells, expressed in varying amounts depending on the cell type. Recently, CK-20 has been shown to be expressed in a number of tumour types and it can be detected in blood samples from patients using a reverse transcriptase polymerase chain reaction technique. This high sensitivity makes CK-20 a potential biomarker for metastases in patients with gastric, pancreatic and colonic carcinomas.

Once the molecular signatures of gastrointestinal cancers and their precursors have been established, it is not unreasonable to expect that molecular lesions could be detected in cells shed into the stool. Already, methods of extracting such DNA from stool specimens have been developed and used to detect mutations, e.g. of codon 12 *ras*. While *ras* is unlikely to be of much use because ACFs and even normal mucosa appear to harbour *ras* mutations, other molecular markers will no doubt prove useful.

## Conclusion

It is clear that several biomarkers of gastrointestinal cancer are relevant, not only because of their potential in epidemiological and clinical use, but also because of the insight they give into the pathogenesis of the tumours themselves. Future work on these markers will hopefully lead to further insights.

## REFERENCES

1 Jass, J.R., Atkin, W.S., Cuzick, J. et al. (1986). The grading of rectal cancer: historical perspectives and a multivariate analysis of 447 cases. *Histopathology*, **10**, 437–59.

2 Yang, K. and Lipkin, M. (1990). AE1 cytokeratin reaction patterns in different differentiation states of squamous cell carcinoma of the esophagus. *American Journal of Clinical Pathology*, **94**, 261–9.

3 Morson, B.C. (1995). Carcinoma arising from areas of intestinal metaplasia in the gastric mucosa. *British Journal of Cancer*, **9**, 377–85.

4 Wright, N.A. (1998). Aspects of biology of regeneration and repair in the human gastrointestinal tract. *Philosophical Transactions of the Royal Society of London – Series B: Biological Sciences*, **353**, 925–33.

5 Moss, S.F. (1998). Cellular markers in the gastric precancerous process. *Alimentary Pharmacology and Therapeutics*, **12**(Suppl 1), 91–109.

6 Shimada, S., Tashima, S., Yamaguchi, K., Matsuzaki, H. and Ogawa, M. (1999). Carcinogenesis of intestinal-type gastric cancer and colorectal cancer is commonly accompanied by expression of brain (fetal)-type glycogen phosphorylase. *Journal of Experimental and Clinical Cancer Research*, **18**, 111–18.

7 Patchett, S.E., Alstead, E.M., Butruk, L., Przytulski, K. and Farthing, M.J. (1995). Ornithine decarboxylase as a marker for premalignancy in the stomach. *Gut*, **37**, 13–16.

8 Fearon, E.R. and Vogelstein, B. (1990). A genetic model for colorectal tumorigenesis. *Cell*, **61**, 759–67.
9 Gerdes, H., Gillin, J.S., Zimbalist, E., Urmacher, C., Lipkin, M. and Winawer, S.J. (1993). Expansion of the epithelial cell proliferative compartment and frequency of adenomatous polyps in the colon correlate with the strength of family history of colorectal cancer. *Cancer Research*, **53**, 279–82.
10 Lipkin, M., Enker, W.E., and Winawer, S.J. (1987). Tritiated-thymidine labeling of rectal epithelial cells in 'non-prep' biopsies of individuals at increased risk of colonic neoplasia. *Cancer Letters*, **37**, 153–61.
11 Risio, M., Lipkin, M., Candelaresi, G., Bertone, A., Coverlizza, S. and Rossini, F.P. (1991). Correlations between rectal mucosa cell proliferation and the clinical and pathological features of nonfamilial neoplasia of the large intestine. *Cancer Research*, **51**, 1917–21.
12 Risio, M., Candelaresi, G. and Rossini, F.P. (1993). Bromodeoxyuridine uptake and proliferating cell nuclear antigen expression throughout the colorectal tumor sequence. *Cancer Epidemiology, Biomarkers and Prevention*, **2**, 363–7.
13 Risio, M. and Rossini, F.P. (1993). Cell proliferation in colorectal adenomas containing invasive carcinoma. *Anticancer Research*, **13**, 43–7.
14 Roncucci, L., Scalmati, A. and Ponz de Leon, M. (1991). Pattern of cell kinetics in colorectal mucosa of patients with different types of adenomatous polyps of the large bowel. *Cancer*, **68**, 873–8.
15 Mills, S.J., Shepherd, N.A., Hall, P.A., Hastings, A., Mathers, J.C. and Gunn, A. (1995). Proliferative compartment deregulation in the non-neoplastic colonic epithelium of familial adenomatous polypsis. *Gut*, **36**, 391–4.
16 Terpstra, O.T., van Blankestein, M., Dees, J. and Eilers, G.A. (1987). Abnormal pattern of cell proliferation in the entire colonic mucosa of patients with colon adenoma or cancer. *Gastroenterology*, **92**, 704–8.
17 Anti, M., Armuzzi, A. and Iascone, E. et al. (1998). Epithelial-cell apoptosis and proliferation in *Helicobacter pylori*-related chronic gastritis. *Italian Journal of Gastroenterology and Hepatology*, **30**, 153–9.
18 Nakamura, S., Kino, I. and Baba, S. (1988). Cell kinetics analysis of background colonic mucosa of patients with intestinal neoplasms by ex vivo autoradiography. *Gut*, **29**, 997–1002.
19 Nakamura, S., Kino, I. and Baba S. (1993). Nuclear DNA content of isolated crypts of background colonical mucosa from patients with familial adenomatous polyposis and sporadic colorectal cancer. *Gut*, **34**, 1240–4.
20 Williams, G.T., Geraghty, J.M., Campbell, F., Appleton, M.A. and Williams, E.D. (1995). Normal colonic mucosa in hereditary non-polyposis colorectal cancer shows no generalised increase in somatic mutation. *British Journal of Cancer*, **71**, 1077–80.
21 Wasan, H.S., Park, H.S. and Liu, K.C. et al. (1998). APC in the regulation of intestinal crypt fission. *Journal of Pathology*, **185**, 246–55.
22 Matthew, J.A., Pell, J.D. and Prior A. et al. (1994). Validation of a simple technique for the detection of abnormal mucosal cell replication in humans. *European Journal of Cancer Prevention*, **3**, 337–44.

23 Green, S.E., Chapman, P. and Burn, J. et al. (1998). Colonic epithelial cell proliferation in hereditary non-polyposis colorectal cancer. *Gut*, **43**, 85–92.
24 Jass, J.R., Ajioka, Y., Radojkovic, M., Allison, L.J. and Lane, M.R. (1997). Failure to detect colonic mucosal hyperproliferation in mutation positive members of a family with hereditary non-polyposis colorectal cancer. *Histopathology*, **30**, 201–7.
25 Park, H.S., Goodlad, R.A. and Wright, N.A. (1997). The incidence of aberrant crypt foci and colonic carcinoma in dimethylhydrazine-treated rats varies in a site-specific manner and depends on tumor histology. *Cancer Research*, **57**, 4507–10.
26 Cameron, I.L., Garza, J. and Hardman, W.E. (1996). Distribution of lymphoid nodules, aberrant crypt foci and tumours in the colon of carcinogen-treated rats. *British Journal of Cancer*, **73**, 893–8.
27 Roncucci, L., Pedroni, M., Fante, R., Di Gregorio, C. and Ponz de Leon, M. (1993). Cell kinetic evaluation of human colonic aberrant crypts. *Cancer Research*, **53**, 3726–9.
28 Takayama, T., Katsuki, S. and Takahashi, Y. et al. (1998). Aberrant crypt foci of the colon as precursors of adenoma and cancer. *New England Journal of Medicine*, **339**, 1277–84.
29 Kinzler, K.W. and Vogelstein, B. (1996). Lessons from hereditary colorectal cancer. *Cell*, **87**, 159–70.
30 Atkin, W.S., Hart, A. and Edwards, R. et al. (1998). Uptake, yield of neoplasia, and adverse effects of flexible sigmoidoscopy screening. *Gut*, **42**, 560–5.
31 Zimmermann, W., Weber, B. and Ortlieb, B. et al. (1988). Chromosomal localization of the carcinoembryonic antigen gene family and differential expression in various tumors. *Cancer Research*, **48**, 2550–4.
32 Loewenstein, M.S. and Zamcheck, N. (1978). Carcinoembryonic antigen (CEA) levels in benign gastrointestinal disease states. *Cancer*, **42**(Suppl), 1412–18.
33 Filella, X., Cases, A. and Molina, R. et al. (1990). Tumor markers in patients with chronic renal failure. *International Journal of Biological Markers*, **5**, 85–8.
34 Moss, S.F., Krivosheyev, V. and de Souza, A. et al. (1999). Decreased and aberrant nuclear lamin expression in gastrointestinal tract neoplasms. *Gut*, **45**, 723–9.
35 Yang, G.Y. and Shamsuddin, A.M. (1996). Gal-GalNAc: a biomarker of colon carcinogenesis. *Histology and Histopathology*, **11**, 801–6.

27

# Markers of malabsorption: coeliac disease

H Julia Ellis, Jocelyn S Fraser, Paul J Ciclitira
Rayne Institute, St. Thomas' Hospital, London, UK

## Introduction

Coeliac disease, or gluten-sensitive enteropathy, is an inflammatory condition that affects mainly the small intestine, resulting in loss of the normal villous architecture. The lesion returns towards normal with a gluten-free diet – that is, the complete avoidance of wheat, rye, barley and possibly oats. In dermatitis herpetiformis, the skin is the organ primarily affected, although some degree of enteropathy is almost always present.

Classically, gluten-sensitive enteropathy consists of a flat small intestinal mucosa, whose greatly reduced surface area results in malabsorption, leading to gross steatorrhoea and multiple loss of nutrients. It is now recognized that there is a spectrum of lesions, some of which are very subtle. The patient may be asymptomatic (silent coeliac disease) or may suffer from deficiency of a single nutrient in the absence of any gastrointestinal symptoms. Thus, unexplained anaemia in a man, or osteoporosis in a premenopausal woman, for example, should set the alarm bells ringing.

Prior to the advent of serological tests or small bowel biopsy, diagnosis had to be made on the basis of measurement of faecal fat. This nonspecific marker indicates only the presence and not the cause of fat malabsorption, which might originate from pancreatic insufficiency among a number of causes. It is now known that many gluten-sensitive individuals have normal fat absorption.

An understanding of the range of gluten-sensitive lesions and advancements in serological screening tests has led to an awareness that gluten sensitivity may affect one in 300 individuals across Europe, rising to one in 100 in Ireland. Serological tests, which are constantly being refined, have not reached 100% sensitivity and specificity. Thus, whether the diagnosis of coeliac disease be suspected on clinical grounds, or indicated by serological screening, it is mandatory to proceed to small intestinal biopsy for confirmation.

## Serological tests for gluten sensitivity

### General considerations

These tests may be used to screen populations or in family studies. Alternatively, they may be used as part of the diagnostic armatorium and, in some cases, to monitor compliance to a gluten-free diet and healing of the mucosa. Some authors have warned that these tests may not be useful when used to investigate individuals in whom a very subtle bowel lesion is present. Where the diagnosis of gluten sensitivity is suspected on clinical grounds, a negative serological test result should not be automatic grounds for abandoning small bowel biopsy. Equally, there are some individuals in whom serological tests prove positive, but subsequent biopsy shows a histologically normal small bowel mucosa. These individuals are now considered to manifest latent coeliac disease and are presently not treated with a gluten-free diet.

Various forms of immunoassay have been developed to detect 'coeliac antibodies'. There are problems associated with them all, not least of which is the absence of any single, standardized procedure for any of the tests discussed below. Thus, different centres will use their own batches of antigen and, in the case of wheat gluten (gliadin), this may vary considerably depending on the species of wheat and the mode of extraction. Secondary antibodies vary by manufacturer and by host species. There are additional variations in detection systems, which may utilize horseradish peroxidase (HRP), alkaline phosphatase (ALP), radiolabel or fluorescein. Until recently, there was no standard reference serum to use as positive control. For all these reasons, the sensitivity and specificity of serological tests is highly dependent on the centre performing the test. The EMRC/ESPGAN (European Medical Research Councils/European Society of Paediatric Gastroenterology and Nutrition) Working Party on 'Serological Screening in Coeliac Disease' has defined robust noncommercial methods to screen for the condition. Reference sera are now available [1].

As will become apparent, the measurement of IgA antibodies to various antigens provides a more sensitive and specific tool than IgG antibodies. However, about 2.5% of coeliac disease patients have selective IgA deficiency, negating the test results in such individuals [2]. Any individual with symptoms suggestive of malabsorption or a family history of coeliac disease should be tested additionally for total serum IgA and individuals with IgA deficiency (less than 0.05 g/l) in the presence of normal IgG should be considered for biopsy.

The results of all the serological tests described below fall rapidly towards normal when a gluten-free diet is commenced. Small bowel biopsy findings may do the same. These trends may make diagnosis difficult if the individual is from a family of coeliac disease patients because the individual, or a parent, may instigate a gluten-free diet prior to investigation by a gastroenterologist.

**Table 27.1.** EMRC/ESPGAN evaluation of serological tests for antigliadin antibodies in coeliac disease

| | Sensitivity | Specificity |
|---|---|---|
| IgA antigliadin | 83% | 82% |
| IgG antigliadin | 86% | 76% |

*Note:*
EMRC/ESPGAN (European Medical Research Councils/European Society of Paediatric Gastroenterology and Nutrition) Working Party methodology was used to evaluate the IgA and IgG antigliadin antibodies.

## Antigliadin antibodies

These antibodies are usually detected by enzyme-linked immunosorbent assay (ELISA) using gliadin coated on a microtitre plate. Gliadin is insoluble in neutral aqueous solution and is therefore usually dissolved in 40–70% ethanol for coating. Gliadin may be obtained commercially (e.g. Sigma Chemicals), but many laboratories make their own pure gliadin. Dilutions of sera are incubated on the plate and antibody binding is detected using a secondary antibody conjugate, usually with HRP or ALP. Many of the published methods have been reviewed by Scott et al. [3]. IgA antigliadin is frequently found in untreated coeliac disease, but rarely in healthy controls. However, elevated levels of IgG antigliadin antibody have been detected in patients with other intestinal disorders and in healthy subjects. Additionally, raised IgG activity to dietary antigens other than gluten is present in the serum of untreated patients with coeliac disease and may reflect antigen passage across the damaged small intestinal mucosa. The EMRC/ESPGAN working party on screening for coeliac disease records the data shown in Table 27.1.

IgA antigliadin, in particular, falls rapidly in response to a gluten-free diet; IgG falls more slowly. These tests have been used to monitor compliance with a gluten-free diet. Measurement of antigliadin antibody is now considered less useful in screening and diagnosis than the other tests described below. However, it was used recently by Hadjivassiliou and colleagues [4] to screen individuals attending a neurology clinic. Antigliadin antibody positivity was prevalent among these patients, but only a proportion were later shown to indeed have gluten-sensitive enteropathy. The authors suggested that there may be a subgroup of gluten-sensitive individuals in whom the brunt of disease is borne by the nervous system. Given that the source of connective tissue antibodies is the small bowel (see below), the authors suggest that antigliadin antibody screening may be more effective in those individuals in whom the nervous system is primarily affected.

### Antireticulin and antijejunal antibodies

Untreated coeliac patients have detectable antibodies to jejunum and to reticulin, which can be demonstrated by immunofluoresence on a number of mammalian tissues. These antibodies are now thought to recognize the same antigen as antiendomysial antibodies.

### Antiendomysial antibodies

Antiendomysial antibodies (EMA) of the IgA type are considered to be highly specific markers of coeliac disease. Using human umbilical cord, the reported sensitivity is 90%, with a specificity of 99% in adults with untreated coeliac disease [1]. IgA antibodies react with endomysium of the smooth muscle bundles of monkey oesophagus and this was the original source of antigen. However, the tamarin monkey is an endangered species, and, in 1994, the EMRC and ESPGAN recommended the use of human umbilical cord, which is nearly as effective as monkey oesophagus. Further deficiencies of monkey tissue are its expense and the absence of smooth muscle bundles from sections taken too near the top of the oesophagus. Monkey oesophagus from a commercial source must be batch tested to check for this.

EMA is detected by indirect immunofluorescence. Serum is incubated on human umbilical cord or monkey oesophagus, and EMA is detected by the addition of fluorescein isothiocyanate (FITC)-conjugated anti-human IgA antibody. EMA positivity results in a typical reticulin-like staining pattern on the smooth muscle bundles of the monkey oesophagus, or, in the case of human umbilical cord, the smooth muscle bundles in the umbilical artery. The value of the test is limited by the subjective interpretation on which the results depend.

So far, EMAs of the IgG type are not used diagnostically. However, 2.5% of patients with coeliac disease have selective IgA deficiency and, in such cases, IgGl EMA may be a useful diagnostic tool. There also appears to be a population of coeliac disease patients with normal quantities of IgA who have high IgGl EMA in the absence of IgA EMA [5]. As with other test results, when a gluten-free diet is introduced, EMA is reduced and returns towards normal in the absence of antigenic stimulus.

### Tissue transglutaminase antibodies

In 1997, Dieterich et al. [6] identified tissue transglutaminase (tTG) as the autoantigen for EMA. tTG is a calcium-dependent enzyme which catalyses crosslink formation between glutamine residues and lysine residues and is expressed in many organs and stored mainly intracellularly. It is possible to detect autoantibodies to tTG by ELISA [7]. Wells are coated with calcium-activated guinea pig liver tTG. Serum, diluted to one in 100, is added and incubated for an hour. Peroxidase-conjugated antihuman IgA or IgG is added, and, after incubation with substrate, the absorbance is read spectrophotometrically. Patients with selective IgA deficiency were found to be tTG IgA negative, but were positive for IgG tTG. However, IgG class tTG antibody was not specific for untreated coeliac disease in patients with normal quantities of serum IgA [7].

The tTG ELISA can be performed using human recombinant tTG [8]. Human tTG can be expressed in the human embryonic cell line 293–EBNA. After purification with streptavidin affinity chromatography, it can be used in the ELISA, giving a specificity of 98.1% and a sensitivity of 98.2%, versus a specificity of 95% and a sensitivity of 93% for guinea pig liver tTG [1].

## Sugar absorption tests

The integrity of the small bowel is impaired in coeliac disease, and intestinal permeability, measured by sugar absorption, reflects this. Sugar absorption tests have been used as a screening tool in coeliac disease and as a tool for follow up of patients taking a gluten-free diet [9]. Their basis is that a specific dose of a marker is given orally, of which a certain amount will permeate across the intestinal mucosa, pass into the circulation and be excreted via the kidneys. The permeability markers can then be measured in the urine.

The passive absorption of molecules >0.5 nm in size (e.g. dissacharides such as lactulose, cellobiose, sucrose or lactose) between the cells is increased by oedema, mucosal inflammation and villous atrophy. Conversely, the proportion of molecules <0.5 nm which are absorbed transcellularly (e.g. monosaccharides such as D-mannitol, L-rhamnose or D-xylose) is unchanged or impaired. Thus, the urinary dissacharide/monosaccharide ratio provides a sensitive, noninvasive index of functional integrity of the small bowel mucosa. The advantage of using differential sugar absorption tests is that both molecules are affected equally by any variation in intestinal or extraintestinal factors. Expressing the results as a ratio negates any external factors.

In practice, the most commonly used test solution contains 5 g of lactulose, 100 g of sucrose and 2 g of mannitol in 450 ml of boiled, demineralized water. This is given after an overnight fast, and the lactulose/mannitol ratio is measured by gas chromatography in urine collected over a period of 5 hours. The measured sensitivity for this test is 100%, with a specificity of 83% [10]. Alternatively, the lactulose and mannitol levels can be measured in the serum, 1 hour after ingestion – a more convenient and less time-consuming process [11]. Other combinations of sugars suitable for intestinal permeability testing include lactulose and rhamnose, and cellobiose and mannitol.

There may be problems with interpreting test results. For instance, the percentage of the oral dose excreted in the urine depends not only on the permeability of the intestinal mucosa, but also on a wide range of other factors, including gastric emptying, intestinal dilution, intestinal transit time, systemic distribution, metabolism and renal clearance.

### Genetic markers of coeliac disease

In northern Europe, around 95% of patients with coeliac disease have the HLA type A1 B8 DR3 (compared with 25% of the normal population). This haplotype is in linkage disequilibrium with the DQ alleles DQA1*0501 and DQB1*0201 which encode HLA DQ2. Here, HLA DQ2 is encoded *cis*, i.e. the encoding alleles are on the same chromosome. However, coeliac disease may also be associated with a heterozygous combination of HLA DR5 and HLA DR7. These individuals can also express DQ2 by encoding the same alleles in *trans* format, i.e. on different chromosomes.

In southern Europe, where the overall prevalence of HLA DQ2 is low, up to 15% of cases may be associated with the DR4 DQ8 haplotype. In a proportion of Italian and Tunisian patients, there also appears to be a significant association with DR53 heterodimers, with maximal risk in those people carrying both DQ2 and DR53 [12].

HLA DQ2- and DQ8-negative patients are extraordinarily rare. It is suggested that the DQ molecules DQ2 and DQ8 alone are capable of binding certain gluten peptides and thus of presenting these peptides to antigen-specific T cells in the gut, initiating the disease process. However, since 25% of the noncoeliac population carry DQ2 in the UK, the primary defect associated with the disease must lie elsewhere.

## Conclusion

A number of screening tests are available for the detection and diagnosis of coeliac disease. A combination of these can give a very high level of suspicion for coeliac disease. However, until a test is available that is 100% sensitive and specific, and not dependent upon continued gluten intake, small bowel biopsy will remain the gold standard for diagnosis.

## REFERENCES

1 Stern, M. for the Working Group on Serological Screening for Coeliac Disease (2000). Comparative evaluation of serological tests for celiac disease: a European initiative towards standardization. *Journal of Pediatric Gastroenterology and Nutrition*, **31**, 513–19.

2 Cataldo, F., Marino, V., Ventura, A., Bottaro, G. and Corazza, G.R. (1998). Prevalence and clinical features of selective immunoglobulin 1 A deficiency in coeliac disease: an Italian multicentre study. *Gut*, **42**, 362–5.

3 Scott, H., Kett, K., Halstensen, T.S., Hvatum, M., Rognum, T.O. and Brandtzaeg, J. (1992). The humoral immune system. In *Coeliac Disease*, ed. M.N. Marsh. Oxford: Blackwell, pp 239–82.
4 Hadjivassiliou, M., Grunewald, R.A. and Davies-Jones, G.A.B. (1999). Gluten sensitivity: a many-headed hydra. *British Medical Journal*, **318**, 1710–11.
5 Sabbatella, L., Di Tola, M., Di Giovambattista, F. et al. (1999). Celiac patients with antiendomysial antibodies of IgGl isotype. *Gastroenterology*, **116**, A919.
6 Dieterich, W., Ehnis, T., Bauer, M. et al. (1997). Identification of tissue transglutaminase as the autoantigen in celiac disease. *Natural Medicines*, **3**, 797–801.
7 Sulkanen, S., Halttunen, T., Laurila, K. et al. (1998). Tissue transglutaminase autoantibody enzyme-linked immunosorbent assay in detecting celiac disease. *Gastroenterology*, **115**, 1322–8.
8 Sardy, M., Odenthal, U., Karpati, S., Paulsson, M. and Smyth, N. (1999). Recombinant human tissue transglutaminase ELISA for the diagnosis of gluten sensitive enteropathy. *Clinical Chemistry*, **45**, 2142–9.
9 Uil, J.J., Van Elburg, R.M., Van Overbeek, F.M., Mulder, C.J.J., VanBerge-Henegouwen, G.P. and Heymans, H.A.S. (1997). Clinical implications of the sugar absorption test: intestinal permeability test to assess mucosal barrier function. *Scandinavian Journal of Gastroenterology*, **32**(Suppl 223), 70–8.
10 Smecuol, E., Vazquez, H., Sugai, E. et al. (1999). Sugar tests detect celiac disease amongst first degree relatives. *American Journal of Gastroenterology*, **94**, 3547–52.
11 Cox, M.A., Lewis, K.O., Cooper, B.T. (1999). Measurement of small intestinal permeability markers, lactulose and mannitol in serum: results in coeliac disease. *Digestive Diseases and Sciences*, **44**, 402–6.
12 Clot, F., Gianfrani, C., Babron, M.C. et al. (1999). HLA-DR53 molecules are associated with susceptibility to celiac disease and selectively bind gliadin-derived peptides. *Immunogenetics*, **49**, 800–7.

# Part 6

# Biomarkers in toxicology

28

# Genomics and biomarkers in toxicology

Jonathan D Tugwood, Katherine M Beckett
AstraZeneca Pharmaceuticals, Macclesfield, UK

## Toxicogenomics

As 'genetics' pertains to the study of genes and their characteristics, so 'genomics' refers to the study of the entire genetic complement (the genome) of organisms and their constituent cells. More importantly, the term genomics encompasses the experimental analysis of the expression patterns of the genes of an organism, with the aim of an improved understanding of how important biological processes are regulated. This has led to the concept of 'gene profiling' or 'transcript profiling', which essentially is the quantitative analysis of many different gene transcripts simultaneously. The ability to perform this type of analysis, together with the availability of sequence information from many thousands of genes from a number of organisms (and often the gene clones themselves), offers an enormous potential benefit to many areas of investigative biology.

Not least is the area of toxicology, where the term 'toxicogenomics' is now becoming widely used. Of the number of important potential benefits offered by the application of genomics to toxicology, perhaps the most significant are: (i) the construction of a large database of gene expression information linked to toxic end-points, and (ii) a more detailed understanding of the molecular mechanisms of compound toxicity. Strategic application of this information should result in the development of more rapid, mechanism-based screens for toxicity.

## Biomarkers in toxicology

The development and use of biomarkers in toxicology is becoming widespread, specifically in the areas of exposure monitoring, and the determination of response and susceptibility to toxins (see [1] for review). There are a number of attributes the ideal toxicology biomarker would possess, but arguably the most important attributes are a relationship to the mechanism of toxicity, a quantitative association with the end-point of toxicity, and ease and reproducibility of measurement. Can large-scale measurement of gene expression *per se*, in whatever experimental system, provide information useful to the toxicological evaluation of a compound and assist in biomarker development? In response to this question, a common assumption is that all toxic responses require modulation of gene expression to some degree and that no gene is regulated in isolation, even in the simplest biological system. This then provides a rationale for evaluating toxicogenomics as a useful tool for the discovery, development and evaluation of biomarkers in toxicology. The following sections describe briefly some of the technologies that have been developed to enable large-scale gene profiling, some examples of how these technologies has been used in the areas of toxicology and/or biomarker development, and, finally, some challenges that need to be overcome before the full potential of toxicogenomics can be realized.

## Transcript profiling technology

Traditionally, researchers have limited their experiments to studying the regulation of relatively small numbers of genes, as the methodology for determining gene expression levels and patterns is not readily applied to large-scale analysis. Recently, however, an enormous amount of resource has been devoted to revolutionizing gene expression analysis, and this has facilitated large-scale gene profiling of the type which is now fairly commonplace. Table 28.1 gives some examples of techniques in use to determine the expression levels of large numbers of genes. These can be essentially divided into 'open' and 'closed' systems – a closed system limits the extent of the analysis to those genes which are physically present as clones or oligonucleotides, e.g. on an array or 'chip', whereas an open system is not limited in this way but may have associated difficulties with data handling and quantitation.

Now that a number of technical hurdles have been overcome, the use of gene 'arrays' and 'microarrays' for obtaining expression profiles has become more commonplace, and an increasing number of 'technical platforms' are becoming available to achieve this. Essentially, a gene array is a large number of gene sequences immobilized on some kind of solid support. The principle involved in its use is simple, in that labelled cDNA is prepared from the cells or tissues of interest and used in hybridization reactions. By this method, the expression levels of many thousands of genes can be determined in a single experiment. Clearly, the volume of data obtained using gene arrays brings problems of its own in terms of data collection, management and interpretation. Development of uniform and 'user-friendly' systems for creating and interrogating gene expression databases, together with some means of integrating these with other types of database, represents a significant challenge.

**Table 28.1.** Examples of 'open' and 'closed' techniques for the analysis of patterns of gene expression

| Method | Open/ closed | Principle | Example of provider or reference |
|---|---|---|---|
| Subtractive hybridization | O | Enrichment of differentially expressed transcripts by a process of subtracting out common sequences between two RNA populations | [2] |
| Differential display | O | Amplification of transcripts from two RNA populations, followed by comparative visualization by gel electrophoresis | [3] |
| Serial amplification of gene ends (SAGE) | O | Isolation, concatenation and sequencing of short 'tagged' cDNA sequences, which give information on the nature and abundance of many genes simultaneously | [4] |
| Filter-based cDNA arrays | C | Cloned cDNAs immobilized on nylon membranes, for hybridization with labelled cDNA probes from cells or tissues | Clontech (www.clontech.com) |
| High-density cDNA microarrays | C | Cloned cDNAs (known genes and 'Expressed Sequence Tags') immobilized on glass support – $\sim$10 000 clones per $cm^2$ | Incyte/Synteni (www.incyte.com) |
| High-density oligonucleotide microarrays | C | Synthetic oligonucleotides immobilized on glass support – $\sim$180 000 per $cm^2$, representing $>$5000 genes | Affymetrix (www.affymetrix.com) |

## Application of genomics technology

Two general approaches to applying array technology to toxicology and safety assessment have evolved, either of which can be deployed depending on the nature of the investigation being carried out. The first ('deductive') approach is to use arrays with a discrete, often limited number of known genes that have been selected on a rational basis as being associated with one or more toxic end-points. The advantage here is that the data analysis is simplified considerably and restricted to those genes that are likely to be 'interesting' and relevant to the toxicological problem in question. The second ('inductive') approach is to use gene arrays with as many genes as possible, both characterized and otherwise, to maximize the amount of information obtainable from a single experiment. This scenario is applicable when novel genetic markers associated with a particular toxicity are sought, or when a gene regulation 'fingerprint' associated with a particular compound or toxic end-point is to be generated.

## Genetic biomarkers of adipogenesis – a deductive approach

The authors have been using the first ('deductive') approach in an effort to identify biomarkers associated with the toxic effects of a class of compounds, the thiazolidinediones (TZDs), developed as therapeutic agents for type II diabetes. Studies with these compounds have indicated undesirable toxic effects – particularly, adipogenesis in bone marrow – which may be explicable in that TZDs are activators of the adipogenesis-related transcription factor peroxisome proliferator activated receptor-gamma (PPAR$\gamma$) [5]. Human bone marrow stromal cell lines have been used as a screen for bone marrow toxicity of TZDs, although the end-point of lipid deposition as detected by Oil Red O staining develops rather slowly over a period of 7–10 days. Our aim with gene expression profiling is to obtain gene biomarkers for TZD-induced bone marrow adipogenesis and, on this basis, to design a more rapid, higher throughput screen for this type of toxicity. With this in mind, a nylon membrane-based gene array was designed with some 800 publicly available human cDNAs. The genes on the array came from a number of toxicologically relevant classes, including stress responses, DNA repair and drug metabolism as well as genes associated with haemopoiesis, osteogenesis and adipogenesis. In order to interrogate these arrays, the human bone marrow stromal cell line was treated separately with a variety of antidiabetic agents (four TZDs and a novel nonTZD) over 7 days. Rabbit serum, rich in PPAR$\gamma$ activators and strongly adipogenic in the stromal cell line, was used as a positive control. Poly(A+) RNA was obtained from these cells after four time points (6 hours, 24 hours, 4 days and 7 days), and radiolabelled cDNA prepared for hybridization to the arrays. For each treatment, it was determined whether all genes were up- or downregulated, or unchanged compared with untreated cells, and genes showing a 2-fold or greater up- or downregulation were included in the analysis. A sample of the results, from the 6-hour and 7-day time points, is shown in Table 28.2.

It is apparent that, even from a small subset of eleven genes, some patterns emerge that are potentially discriminatory of classes of compounds, or of individual compounds. With a more comprehensive data set and analysis, including more representatives of the compound classes, it is anticipated that a 'miniarray' of perhaps 20–30 'biomarker' genes can be designed that would function as a predictive tool for the adipogenic potential of compounds in this stromal cell system. The key test will be whether such an array is predictive of the toxicity of these compounds in vivo. If this were the case, then such a tool would be a valuable addition to the screening process for the identification of efficacious and nontoxic compounds for diabetes therapy.

**Table 28.2.** Sample data set from 'transcript profiling' experiments

| | Compound | | | | | | | | | | | | |
|---|---|---|---|---|---|---|---|---|---|---|---|---|---|
| | TZD 1 | | TZD 2 | | TZD 3 | | TZD 4 | | Rabbit serum | | Non TZD 1 | | Possible biomarker for? |
| | Time point | | | | | | | | | | | | |
| Gene | 6 h | 7 d | 6 h | 7 d | 6 h | 7 d | 6 h | 7 d | 6 h | 7 d | 6 h | 7 d | |
| Raf | | ↓ | | | | ↓ | | ↓ | | | | | TZD |
| G3PDH | | ↓ | | ↓ | | | | ↓ | | | | | |
| Fas ligand | | | | | | | | | ↑ | | | | |
| E-cadherin | | ↓ | | ↓ | | ↓ | | ↓ | | ↓ | | | TZD |
| $\alpha$-actin | | ↓ | | ↓ | | ↓ | | ↓ | | ↓ | | | |
| c-fos | | ↓ | | ↓ | | | | ↓ | | ↓ | | | |
| TransGolgi network glycoprotein 51 | | ↓ | | ↓ | | ↓ | | ↓ | | ↓ | | ↓ | |
| Angiotensin | ↑ | | ↑ | | ↑ | | | | ↑ | | ↑ | | |
| Paxillin B | ↑ | | | | | | | | | | ↑ | | |
| Interleukin 1$\beta$ | | | ↑ | | | | | | | | ↑ | | |
| Retinoid X receptor $\gamma$ | | | | | | | | | | | ↑ | | NonTZD |

*Note:*
Human bone marrow stromal cells were exposed to four different TZDs, a nonTZD or rabbit serum at a single concentration for the indicated period of time (6 hours or 7 days). Levels of gene expression were determined using a custom designed human cDNA array. Up or down arrows indicate that the compounds altered the mRNA abundance greater than 1.5-fold relative to untreated control cultures, for each of three duplicate experiments.

## Isolation of new genetic biomarkers and elucidation of molecular mechanisms – an inductive approach

Although gene arrays are a relatively new innovation, there are already several good examples in the scientific literature of the use of this technology in biomarker identification and interrogation of molecular mechanisms underlying a toxic event. One such publication is by Voehringer et al. [6] on an analysis of the molecular determinants of sensitivity to apoptosis of mouse lymphoma cells. A pair of lymphoma cell lines, one resistant to radiation-induced apoptosis and the other sensitive, were analysed by transcript profiling both before and after radiation exposure. The expression levels of some 11 000 genes were monitored using an oligonucleotide-based array, and, unsurprisingly, a large number of differences were observed between the two cell lines both before and after irradiation. This highlights a general difficulty with this type of approach, in that, because such large data sets are generated, specialized bioinformatic analyses are required to tease out the important information. One such type of analysis is 'clustering' (see, for example, [7]), which groups together genes that show similar change profiles within the same treatment categories. In this way, Voehringer et al. were able to identify classes of differentially regulated genes, and to propose a hypothetical model for apoptosis regulation in the lymphoma cells based on modulation of glutathione homoeostasis and intracellular redox potential. In addition, the experimenters identified two good genetic biomarkers of radiation sensitivity, the uncoupling protein (UCP-2) and the voltage-dependent anion channel component (VDAC-1), both of which disrupt normal mitochondrial function. Interestingly, whereas both these genes are strongly upregulated by radiation in the sensitive cell type but not the resistant, this differential regulation is not seen with closely related genes of the same family, e.g. UCP-1 and VDAC-2. This implies a specific role for these proteins and enhances their credibility as biomarkers.

A further illustration of the utility of gene arrays in biomarker development, albeit not directly related to toxicology, is provided by a study aimed at obtaining biomarkers of clinical importance in renal cell carcinomas (RCCs) [8]. The initial step involved the use of a nylon filter-based gene array with 5184 human cDNAs to obtain transcript profiles from normal human kidney and from an RCC cell line. A simple visual comparison of the array images identified 89 cDNAs as being differentially expressed between the two samples. One of these was vimentin, a good candidate for further investigation as this had been shown previously to be a good prognostic indicator in breast and cervical cancer. To evaluate the usefulness of vimentin as a prognostic biomarker, another type of array was used, comprising 532 0.6 mm-diameter RCC tissue samples arrayed on microscope slides. Immunohistochemistry was used to determine levels of vimentin protein in the samples, and correlations sought with tumour histology and patient survival. It was found that vimentin expression in clear cell RCC correlated with poorer survival (relative risk of 1.6), and illustrated the usefulness of array technology in the rapid identification and evaluation of biomarkers.

Another recent example has employed gene profiling to identify a 50-gene set that can be used to discriminate between acute myeloid leukaemia and acute lymphoblastic leukaemia [9]. Each of the genes within this set can be considered to be an individual biomarker, and the confidence in the accuracy of these will be enhanced or diminished as more samples are analysed.

## Applying genomics to toxicology – some challenges and the way forward?

From the preceding examples, it is clear that in the field of genomics the selection of a biological system around which clear questions can be posed is critical. It is no coincidence that the vast majority of good quality published data on transcript profiling has been derived from in vitro systems, or from simpler in vivo systems such as bacteria and yeast. This poses a problem for toxicologists, since there is the dual issue of the complexity of toxic responses in animal models, and the need for obtaining information on potential human toxicity from these models. Clearly, in vitro models of toxicity do exist and have been widely used, but, with the possible exception of in vitro genotoxicity assays, there is continual debate and controversy about the usefulness of these.

As mentioned previously, the hope for toxicogenomics is that it will provide both information on toxic mechanisms and allow the development of more accurate and rapid screens for toxicity. With this in mind, many laboratories, both individually and collectively as collaborations and consortia, are embarked on a 'proof of concept' exercise that hopefully will demonstrate that transcript profiling will prove useful to toxicology and to biomarker development. Most initial experiments are focused on systems where there is a wealth of available information on toxic responses, and where a good selection of test compounds is available. Favoured experimental systems are liver, both rat/mouse in vivo and in primary hepatocyte culture, and in vitro genotoxicity systems such as L5178Y $TK^{+/-}$ mouse lymphoma cells.

An obvious caveat is that, while transcript profiling provides information on mRNA abundance, the mediators of biological effects are the proteins they specify and mRNA abundances are not always representative of protein abundances. The corresponding field of 'proteomics' has evolved to study protein identity and abundance on a large scale, but, compared with genomics/transcript profiling, progress has been slow due to limitations of the technology with regard to throughput and sensitivity. The recent advent of 'protein chips' should go a long way towards resolving these issues.

Finally, and critically, the usefulness or otherwise of toxicogenomics will depend on how the enormous volume of data generated is collated, analysed and interpreted. A large amount of resource is already being ploughed into database construction and the development of analytical tools. Importantly, the interaction of cells and tissues with xenobiotic compounds will result in changes in patterns of gene expression, not all of which will be of significance in terms of toxicity. The ultimate challenge will be for those whose task it is to exercise their toxicological experience and judgement in interpreting the data, and in deriving novel biomarkers and mechanistic information.

## REFERENCES


1 Timbrell, J.A. (1998). Biomarkers in toxicology. *Toxicology*, **129**, 1–12.

2 Diatchenko, L., Lau, Y.F.C., Campbell, A.P. et al. (1996). Suppression subtractive hybridization – a method for generating differentially regulated or tissue-specific cDNA probes and libraries. *Proceedings of the National Academy of Sciences USA*; **93**, 6025–30.

3 Liang, P., Pardee, A.B. (1992). Differential display of eukaryotic messenger RNA by means of the polymerase chain reaction. *Science*, **257**, 967–71.

4 Velculescu, V.E., Zhang, L., Vogelstein, B. and Kinzler, K.W. (1995). Serial analysis of gene expression. *Science*, **270**, 484–7.

5 Lambe, K.G. and Tugwood, J.D. (1996). A human peroxisome-proliferator-activated receptor-gamma is activated by inducers of adipogenesis, including thiazolidinedione drugs. *European Journal of Biochemistry*, **239**, 1–7.

6 Voehringer, D.W., Hirschberg, D.L., Xiao, J. et al. (2000). Gene microarray identification of redox and mitochondrial elements that control resistance or sensitivity to apoptosis. *Proceedings of the National Academy of Sciences USA*, **97**, 2680–5.

7 Eisen, M.B., Spellman, P.T., Brown, P.O. and Botstein, D. (1998). Cluster analysis and display of genome-wide expression patterns. *Proceedings of the National Academy of Sciences USA*, **95**, 14863–8.

8 Moch, H., Schraml, P., Bubendorf, L. et al. (1999). High-throughput tissue microarray analysis to evaluate genes uncovered by cDNA microarray screening in renal cell carcinoma. *American Journal of Pathology*, **154**, 981–6.

9 Golub, T.R., Slonim, D.K., Tamayo, P. et al. (1996). Molecular classification of cancer: class discovery and class prediction by gene expression monitoring. *Science*, **286**, 531–7.

29

# Protein profiling and proteomic databases

Julio E Celis, Pavel Gromov, Morten Østergaard, Hildur Pálsdóttir, Irina Gromova

University of Aarhus, Aarhus C, Denmark

## Introduction

With the completion of the human genome sequencing project [1, 2], the next great challenge in the life sciences in this millennium will be in decoding of the genome information in terms of regulation and function. Gradually, the emphasis of the Human Genome Project is starting to shift towards functional genomics, an area of the post-genomic era that aims to identify and fuctionally characterize proteins, the main effectors of cellular function. Indeed, today we are experiencing a rapid explosion of technology for the high throughput expression analysis of genes and their products.

Proteomics is a key area of research within functional genomics. First derived by Wilkins and colleagues in 1996 [3], the term proteomics describes an emerging technology making use of a plethora of protein analysis techniques (high-resolution two-dimensional polyacrylamide gel electrophoresis [2D PAGE], mass spectrometry, bioinformatics, etc.) to resolve, quantitate and identify proteins as well as to reveal their interacting partners. This information, together with protein behavioural data generated in various cell types and tissues, has speeded up the establishment of comprehensive 2D PAGE databases that aim to link protein information with DNA sequencing and mapping data from genome projects [4–7; http:/biobase.dk/cgi-bin/celis]. 2D PAGE databases play an important role in the functional annotation of genes and, in particular, human proteomic databanks are expected to expedite drug discovery by pinpointing candidate drug targets on the basis of changes in the proteome expression profile of biopsies obtained from patients and controls [8].

Today, there is a great deal of interest in proteomics, particularly in the study of diseases, as these technologies are expected to reveal gene regulation events involved in disease progression as well as generating potential targets for drug discovery and diagnostics. Furthermore, the technology is expected to have a great impact on agriculture, toxicology and the biotechnology industry in general.

## How many proteins are expressed in a human cell type?

The usefulness of the 2D PAGE technique for large-scale proteomic projects depends very much on the number of proteins to be resolved in a complex protein mixture – for example, a human cell. Current estimates indicate that the human genome may contain no more than 30000 protein coding genes [1, 2]. Data from a few laboratories, including the authors', have shown that only a fraction of these genes are switched on in a given cell type [6, 9]. Single cells typically express in the range of 5000–6000 different primary translation products plus their modifications, which can be extensive in some cases [6, 8]. The number of proteins in a human cell is significantly higher than that reported for yeast cells, which are known to express about 4500 proteins under growing conditions and to contain far fewer modifications. In humans, as much as 80–90% of the proteins may represent housekeeping genes (i.e. components of metabolic pathways, cytoarchitectural elements, etc.) that are expressed, albeit in variable amounts, by most cell types [6, 8 and references therein]. Considering that there are at least 250 different cell types in the human body, each expressing on average about 100–200 proteins unique to the cell type, one ends up with a total number of polypeptides that is reasonably close to the estimated number of genes.

At present, only a small proportion of the human proteome [3] – that is, the protein equivalent of the human genome – has been identified and, for most of these proteins, their physiological roles remain unknown. Proteins orchestrate most cellular functions and, therefore, it is of importance to develop technology to resolve, quantitate and identify them, as well to reveal their interacting partners.

## Human 2D PAGE databases

### The 2D gel technology

For the last 25 years, high-resolution 2D PAGE has been the technique of choice to analyse the protein composition of a given cell type and for monitoring changes in gene activity through the quantitative and qualitative analysis of the thousands of proteins that orchestrate various cellular functions. Proteins are frequently the functional molecules and, therefore, the most likely to reflect differences in gene expression. Genes may be present, they may be mutated, but they are not necessarily transcribed. Some genes are transcribed but not translated, and the number of mRNA copies per cell does not necessarily reflect the number of functional protein molecules [10 and references therein]. Consequently, because of these uncertainties, focusing on the proteins eliminates many of these constraints.

The technique, which was originally described by O'Farrell [11, 12] and Klose [13], separates proteins both in terms of their isoelectric point (pI) and molecular weight (Mr). Usually, one chooses a condition of interest and lets the cell reveal the global protein behavioural response because all detected proteins can be analysed (relative abundance, post-translational modifications, coregulated proteins, etc.) both qualitatively and quantitatively in relation to each other. As an example, Figure 29.1 depicts changes in the proteome expression profile of primary human keratinocytes induced to differentiate by the addition of 10% fetal calf serum. Clearly, the expression of many polypeptides, known and unknown, changes in response to factors in the serum.

A systematic analysis of the human proteome by 2D PAGE requires a reproducible gel system to resolve the proteins as well as computer-assisted technology to scan the gels, make synthetic images, assign numbers to individual spots and match gel spots. In addition, one needs functions to enter and retrieve qualitative and quantitative information. To date, the authors' work has been based entirely on the use of carrier ampholytes as originally described by O'Farrell [11, 12]. Although gels run with carrier ampholytes are difficult to reproduce, we have standardized the technology to a level at which it is possible to obtain reproducible separations routinely of nearly 4000 [$^{35}$S]-methionine-labelled polypeptides from whole cell extracts (see procedures and videos at http:/biobase.dk/cgi-bin/celis). The lowest levels of detection for [$^{35}$S]-methionine-labelled proteins correspond to polypeptides that are present in about 40 000 molecules per cell, while immunoblotting in combination with enhanced chemoluminescence (ECL) detection can reveal components that are present in very few copies per cell. The latter is exemplified in Figure 29.2 which shows an ECL-developed 2D gel Western blot of crude keratinocyte extracts reacted with antibodies raised against Ha-ras, a protein that is known to be expressed in about 20 000 molecules per cell. We surmise from the intensity of the various spots in Figure 29.2 that the lower abundancy product indicated with an arrow may be present in no more than 1000 molecules per cell. Thus, by increasing the sensitivity of the detection procedures, and/or by analysing subcellular fractions, it should be possible to detect very low abundancy.

Bjellqvist and others [14–16] have introduced the use of immobilized pH gradients (IPGs), which are integrated into the polyacrylamide matrix and, therefore, provide a more reproducible focusing pattern. By avoiding some of the problems associated with carrier ampholytes, such as cathodic drift and endosmosis, IPGs allow a higher loading capacity for micropreparative runs, and provide increased charge resolution when narrow pH gradients are used (1 pH unit/18 cm) [17, 18 and references therein]. In the authors' hands, however, carrier ampholytes (3.5–10) and broad range IPGs resolve similar number of [$^{35}$S]-methionine-labelled polypetides. Narrow range IPG gradients, however, offer higher resolution: proteins with a difference of less than 0.025 pI units can be resolved [17, 18]. It has been proposed that narrow range, overlapping IPG gradients viewed side by side

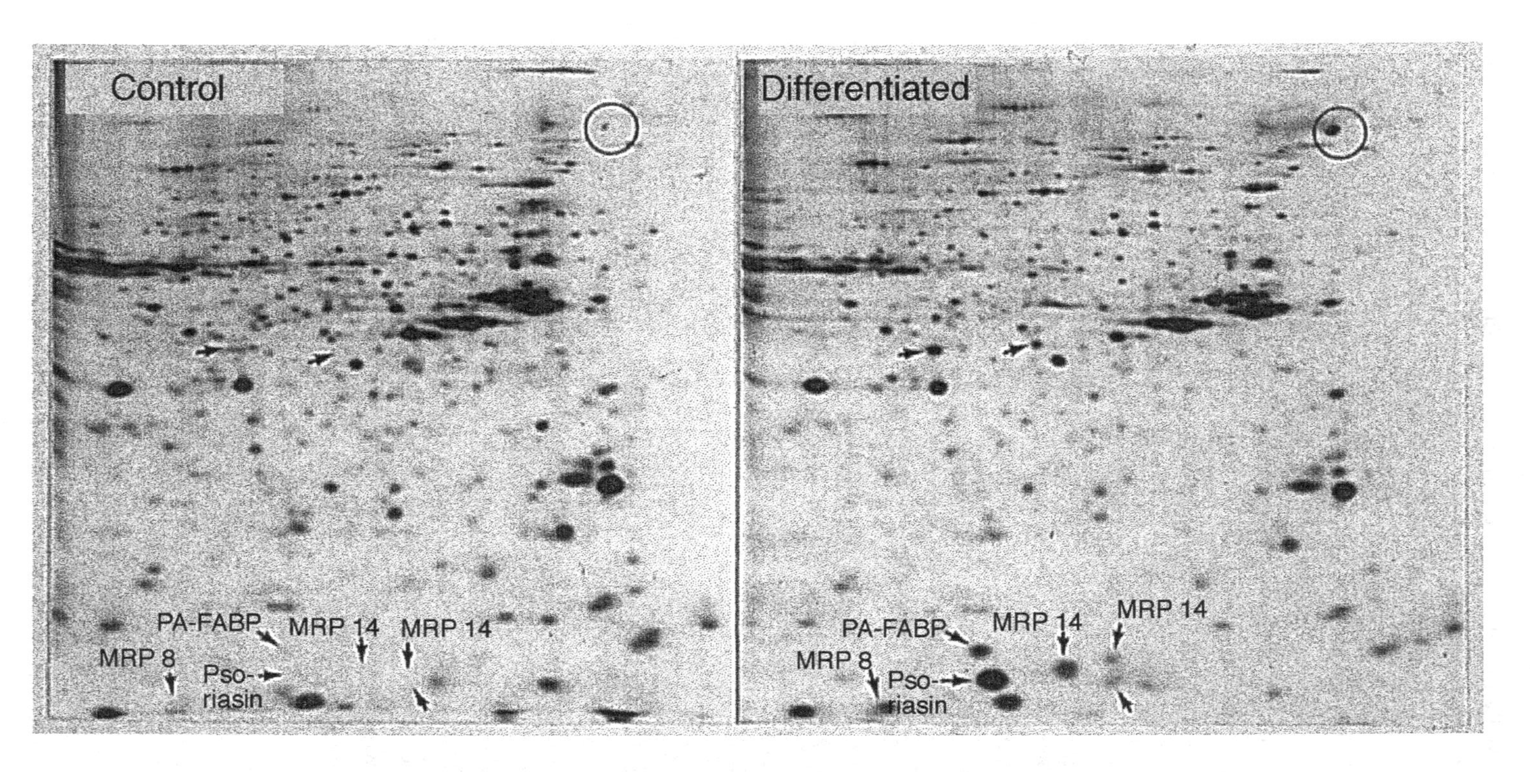


Figure 29.1 [$^{35}$S]-methionine-labelled keratinocyte proteins from control and 10% fetal calf serum-treated noncultured keratinocytes.

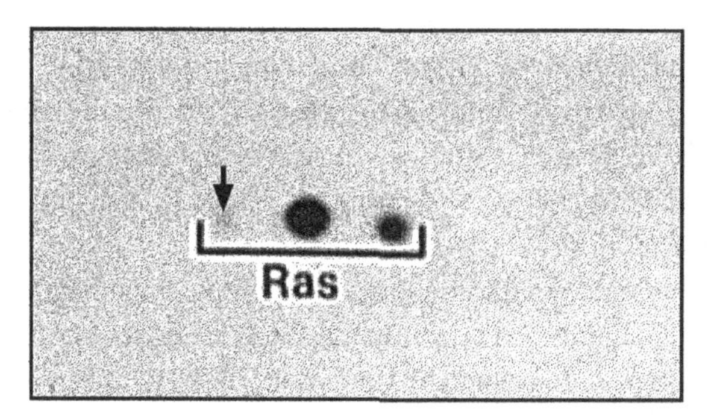


Figure 29.2 2D gel immunoblot of total keratinocyte proteins reacted with Ha-ras antibodies and developed using the ECL procedure.

may provide a solution to the problem of resolving and depicting the proteome of a given cell type. Recently, however, Corthals et al. found this solution unrealistic as it will require running a large number of gels [18].

## Making a 2D PAGE database

The first step in making a comprehensive 2D PAGE database is to prepare a synthetic image (i.e. a digital form of the gel image) of the chosen standard or master reference gel (autoradiogram, fluorogram, Coomassie blue- or silver-stained gel). This can be done with laser scanners, charge couple devices (CCDs), array scanners, television cameras, rotating drum scanners and multiwire chambers [19]. Various software for 2D PAGE analysis have been developed that work on different hardware; these include Tycho, Gellab, Kepler, Quest, PDQUEST, Lips, Elsie, Hermes Gemini and Melanie [19 and references therein].

In Aarhus, fluorograms are scanned using a laser scanner and the data are analysed with the use of the PDQUEST II software (formerly Protein Databases Inc., now bought by Bio-Rad) running on a SPARK station computer from SUN Microsystems, Inc. The scanner measures intensity in the range of 0–2.0 absorbance. A typical scan of a 17 cm × 17 cm fluorogram takes about 2 min. Steps in the image analysis include: image acquisition, initial smoothing, background subtraction, final smoothing, spot detection and fitting of ideal Gaussian distributions to spot centres. Spot intensity is calculated as the integration of a fitted Gaussian. If calibration strips containing individual segments of known amount of radioactivity are used, it is possible to merge multiple exposures of the sample image into a single data image of greater dynamic range. Functions that can be used to edit the images include: *cancel* (e.g. to erase scratches that may have been interpreted as spots by the computer and to cancel streaks or low dpm [decays per min] spots), *combine* (sometimes a spot may be resolved into several closely packed spots), *restore*, *uncombine* and *add spot* to the gel. The editing process is time consuming and can only be performed by operators who have experience in the analysis of gel samples. Today, more accurate quantitative data can be obtained by using a phosphorimager. Once the synthetic image is created, it can be stored on disk and displayed directly on the monitor.

Each polypeptide is assigned a number by the computer, a fact that facilitates the entry and retrieval of qualitative and quantitative information for any given spot in the gel. The standard image can be matched automatically by the computer to other standard or reference gels provided a few landmark spots are given manually as reference to initiate the process. Proteins are matched according to their gel position and, therefore, additional ways to verify their relatedness are needed before one can take full advantage of the data. Once a standard map of a given protein sample is made, one can enter qualitative or quantitative information to establish a reference or master database (see, for example, http:/biobase.dk/cgi-bin/celis) [6, 8]. Categories or entries are created so as to gather information on physical, chemical, biochemical, physiological, genetic, architectural and biological properties of proteins. In general, entries reflect the type of biological problem – for example, normal versus disease sample – that is being studied using the database approach.

## Aarhus 2D PAGE databases on the Internet: http://biobase.dk/cgi-bin/celis

As a result of a long-term and systematic effort to analyse the human proteome in bladder cancer and skin ageing, the authors have established comprehensive databases of noncultured keratinocytes [6], tissues (bladder transitional cell carcinomas [TCCs] and squamous cell carcinomas [SCCs]; http://biobase.dk/cgi/bin/celis) and fluids (urine) that can be accessed in part in the World Wide Web using a custom made software developed by Protein Databases Inc. Of the databanks, the keratinocyte database is by far the most extensive, although the TCC database is expected to be our major database in the future.

Figure 29.3 shows the synthetic master 2D PAGE image (isoelectric focusing, IEF) of TCC proteins as depicted on the World Wide Web (http://biobase.dk/cgi-bin/celis). Proteins flagged with a cross correspond to known polypeptides. About 600 polypeptides have been identified in this database (IEF and NEPHGE) of the nearly 3200 that have been resolved and catalogued [8]. Proteins have been identified by one or a combination of techniques that include: (i) mass spectrometry of tryptic peptides; (ii) 2D gel immunoblotting using specific antibodies and ECL detection; (iii) microsequencing of Coomassie Brilliant Blue-stained proteins; (iv) comigration with known human proteins (individual proteins and organelle components); (v) vaccinia virus and COS-1 expression of full length cDNAs; (vi) in vitro coupled

Transitional Cell Carcinomas-IEF database

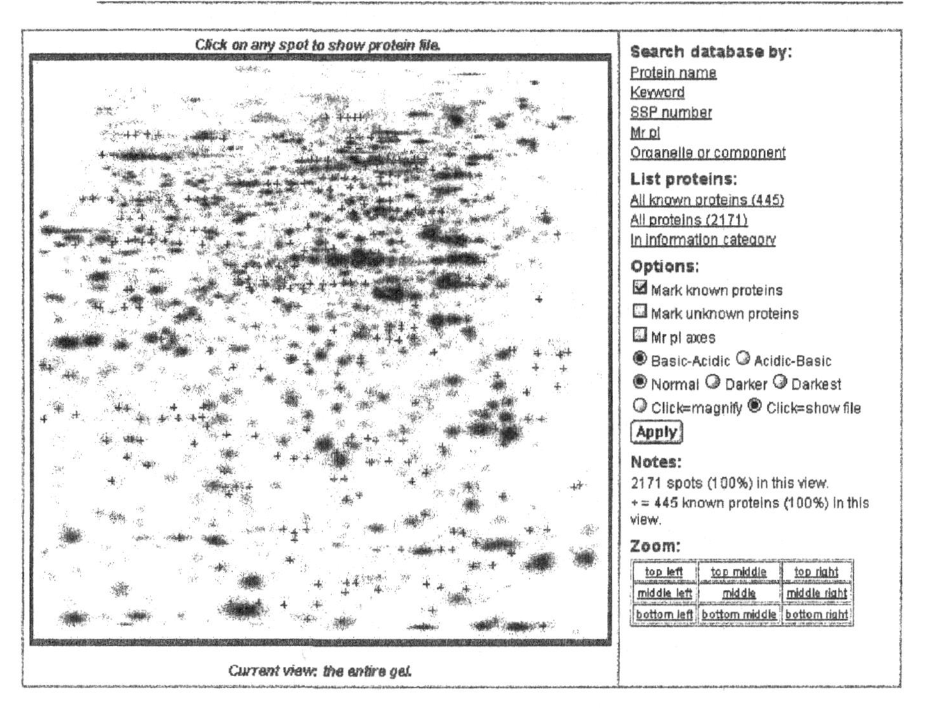


Figure 29.3 Master synthetic image of human TCC proteins separated by IEF 2D PAGE as depicted in the World Wide Web. Proteins flagged with a cross correspond to known proteins. By clicking on any spot, it is possible to obtain a file containing information about the protein as well as links to other sites in the World Wide Web.

transcription–translation; (vii) blot overlay techniques; and (viii) labelling of post-translational modifications. Both microsequencing and mass spectrometry are instrumental for the identification of novel proteins as they provide sequence information for molecular cloning of the corresponding mRNAs. Today, mass spectrometry is the technique of choice for protein identification as it requires picomole levels of the proteins and, therefore, many low abundancy proteins have become amenable to analysis [20]. In addition, peptide sequencing is also possible using mass spectrometry – a fact that will greatly facilitate the identification of unknown proteins on 2D gels as well as the cloning of genes coding for novel proteins [20, 21].

Information gathered on any given polypeptide, known or unknown, can be easily retrieved by clicking on the corresponding spot – in this case, to the adipocyte fatty acid-binding protein, A-FABP, which exhibits a very narrow tissue distribution and is downregulated during tumour progression [22]. Files for known proteins contain links to a subset of Medline (http:// www.ncbi.nlm.nih.gov/PubMed/), Swiss-Prot (http://expasy.hcuge.ch/sprot/sprot-top.html) and PDB (http://www.embl-heidelberg.de/pdb/). Other links include OMIM (http://www.ncbi.nlm.nih.gov/Omim/), GeneCards (http://bioinformatics.weizmann.ac.il/cards), UniGene (http://www.ncbi.nlm.nih.gov/UniGene/index.html) and other Web sites such as CySPID (cytoskeletal protein database; http://paella.med.yale.edu/~panzer/cytoskdb/index.html), metabolic pathways (compiled by KEGG; http://www.genome.ad.jp/kegg/), the cytokine explorer (http://kbot.mig.missouri.edu:443/cytokines/explorer.html), histology images (http://biosun.biobase.dk/~pdi/jecelis/micrographs.html), etc. In the future, as new databases and related Web sites become available, it will be possible to navigate throughout various databanks containing complementary information (i.e. nucleic acid and protein sequence, genome mapping, diseases, protein structure, post-translational modifications, antibodies, signalling pathways, etc.). 2D PAGE databases are expected to annotate DNA sequences, and will be instrumental in linking protein and DNA sequencing with mapping information, offering a global approach to the study of cell regulation [6, 8].

Database queries can be by name, protein number or keywords (Figure 29.3) as well as by organelle or cellular component. By clicking on any of the organelles, cellular structures and components, it is possible to get a protein list as well as their relative position in the master image. In addition, one can retrieve a list of all known proteins recorded in the database.

Today, about 100 information categories are available in the World Wide Web version of the TCC database. To name a few, these include protein name, cellular localization, gene map, gene name, proteins expressed in other cell types, heat shock proteins, transformation-sensitive proteins, levels in fetal human tissues, etc. Categories, or entries, are created so as to gather information on physical, chemical, biochemical, physiological, genetic, architectural and biological properties of proteins. In general, entries reflect the type of specific biological problem that is being studied using the database approach.

## Application of the keratinocyte 2D PAGE database to the study of bladder SCCs

SCCs are composed of a single cell type closely resembling keratinocytes both in morphology and proteome expression profile [22]. Grading of SCCs is subjective and takes into consideration the degree of nuclear polymorphism, nuclear to cytoplasmic ratio and chromatin clumping, as well as the number of mitotic cells. The lack of objective histopathological criteria for the grading of SCCs, as well as of

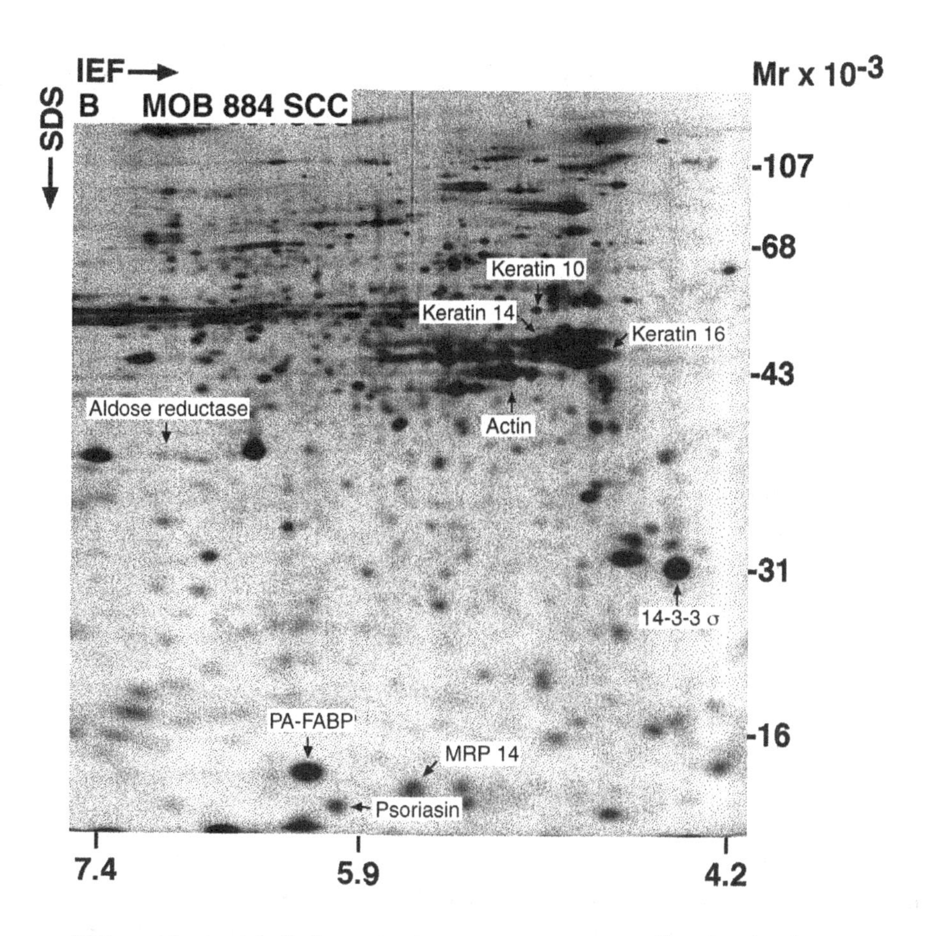


Figure 29.4 [$^{35}$S]-methionine-labelled proteins from an SCC expressing abundant levels of psoriasin.

markers for the early detection of individuals at risk, has presented major obstacles for the early preventive intervention of these lesions. SCCs have a bad prognosis and, therefore, it is of paramount importance to reveal biomarkers that may identify individuals at risk.

In the authors' laboratory, SCC biomarkers are being searched for, taking advantage of the fact that these lesions share most of their proteins in common with keratinocytes (compare Figures 29.1 and 29.4), a cell type for which we know approximately 35% of all resolved proteins and for which we have established a comprehensive 2D PAGE database that contains substantial information on proteins, including tissue distribution and whether they are secreted or externalized to the culture media. The latter information is invaluable when searching for specific SCCs markers that are externalized to the urine. Proteome analysis of fresh SCCs

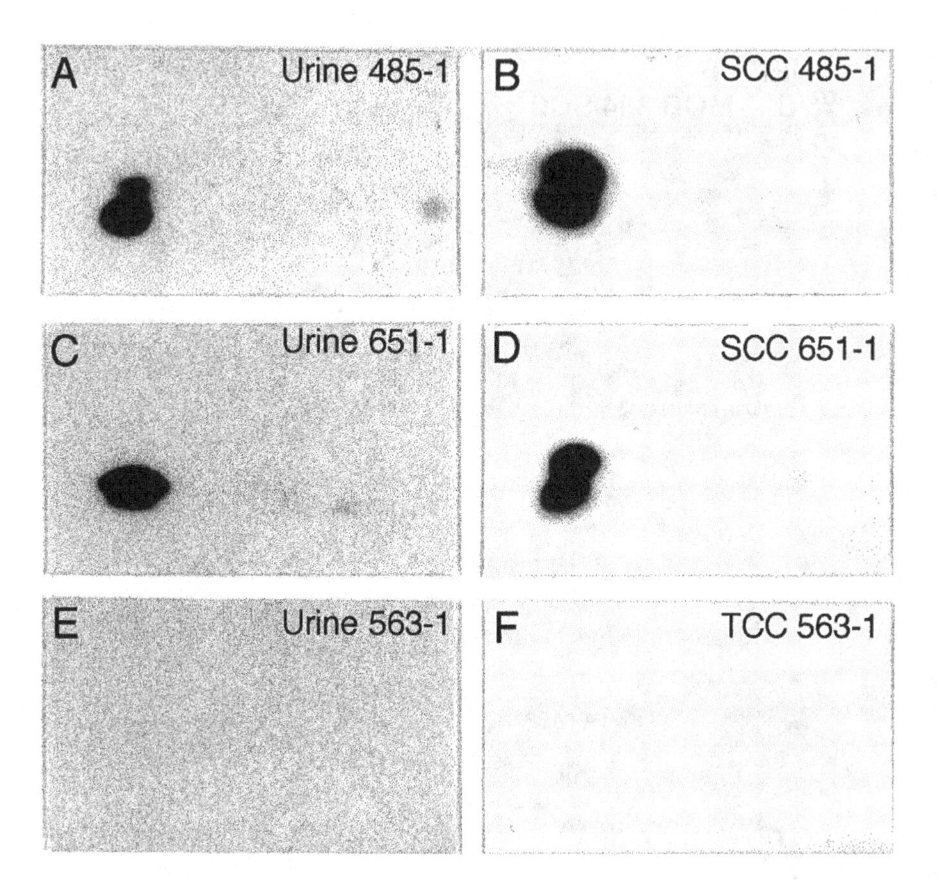


Figure 29.5 2D gel Western blots reacted with antibodies against psoriasin. (A) urine SCC 485–1, (B) SCC 485–1, (C) urine SCC 651–1, (D) SCC 651–1, (E) urine TCC 563–1, (F) TCC 563–1. The equivalent of 1 ml of urine was applied to the first dimension gels.

as well as of the patient's urine using 2D PAGE electrophoresis, in conjunction with protein identification techniques, has so far revealed a single biomarker, the calcium-binding protein psoriasin [23, 24] (Figure 29.4), which is synthesized by some differentiated cells in the SCCs and that can be detected in the urine of SCC patients by immunoblotting (Figure 29.5). With the exception of urothelial tissue exhibiting squamous metaplasia, psoriasin is not expressed by any other cell type in the urinary tract [24]. Psoriasin, alone or as part of a biomarker profile, may prove valuable for the noninvasive follow up of SCC-bearing patients, particularly in males as the authors have previously shown that the frequent presence of stratified squamous epithelia in the female trigone may lead to false positives [23, 24].

At present, the authors are carrying out a systematic analysis of hundreds of fresh tumours (TCCs and SCCs) [22–24] and random biopsies in an effort to identify markers that may subdivide histopathological types and that will provide specific probes (antibodies for immunohistochemical analysis of paraffin-embedded tissue samples and for urine-based enzyme-linked immunosorbent assays) for the objective diagnosis, prognosis and treatment of these lesions. In addition, these putative biomarkers will serve as landmarks for forthcoming research aimed at dissecting the various stages of tumour progression. The aim of these studies is to identify signalling pathways and components that are deregulated in bladder cancer and that may provide novel leads for drug discovery.

## Concluding remarks and perspectives

Advances in mass spectrometry have made possible the establishment of comprehensive proteomic 2D PAGE databases (http://biobase.dk/cgi-bin/celis) that aim to link protein information with existing and forthcoming DNA mapping and sequence data from the Human Genome Project (Figure 29.6). These databases may provide an alternative route to drug discovery by pinpointing signalling pathways and protein components that are deregulated in particular diseases. With the integrated approach offered by 2D PAGE databases, it is now possible to reveal phenotype-associated proteins, assess their identity, search for homology with previously identified proteins, clone the cDNAs, assign partial protein sequences to genes for which the full DNA sequence and the chromosome location are known, and to study the regulatory properties and function of groups of proteins that are co-ordinately expressed in a given biological process. Human 2D PAGE databases are expected to provide an integrated picture of the expression levels and properties of the thousands of protein components of organelles, pathways and cytoskeletal systems. Obtained under both physiological and abnormal conditions, these will complement the current efforts to map and sequence the entire human genome (see also http://expassy.hcuge.ch/www/expassy-top.html). In addition, 2D PAGE databases of target tissues are expected to be instrumental in toxicology studies. Most importantly, proteomic databases and associated technologies are expected to address problems that cannot be approached by DNA analysis – namely, relative abundancy of the protein product, post-translational modification, subcellular localization, turnover, interaction with other proteins and functional aspects [8].

At present, there is a great deal of effort being devoted to the development of high throughput technology to resolve, visualize, quantitate and characterize proteins in an effort to find correlations between protein expression profiles and biological function both in health and disease [8]. In addition, there is much work being devoted to the study of protein/protein interactions as well as to determining the functional meaning of post-translational modifications. While there is still much to be done about the complexity of tissue samples, which complicates the interpretation

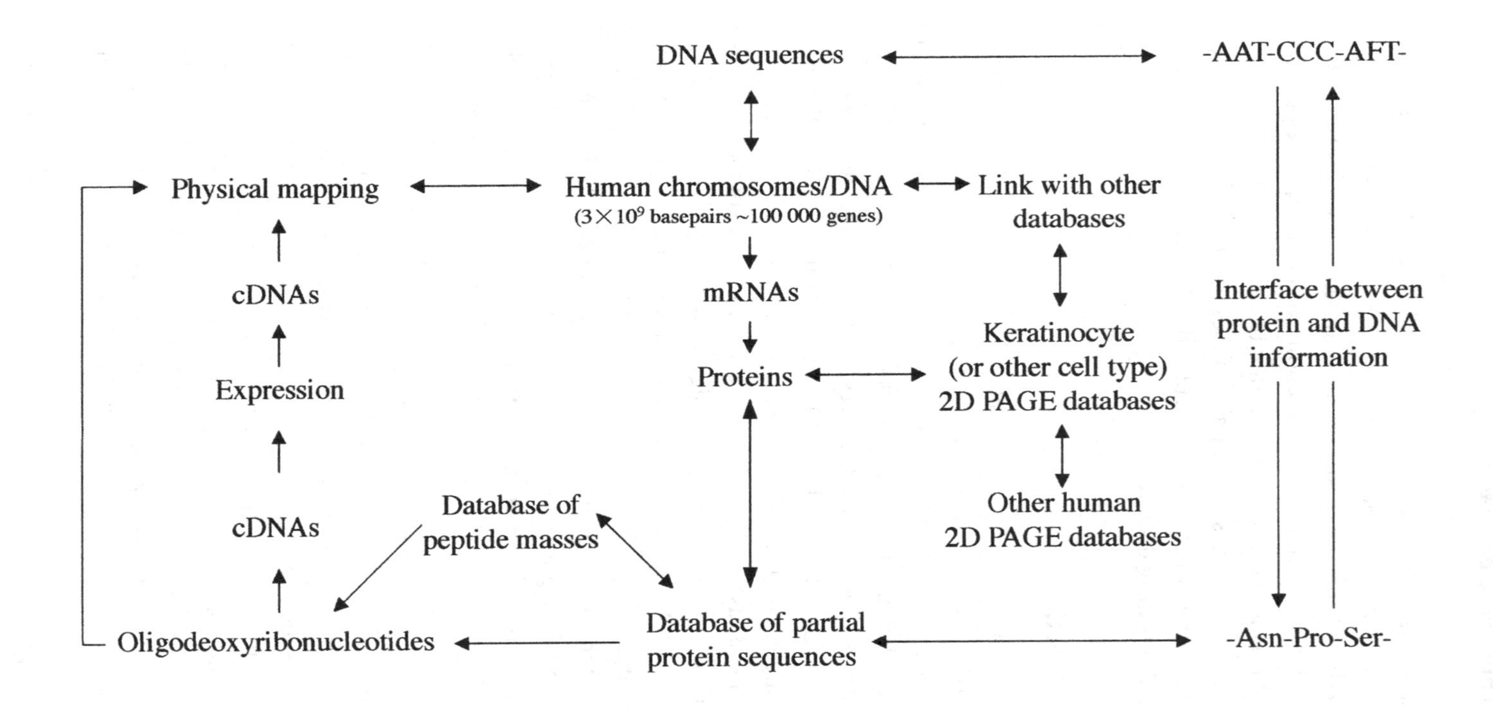


Figure 29.6 Interface between human 2D PAGE proteomic databases and the human genome mapping and sequencing programmes.

of the results, we believe that the current technology has reached a stage in which it is possible to contemplate the translation of basic discoveries into meaningful clinical applications.

## Acknowledgements

We would like to thank Pamela Celis, Bodil Basse, Jette B. Lauridsen, Inger Andersen and Gitte Ratz for expert technical assistance and other colleagues in the group for helpful discussions. The work was supported by grants from the Danish Biotechnology Programme, the Danish Cancer Society and the Danish Centre for Molecular Gerontology.

## REFERENCES

1 Venter, J.C., Adams, M.D., Myers, E.W. et al. (2001). The sequence of the human genome. *Science*, **29**, 1304–51.

2 Lander, E.S., Linton, L.M., Birren, B. et al. (2001). Initial sequencing and analysis of the human genome. *Nature*, **409**, 860–921.

3 Wilkins, M.R., Sanchez, J.-C., Gooley, A.A. et al. (1996). Progress with proteome projects: why all proteins expressed by a genome should be identified and how to do it. *Biotechnology of Gene Engineering Reviews*, **13**, 19–50.

4 Anderson, N.G. and Anderson, L. (1982). The Human Protein Index. *Clinical Chemistry*, **28**, 739–48.

5 Bravo, R. and Celis J.E. (1982). Up-dated catalogue of HeLa cell proteins: percentages and characteristics of the major cell polypeptides labeled with a mixture of 16 $^{14}C$-labeled amino acids. *Clinical Chemistry*, **28**, 766–81.

6 Celis, J.E., Rasmussen, H.H., Leffers, H. et al. (1991). Human cellular protein patterns and their link to genome DNA sequence data: usefulness of two-dimensional gel electrophoresis and microsequencing. *FASEB Journal*, **5**, 2200–8.

7 Humphery-Smith, I. (1997). Microbial proteomes. *Electrophoresis*, **18**, 1205–498.

8 Celis, J.E., Gromov, P., Østergaard, M. et al. (1996). Human 2-D PAGE databases for proteome analysis in health and disease: http://biobase.dk/cgi-bin/celis. *FEBS Letters*, **398**, 129–34.

9 Duncan, R. and McConkey, E.H. (1982). How many proteins are there in a typical mammalian cell? *Clinical Chemistry*, **28**, 749–55.

10 Anderson, L. and Seilhamer, J. (1997). A comparison of selected mRNA and protein abundances in human liver. *Electrophoresis*, **18**, 533–7.

11 O'Farrell, P. (1975). High resolution two-dimensional electrophoresis of proteins. *Journal of Biological Chemistry*, **250**, 4007–21.

12 O'Farrell, P.Z., Goodman, H.M. and O'Farrell, P. (1977). High resolution two-dimensional electrophoresis of basic as well as acidic proteins. *Cell*, **12**, 1133–41.

13 Klose, J. (1989). Systematic analysis of the total proteins of a mammalian organism: principles, problems and implications for sequencing the human genome. *Electrophoresis*, **10**, 140–52.
14 Bjellqvist, B., Ek, K., Righetti, P.G. et al. (1982). Isoelectric focusing in immobilized pH gradients: principle, methodology and some applications. *Journal of Biochemical and Biophysical Methods*, **6**, 317–39.
15 Görg, A., Postel, W. and Günther, S. (1988). The current state of two-dimensional electrophoresis with immobilized pH gradients. *Electrophoresis*, **9**, 531–46.
16 Righetti, P.G. (1990). *Immobilized pH Gradients: Theory and Methodology*. Amsterdam: Elsevier.
17 Görg, A., Obermaier, C., Boguth, G. et al. (2000). The current state of two-dimensional electrophoresis with immobilized pH gradients. *Electrophoresis*, **21**, 1037–53.
18 Corthals, G.L., Wasinger, V.C., Hochstrasser, D.F. and Sanchez, J.C. (2000). The dynamic range of protein expression: a challenge for proteomic research. *Electrophoresis*, **21**, 1104–15.
19 Miller, M.J. (1989). Computer-assisted analysis of two-dimensional gel electrophoretograms. *Advances in Electrophoresis*, **3**, 182–217.
20 Pappin DJ. (1997). Peptide mass fingerprinting using MALDI-TOF mass spectrometry. *Methods in Molecular Biology*, **64**, 165–73.
21 Mann, M. A shortcut to interesting human genes: peptide sequence tags, expressed-sequence tags and computers. *Trends in Biochemical Sciences*, **21**, 494–5.
22 Celis, J.E., Østergaard, M., Basse, B. et al. (1996). Loss of adipocyte-type fatty acid binding protein and other protein biomarkers is associated with progression of human bladder transitional cell carcinomas. *Cancer Research*, **56**, 4782–90.
23 Celis, J.E., Rasmussen, H.H., Vorum, H. et al. (1996). Bladder squamous cell carcinomas express psoriasin and externalize it to the urine. *Journal of Urology*, **155**, 2105–12.
24 Østergaard, M., Rasmussen, H.H., Nielsen, H.V. et al. (1997). Proteome profiling of bladder squamous cell carcinomas: identification of markers that define their degree of differentiation. *Cancer Research*, **57**, 4111–17.

30

# Biomarkers for evaluating the safety of genetically modified foods

Ad ACM Peijnenburg, Hubert PJM Noteborn, Harry A Kuiper
RIKILT (National Institute for Quality Control of Agricultural Products), Wageningen, The Netherlands

Recombinant DNA technology has found many interesting applications within the area of plant breeding. Food crops have been developed with altered characteristics like improved pest and disease resistance and prolonged shelf-life. Although the registration of new varieties of existing crop plants has not resulted in adverse effects in humans, the conventional assessments by plant breeders were not considered to be sufficient to ensure food safety of the genetically modified (GM) crop plants. Therefore, the European Community has established a legal framework for the introduction of novel foods with the Regulation on Novel Foods and Novel Food Ingredients (EC/258/97), which came into force in May 1997. The accompanying guidelines for the safety assessment of GM foods are centred around the so-called 'Concept of Substantial Equivalency'. According to this principle, the evaluation should be based on a comparison with conventionally bred products, assuming that these traditional foods have a long history of safe use. The guidelines indicate the type of agronomical, (bio)chemical and nutritional parameters to be considered for the safety assessment and deal with three main food safety issues:

- the nutritional and toxicological consequences of inserted gene products;
- the allergenicity of expressed proteins and novel foodstuffs; and
- the potential of pleiotropic (unintended) effects in the host organism due to the insertion event.

This chapter describes existing methods for the safety evaluation of transgenic crop plants in general. Emphasis is given to advanced new technologies that might be used not only to identify new functional biomarkers for food safety testing but also to establish the similarities and differences between GM plants and their conventionally bred counterparts.

## Nutritional and toxicological effects of inserted gene products

### General principles

If substantial equivalence can be established apart from the newly introduced trait of the transgenic crop plant, any further assessment of safety should focus specifically on the introduced trait itself. A principal drawback is that the new proteins in GM crop plants are often present at levels which are too low for extensive testing. For that reason, many toxicity studies so far performed have used proteins obtained from cultures of GM bacteria. This, however, carries the risk that toxic impurities may be present, which are uncommon for the crop plant, and that protein processing, like glycosylation, may be different in plants and bacteria. Nowadays, more appropriate systems are available for the heterologous expression of eukaryotic proteins. However, even when a protein is produced in baculovirus-infected insect cells or in the yeast *Pichia pastoris*, it is crucial to confirm that this protein is equivalent to the newly expressed protein in the plant – for instance, by comparing the behaviour on two-dimensional gels, immunoreactivity to poly- and/or monoclonal antibodies and through its intrinsic functional characteristics. As an example, the functional integrity of Cry proteins, which are often introduced in transgenic crops for their insecticidal properties, can easily be tested in the larvae of target herbivoral insect species [1]. These proteins, also referred to as Bt toxins, are originally produced as protoxins in the bacterium *Bacillus thuringiensis.* After ingestion by the target insect, the active toxin is liberated by enzymatic cleavage in the gut. Epithelial cells in the midgut of susceptible insect larvae possess receptors specific for a particular Cry protein. Receptor binding leads to pore formation, cell lysis, disintegration of the epithelium lining in their midgut and eventually to death of the insect larvae due to starvation.

Following extensive characterization of the recombinant protein, its safety can be demonstrated using a case-by-case strategy within a tiered approach in which the function and mode of action should be taken into account if known [2]. Where the mechanism of action is not known, a search for sequence and structural homology to known toxic proteins might provide useful parameters to be included in the safety testing procedure. Several protocols have been described for the safety assessment of food additives by the Joint Food and Agriculture Organization (FAO)/World Health Organization Expert Committee on Food Additives (JECFA), Food and Drug Adminstration (FDA) and Organization for Economic Cooperation and Development (OECD). A number of aspects in these protocols can also be implemented in the safety evaluation of GM food. The following issues may be addressed in the safety testing of novel food proteins:

- Digestibility and stability of the protein in simulated gastric and intestinal fluids, ex vivo gastric fluid and in vivo models to reflect the physiological situation in the digestive tract.
- Acute oral toxicity study in a rodent species using the protein at one high dose level (the fixed-dose method as described in the OECD guideline 420). The protocol involves isolation of animal tissue 2 hours after protein administration and histopathological investigations for characteristics related to its mode of

action. In the case of Cry proteins mentioned above, candidate biomarkers would be recognized by an ability to bind to the intestinal epithelium and subsequent induction of cell lysis.

- Subacute repeated dose feeding study (28 days; OECD guideline 407) to establish dose response characteristics (i.e. the Lowest Observed Adverse Effect Level [LOAEL] and/or the No Observed Effect Level [NOEL]) which may be used in the design of a 90-day animal study.
- Subchronic 90-day feeding study (OECD guideline 408) with both the individual recombinant protein and transgenic crop plant in order to yield additional information – on the occurrence of unintended effects, for example (see below).

## Specific investigations

In practice, feeding studies with whole foods are difficult to carry out and one may have to deal with a number of experimental shortcomings, including limited dose ranges in order to prevent unbalanced diets, which may make the results difficult to interpret and of limited value. Besides the feeding experiments, in vitro studies with organs and tissues from animals and various types of cell lines have already been successfully used for screening the toxic potential of food additives. It may be that this approach will lead to a redesign, refinement or even replacement of the rodent feeding trials. As the mucosa of the gastrointestinal tract is clearly a primary site for interaction with food components, intestinal cell lines grown on microporous membrane filters provide suitable systems for toxicity testing. Upon exposure, the cells can be tested for transmucosal transport parameters such as the transport of water-insoluble probes (e.g. $^{14}$C-polyethylene glycol) and for cytological parameters such as leakage of lactate dehydrogenase (LDH), conversion of 3–[4,5-dimethylthiazol-2-yl]-2,5-diphenyltetrazolium-bromide (MTT), uptake of neutral red (NR) and glutathione-*S*-transferase (GST) activity [3]. However, most of these parameters concern in vitro end-points for cytotoxicity and toxicity studies obviously would benefit from the identification of early warning markers. The recent developments in genomics and its application to toxicology, also referred to as toxicogenomics [4], provide an opportunity to find new informative leads and to enhance the predictive power of in vitro model systems. The DNA microarray technology has come into view as a powerful method for gene expression profiling [5]. The major advantage of this technology over conventional techniques for gene expression analysis (e.g. Northern blotting, differential display) is that it allows small-scale analysis of expression of a large number of genes at a time in a sensitive and quantitative manner. Furthermore, it allows comparison of gene expression profiles under various experimental conditions. For a comprehensive overview of the microarray technology, see Van Hal et al. [6].

## Potential allergenicity

The potential allergenicity of the newly introduced protein is one of the major safety concerns in the evaluation of GM foods. Where the candidate crop has acquired allergenic properties by genetic modification, it will, in general, not be approved for market introduction. Food allergy is an adverse reaction of the immune system to food components and manifests itself by symptoms such as dermatitis, oedema, vomiting, diarrhoea and fatal anaphylaxis in the extreme. It is generally accepted that the body is sensitized on initial contact with the allergen and that subsequent exposures elicit an allergic reaction. During the the sensitization process, the allergen stimulates B lymphocytes to synthesize allergen-specific IgE antibodies, probably with the aid of IL-4-producing helper CD4 T cells. The IgE binds to mast cells via the high affinity Fcε receptor (FcεRI). Upon subsequent exposure, the allergen crosslinks two surface-bound IgE molecules and triggers the mast cell to release mediators of allergic reactions, such as histamines and leukotrienes. Exposure to a small amount of plant-derived allergen can induce the release of allergic mediators. Consequently, the tolerance level for food allergens may be quite low in many cases.

In an attempt to define a predictive allergenicity assessment protocol for novel food proteins, a decision tree has been set up by the the International Food Biotechnology Council (IFBC) together with the International Life Sciences Institute (ILSI) [7]. The recommended strategy is focused on: (i) the source of the gene; (ii) the immunochemical reactivity of the newly introduced protein with IgE from blood serum of individuals known to be allergic to the source of the transferred genetic material; (iii) the sequence homology of the newly introduced protein to known allergens; and (iv) the physicochemical properties of the newly introduced protein. With respect to the source of the gene, the three possibilities considered are that the introduced gene is: (i) derived from commonly allergenic food such as peanuts, soybeans, tree nuts, wheat, milk, eggs, fish and crustacea; (ii) derived from a less commonly allergenic source; or (iii) from a source without a history of allergenicity.

### Common allergenic sources

Where the introduced gene is from a common allergenic source, the first step in the assessment process is a radioallergosorbent test (RAST, ELISA) with the detection of IgE antibodies in the serum from at least 14 individuals confirmed to be allergic to the gene source. When these immunoassays generate negative or equivocal data, the purified protein as well as a protein extract of the GM crop plant should be tested further using skin-prick tests involving at least 14 allergic individuals. A positive result in these latter tests indicates that the protein and/or transgenic food extract is able to elicit histamine release from skin mast cells in vivo. Finally, the IFBC/ILSI decision tree suggests that the assessment be confirmed using double-blind, placebo-controlled food challenges (DBPCFC) involving at least 14 allergic individuals. A positive result in either of these in vivo tests would provide sufficient evidence for the allergenicity of the transgenic food.

### Infrequent allergenic sources

If the introduced genetic material is from a less common allergenic source, the same type of tests should be performed except that at least five rather than 14 subjects are examined, recognizing that it will be much more difficult to identify individuals with these allergies. Although the degree of assurance that a major allergen from the donor organism has not been introduced in the transgenic crop plant is less than with the proteins obtained from a common allergenic source, the size of the affected population and thus the risk to the consumers is also much less. If less than five allergic individuals are included in the studies, the decision tree approach advocates two additional steps: (i) a structural comparison of the protein with known allergens; and (ii) an investigation of the stability of the protein (see below).

### Unknown allergenic sources

The most difficult assessment involves genetic material obtained from (nonfood) sources of unknown allergenic potential. A merely predictive approach for allergenicity testing could be followed. According to the IFBC/ILSI, this testing should include at least a comparison of the amino acid sequence of the novel protein with the sequence of known allergens and an analysis of the stability of the protein. Comparison of the amino acid sequence of the novel protein with the amino acid sequences of known allergens may be done using protein structure databases such as GenBank, SWISS-PROT and PIR, and alignment programs such as the FASTA and BLAST algorithms. The amino acid sequences of 198 major allergens including about 30 food allergens from plant origin are known [7]. So far, a relationship, if any, between allergenicity and (putative) biological function and/or protein structure has not been established. Although helper T-cell epitopes and B-cell epitopes of various allergens have been identified, comparison of the amino acid sequence of these allergenic proteins has not yet led to the identification of a consensus pattern specifically related to the ability to induce an IgE response.

One of the criteria currently used to predict the presence of potential antigenic determinants is a match of at least eight contiguous, identical amino acids between the novel protein and the known allergen and is based on the minimal peptide length for binding to helper T cells which would be required for allergen sensitization [7]. However, as long as the search is focused on linear sequences rather than discontinuous or conformational epitopes, the value of the approach is limited.

This may change in the near future as protein modelling software has become available recently, allowing for the prediction of three-dimensional structures from linear sequences; for an overview, see: http//www.embl-heidelberg.de/~rost/Papers/sisyphus.html. Since many food allergens tend to be stable to digestion and (acid and heat) processing, resistance of the novel protein to proteolysis is used as another criterion for potential allergenicity. The digestive stability of the protein can be analysed in simulated gastric and intestinal fluids. Stability to processing should be approached on a case-by-case basis depending on the processes applied to the food of interest. It cannot be emphasized enough that both sequence comparison and stability experiments should be interpreted carefully and that they only give an impression of the potential allergenicity. Stability of the protein and/or lack of structural homology do not automatically imply absence of allergenicity. For example, the oral allergy syndrome (OAS) is known to be caused by labile allergens. Moreover, little is known about the possible generation of new allergenic epitopes upon denaturation of the protein by heat or proteolysis.

## Additional evaluations of allergenicity

In addition to the assays mentioned in the IFBC/ILSI decision tree, animal experiments should be part of the evaluation in case of insufficient evidence for nonallergenicity. An animal model has been developed to test the potential allergenicity of food components in which Brown Norway rats (high IgE responders) are sensitized with or without an adjuvant prior to intraperitoneal and oral exposure to the test compound [8]. In order to avoid the induction of tolerance, these rats are reared for at least two generations on an allergen-free or test protein diet prior to challenge with the test compound. The outcome of such experiments should be carefully evaluated and their deficiencies recognized, e.g. a rat experiment failed to demonstrate the allergenicity of the 2S albumin from Brazil nut transferred into soybean, whereas individuals allergic to Brazil nut reacted positively to the novel product [9].

An additional issue with respect to the evaluation of potential allergenicity of a novel plant protein is the glycosylation pattern. Many food allergens are proteins that have been modified post-translationally by glycosylation, i.e. covalent linkage of oligosaccharide chains (glycans) to the protein backbone. In particular, *N*-glycosylation involving $\alpha$-1,3 fucose and $\beta$-1,2 xylose residues is suggested to be a key feature responsible for the high allergenicity and nonspecific cross-reactivity of various pollen and vegetable proteins [10]. Furthermore, protein glycosylation might be indirectly responsible for the antigenic character of a glycoprotein. Particularly contributing to an increased allergenicity of the glycoprotein counterparts of unglycosylated proteins are their increased solubility and proteolytic and thermal stability resulting from the presence of the glycans. A generic technology for the isolation, purification and characterization of *N*-linked oligosaccharides in plant tissues has been developed, taking Bt transgenic tomatoes, expressing the cryIAb5 gene, as a model [11].

It may be envisaged that genetic modification increases a crop plant's allergenicity not only by the introduction of a new gene encoding an allergenic protein but also by modification of endogenous proteins and by raising the endogenous levels of allergens. Taking these potential unforeseen allergenic effects into account, the assessment should be focused not only on the protein encoded by the inserted gene but also, as much as possible, on the whole transgenic crop plant. Even if there is no indication for a novel protein to be allergenic and it is considered for market approval, postmarketing surveillance should be recommended.

## Detection and characterization of unintended effects

Another key issue in the safety evaluation of GM foods is whether unexpected changes may have taken place in the transgenic crop plant as a result of the genetic modification process. If so, do these unintended (secondary) changes affect the safety or nutritional status of the modified organism? Unexpected changes in agronomical traits or composition may result from insertional mutagenesis or from the metabolic effects of the new gene product(s). In order to identify such changes, a systematic comparison should be made between the genetically modified organism and its parent grown under conditions that are as near identical as possible, since environmental factors may interact with the genotype of the crop, possibly resulting in alterations in phenotypic traits and composition. Compositional analysis has until now been performed on single constituents, such as key nutrients and natural plant toxins, as a screening strategy for unintended effects. A subchronic animal feeding study of 90 days' duration according to OECD guideline 408 is considered to be the minimum requirement to demonstrate the safety of repeated consumption of a novel crop plant in the diet. In order to design such a long-term toxicity study properly, compositional analysis (micro- and macronutrients) of the non-transgenic comparator crop and a pilot study of short duration (28 days; OECD guideline 407) are prerequisites. These are to ensure that the semisynthetic diet containing the whole food is palatable to the laboratory animals, to avoid the complications of possible nutritional inadequacy of the diets and to establish a LOAEL/NOEL of the novel plant protein. The highest dose level used in any animal feeding trials should be the one that is maximally achievable without causing nutritional imbalance while the lowest level used should be comparable to the anticipated human intake.

The 90-day subchronic study should be composed of at least the following test groups:

(1) semisynthetic human-type diet containing the maximal achievable amount (MTD) of the nontransgenic crop plant;
(2) as (1) but spiked with the recombinant protein of interest at LOAEL concentration;
(3) semisynthetic human-type diet containing the MTD of the transgenic crop plant; and
(4) as (3) but spiked with the recombinant protein of interest at LOAEL concentration.

Inclusion of these test groups allows distinction between toxicity caused by the novel protein *per se* or by the occurrence of unintended changes in the transgenic crop plant.

A current trend in plant biotechnology is the modification of crops with multiple genes which allows for the introduction of new metabolic pathways. This will make the assessment of substantial equivalence even more complicated. Therefore, alternative methods which allow for the simultaneous screening of many components are needed. Recent developments in molecular biology and analytical chemistry provide new opportunities to detect and evaluate unintended effects at different integration levels, i.e. transcriptome (mRNA), proteome (protein) and metabolome (secondary metabolites, macro- and micronutrients). RIKILT and 10 laboratories across Europe are currently testing different methods to be used for the detection and characterization of unintended effects as a result of genetic modification. Within the EU 5FP RTD project QLK1–1999–00765, called GMOCARE, the consortium plans to exploit DNA microarrays, 2D PAGE/mass spectrometry, and metabolite profiling techniques (liquid chromatography – nuclear magnetic resonance [LC–NMR]). Furthermore, RIKILT is presently testing within the framework of another multipartner EU project (QLK1–1999–00651; SAFOTEST) whether application of the DNA microarray technology is a useful approach to identifying functional biomarkers which can be used to improve the sensitivity and specificity of standard OECD guidelines 407 and 408 toxicity tests. Upon evaluation of classical and new methodologies, the challenge for regulators will be to harmonize rapidly the international safety assessment of GM food.

## REFERENCES

1 Van Rie, J., McGaughey, W.H., Johnson, D.E., Barnett, B.D. and Van Mellaert, H. (1990). Mechanism of insect resistance to the microbial insecticide Bacillus thuringiensis. *Science*, 247, 72–4.

2 Noteborn, H.P.J.M., Bienenmann-Ploum, M.E., van den Berg, J.H.J. et al. (1995). Safety assessment of the Bacillus thuringiensis insecticidal crystal protein CryIA(b) expressed in

transgenic tomatoes. In *Genetically Modified Foods*, ed. K.-H. Engel, G.R. Takeoka and R. Teranishi. Washington, DC: ACS Symposium Series 605, pp. 134–47.

3 Noteborn, H.P.J.M., Lommen, A., Weseman, J.M., Van der Jagt, R.C.M. and Groenendijk, F.P.F. (1998). Chemical fingerprinting and in vitro toxicological profiling for the safety evaluation of transgenic food crops. In *Report of the Demonstration Programme on Food Safety Evaluation of Genetically Modified Foods as a Basis for Market Introduction*, ed. M. Horning. The Hague, The Netherlands. Information and News Supply Department, Ministry of Economic Affairs, pp. 51–79.

4 Pennie, W.D., Tugwood, J.D., Oliver, G.J.A. and Kimber, I. (2000). The principles and practice of toxicogenomics: applications and opportunities. *Toxicological Sciences*, **54**, 277–83.

5 Schena, M., Shalon, D., Davis, R.W. and Brown, P.O. (1995). Quantitative monitoring of gene expression patterns with a complementary DNA microarray. *Science*, **270**, 467–70.

6 Van Hal, N., Vorst, O., Van Houwelingen, A.M. et al. (2000). The application of DNA microarrays in gene expression analysis. *Journal of Biotechnology*, **78**, 271–80.

7 Metcalfe, D.D., Astwood, J.D., Townsend, R., Sampson, H.A., Taylor, S.L. and Fuchs, R.L. (1996). Assessment of the allergenic potential of foods derived from genetically engineered crop plants. *Critical Reviews in Food Science and Nutrition*, **36**, S165–86.

8 Atkinson, H.A.C., Johnson, I.T., Gee, J.M., Grigoriadou, F. and Miller, K. (1996). Brown Norway rat model of food allergy. Effect of plant components on the development of oral sensitization. *Food and Chemical Toxicology*, **34**, 27–32.

9 Melo, V.M.M., Xavier-Filho, J., Silva-Lima, M. and Prouvost-Danon, A. (1994). Allergenicity and tolerance to proteins from Brazil nut (Bertholleria excelsa H.B.K). *Food and Agricultural Immunology*, **6**, 185–95.

10 Garcia-Casado, G., Sanchez-Monge, R., Chrispeels, M.J., Armentia, A., Salcedo, G. and Gomez, L. (1996). Role of complex asparagine-linked glycans in the allergenicity of plant glycoproteins. *Glycobiology*, **6**, 471–7.

11 Zeleny, R., Altmann, F. and Praznik, W. (1999). Structural characterisation of the *N*-linked oligosaccharides from tomato fruit. *Phytochemistry*, **51**, 199–210.

Part 7

# Biomarkers of cardiovascular disease and dysfunction

31

# The impact of biochemical tests on patient management

Paul O Collinson

St George's Hospital, London, UK

## Introduction

Biochemical testing is the 'Gold standard' for the diagnosis of acute myocardial infarction (AMI). Although the electrocardiogram (ECG) is the first test used for the differential diagnosis of patients who present with suspected acute coronary syndromes (ACS), the diagnostic accuracy of the ECG is only 55–75%. ST segment elevation has excellent specificity for AMI and identifies patients who will respond to thrombolytic therapy [1]. Biochemical tests form part of the World Health Organization criteria for AMI and have a diagnostic sensitivity of 95–100%.

The breakdown of diagnoses in patients with chest pain presenting to a typical district general hospital is shown in Figure 31.1. Only 10% have ST segment elevation AMI and the majority of patients have either low-risk ischaemic heart disease (IHD) or no IHD. Admission of these patient groups represents a significant waste of scarce healthcare resources. As only 10% of the patients present with characteristic ECG changes, biochemical diagnosis to confirm or exclude a diagnosis of AMI is required for 90% of the patients who present with suspected ACS. The role of the ECG is, therefore, to select patients for thrombolysis, not as the definitive diagnostic test for AMI. Creatinine kinase (CK)-MB is the best of the current enzymes and CK-MB mass the best test for the diagnosis of AMI according to current criteria [2]. Diagnosis can be achieved reliably within 12 hours from admission by serial measurement of myoglobin, CK or CK-MB. The use of rate of change of these markers (delta values) allows very early and accurate diagnostic categorization of patients with suspected ACS into those with or without AMI, within as little as 2–4 hours from admission [3, 4].

Immunoassays for the cardiac troponins (cTn) T and I provide the laboratory with a marker that is completely specific for cardiac damage. Measurements of cTnT and cTnI are 100% sensitive for the diagnosis of AMI. The specificity of both markers in nonAMI populations with unstable angina has been found to be less than 100%. This is due to the use of CK-MB mass as the reference standard. If a CK-MB result of more than twice the upper reference limit (URL) is used to define

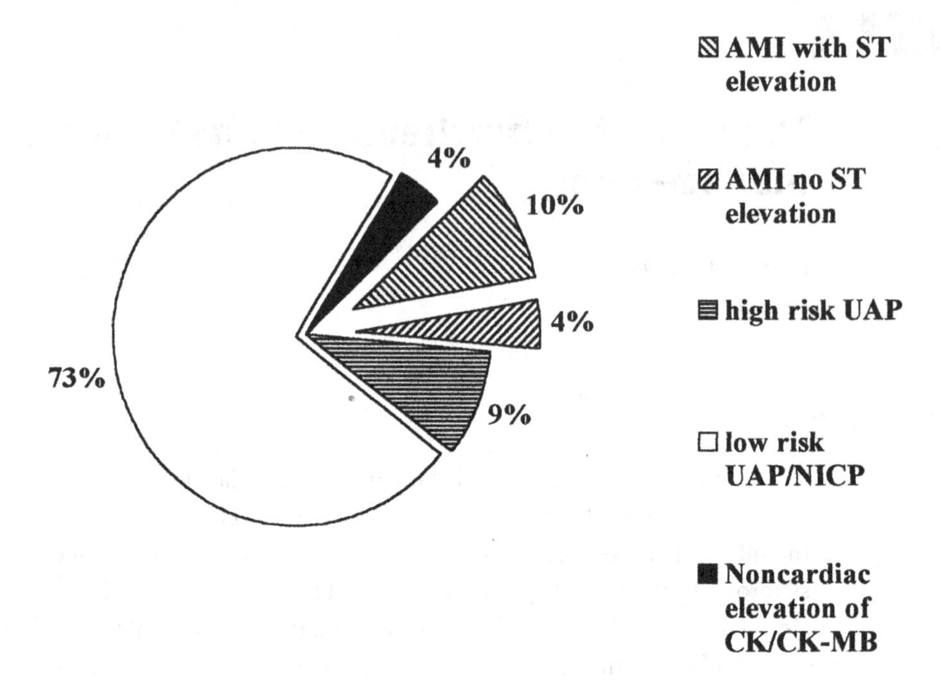


Figure 31.1 Distribution of patients admitted with suspected acute coronary syndromes according to final diagnosis. UAP, unstable angina pectoris; NICP, nonischaemic chest pain.

AMI, then any test evaluation including CK-MB will de facto assign 100% sensitivity and specificity to CK-MB. The classification of patients as suffering AMI becomes circular. AMI is diagnosed when CK-MB mass is elevated, making CK-MB mass measurement, by definition, 100% sensitive and specific. Cardiac TnT has not been detected in the serum of a large normal population. In patients with a final diagnosis of unstable angina, detectable cTnT and cTnI in the serum predicts the risk of cardiac death or recurrent cardiac events. Risk is proportional to the degree of both cTnT and cTnI elevation [5, 6]. This explains the apparent reduction in specificity of cTns. Measurement of cTnT and cTnI detects small degrees of myocardial necrosis not detected by conventional markers. In this respect, it has been noted that small elevations of CK-MB do occur in patients with suspected ACS and that these are associated with a poor prognosis. Histologically, microinfarcts due to plugging of small arteries by platelet aggregates have been demonstrated. These platelet microemboli arise as a consequence of platelet aggregation on a ruptured unstable plaque, which then breaks off and lodges distally. The presence of cTnT and cTnI in the serum is, therefore, a surrogate marker for an active unstable plaque. Recent consensus documents recommend the use of two markers, one with a short time window, and the other a cTn. The use of a delta value for CK, CK-MB mass or myoglobin, plus cTnT or cTnI concentration, would seem to be the optimal combination. The impact of cardiac markers on management can then be considered in two general areas: cardiospecific diagnosis, risk stratification and therapeutic targeting, and cost-effective management strategies.

## Cardiospecific diagnosis, risk stratification for management and therapeutic targeting

### Cardiospecific diagnosis

Cardiospecific diagnosis is possible by the measurement of cTnT or cTnI. Elevation of cTn does not occur when there is elevation of CK or CK-MB of musculoskeletal origin, as occurs in arduous physical training [7] or following direct current (DC) cardioversion. Measurement of cTn can, therefore, be used to detect perioperative AMI and to monitor cardiothoracic surgery. Elevation of cTn in cardiothoracic surgery is proportional to the degree of damage, either due to perioperative AMI, crossclamp time or the type of surgical procedure. Elevations of cTn do not occur in minimally invasive cardiac surgery [8].

### Risk stratification for management and therapeutic targeting

For management by risk stratification and therapeutic targeting, patients can be divided into three categories: ST segment elevation ACS; clinical high-risk nonST segment elevation suspected ACS; and clinically low-risk nonST segment elevation ACS. In all these patients, cTnT and cTnI can be combined with other diagnostic modalities.

#### ST segment elevation ACS

In patients with ST segment elevation, measurement of cTnT allows further risk stratification in addition to the ECG and CK-MB measurement. Patients who have detectable cTnT on admission have a worse prognosis than those who have no detectable cTnT [9]. An elevated cTnT on admission correlates with failure to achieve TIMI-3 (thrombolysis in myocardial infarction) flow but does not correlate with the time of onset of symptoms. This has implications for therapy. A more aggressive approach to revascularization, either by invasive intervention or the use of an aggressive thrombolytic regimen, combined with a glycoprotein IIb/IIIa antagonist may be more appropriate in this group.

#### Clinical high-risk nonST segment elevation suspected ACS

Patients considered on clinical grounds to be at high risk of ACS as a cause of their symptoms require admission to an appropriate observation facility for risk stratification and investigation. This will require a minimum of 12 hours as the kinetics

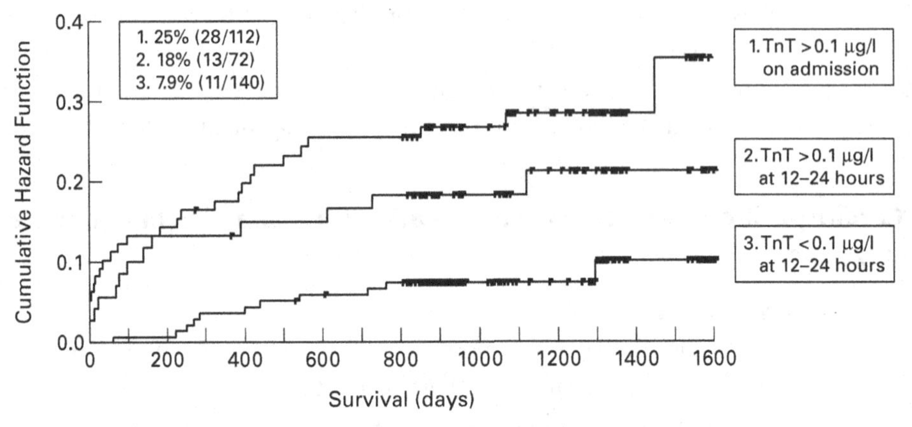


**Figure 31.2 Cardiac death as first event according to timing of troponin T measurement and concentration in patients admitted with chest pain without ST segment elevation. Reproduced with permission [11].**

of cTn release mean that 100% sensitivity for exclusion of prognostically significant damage is not achieved until this time point. Measurement of cTn on admission and at 12–24 hours from admission allows separation into high-, intermediate- and low-risk categories (Figure 31.2) [10]. Both admission and peak cTnT values can be combined with ECG category (ST depression, T wave inversion, normal ECG or uncodeable ECG) to allow further prognostic risk stratification. Measurement of cTnT can be combined with the stress ECG [11]. Patients with undetectable cTnT and a negative stress test are at very low risk.

Measurement of cTn can be used to guide the need for therapy. Revascularization of patients with unstable angina who are cTnT positive results in reduction of cardiac events. Measurement of cTnT can be used to identify patients with a reduced ejection fraction in whom ACE inhibitors will be beneficial. As detectable cTnT and cTnI in the serum are surrogate markers for platelet aggregation and platelet emboli, it would be expected that drugs acting on the thrombotic pathway would be maximally effective in this group. It has been found that the response to low molecular weight heparin and to inhibitors of the glycoprotein IIb/IIIa receptor are only effective in patients in whom cTn can be detected [12, 13]. Measurement of cTn can, therefore, be used to guide therapy.

## Clinically low-risk nonST segment elevation ACS

Patients in whom the suspicion of ACS is low can be investigated and managed differently. The objective is to identify those patients who have ACS with myocar-

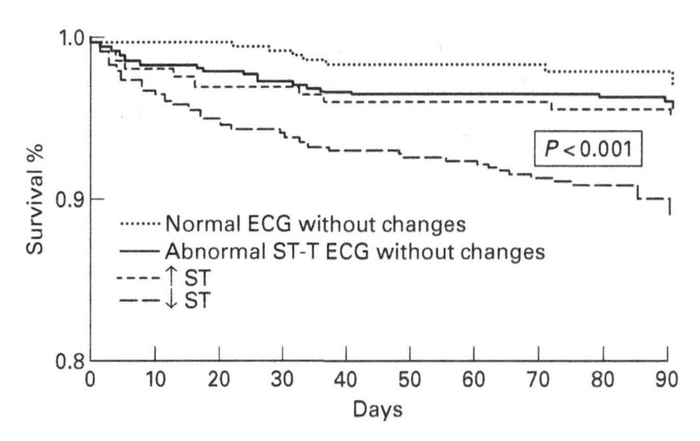


Figure 31.3 ECG changes and incidence of cardiac events. Reproduced with permission [15].

dial necrosis, but to avoid inappropriate admissions. Current strategies fail to identify this patient group. Of patients who are considered not to have IHD and who are discharged home from the emergency department, 6% have had an AMI.

Initial clinical assessment should identify patients with an atypical history and a normal ECG. Measurement of cTn on hospital attendance and at 4 hours (at least 6 hours from the onset of symptoms) allows risk categorization [14]. A normal ECG (Figure 31.3) and undetectable cTn at 4 hours identifies a very low risk group who can be discharged from hospital for outpatient investigation or immediately undergo noninvasive testing to exclude IHD [15]. This strategy is only appropriate for low-risk patients where the prior probability of disease is 5–10%. Not all patients will be detected but those discharged will be at low risk and can be identified by subsequent investigation.

## Cost economics

The cost economics of cardiac marker testing utilize the ability to combine rapid diagnosis with other investigations as part of an integrated clinical decision-making strategy. This can be divided into two pathways corresponding to the clinical classification previously discussed.

Typical test costs for different strategies are summarized in Table 31.1. It is apparent that the unit test cost of cTn measurement is more expensive than that of a single measurement of CK-MB mass, but the substitution of a single cTn measurement for multiple measurements of CK-MB mass in patients admitted with ST segment elevation would be cost neutral.

In patients with clinical high-risk nonST segment elevation suspected ACS, the combination of cardiac markers, the ECG and stress ECG can be used to improve

**Table 31.1.** Costs of individual tests and test strategies

| Test costs | Fixed | Variable | Full |
|---|---|---|---|
| AST | 0.70 | 0.37 | 1.07 |
| CK | 1.47 | 0.50 | 1.97 |
| CK-MB | 6.58 | 3.28 | 9.86 |
| LDH | 0.66 | 0.34 | 0.99 |
| cTnT | 1.65 | 4.80 | 6.45 |
| cTnI | 1.65 | 4.80 | 6.45 |
| Strategy costs | | | |
| CK + AST + LDH | 2.83 | 1.20 | 4.03 |
| CK + AST + LDH × 3 | 8.50 | 3.60 | 12.10 |
| CK × 3 | 4.42 | 1.49 | 5.90 |
| CK + CK-MB | 8.05 | 3.78 | 11.82 |
| CK + CK-MB × 3 | 24.14 | 11.33 | 35.47 |
| CK-MB × 3 | 19.73 | 9.84 | 29.57 |
| cTnT | 1.65 | 4.80 | 6.45 |
| cTnT × 2 | 3.30 | 9.60 | 12.90 |
| CK × 2 + cTnT | 4.59 | 5.79 | 10.38 |

*Note:*
All costs in Euros. Indicative costs based on internal hospital data.

substantially efficiency by process re-engineering. Currently, typical strategies utilize enzyme measurement over 2–3 days with a subsequent stress ECG. This produces a total average length of stay of 1.5 days on the coronary care unit (CCU) and 2.5 days on a step-down/general ward. If a rapid rule-out strategy is introduced, which reduces CCU stay to 1 day with rule out of myocardial necrosis at 12 hours from admission, and 1 day in a step-down bed with a stress test, total duration of stay is reduced by 2 days. For a typical case load, this offers substantial savings. This is illustrated in Table 31.2, in which reduction in length of stay generates a reduction of 1.1 million Euros in total patient episode cost. In addition, high-risk patients are identified who can then be managed appropriately. Using the same data, the impact of various treatment strategies can also be costed (Table 31.3). Two different treatment options have been considered, based on treatment with a low molecular weight heparin or a glycoprotein IIb/IIIa antagonist for 5 days. If all patients presenting with chest pain and suspected ACS are treated, costs are unrealistic. Treatment solely of nonQ-wave AMI patients detected by CK-MB measurement is less costly, but misses patients at risk. It is apparent that treatment on clinical grounds alone is not cost-efficient. Treatment of patients based on cTn measurement will, however, be more cost-effective, as only those known to benefit will receive therapy.

**Table 31.2.** Costs of test strategies on shifting from conventional diagnosis (4-day stay, 1.5 days CCU, 2.5 days step-down) to an accelerated rule-out protocol (2-day stay, conventional 1 day CCU, 1 day step-down). This assumes 4500 patients per annum with 2608 eligible for fast-track rule out

| | Test cost | Total laboratory cost | Total patient episode cost | Change |
|---|---|---|---|---|
| CK (conventional management) | 5.90 | 16605 | 4107412 | 0 |
| CK plus CK-MB ×3 (conventional management) | 35.47 | 99765 | 4135882 | 111630 |
| CK ×2 plus cTnT (conventional management) | 10.38 | 29205 | 4161378 | 166566 |
| CK plus CK-MB ×3 (rapid rule-out strategy) | 35.47 | 99765 | 2839733 | −1184519 |
| CK ×2 plus cTnT (rapid rule-out strategy) | 10.38 | 29205 | 2935149 | −1159664 |

*Note:*
All costs in Euros.

**Table 31.3.** Cost of alternative therapeutic strategies assuming 5 days' treatment with low molecular weight heparin or a glycoprotein IIb/IIIa antagonist for a case load of 4500 admissions. Test costs are based on point-of-care testing

| | | | Cost of treatment strategies | |
|---|---|---|---|---|
| Patient category | Number | Laboratory cost | Low molecular weight heparin | Glycoprotein IIb/IIIa antagonist |
| All patients with clinically suspected ACS (diagnosis based on CK) | 3074 | 16232 | 491892 | 15648649 |
| Patients with cTn-positive ACS | 466 | 102857 | 74595 | 2797297 |
| Patients with nonQ-AMI by CK-MB | 163 | 52670 | 26108 | 979054 |

*Note:*
All costs in Euros.

## Patients considered at low risk for ACS scheduled for discharge from the emergency department

The use of rapid rule-out strategies for the Emergency Department (ED) requires initial clinical identification of low-risk patients and a normal ECG with subsequent biochemical testing to exclude those patients with myocardial necrosis currently missed. In a series of studies, the author and colleagues have evaluated several strategies for the rule out of AMI in the ED. Use of serial enzyme measurement over a 4-hour period allows the accurate identification of AMI patients currently missed

**Table 31.4.** Cost of different strategies for Emergency Department rule out based on 990 patients per annum

| Test strategy | Total test cost | Bed cost | Total cost | Saving | Extra laboratory cost |
|---|---|---|---|---|---|
| CK | 3897 | 159 152 | 163 049 | 0 | 0 |
| CK + CK-MB | 4843 | 76 328 | 81 171 | 81 877 | 947 |
| CK + cTnT | 4434 | 0 | 4434 | 158 614 | 538 |
| cTnT | 6384 | 0 | 6384 | 156 665 | 6384 |
| cTnT × 2 | 12 767 | 0 | 12 767 | 150 282 | 12 767 |

*Note:*
Incidence of raised CK 120/990 (12.1%) with incidence of AMI 22/990 (2.2%). This assumes all patients with elevated cardiac enzymes are admitted. (All costs in Euros.)

and discharged home. The use of serial measurement of delta CK is as efficient as delta CK-MB, but less specific. Both strategies detect missed AMI, but lead to false-positive admissions due to elevated CK/CK-MB of musculoskeletal origin. The use of a cardiac-specific second test in those with elevated CK or CK-MB will improve diagnostic accuracy by excluding these false-positive admissions. The costs and outcomes of this strategy are illustrated in Table 31.4 for a population of 990 patients attending the ED. However, this strategy will miss some patients with myocardial damage. Comparison of a detection strategy based on CK, CK-MB mass and cTnT measurement in a total of 432 patients (247 males, 179 females; age range 26.3–88.9 years, median 53.5 years) found 6.4% had myocardial damage missed clinically, 10.4% low-risk angina and 83% nonischaemic chest pain. The miss rate for high-risk myocardial damage by each strategy was as follows: CK 2.1%, CK-MB mass 1.2% and CK/CK-MB ratio 2.1%. Sequential measurement of cTn will reduce both inappropriate admissions and inappropriate discharges.

In conclusion, the use of current available biochemical markers in a structured manner for clinical decision making improves clinical outcomes and cost-efficiency.

## REFERENCES

1 Fibrinolytic Therapy Trialists Collaborative Group. (1994). Indications for fibrinolytic therapy in suspected acute myocardial infarction: collaborative overview of early mortality and major morbidity results from all randomised trials of more than 1000 patients. *Lancet*, **343**, 311–22.

2 Gerhardt, W., Katus, H., Ravkilde, J. et al. (1991). S-troponin T in suspected ischaemic myocardial injury compared with mass and catalytic concentrations of S-creatine kinase isoenzyme B. *Clinical Chemistry*, **37**, 1405–11.
3 Collinson, P.O., Rosalki, S.B. Kuwana, T. et al. (1992). Early diagnosis of acute myocardial infarction by CK-MB mass measurements. *Annals of Clinical Biochemistry*, **29**, 43–7.
4 Young, G.P., Gibler, W.B., Hedges, J.R. et al. for the EMREG II study group. (1997) Serial creatine kinase-MB results are a sensitive indicator of acute myocardial infarction in chest pain patients with non-diagnostic electrocardiograms: the second Emergency Medicine Cardiac Research Group study. *Academic Emergency Medicine*, **4**, 869–77.
5 Lindahl, B., Venge, P. and Wallentin, L. (1996). Relation between troponin T and the risk of subsequent cardiac events in unstable coronary artery disease. *Circulation*, **93**,1651–7.
6 Antman, E.M., Tanasijeevic, M.J., Thompson, B. et al. (1996). Cardiac specific troponin I levels to predict the risk of mortality in patients with acute coronary syndromes. *New England Journal of Medicine*, **335**, 1342–9.
7 Collinson, P.O., Chandler, H.A., Stubbs, P.J., Moseley, D.S., Lewis, D. and Simmons, M.D. (1995). Cardiac troponin T and CK-MB concentration in the differential diagnosis of elevated creatine kinase following arduous physical training. *Annals of Clinical Chemistry*, **32**, 450–3.
8 Braun, S.L., Barankay, A. and Mazzitelli, D. (2000). Plasma troponin T and troponin I after minimally invasive coronary bypass surgery. *Clinical Chemistry*, **46**, 279–81.
9 Stubbs, P., Collinson, P., Moseley, D., Greenwood, T. and Noble, M. (1996). Prognostic significance of admission troponin T concentrations in patients with myocardial infarction. *Circulation*, **94**,1291–7.
10 Collinson, P.O. (1998). Troponin T or troponin I or CK-MB (or none?). *European Heart Journal*, **19**(Suppl) N16–24.
11 Lindahl, B., Andren, B., Ohlsson, J., Venge, P. and Wallentin, L. (1997). Risk stratification in unstable coronary artery disease. Additive value of troponin T determinations and predischarge exercise tests. FRISK Study Group. *European Heart Journal*, **18**, 762–70.
12 Hamm, C.W., Heeschen, C., Goldman, B. et al. (1999). Benefit of acbiximab in patients with refractory unstable angina in relation to serum troponin T levels. *New England Journal of Medicine*, **340**, 1623–9.
13 Heescheen, C., Hamm, C.W., Goldmann, B., Deu, A., Langenbrink, L. and White, H.D. for the PRISM Study Investigators. (1999). Troponin concentrations for stratification of patients with acute coronary syndromes in relation to therapeutic efficacy of tirofiban. *Lancet*, **354**, 1757–62
14 Hamm, C.W., Goldmann, B.U., Heeschen, C., Kreymann, G., Berger, J. and Meinertz, T. (1997). Emergency room triage of patients with acute chest pain by means of rapid testing for cardiac troponin T or troponin I. *New England Journal of Medicine*, **337**, 1648–53
15 López de Sá, E., López-Sendón, J., Bethencourt, A. and Bosch, X. (1998). Prognostic value of ECG changes during chest pain in patients with unstable angina. Results of the Proyecto de Estudio del Pronostico de la Angina (PEPA). *Journal of the American College of Cardiology*, 31(Suppl A): 79A.

32

# Cardiac natriuretic peptides in risk assessment of patients with acute myocardial infarction or congestive heart failure

Johannes Mair
University of Innsbruck, Innsbruck, Austria

## Introduction

In the evidence-based medicine era, risk assessment has become an important issue in the current debate on new laboratory markers in cardiology and other fields of diagnostic applications. This is because risk stratification is an important part of the medical decision-making process for the care of patients. Diagnostic tests are included in this process as important tools. Since the discovery of atrial natriuretic peptide (ANP) 20 years ago [1], it is apparent that the heart is also an endocrine gland which releases natriuretic peptides (NP). The NP system has emerged as one of the most important hormonal systems in the control of cardiovascular function. The prognostic significance of plasma concentrations of circulating NP for risk stratification in heart failure patients has been investigated intensively during the last two decades. The modulation of the NP system by pharmacological interventions continues to be an important goal of the therapeutics of congestive heart failure (CHF). This chapter will critically review the current knowledge on the use of cardiac NP for risk stratification in acute myocardial infarction (AMI) and in CHF.

## Pathophysiology of natriuretic peptides

ANP and brain natriuretic peptide (BNP) are the naturally occurring antagonists of the renin–angiotensin–aldosterone system (RAAS) and the sympathetic nervous system. They act as a dual natriuretic system in regulating blood pressure and fluid balance, and have an important role in the body's defence against mineralocorticoid- and salt-induced hypertension and plasma volume expansion [2]. NPs

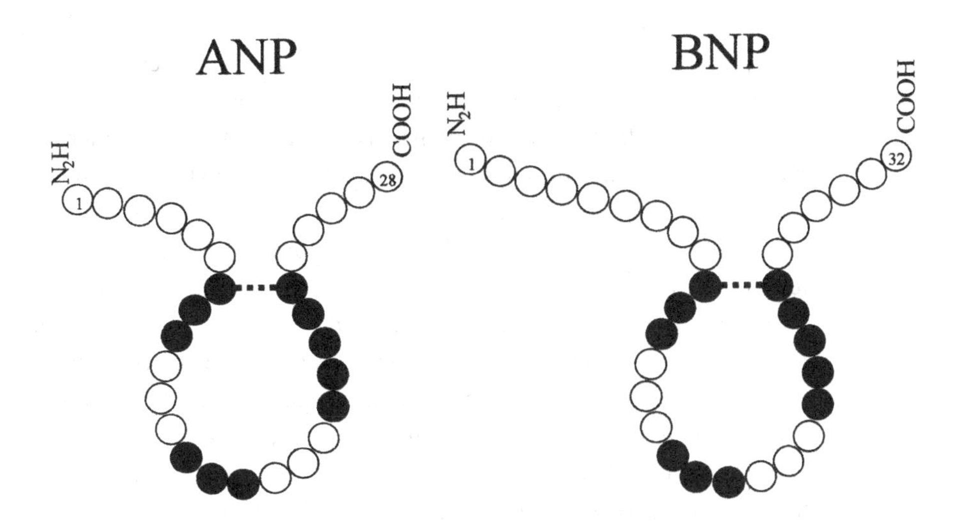


Figure 32.1 Structures of human atrial and brain natriuretic peptides. Homologous amino acids of both peptides, in the ring structure which is essential for receptor binding, are plotted in black. The disulphide bond between two cysteine residues is indicated by the broken line.

decrease systemic and pulmonary vascular resistance and systemic and pulmonary blood pressure. They decrease preload and venous return, and reduce atrial and ventricular end diastolic pressure. They increase coronary and renal blood flow, glomerular filtration rate and urine output. The name 'brain natriuretic peptide' is misleading because circulating BNP originates from the heart and the highest tissue concentrations of this peptide are found in myocardium. However, BNP was first isolated from porcine brain tissue in 1988 [3], which explains the name of this peptide. It soon became clear, however, that the main source of BNP was the cardiac ventricles rather than brain and, therefore, BNP is also more correctly called B-type natriuretic peptide. ANP and BNP are peptide hormones with high homology within the common 17 amino acid ring structure (Figure 32.1). This central loop is formed by a disulphide bridge between two cysteine amino acid residues. The intact ring structure is essential for receptor binding. Physiological actions of NP are mediated through guanylate cyclase-linked receptors (NPR-A and NPR-B) [4]. These receptors are transmembrane proteins composed of an extracellular binding site, a single transmembrane-spanning region and an intracellular domain which contains a protein kinase-like homology region and a guanylyl cyclase catalytic site that catalyses the conversion of guanosine triphosphate to cyclic guanosine monophosphate, the second messenger of NP. Another NP receptor (NPR-C) promotes the clearance of NP from the blood by binding it and removing it from the circulation by receptor-mediated endocytosis with subsequent lysosomal degradation.

NPs are also degraded through enzymatic cleavage by neutral endopeptidase, a membrane-bound nonspecific enzyme present in the kidneys and vascular beds which opens the ring structure and inactivates the molecules.

ANP and BNP are produced primarily in the cardiac atria under normal conditions. In left ventricular dysfunction (LVD), ventricular hypertrophy and other cardiac pathologies with chronic haemodynamic pressure or volume overload, ventricular myocytes undergo phenotypic modifications and re-express several fetal genes including ANP and BNP. Ventricles become an important source of NP, particularly BNP [5]. Under these conditions, the increased production of BNP changes the normal plasma concentration ratio of ANP to BNP. Atrial stretch, probably via G protein-coupled mechanisms, is the dominant stimulus for ANP release. Additionally, neurohormones, such as endothelin, arginine-vasopressin and phenyladrenaline, stimulate NP gene expression and release after binding to G protein-coupled receptors. ANP is stored in granules within atrial cardiomyocytes, thereby providing a source for rapid release. Regulation of ANP release occurs mainly at the level of hormone secretion. A critical decrease in ANP stores must be reached before transcriptional activation occurs. Members of the family of zinc finger proteins, such as GATA-4, appear to be important transcription factors of cardiac NP genes. ANP is primarily a measure of extracellular volume status with very rapid response. Upon secretion, proANP is split by a membrane-bound protease (atriopeptidase) into a C-terminal and N-terminal portion on an equimolar basis. The CT–ANP is the physiologically active hormone. NT–proANP occurs mainly in the high molecular mass form proANP 1–98. NT–proANP may be cleaved into smaller fragments. In contrast to ANP, it is believed that acute regulation of BNP synthesis and excretion occurs mainly at the level of gene expression, although BNP coexists in some of the secretery granules of human atrial and ventricular myocardium. However, cardiac BNP does not appear to be stored to the same extent as ANP and, hence, increased BNP release requires increased synthesis. Mechanical stretch-induced BNP gene expression depends at least in part on activation of zinc finger transcription factor GATA-4. The induction of BNP gene expression in cardiomyocytes is an early marker of haemodynamic overload. Whether proBNP is split into BNP-32 and NT–proBNP upon secretion, or whether proBNP is split later in serum into the physiologically active hormone and the NT-portion, is currently not known definitively. The major origin of BNP under pathophysiological conditions is ventricles. Left ventricular stretch or wall tension are the primary regulators of BNP release. Therefore, BNP is primarily a marker of LVD and cardiac hypertrophy. ANP and BNP form a dual, integrated NP system, and BNP is the slower reacting NP and may be a backup hormone activated only after prolonged ventricular overload.

### NP in heart failure

NPs have an important role in maintaining the compensated state of heart failure and delaying the progression to overt heart failure [6]. Increased plasma concentrations of NPs and their NT prohormone fragments are frequently found in symptomless LVD. Although the activity of the sympathetic nervous system is increased, plasma levels of noradrenaline are within reference limits during this stage of the disease. Similarly, the plasma levels of the hormones of the RAAS system are not increased in untreated patients with symptomless LVD because this system is completely suppressed by NP. With the progression of the disease to overt heart failure, the NP cannot sufficiently counteract and suppress the sympathetic nervous system and the RAAS system. As a result of this, vasoconstrictor and sodium- and water-retaining effects predominate and, consequently, symptoms develop. In addition, in overt heart failure, the hormones of the RAAS system and noradrenaline are increased in plasma. In contrast to mild heart failure, diagnosing moderate or severe heart failure by clinical examination is not very difficult. Depending on the cause of heart failure, with predominant overload of the atria or the ventricles, the secretion pattern of ANP and BNP in heart failure patients varies, since BNP predominantly reflects the degree of ventricular and ANP the degree of atrial overload.

ANP and BNP are activated to their greatest extent in heart failure; however, they are also markedly increased in all oedematous disorders with a volume overload which leads to an increase in atrial or ventricular tension, or the central blood volume. Examples include systemic and pulmonary hypertension, renal failure and increases in NP found in patients with ascitic liver cirrhosis and some endocrine disorders [7].

## Prognostic role of natriuretic peptides after acute myocardial infarction

Apart from myocardial damage, AMI leads to a pronounced transient activation of all major neurohormonal systems controlling vascular tone and fluid balance (i.e. sympathetic nervous system, RAAS, NP) during the first days after AMI, in response to acute ventricular dysfunction which is associated with myocardial necrosis [8]. Neurohormonal activation persists in case heart failure develops. NP increase in AMI patients is due to several factors, including alterations in haemodynamics, ischaemia, increased synthesis especially in the peri-infarct zone or release from necrotic myocardium. Although it is known from experimental studies and clinical investigations in humans that NPs are released from ischaemic cardiomyocytes, and that hypoxia *per se* stimulates the release of NP from myocardium [9], the available data suggest that myocardial ischaemia does not increase peripheral venous NP concentrations in patients, as long as ventricular or atrial dysfunction are not present [10]. The highest ANP values are found early after

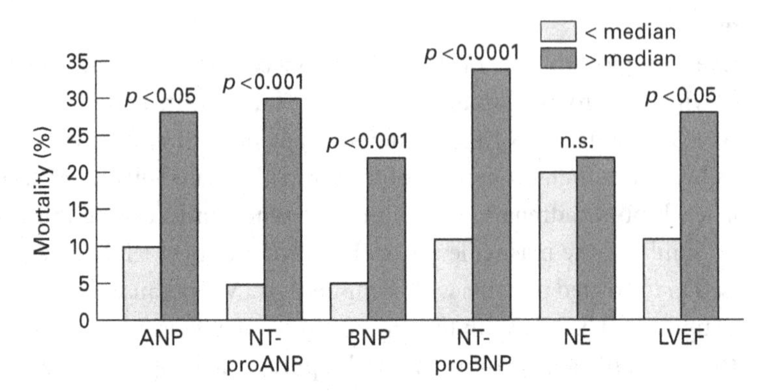


Figure 32.2 Predictors of 2-year prognosis after acute myocardial infarction. All markers except the catecholamines (NE) were significant predictors of 2-year mortality in univariate analyses. In a comparison of atrial natriuretic peptide (ANP), N-terminal proANP (NT–proANP), brain natriuretic peptide (BNP) and N-terminal proBNP (NT–proBNP), the measurement of BNP and NT–proBNP, 2–4 days after myocardial infarction, best reflected ventricular function; BNP and NT–proBNP were the best predictors of prognosis in multivariate analyses [19]. Left ventricular ejection fraction (LVEF).

AMI, usually on admission, with a subsequent decline and increase. ANP concentrations on day 2 or 3 are related to left ventricular ejection fraction (LVEF) and the presence of clinical heart failure symptoms [11]. By contrast, BNP values peak about 16 hours after admission and decrease thereafter. In patients with LVD, a second peak is found during the subacute phase several days after AMI [12]. The biphasic pattern was more frequently found with anterior Q-wave AMI, heart failure, low LVEF and high creatinine kinase (CK)-MB peaks. Interestingly, it could be demonstrated that the majority of patients in whom angiotensin-converting enzyme (ACE) inhibitor treatment was started early after infarction belonged to the group of patients with a single early BNP peak [13].

A consistent finding of the vast majority of studies which compared the prognostic power of neurohormones for the prediction of morbidity and mortality during the subacute phase of AMI is that NPs were the best prognostic markers for long-term cardiovascular mortality [14–20]. Plasma BNP, measured 3 days after AMI, was the only significant independent predictor of cardiovascular mortality in multivariate analysis when compared with ANP or NT–proANP, a risk that could not be identified clinically or by the measurement of LVEF [16, 17]. In a recently published study, Richards and coworkers compared all NPs with noradrenaline, and LVEF for risk stratification after AMI [19]. All markers except the catecholamines were significant predictors of 2-year mortality in a univariate analysis (see Figure 32.2). In a comparison of ANP, NT–proANP, BNP and NT–proBNP, the measurement of BNP and NT–proBNP, 2–4 days after infarction, best reflected ventricular function; these were the best predictors of prognosis in AMI patients. In a multivariate analysis, BNP and NT-BNP, measured 2–4 days after AMI, were comparably useful and independent, additive, prognostic markers to LVEF for prediction of 2-year survival. Thus, the measurement of the ventricular markers BNP or NT–proBNP for risk stratification after AMI may well prove to be an inexpensive and reliable screening tool.

## Natriuretic peptides as diagnostic tools and markers of prognosis in CHF

The utility of NP measurements is not only restricted to the postAMI patient – they are also useful diagnostic and prognostic tools in patients with CHF. Patients with suspected CHF should, ideally, be investigated with either echocardiography (ECG) or radionuclide ventriculography. This may be feasible in teaching hospitals but not necessarily in district general hospitals or if the patient is seen by a general practitioner. There has been enormous interest in the use of NP to detect LVD – in particular, BNP has been proposed as a biochemical marker that might provide a useful screening test to select patients for further cardiac investigations [21–27]. Recently, Nielsen and coworkers [27] reported on the usefulness of NT–proANP measurements for the assessment of risk of heart failure in the primary care setting. The results of their multivariate analysis provide three simple key questions which are very helpful in identifying patients who should be referred for echocardiography: (i) Are there QRS or ST-T changes, or both on the ECG? (ii) Is resting supine heart rate (beats/minute) greater than diastolic blood pressure (mm Hg)?; and (iii) Is NT–proANP increased? If the answer to all three questions is yes, the patient has LVD with near certainty; two positive answers indicate a moderate risk, one a low risk and three negative answers exclude the presence of CHF with near certainty. A limitation of this study is that BNP was not measured. Previously, in an unselected population of more than 1000 outpatients, McDonagh et al. [24] could demonstrate that BNP was superior to NT–proANP in the diagnosis of systolic LVD, which was defined as an echocardiographically determined LVEF of $<30\%$. They concluded that BNP was useful for screening for CHF in symptomatic or high-risk individuals and in helping primary care physicians in identifying which patients require further investigation. Thus, BNP emerges as the marker of first choice for the detection of systolic or diastolic LVD. First results on NT–proBNP are promising [19, 26], but whether NT–proBNP is superior to BNP cannot be definitively judged at the moment because sufficient comparative data are still lacking.

However, NP concentrations may be even more useful as prognostic markers [28], because reliable prognosis and adequate risk stratification are of primary importance in CHF. Elevated BNP concentrations are an independent predictor of

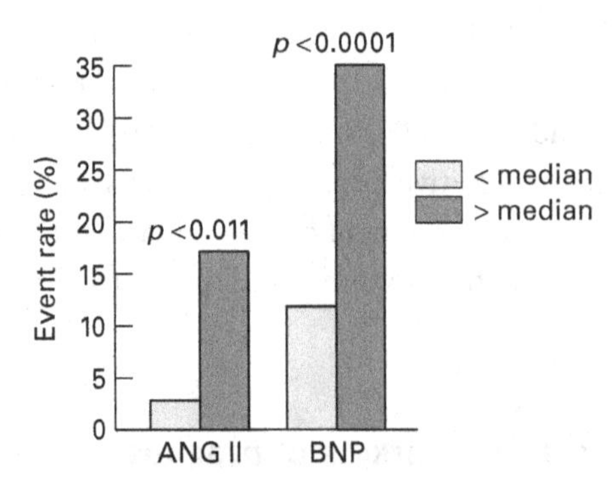


Figure 32.3 Brain natriuretic peptide as a biochemical marker of long-term morbidity and mortality in patients with asymptomatic or minimally symptomatic left ventricular dysfunction. In a univariate analysis, neurohormonal factors such as angiotensin II (ANG II) and brain natriuretic peptide (BNP) were significant predictors of cardiac events. However, according to a stepwise multivariate analysis, only high concentrations of BNP were a significant independent predictor ($p < 0.0001$). The median follow-up period was 2.2 years. Data adapted from reference [30].

the severity of heart failure and of mortality in CHF patients [29, 30]. Moreover, BNP measurements in the general elderly population can predict total mortality and reflect cardiac diseases, such as CHF and ischaemic heart disease [31]. Tsutamoto and coworkers [29] demonstrated that of noradrenaline, ANP and BNP, BNP was the best neurohormonal marker for assessing prognosis in overt heart failure patients. BNP was the only marker that provided prognostic information independently from haemodynamic markers. Later, the same group confirmed this finding in asymptomatic or mildly symptomatic CHF patients [30]. Of all tested neurohormones, BNP was the best predictor of morbidity and mortality in these patients (see Figure 32.3). BNP was the only marker that provided independent prognostic information on morbidity and mortality, distinct from haemodynamic markers (such as LVEF, left ventricular end diastolic volume index and pressure, and pulmonary capillary wedge pressure). Patients with low plasma BNP had an excellent long-term prognosis, while high BNP concentrations were related to high mortality (approximately 60% in 3-years' follow up). Thus, the measurement of BNP is a simple and useful addition to the standard clinical investigation of patients with asymptomatic or minimally symptomatic LVD. Mild and moderate CHF patients are mostly in the care of primary care physicians – therefore, the need for simple and readily available prognostic indicators, such as BNP, is imperative. BNP changes over time can be at least as important in prognostication as a BNP value measured at a single time point.

## Utility and applications of cardiac markers in the management of patients with myocardial infarction or heart failure

Cardiac troponins as well as BNP and NT–proBNP are going to fulfil the criteria of an evidence-based medicine approach for the introduction of new laboratory parameters into routine use. Similar to troponins, BNP and NT–proBNP have the advantages of stability in whole blood and ease of storage and measurement [32]. Both parameters have evolved from diagnostic tool to powerful prognostic marker. For risk assessment in AMI patients, a combination of the troponin concentration on admission with a BNP or NT–proBNP concentration on day 3 is suggested. BNP values during the subacute phase of AMI can distinguish patients with low LVEF from those with relatively preserved LVEF (>40%) and have been proposed to guide therapy with ACE inhibitors after AMI [33]. In patients with CHF, BNP and NT–proBNP are currently the most promising markers for risk assessment. In this clinical setting, a combination of troponin and NP measurements makes sense because, according to recently published studies, CHF patients with detectable troponin concentrations have a significantly lower LVEF and a worse prognosis than patients without [34, 35]. In addition to their diagnostic and prognostic significance, current ongoing studies are focusing on tailoring treatment according to troponin and BNP or NT–proBNP test results. In CHF patients the first studies have been published recently, comparing an intensive clinically guided treatment protocol with a strategy that tailors intensity of treatment to clinical signs and symptoms together with BNP or NT–proBNP test results, with the aim of lowering these markers towards the normal range [36, 37]. The greater benefit with NT–proBNP monitoring (reduced total cardiovascular events and delayed time to first event) observed by Troughton et al. [37] than in a previous study [36] of BNP-guided therapy may reflect a more advanced stage of CHF than in the earlier work, in which most patients had filling pressures within the normal range at the start of the trial. It remains to be seen whether BNP may be used as a surrogate end-point for the treatment of CHF, but BNP may lay the foundation of a bridge between the haemodynamic and neurohormonal approaches to CHF treatment.

In summary, the diagnostic utility and impact of NP on patient management have been convincingly demonstrated and are better documented than for several widely used routine laboratory parameters. The use of these novel markers in daily clinical practice is encouraged by the clinical findings.

## REFERENCES

1 de Bold, A.J., Borenstein, H.B., Veress, A.T. et al. (1981). A rapid and potent natriuretic response to intravenous injection of atrial myocardial extracts in rats. *Life Sciences*, **28**, 89–94.

2 Levin, E.R., Gardner, D.G. and Samson, W.K. (1998). Natriuretic peptides. *New England Journal of Medicine*, **339**, 321–8.

3 Sudoh, T., Kangawa, K., Minamino, N. and Matsuo, H. (1988). A new natriuretic peptide in porcine brain. *Nature*, **332**, 78–81.

4 Koller, K.J. and Goeddel, D.V. (1992). Molecular biology of the natriuretic peptides and their receptors. *Circulation*, **86**, 1081–8.

5 Thibault, G., Amiri, F. and Garcia, R. (1999). Regulation of natriuretic peptide secretion by the heart. *Annual Review of Physiology*, **61**, 193–217.

6 Schrier, R.W. and Abraham, W.T. (1999). Hormones and hemodynamics in heart failure. *New England Journal of Medicine*, **341**, 577–85.

7 Mair, J., Friedl, W., Thomas, S. and Puschendorf, B. (1999). Natriuretic peptides in assessment of left-ventricular dysfunction. *Scandinavian Journal of Clinical and Laboratory Investigation*, **59**(Suppl 230), 132–42.

8 Rouleau, J.L., Packer, M., Moye, L. et al. (1994). Prognostic value of neurohumoral activation in patients with an acute myocardial infarction: effect of captopril. *Journal of the American College of Cardiology*, **24**, 583–91.

9 Mair, P., Mair, J., Bleier, J. et al. (1997). Augmented release of brain natriuretic peptide during reperfusion of the human heart after cardioplegic cardiac arrest. *Clinica Chimica Acta*, **261**, 57–68.

10 Ikaheimo, M.J., Ruskoaho, H.J., Airaskinen, K.E.J. et al. (1989). Plasma levels of atrial natriuretic peptide during myocardial ischemia induced by percutaneous transluminal coronary angioplasty or dynamic exercise. *American Heart Journal*, **117**, 837–41.

11 Wencker, M., Lechleitner, P., Dienstl, F. et al. (1987). Early decrease of atrial natriuretic peptide in acute myocardial infarction. *Lancet*, **1**, 1369.

12 Morita, E., Yasue, H., Yoshimura, M. et al. (1993). Increased plasma levels of brain natriuretic peptide in patients with acute myocardial infarction. *Circulation*, **88**, 82–91.

13 Mizuno, Y., Yasue, H., Oshima, S. et al. (1997). Effects of angiotensin-converting enzyme inhibitor on plasma B-type natriuretic peptide levels in patients with acute myocardial infarction. *Journal of Cardiac Failure*, **3**, 287–93.

14 Omland, T., Aarsland, T., Aakvaag, A. et al. (1993). Prognostic value of plasma atrial natriuretic factor, norepinephrine and epinephrine in acute myocardial infarction. *American Journal of Cardiology*, **72**, 255–9.

15 Hall, C., Rouleau, J.L., Moye, L. et al. (1994). N-terminal proatrial natriuretic factor – an independent predictor of long-term prognosis after myocardial infarction. *Circulation*, **89**, 1934–42.

16 Omland, T., Aakvaag, A., Bonarjee, V.V.S. et al. (1996). Plasma brain natriuretic peptide as an indicator of left ventricular systolic function and long-term survival after acute myocardial infarction – comparison with plasma atrial natriuretic peptide and N-terminal proatrial natriuretic peptide. *Circulation*, **93**, 1963–9.

17 Dabar, D., Davidson, N.C., Gillespie, N. et al. (1996). Diagnostic value of B-type natriuretic peptide concentrations in patients with acute myocardial infarction. *American Journal of Cardiology*, **78**, 284–7.

18 Arakawa, N., Nakamura, M., Aoki, H. and Kiramori, K. (1996). Plasma brain natriuretic

peptide concentrations predict survival after acute myocardial infarction. *Journal of the American College of Cardiology*, **27**, 1656–61.
19 Richards, A.M., Nicholls, M.G., Yandle, T.G., et al. (1998). Plasma N-terminal pro-brain natriuretic peptide and adrenomedullin: new neurohormonal predictors of left ventricular function and prognosis after myocardial infarction. *Circulation*, **97**, 1921–9.
20 Richards, A.M., Nicholls, M.G., Yandle, T.G. et al. (1999). Neuroendocrine prediction of left ventricular function and heart failure after acute myocardial infarction. *Heart*, **81**, 114–20.
21 Friedl, W., Mair, J., Thomas, S. et al. (1996). Natriuretic peptides and cyclic guanosine 3',5'-monophosphate in asymptomatic and symptomatic left ventricular dysfunction. *Heart*, **76**, 129–36.
22 Lerman, A., Gibbons, R.J., Rodeheffer, R.J. et al. (1993). Circulating N-terminal atrial natriuretic peptide as a marker for symptomless left-ventricular dysfunction. *Lancet*, **341**, 1105–9.
23 Cowie, M.R., Struthers, A.D., Wood, D.A. et al. (1997). Value of natriuretic peptides in assessment of patients with possible new heart failure in primary care. *Lancet*, **350**, 1347–51.
24 McDonagh, T.A., Robb, S.D., Murdoch, D.R. et al. (1998). Biochemical detection of left-ventricular dysfunction. *Lancet*, **351**, 9–13.
25 Yamamoto, K., Burnett, J.C., Jougasaki, M. et al. (1996). Superiority of brain natriuretic peptide as a hormonal marker of ventricular systolic and diastolic dysfunction and ventricular hypertrophy. *Hypertension*, **28**, 988–94.
26 Talwar, S., Squire, B., Davies, J.E. et al. (1999). Plasma N-terminal pro-brain natriuretic peptide and the ECG in the assessment of left-ventricular systolic dysfunction in a high risk population. *European Heart Journal*, **20**, 1736–44.
27 Nielsen, O.W., Hansen, J.F., Hilden, J. et al. (2000). Risk assessment of left ventricular systolic dysfunction in primary care: cross sectional study evaluating a range of diagnostic tests. *British Medical Journal*, **320**, 220–4.
28 Swedberg, K., Eneroth, P., Kjekshus, J. et al. (1990). Hormones regulating cardiovascular function in patients with severe congestive heart failure and their relation to mortality. *Circulation*, **82**, 1730–6.
29 Tsutamoto, T., Wada, A., Maeda, K. et al. (1997). Attenuation of compensation of endogenous cardiac natriuretic peptide system in chronic heart failure – prognostic role of plasma brain natriuretic peptide concentration in patients with chronic symptomatic left ventricular dysfunction. *Circulation*, **96**, 509–16.
30 Tsutamoto, T., Wada, A., Maeda, K. et al. (1999). Plasma brain natriuretic peptide level as a biochemical marker of morbidity and mortality in patients with asymptomatic or minimally symptomatic left ventricular dysfunction – comparison with angiotensin II and endothelin-1. *European Heart Journal*, **20**, 1799–807.
31 Wallen, T., Landahl, S., Hedner, T. et al. (1997). Brain natriuretic peptide predicts mortality in the elderly. *Journal of Internal Medicine*, **77**, 264–7.
32 Buckley, M.G., Marcus, N.J. and Yacoub, M. (1999). Cardiac peptide stability, aprotinin and room temperature: importance for assessing cardiac function in clinical practice. *Clinical Sciences*, **97**, 689–95.
33 Motwani, J.G., McAlpine, H., Kennedy, N. et al. (1993). Plasma brain natriuretic peptide as an indicator of ACE inhibition after myocardial infarction. *Lancet*, **341**, 1109–13.

34 Missov, E., Calzolari, C. and Pau, B. (1997). Circulating cardiac troponin I in severe congestive heart failure. *Circulation*, **96**, 2953–8.

35 Missov, E. and Mair, J. (1999). A novel biochemical approach to congestive heart failure: cardiac troponin T. *American Heart Journal*, **138**, 95–9.

36 Murdoch, D.R., McDonagh, T.A., Byrne, J. et al. (1999). Titration of vasodilator therapy in chronic heart failure according to plasma brain natriuretic peptide concentration: randomized comparison of the hemodynamic and neuroendocrine effects of tailored versus empirical therapy. *American Heart Journal*, **138**, 1126–32.

37 Troughton, R.W., Frampton, C.M., Yandle, T.G. et al. (2000). Treatment of heart failure guided by plasma aminoterminal brain natriuretic peptide concentrations. *Lancet*, **355**, 1126–30.

33

# Serum markers of inflammation and cardiovascular risk

Juan Carlos Kaski

St George's Hospital Medical School, London, UK

Coronary artery disease (CAD) is associated with risk factors such as smoking, diabetes mellitus, dyslipidaemia and hypertension, and represents the major cause of morbidity and mortality in the Western world [1]. In recent years, it has become apparent that inflammation plays a major pathogenic role in both atherogenesis and rapid CAD progression [2,3].

## Inflammation and atherosclerosis

Despite important observations regarding an association between inflammation and atherosclerosis in the past, it has been only in the last two decades that scientists have specifically focused their attention on inflammation as a major player in the development of atherosclerosis. As reviewed recently by Ross [3], the atherogenic process involves inflammatory mechanisms with pro- and anti-inflammatory cytokine production, and increased blood concentrations of acute phase reactants. Acute phase reactants, such as fibrinogen, C-reactive protein (CRP), serum amyloid A protein, syalic acid, caeruloplasmin and albumin, have been noted as markers of coronary disease activity, similar to other inflammatory disease processes. Typically, cells involved in chronic inflammation include macrophages, lymphocytes, mast cells and plasma cells. Different cell types, i.e. endothelial cells, vascular smooth muscle cells, macrophages and lymphocytes, as well as numerous families of cytokines and growth factors, are involved in the atherosclerotic process [3]. Cytokines and growth factors have effects on the vasculature, inducing a variety of responses by activated endothelial cells. Endothelial cell activation and endothelial dysfunction may be triggered by 'established' risk factors, i.e. hypercholesterolaemia, diabetes mellitus, hypertension, etc., and others, more recently identified, such as homocysteine and chronic inflammation, which trigger the expression of specific adhesive glycoproteins on the surface of the endothelial cells [3]. These glycoproteins are members of the selectin group of molecules (E-selectin and P-selectin) and the immunoglobulin superfamily (platelet-endothelial cell adhesion molecule-1 [PECAM-1], intercellular adhesion molecules [ICAMs] and vascular cell adhesion molecule-1 [VCAM-1]) [4]. Adhesive glycoproteins on the surface of endothelial cells are recognized by integrins in the surface of monocytes and T lymphocytes which facilitate the attachment of white blood cells to the endothelial surface [5]. Monocytes and T lymphocytes subsequently migrate to the interior of the vessel wall through endothelial cell junctions – a process that is influenced by growth regulatory molecules and chemoattractants released by both the altered endothelial cells and the adherent leucocytes, i.e. interleukins (IL), leukotrienes, platelet-derived growth factor (PDGF), monocyte chemotactic protein-1 (MCP-1) and PECAM-1 [3, 6]. MCP-1 also attracts CD4+ and CD8+ lymphocytes, and stimulates the release of inflammatory cytokines such as IL-1 and IL-6 by monocytes [3]. Cytokines increase free radical and enzyme production by endothelial cells and macrophages which contribute to low-density lipoprotein (LDL) oxidation. Oxidized LDL constitutes a ligand for a specific receptor which is expressed in cytokine-activated macrophages. This receptor binds to LDL and internalizes oxidized LDL which leads to the accumulation of foam cells within atherosclerotic plaques. These cells present antigens to lymphocytes and trigger an immune response. Moreover, activated macrophages produce proinflammatory cytokines, i.e. IL-12 and IL-18, which induce the production of interferon-$\gamma$ (IFN-$\gamma$) by T lymphocytes which, in turn, activates the macrophages thus establishing a positive feedback loop. T lymphocytes can also produce IL-10 which may limit the proinflammatory process [7].

## Inflammation, atherosclerotic plaque disruption and thrombogenesis

Inflammation is also involved in the active process of plaque rupture. While certain cytokines and growth factors are implicated in the synthesis of collagen in the fibrous cap of atherosclerotic plaques, i.e. transforming growth factor-$\beta$ (TGF-$\beta$) and PDGF, others, such as IFN-$\gamma$, impair collagen synthesis by the smooth muscle and inhibit smooth muscle cell proliferation [6]. Macrophages in active plaques express matrix metalloproteinases that induce degradation of collagen in vulnerable regions of the plaque [8]. This results in the weakening of the fibrous cap, cap fissuring, plaque disruption and thrombogenesis [9]. There is a clear link between inflammation and thrombosis [10, 11]. Upon stimulation by cytokines, endothelial cells produce procoagulant molecules such as von Willebrand factor, tissue factor and plasminogen activator inhibitor-1 and -2. Moreover, activated inflammatory cells synthesize a variety of molecules which modulate the thrombotic cascade, such as tissue factor and thrombin [12]. Fibrinogen, an acute-phase reactant and a key factor in the thrombogenic process, also plays a major role in platelet adhesion and aggregation [5, 13, 14]. Thrombosis and inflammation are also linked through the regulatory roles that cytokines and growth factors, such as IL-1 and IL-4, exert on the thrombotic–thrombolytic balance. IL-1 can stimulate endothelial cells to produce plasminogen activator inhibitor-1 (PAI-1) [15] and IL-4 induces the production of tissue plasminogen activator (t-PA) by the monocytes [16].

## Markers of inflammation in atherosclerosis

### Cytokines

As discussed earlier, cytokines are signalling peptides that stimulate inflammatory responses acting through high affinity receptors in the cell surface. Cytokines act locally, in both autocrine and paracrine fashion, and are produced as a response to tissue injury. Most of the cytokines are multifunctional molecules that have different actions and, usually, the effects of these different cytokines tend to overlap. Among proinflammatory cytokines that trigger the inflammatory response are IL-1, IL-6 and tumour necrosis factor alpha (TNF-$\alpha$). Their actions are modulated by other cytokines such as IL-8, IL-10, IL-11, IL-12, IL-18 and IFN-$\gamma$. Cytokines induce the production of important inflammation mediators such as C5a complement fragment and platelet activator factor [6, 12].

Large epidemiological studies have shown a correlation between atherosclerosis and cytokine plasma concentrations [17, 18]. These studies involving patients from the CARE (Cholesterol and Recurrent Events) database and the Physician's Health Study have reported a significant association between elevation of TNF-$\alpha$ plasma levels and increased risk of recurrent coronary events and between raised levels of IL-6 and cardiovascular risk, respectively. However, cytokine plasma levels may not represent a very accurate measure of the local inflammation that takes place within the coronary artery wall. These markers, however, may give a rough indication of the total atherosclerotic burden. Measurement of plasma cytokine levels may not represent an ideal marker of risk. Cytokine plasma levels may vary rapidly and, therefore, single measurements of these markers may not give an accurate picture of the ongoing inflammatory process that leads to plaque formation or plaque disruption.

### Neopterin

Neopterin is a pterydine derivative secreted by macrophages following stimulation by IFN-$\gamma$ [19, 20]. Neopterin modulates the intracellular redox state, probably through stimulation of the activity of both constitutive and inducible nitric oxide synthase [20–22]. As a consequence of the modification of the redox state, neopterin activates the translocation of NFkB subunits to the nucleus [23], which in turn upregulates proinflammatory genes responsible for the production of IL-1, IL-6,

IL-8, IFN-$\gamma$, macrophage chemotactic protein-1, colony-stimulating factors, c-myc, VCAM-1, ICAM-1, E-selectin, tissue factor, TNF-$\alpha$ and other factors [24–26]. Studies have reported an association between plasma neopterin and both peripheral vascular occlusion and the extent of carotid atherosclerosis [19, 27]. Other studies have shown that plasma neopterin is elevated in patients with unstable angina or myocardial infarction compared with patients with stable angina [28, 29]. Observations from the author's group showed that neopterin levels had a significant relationship with the number of complex lesions as assessed by coronary angiography [30]. Recently, the author and coworkers have also shown that baseline plasma neopterin is a marker of cardiovascular risk in women with unstable angina and in those with chronic stable angina [31]. Renal function is a major determinant of neopterin levels in blood [32] and correction for creatinine levels is therefore necessary. Neopterin appears to be a promising marker of risk in patients with angina. The finding that a marker of immune activation is also a marker of cardiovascular risk is consistent with previous reports that coronary disease progression is associated with inflammatory mechanisms [33].

### Acute phase reactants and CAD – C-reactive protein

Acute phase reactants are sensitive, albeit nonspecific, markers of ongoing inflammation. These are proteins synthesized by cytokine-stimulated hepatocytes [34]. Acute phase reactant concentrations are raised in patients with CAD who are at a high risk of adverse cardiovascular events [35–37]. Among acute phase reactant proteins, fibrinogen [35, 38, 39], serum amyloid-A protein [40, 41], sialic acid [42], albumin [43], caeruloplasmin [44] and CRP have been shown to be associated with cardiovascular risk. However, for several reasons, CRP is currently the focus of attention of scientists worldwide.

## CRP and cardiovascular risk

High CRP concentrations have been shown to be an independent prognostic marker in apparently healthy individuals, patients with CAD and patients with peripheral vascular disease.

### CRP and cardiovascular risk in healthy individuals

Several large epidemiological studies have assessed the role of CRP measurements for the identification of individuals at risk of cardiovascular events. A substudy involving individuals enrolled in the Multiple Risk Factor Intervention Trial (MRFIT) study [45] showed for the first time an association between CRP plasma levels and cardiovascular risk in apparently healthy men. In this investigation, 200 cases and 400 controls from the initial 12 866 apparently healthy men recruited and followed up for up to 17 years were entered in a nested study. Cases comprised the first 50 nonsmoker and the first 100 smoker CAD deaths, as well as the first 35 non-smoker and the first 65 smoker myocardial infarction patients. Two controls per case, matched by age and smoking status, were studied. Men without events during follow up had a baseline CRP concentration of 2.0 mg/l, compared with 2.7 mg/l in men who developed myocardial infarction and 3.34 mg/l in those with cardiac death. The odds ratio for men in the highest CRP quartile compared with those in the lowest CRP quartile was 2.8 (1.4–5.4). In another important nested study involving individuals who took part in the Physician's Health Study (22071 apparently healthy male physicians in the USA), Ridker et al. [46] selected 543 men with cardiovascular events during follow up and 543 controls without events. Both groups had been previously randomized to aspirin or placebo. Baseline CRP was 1.13 mg/l in the control group, compared with 1.40 mg/l in men with cardiovascular events ($p<0.001$). Men who developed myocardial infarction, haemorrhagic stroke and ischaemic stroke had significantly higher CRP levels than men without events (1.51 mg/l, 1.36 mg/l and 1.38 mg/l, respectively). Men in the highest CRP quartile had a 2.9-fold higher relative risk for myocardial infarction than those in the lowest CRP quartile. Aspirin was associated with a significant reduction of events in men within the highest CRP quartile (55.7% risk reduction, $p=0.02$). However, risk reduction in men within the lowest CRP quartile was not significant (13.9%, $p=0.77$).

## CRP and cardiovascular risk in unstable angina patients

Liuzzo et al. [40] studied the prognostic role of CRP levels in 32 patients with stable angina, 31 unstable angina patients and 29 myocardial infarction patients. A CRP threshold value of 3 mg/l was used, based on the 90th percentile of the normal distribution observed in the study. Unstable angina patients had higher CRP concentrations than stable angina patients. Moreover, unstable angina patients with high CRP concentrations had more ischaemic episodes ($4.8 \pm 2.5$ versus $1.8 \pm 2.4$; $p=0.004$), and were more likely to die (nonsignificant trend) during coronary care unit admission than those with the lowest concentrations of CRP. These findings were later confirmed by a larger study of 965 patients with unstable angina or nonQ-wave myocardial infarction (FRISC [Fast Revascularization during Instability in Coronary artery disease]) [47]. However, only a trend of increased CRP was found in the group who died and/or had a new myocardial infarction compared with patients with no events during follow up (7.5 [1–17] versus 5 [0–14], respectively; $p=0.067$). In this study, logistic multiple regression analysis showed that CRP was not an independent risk factor for death or myocardial infarction. A recent study by Ferreiros et al. [48] showed that CRP measured at hospital admission was not predictive of events at discharge.

However, they found that CRP measured at hospital discharge was highly predictive of coronary risk at 90 days [48].

### CRP and cardiovascular risk in stable angina patients

In the European Concerted Action on Thrombosis (ECAT) group's studies [11, 36, 49] involving over 3000 patients with unstable angina, stable angina pectoris and atypical chest pain, high plasma CRP was found to be associated with cardiovascular events during follow up. CRP was higher in patients with events than in those without events (2.15 mg/l versus 1.61 mg/l; $p=0.05$). Thompson et al. [49] found that CRP lost its predictive value after adjusting for fibrinogen. Independent predictors of events in this study were fibrinogen, von Willebrand factor antigen and t-PA antigen. However, Haverkate et al. [36] showed a significant difference in CRP concentrations between those patients with and those without a history of myocardial infarction (1.82 mg/l versus 1.65 mg/l, respectively). Patients with zero vessel disease had a CRP concentration of 1.43 mg/l, compared with: those with one vessel disease, 1.73 mg/l; two vessel disease, 1.90 mg/l; and three or more vessel disease, 1.86 mg/l. Differences between groups were statistically significant ($p=0.01$). In this study, patients with high CRP levels (the highest fifth) had a greater risk, more than twofold, of developing a cardiovascular event, compared with the remaining patients.

### CRP and cardiovascular risk in women

Most of the studies described earlier were carried out mainly or, exclusively, in men. Recently, however, the prognostic value of CRP was assessed in a nested study (122 women who suffered a first cardiovascular event – myocardial infarction, stroke, PTCA (percutaneous transluminal coronary angioplasty), CABG (coronary artery bypass graft) or cardiovascular death – and 244 age- and tobacco-matched controls) in apparently healthy postmenopausal women [50]. The authors observed that, after adjusting for other risk factors, women with cardiovascular events had significantly higher CRP levels than women without events at follow up – a result similar to that obtained in previous studies in apparently healthy men [45]. Recent observations from the author's group [51] have shown that women with chronic stable angina have significantly higher CRP concentrations than men, even after adjustment for other variables associated with gender. In agreement with the findings of other groups [52, 53], the author and coworkers observed that women who received hormone replacement treatment (HRT) had significantly higher levels of CRP than women who received no HRT.

### CRP combined with other parameters and cardiovascular risk

Recent reports have discussed the additional information provided by CRP measurements, over and above that supplied by established risk markers, i.e. cardiac troponin T and cholesterol concentrations [54, 55]. In studies by Ridker et al. [55, 56], the assessment of CRP added significantly to the predictive value of lipid parameters in determining risk of first myocardial infarction. In another study [54], elevated CRP at admission ($\geq$15.5 mg/l) was a predictor of early mortality, both independently and in combination with troponin T, in acute coronary syndromes.

## Conclusions

In large epidemiological studies, markers of inflammation, in particular CRP, are predictors of cardiovascular risk. CRP appears to add prognostic information over and above that obtained by assessing conventional risk factors. However, the extrapolation of results obtained in large epidemiological studies to the individual patient is not straightforward. Further studies in 'real life' patients are required, together with the definition of clinically useful cut-off points, before markers such as CRP can be confidently and successfully used in the clinical setting.

## REFERENCES

1 American Heart Association (1996). *Heart and Stroke Facts: 1996. Statistical Supplement.* Dallas: American Heart Association, pp. 1–23.

2 Neri Serneri, G.G., Prisco, D., Martini, F. et al. (1997). Acute T-cell activation is detectable in unstable angina. *Circulation*, **95**, 1806–12.

3 Ross, R. (1999). Atherosclerosis – an inflammatory disease. *New England Journal of Medicine*, **340**, 115–26.

4 Mehta, J.L., Saldeen, T.G.P. and Rand, K. (1998). Interactive role of infection, inflammation and traditional risk factors in atherosclerosis and coronary artery disease. *Journal of the American College of Cardiology*, **31**, 1217–25.

5 Loscalzo, J. (1992). The relation between atherosclerosis and thrombosis. *Circulation*, **86**(Suppl), III95–9.

6 Libby, P. (1995). Molecular bases of the acute coronary syndromes. *Circulation*, **91**, 2844–50.

7 de Waal Malefyt, R., Abrams, J., Bennett, B., Figdor, C.G. and de Vries, J.E. (1991). Interleukin 10 (IL-10) inhibits cytokine synthesis by human monocytes: an autoregulatory role of IL-10 produced by monocytes. *Journal of Experimental Medicine*, **146**, 3444–51.

8 Galis, Z.S., Sukhova, G.K., Lark, M.W., et al. (1994). Increased expression of matrix metalloproteinases and matrix degrading activity in vulnerable regions of human atherosclerotic plaques. *Journal of Clinical Investigation*, **94**, 2493–503.

9 Davies, M.J. (1996). Stability and instability: two faces of coronary atherosclerosis. *Circulation*, **94**, 2013–202.

10 Grau, A.J., Buggle, F., Becher, H., Werle, E. and Hacke, W. (1996). The association of leukocyte count, fibrinogen and C-reactive protein with vascular risk factors and ischemic vascular diseases. *Thrombosis Research*, **82**, 245–55.

11 Haverkate, F., Thompson, S.G. and Duckert, F. (1995). Haemostasis factors in angina pectoris; relation to gender, age and acute-phase reaction. Results of the ECAT Angina Pectoris Study Group. *Thrombosis and Haemostasis*, **73**, 561–7.

12 Hansson, G., Stemme, V. and Yokota, T. (1997). Cytokines and the cardiovascular system. In *Cytokines*, eds. D.G. Remick and J.S. Friedland. New York: Marcel Decker, pp. 507–18.

13 Danesh, J., Collins, R., Appleby, P. and Peto, R. (1998). Association of fibrinogen, C-reactive protein, albumin, or leukocyte count with coronary heart disease. Meta-analyses of prospective studies. *Journal of the American Medical Association*, **279**, 1477–88.

14 Danesh, J., Muir, J., Wong, Y.K., Ward, M., Gallimore, J.R. and Pepys, M.B. (1999). Risk factors for coronary heart disease and acute-phase proteins. A population-based study. *European Heart Journal*, **20**, 954–9.

15 Emeis, J. and Kooistra, T. (1986). Interleukin-1 and lypopolysaccharide induce an inhibitor of t-PA in vivo and in cultured endothelial cells. *Journal of Experimental Medicine*, **163**, 1860–6.

16 Hait, P.H., Burgess, D.R., Vitti, G.F. and Hamilton, J.A. (1989). Interleukin-4 stimulates human monocytes to produce tissue-type plasminogen activator. *Blood*, **74**, 1282–5.

17 Ridker, P.M., Rifai, N., Pfeffer, M. et al. (2000). Elevation of tumor necrosis factor-alpha and increased risk of recurrent coronary events after myocardial infarction. *Circulation*, **101**, 2149–53.

18 Ridker, P.M., Rifai, N., Meir, J. et al. (2000). Plasma concentration of interleukin-6 and the risk of future myocardial infarction among apparently healthy men. *Circulation*, **101**, 1767–72.

19 Tatzber, F., Rabl, H., Koriska, K. et al. (1991). Elevated serum neopterin levels in atherosclerosis. *Atherosclerosis*, **89**, 203–8.

20 Huber, C., Batchelor, R., Fuchs, D. et al. (1984). Immune response-associated production of neopterin. Release from macrophages primarily under control of interferon-gamma. *Journal of Experimental Medicine*, **160**, 310–16.

21 Schobersberger, W., Hoffmann, G., Grote, J. et al. (1995). Induction of inducible nitric oxide synthase expression by neopterin in vascular smooth muscle cells. *FEBS Letters*, **377**, 461–4.

22 Werner-Felmayer, G., Werner, E.R., Fuchs, D. et al. (1993). Pteridine biosynthesis in human endothelial cells. Impact on nitric oxide-mediated formation of cyclic GMP. *Journal of Biological Chemistry*, **268**, 1842–6.

23 Wirleitner, B., Baier-Bitterlich, G., Hoffmann, G. et al. (1996). Neopterin activates transcription factor nuclear factor-kappa B in vascular smooth muscle cells. *FEBS Letters*, **391**, 181–4.

24 Brand, K., Page, S., Rogler, G. et al. (1996). Activated transcription factor nuclear factor-kappa B is present in the atherosclerotic lesion. *Journal of Clinical Investigation*, **97**, 1715–22.

25 Collins, T. (1993). Endothelial nuclear factor-kappa B and the initiation of the atherosclerotic lesion. *Laboratory Investigation*, **68**, 499–508.

26 Barnes, P.J. and Karin, M. (1997). Nuclear factor – $\kappa$B – a pivotal transcription factor in chronic inflammatory diseases. *New England Journal of Medicine*, **336**, 1066–71.

27 Weiss, G., Willeit, J., Kiechl, S. et al. (1994). Increased concentrations of neopterin in carotid atherosclerosis. *Atherosclerosis*, **106**, 263–71.
28 Gupta, S., Fredericks, S., Schwartzman, R.A. et al. (1997). Serum neopterin in acute coronary syndromes. *Lancet*, **349**, 1252.
29 Schumacher, M., Halwachs, G., Tatzber, F. et al. (1997). Increased neopterin in patients with chronic and acute coronary syndromes. *Journal of the American College of Cardiology*, **30**, 703–7.
30 Garcia-Moll, X., Coccolo, F., Cole, D. and Kaski, J.C. (2000). Serum neopterin and complex stenosis morphology in patients with unstable angina. *Journal of the American College of Cardiology*, **35**, 956–62.
31 Garcia-Moll, X., Cole, D., Zouridakis, E. and Kaski, J.C. (2000). Increased serum neopterin: a marker of coronary artery disease activity in women. *Heart*, **83**, 346–50.
32 Werner, E.R., Bichler, A., Daxenbichler, G. et al. (1987). Determination of neopterin in serum and urine. *Clinical Chemistry*, **33**, 62–6.
33 Neri Serneri, G.G., Abbate, R., Gori, A.M. et al. (1992). Transient intermittent lymphocyte activation is responsible for the instability of angina. *Circulation*, **86**, 790–7.
34 Steel, D.M. and Whitehead, A.S. (1994). The major acute phase reactants: C-reactive protein, serum amyloid P component and serum amyloid A protein. *Immunology Today*, **15**, 81–8.
35 Kannel, W.B., D'Agostino, R.B. and Belanger, A.J. (1987). Fibrinogen, cigarette smoking, and risk of cardiovascular disease: insights from the Framingham Study. *American Heart Journal*, **113**, 1006–10.
36 Haverkate, F., Thompson, S.G., Pyke, S.D.M., Gallimore, J.R. and Pepys, M.B., for the European Concerted Action on Thrombosis and Disabilities Angina Pectoris Study Group (1997). Production of C-reactive protein and risk of coronary events in stable and unstable patients. *Lancet*, **349**, 462–6.
37 de Beer, F.C., Hind, C.R., Fox, K.M., Allan, R.M., Maseri, A. and Pepys, M.B. (1982). Measurement of serum C-reactive protein concentration in myocardial ischaemia and infarction. *British Heart Journal*, **47**, 239–43.
38 Meade, T.W. (1995). Fibrinogen in ischaemic heart disease. *European Heart Journal*, **16**(Suppl A), 31–4.
39 Kannel, W.B., Wolf, P.A., Castelli, W.P. and D'Agostino, R.B. (1987). Fibrinogen and risk of cardiovascular disease. *Journal of the American Medical Association*, **258**, 1183–6.
40 Liuzzo, G., Biasucci, L.M., Gallimore, R. et al. (1994). The prognostic value of C-reactive protein and serum amyloid A protein in severe unstable angina. *New England Journal of Medicine*, **331**, 417–24.
41 Pepys, M.B. and Baltz, M.L. (1983). Acute phase proteins with special reference to C-reactive protein and related proteins (pentaxins) and serum amyloid A protein. *Advances in Immunology*, **34**, 141–212.
42 Lindberg, G., Eklund, G.A., Gullberg, B., Grandits, G.A., McCallum, L. and Tracy, R.P. (1991). Serum sialic acid concentration and cardiovascular mortality. *British Medical Journal*, **302**, 143–6.
43 Kuller, L.H., Eichner, J.E., Orchard, T.J. et al. (1991). The relation between serum albumin levels and risk of coronary heart disease in the Multiple Risk Factor Intervention Trial. *American Journal of Epidemiology*, **134**, 1266–77.

44 Reunanen, A., Knekt, P. and Aaran, R.-K. (1992). Serum ceruloplasmin and the risk of myocardial infarction and stroke. *American Journal of Epidemiology*, **136**, 1082–90.

45 Kuller, L.H., Tracy, R.P., Shaten, J. and Meilahn, E.N. (1996). Relation of C-reactive protein and coronary heart disease in the MRFIT nested case-control study. Multiple Risk Factor Intervention Trial. *American Journal of Epidemiology*, **144**, 537–47.

46 Ridker, P.M., Cushman, M., Stampfer, M.J., Russell, P.T. and Hennekens, C.H. (1997). Inflammation, aspirin, and the risk of cardiovascular disease in apparently healthy men. *New England Journal of Medicine*, **336**, 973–9.

47 Toss, H., Lindahl, B., Siegbahn, A. and Wallentin, L. for the FRISC Study Group (1997). Prognostic influence of increased fibrinogen and C-reactive protein levels in unstable coronary artery disease. *Circulation*, **96**, 4204–10.

48 Ferreiros, E.R., Boissonnet, C.P., Pizarro, R. et al. (1999). Independent prognostic value of elevated C-reactive protein in unstable angina. *Circulation*, **100**, 1958–63.

49 Thompson, S.G., Kienast, J., Pyke, S.D.M., Haverkate, F. and van de Loo, J.C.W. (1995). Hemostatic factors and the risk of myocardial infarction or sudden death in patients with angina pectoris. *New England Journal of Medicine*, **332**, 635–41.

50 Ridker, P.M., Buring, J.E., Shih, J., Matias, M. and Hennekens, C.H. (1998). Prospective study of C-reactive protein and the risk of future cardiovascular events among apparently healthy women. *Circulation*, **98**, 731–3.

51 Garcia-Moll, X., Zouridakis, E., Cole D. and Kaski, J.C. (2000). C-reactive protein in chronic stable angina: differences in baseline serum concentration between women and men. *European Heart Journal*, **21**, 1598–606.

52 Ridker, P.M., Hennekens, C.H., Rifai, N., Buring, J.E. and Manson, J.A.E. (1999). Hormone replacement therapy and increased plasma concentration of C-reactive protein. *Circulation*, **100**, 713–16.

53 Cushman, M., Legault, C., Barrett-Connor, E. et al. (1999). Effect of postmenopausal hormones on inflammation-sensitive proteins: the Postmenopausal Estrogen/Progestin Interventions (PEPI) study. *Circulation*, **100**, 717–22.

54 Morrow, D.A., Rifai, N., Antman, E.M. et al. (1998). C-reactive protein is a potent predictor of mortality independently of and in combination with troponin T in acute coronary syndromes: a TIMI 11A substudy. *Journal of the American College of Cardiology*, **31**, 1460–5.

55 Ridker, P.M., Glynn, R.J. and Hennekens, C.H. (1998). C-reactive protein adds to the predictive value of total and HDL cholesterol in determining risk of first myocardial infarction. *Circulation*, **97**, 2007–11.

56 Ridker, P.M., Hennekens, C.H., Buring, J.E. and Rifai, N. (2000). C-reactive protein and other markers of inflammation in the prediction of cardiovascular disease in women. *New England Journal of Medicine*, **342**, 836–43.

# The clinical significance of markers of coagulation in acute coronary syndromes

Lina Badimon, Antonio Bayés-Genís
Cardiovascular Research Center, Barcelona, Spain

## Introduction

Arterial thrombus formation seems to be an important factor in the conversion of chronic to acute atherosclerotic coronary events after plaque rupture, in the progression of coronary disease and in the acute phase of revascularization interventions. The presence of intraluminal thrombi, both in unstable angina and in acute myocardial infarction, has been documented in pathological, angiographic, angioscopic and intravascular ultrasound studies. In contrast with the very high incidence of thrombi in acute myocardial infarction, its incidence in unstable angina varies significantly among different studies, related, in part, to the interval between the onset of symptoms and the angiographic study. Presumably, the thrombus is occlusive at the time of anginal pain and later may become subocclusive and slowly lysed or digested. Local and systemic thrombogenic risk factors at the time of coronary plaque disruption may influence the type of thrombus and, hence, the different pathological and clinical syndromes [1].

## Thrombosis: platelets and coagulation

In severe injury, with exposure of components of the plaque, as in spontaneous plaque rupture or in angioplasty, marked platelet aggregation with mural thrombus formation follows. Vascular injury of this magnitude stimulates thrombin formation through both the intrinsic (surface-activated) and extrinsic (tissue factor-dependent) coagulation pathways, in which the platelet membrane facilitates interactions between clotting factors. This concept of vascular injury as a trigger of the platelet coagulation response is important in understanding the pathogenesis of the various vascular diseases associated with atherosclerosis and coronary artery disease. Growing thrombi may locally occlude the lumen, or embolize and be washed away by the blood flow to occlude distal vessels. However, thrombi may be physiologically and spontaneously lysed by mechanisms that block thrombus propagation. Thrombus size, location and composition is regulated by haemodynamic forces (mechanical effects), thrombogenicity of exposed substrate (local molecular effects), relative concentration of fluid phase and cellular blood components (local cellular effects) and the efficiency of the physiological mechanisms of control of the system – mainly fibrinolysis [2–5].

## Platelets

The initial recognition of damaged vessel wall by platelets involves: (i) adhesion, activation and adherence to recognition sites on the thromboactive substrate (extracellular matrix proteins; e.g. von Willebrand factor, collagen, fibronectin, vitronectin, laminin); (ii) spreading of the platelets on the surface; and (iii) aggregation with each other to form a platelet plug or white thrombus. The efficiency of the platelet recruitment will depend on the underlying substrate and local geometry. A final step of recruitment of other blood cells also occurs; erythrocytes, neutrophils and, occasionally, monocytes are found on evolving mixed thrombus. However, after plaque rupture, the exposed atherosclerotic vessel structures induce platelet aggregation and thrombosis by mechanisms, in some instances, different from those prevalent in haemostatic plug formation. The ulcerated atherosclerotic plaque may contain a disrupted fibrous cap, a lipid-rich core, abundant extracellular matrix and inflammatory cells. Such structures exhibit a potent activating effect on platelet aggregation and the triggering of coagulation.

Thrombin plays an important role in the pathogenesis of arterial thrombosis. It is one of the most potent known agonists for platelet activation and recruitment. Thrombin is a critical enzyme in early and late thrombus formation, cleaving fibrinopeptide A and B from fibrinogen to yield insoluble fibrin which, effectively, anchors the evolving thrombus. Both free and fibrin-bound thrombin is able to convert fibrinogen to fibrin, allowing propagation of the thrombus at the site of injury. Therefore, platelet activation triggers intracellular signalling and expression of platelet membrane receptors for adhesion, and initiation of cell contractile processes that will induce shape change and secretion of the granular contents. The expression of the integrin IIb/IIIa ($\alpha$IIb$\beta_3$) receptors for adhesive glycoprotein ligands (mainly fibrinogen and von Willebrand factor) in the circulation initiate platelet to platelet interaction. The process becomes perpetuated by the arrival of platelets brought by the circulation. Most of the glycoproteins in the platelet membrane surface are receptors for adhesive proteins or mediate cellular interactions.

## Coagulation system

During plaque rupture, in addition to platelet deposition in the injured area, the clotting mechanism is activated by the exposure of the atherosclerotic plaque matrix. The activation of the coagulation cascade leads to the generation of thrombin, which, as mentioned before, is a powerful platelet agonist that contributes to platelet recruitment, in addition to catalysing the formation and polymerization of fibrin. Fibrin is essential in the stabilization of the platelet thrombus and its ability to withstand removal by flow forces, shear and high intravascular pressure. These basic concepts have clinical relevance in the context of the acute coronary syndromes, in which plaque rupture exposes vessel wall matrix and plaque core materials which, by activating platelets and the coagulation system, results in the formation of a fixed and occlusive platelet–fibrin thrombus. The efficacy of fibrinolytic agents pointedly demonstrates the importance of fibrin-related material in the thrombosis associated with myocardial infarction. The regulation of blood coagulation is dependent on processes which take place on membrane surfaces. The proteins which compose the clotting enzymes do not collide and interact in a random manner in the plasma, but interact in complexes in a highly efficient manner on platelet and vascular surfaces. The major regulatory events in coagulation (activation, inhibition, generation of anticoagulant proteins) occur in membrane surfaces.

Tissue factor appears to be a major procoagulant factor in the vascular space immediately underlying the endothelial lining of arteries, a site which might be readily accessible upon local injury or upon rupture of an atherosclerotic plaque [6–9].

## Haemostatic markers and risk stratification

Clinical decision making requires appropriate tools to help in the diagnosis and to measure future risk in patients. Risk estimations of a patient having cardiovascular disease or of developing complications, such as myocardial infarction, sudden death or stroke, are necessary as they will determine which type of intervention to perform. The need for clinically useful markers, to help in the evaluation of risk, is, therefore, one of the primary targets of research. Biochemical parameters have been, and still are, sought to cover what epidemiological research has signalled as risk markers. Within the scope of this chapter, coagulation markers are noted which are good candidates to help with patient risk stratification and prognosis. Cardiac troponins have emerged recently as good tools for the detection of myocardial damage associated with thrombotic complications of the coronary atherosclerotic plaques.

Several markers of thrombus formation and lysis have been identified, both in stable and unstable patients [10]. Fibrinogen, activated factor VII, D-dimer, thrombin–antithrombin complexes, prothrombin fragments 1 and 2 and plasmin-$\alpha 2$–antiplasmin complexes are substrates and end-products of the coagulation and fibrinolysis cascades in ischaemic coronary patients [11–13]. However, little is known of their potential role in the triage and risk stratification of chest pain patients in the emergency room (ER).

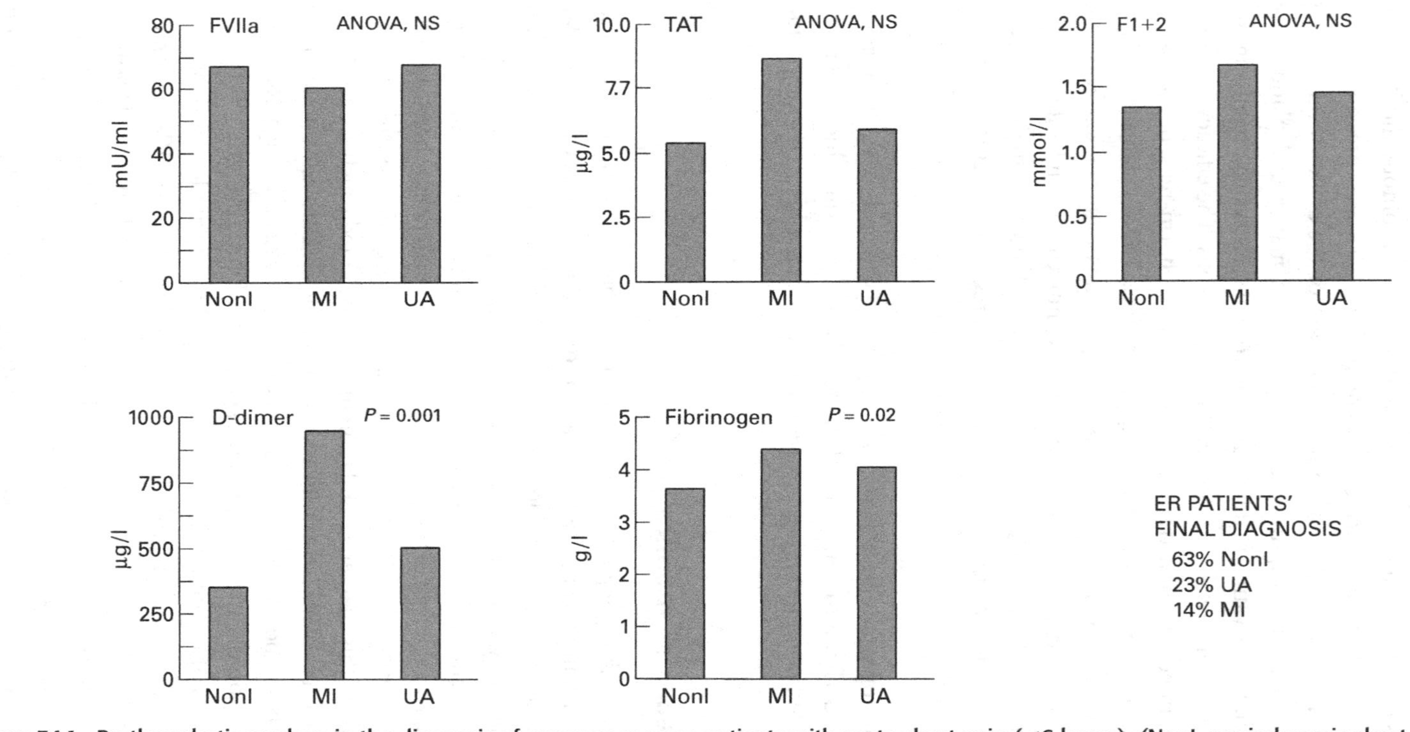


Figure 34.1 Prothrombotic markers in the diagnosis of emergency room patients with acute chest pain (<6 hours). (NonI, nonischaemic chest pain; MI, myocardial infarction; UA, unstable angina.)

Recently, the authors have performed a study in which plasma D-dimer levels were determined in order to assess their independent diagnostic value over conventional tests for the early diagnosis of acute coronary syndromes in the ER. Measurements of thrombin–antithrombin complex (TAT), prothrombin fragments 1 and 2 ($F_{1+2}$), activated factor VII, fibrinogen and D-dimer were also determined and compared. This was a prospective study that enrolled 300 consecutive patients admitted to a teaching hospital ER for acute chest pain. All patients were 25 years or older and came to the ER with the principal complaint of central or left-sided chest pain. Only patients with an onset of symptoms <6 hours before admission were eligible for study [14].

D-dimer is the end-product of crosslinked fibrin breakdown and indicates active thrombus formation and lysis. In the past decade, D-dimer testing has been established as a useful aid in the diagnosis of deep-vein thrombosis of the lower limbs and pulmonary embolism [15, 16]. Few data exist on D-dimer values in coronary artery disease, and these data are controversial. Small studies show that D-dimer levels are higher in patients with myocardial infarction (MI) [10, 12, 17]. Francis et al. [12] provided the first demonstration of increased plasma concentration of crosslinked fibrin polymers in patients with MI using a laborious inhouse electrophoretic technique, which was not suited for ER use. Lee et al. [17] showed that marked increases in circulating D-dimer are indicative of thrombotic complications in patients with MI, suggesting that D-dimer, besides being useful as a marker for early diagnosis, is also a risk factor for the development of MI complications. Other series, however, found normal D-dimer concentrations both in patients with MI and in healthy volunteers [18]. These studies determined D-dimer several hours after the onset of pain in hospitalized patients with a well-established diagnosis of MI.

In the authors' study, ischaemic patients had significantly higher plasma D-dimer concentrations (mean [95% CI] 656 μg/l [484–865]) than nonischaemic patients (338 μg/l [305–375]; $p<0.01$), even after adjustment for the covariates age, previous history of ischaemic heart disease and fibrinogen. Both MI (925 μg/l [635–1349]) and unstable angina (UA) (484 μg/l [398–589]) showed significantly higher levels than nonischaemic patients ($p<0.001$ and $p<0.05$, respectively) (Figure 34.1). Levels in MI were significantly higher than in UA ($p<0.02$). No statistically significant differences in D-dimer levels were found for Q-wave and nonQ-wave MI. Mean and 95% confidence interval for Q and nonQ-wave MI were 1047 μg/l (562–1950 μg/l) and 794 μg/l (550–1122 μg/l), respectively (not significant). Among the MI patients with D-dimer levels<500 μg/l, 58% of patients evolved to Q-wave MI and 42% of patients to nonQ-wave MI [14].

Receiver operating characteristic (ROC) curve analysis was performed at different D-dimer levels ranging from 400 μg/l (the upper normal range in our control group) to 700 μg/l. The discriminate power of the ROC curve analysis (combined sensitivity and specificity) for MI was best at levels of 500 μg/l (area under the curve, 0.81; standard error, 0.04), allowing the study of MI patients with D-dimer values above 500 μg/l (DD>500 μg/l). The sensitivity and specificity of D-dimer values >500 μg/l to identify MI were 65% and 80%, respectively, and the positive and negative predictive values were 36% and 93%, respectively. Moreover, D-dimer levels over 1000 μg/l were found in 46% of MI, in 17.2% of UA patients and in 3.7% of nonischaemic patients ($p<0.001$) [14].

Plasma levels of activated factor VII were not significantly different when the three patient groups (nonischaemic, MI, UA) were compared. TAT was significantly higher in MI patients than in the other two groups, and $F_{1+2}$ was higher in MI patients than in nonischaemic patients before adjusting for age and previous ischaemic heart disease. However, after adjustment for these variables, the significance for both markers was lost ($p=0.185$ and $p=0.456$, respectively). Fibrinogen levels were higher among patients with MI and UA than in nonischaemic patients ($p=0.02$) (Figure 34.1).

Logistic regression modelling identified four variables containing independent prognostic information for MI: D-dimer >500 μg/l, clinical assessment, positive ECG and plasma CK levels >180 U/l. Using a stepwise approach similar to that currently used in ER evaluation, a model that included clinical assessment, positive ECG and first CK>180 U/l had a diagnostic sensitivity for MI of 73% and a specificity of 98%. The addition of D-dimer >500 μg/l to the analysis resulted in a 19% increase of the diagnostic sensitivity for MI, from 73% to 92%, without losing specificity (97%). None of the eliminated variables improved this model significantly when included one at a time (all $p>0.2$). The multivariate analysis did not identify D-dimer levels as independent predictors for UA, although they were significantly raised in UA.

Many studies suggest that 2–4% of patients discharged from the ER may develop MI within 48 hours [19]. In the authors' series, five patients who ultimately had MI confirmed arrived to the ER with nonspecific ECG and initial cardiac enzymes in the normal range. D-dimer levels were raised (>500 μg/l) in 60% of these patients. Troponin assays are highly selective for myocardial injury, but, for early triage, have the limitation of being measurable only after 3–4 hours after the onset of pain [20]. D-dimer levels rise earlier than cardiac injury markers (including myoglobin) in acute ischaemic events, since they represent an earlier stage in the pathophysiology of MI and UA. The most appealing potential of D-dimer relates to detecting ongoing thrombus formation/dissolution in patients with acute ischaemic syndromes that are undetectable by conventional methods. This could result in a more cost-effective use of hospital facilities and, thereby, reduce overall costs. It is possible that elevated D-dimer levels can identify those patients most suitable for lytic and/or antiplatelet therapy, while those without such elevations are more suitable for interventional therapies.

In addition to the diagnostic utility of D-dimer for MI, this marker may also be of potential prognostic utility. D-dimer prognostic information may be particularly useful in UA patients with less elevated basal D-dimer levels. Some reports have found higher D-dimer concentrations in atherosclerotic patients than in control subjects [21], and the Physicians' Health Study [22] found that raised D-dimer levels were a marker for future MI among healthy men. Likewise, a recent prospective study shows that D-dimer is a predictor of recurrent coronary events [23], with implications for the prevention of secondary coronary events. These data fit with the concept of a hypercoagulable state that has been hypothesized to precede clinical events [24]. These results and others [17] suggest that D-dimer increases probably reflect enhanced physiological fibrinolysis as a result of ongoing thrombosis or a more pronounced fibrinolytic response to coronary thrombosis.

In another study, the authors also studied hospitalized patients with UA, subclassified into three groups: group A included patients with postMI angina (from days 3 to 30 after MI); group B comprised patients with new-onset angina ($<$2 months from onset, without previous MI during the last month) and no history of coronary heart disease (CHD); and group C included patients with crescendo angina ($>$2 months from onset, with no history of MI during the last month) and a history of CHD [13]. TAT, $F_{1+2}$ and plasminogen activator inhibitor (PAI-1) were not different among these groups of patients. Only D-dimer and plasmin–antiplasmin (PAP) showed significant changes and were higher than levels in healthy controls. Increased levels of PAP and D-dimer were found, particularly in the group with postinfarct angina, and both parameters were significantly correlated (Figures 34.2 and 34.3). This relationship is not found in new-onset angina or progressive angina. However, in this series, PAP levels did not predict an uneventful outcome, either in the acute phase or long term (6 months) [13].

## Conclusions

The results of the authors' study suggest that D-dimer measurement can improve clinical decision models to facilitate risk stratification. Substantially increased D-dimer levels in MI and UA patients were found. Our findings also provide a clinically useful D-dimer threshold in MI, $>$500 μg/l, which, added to the conventional diagnosis in the ER, increased the diagnostic sensitivity for MI from 73% to 92%. TAT, $F_{1+2}$ and activated factor VII provided no additional diagnostic information

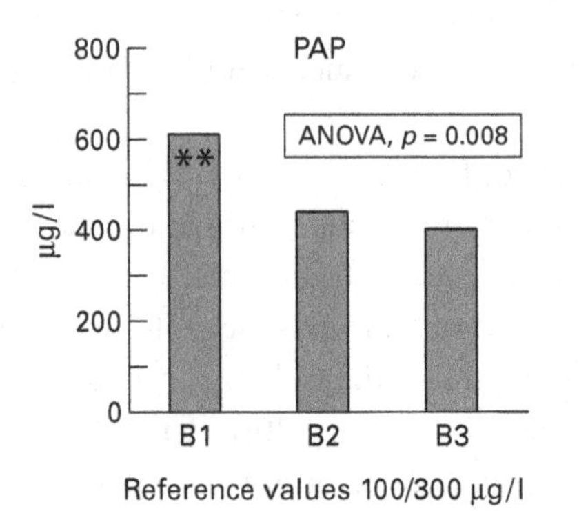


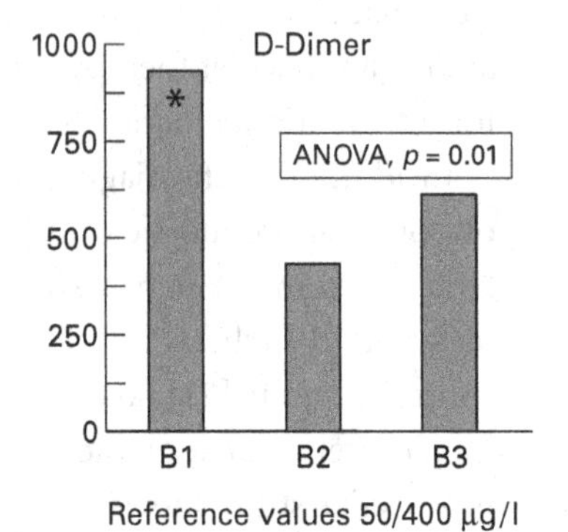


Figure 34.2 Unstable angina patients and prothrombotic markers. PAP, plasmin–antiplasmin complexes. B1, post MI angina patients; B2, new-onset angina patients; and B3, crescendo angina patients. TAT (range, 4.4–10.4 µg/l); F1.2 (1.1–2.1 nmol/l) and PAI-1 (17.1–35.8 ng/ml) were similar among UA patients and to reference normal values.

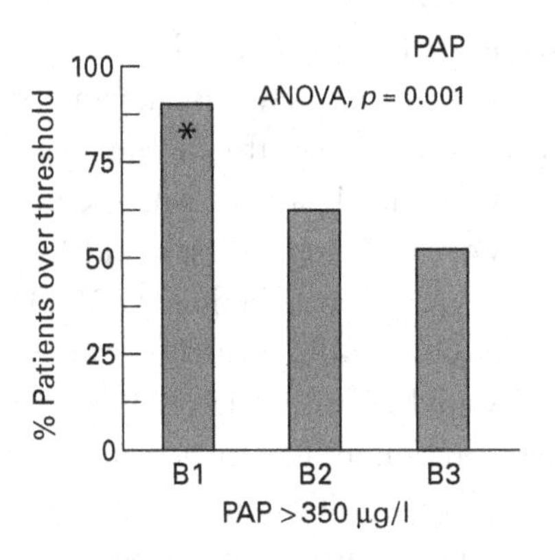


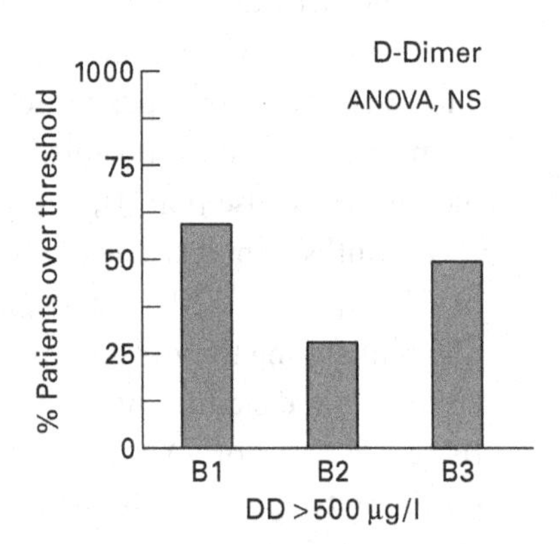


Figure 34.3 Percentage of unstable angina patients over threshold for the three groups: B1, post MI angina patients; B2, new-onset angina patients; and B3, crescendo angina patients.

for acute ischaemic patients in the ER. Although fibrinogen levels were raised in ischaemic patients, they do not represent a marker of coagulation activation but, instead, a constitutive zymogen of the coagulation cascade and an independent risk factor for ischaemic events.

Evaluation of cardiac troponins has also emerged as a very good marker of ischaemia. In the future, the association of troponins with D-dimer may improve our capacity to identify patients at risk.

## REFERENCES

1 Badimon, L. and Badimon, J.J. (1996). Interaction of platelet activation and coagulation. In *Atherosclerosis and Coronary Artery Disease*, eds. V. Fuster, R. Ross, and E.J. Topol. Philadelphia: Lippincott-Raven Publishers, pp. 639–56.

2 Fuster, V., Badimon, L., Badimon, J.J. and Chesebro, J.H. (1992). The pathogenesis of coronary artery disease and the acute coronary syndromes (Part I). *New England Journal of Medicine*, **326**, 242–50.

3 Fuster, V., Badimon, L., Badimon, J.J. and Chesebro, J.H. (1992). The pathogenesis of coronary artery disease and the acute coronary syndromes (Part II). *New England Journal Medicine*, **326**, 310–18.

4 Badimon, L., Chesebro, J.H. and Badimon, J.J. (1992). Thrombus formation on ruptured atherosclerotic plaques and rethrombosis on evolving thrombi. *Circulation*, **86**(Suppl III), III74–85.

5 Badimon, J.J., Fuster, V., Chesebro, J.H. and Badimon, L. (1993). Coronary atherosclerosis. *Circulation*, **87**(Suppl II), II3–16.

6 Badimon, L., Badimon, J.J., Lassila, R., Heras, M., Chesebro, J.H. and Fuster, V. (1991). Thrombin regulation of platelet interaction with damaged vessel wall and isolated collagen type I at arterial flow conditions in a porcine model. Effects of hirudins, heparin and calcium chelation. *Blood*, **78**, 423–34.

7 Fernández-Ortiz, A., Badimon, J.J., Falk, E, et al. (1994). Characterization of the relative thrombogenicity of atherosclerotic plaque components: implications for consequences of plaque rupture. *Journal of the American College of Cardiology*, **23**, 1562–9.

8 Toschi, V., Gallo, R., Lettino, M. et al. (1997). Tissue factor modulates the thrombogenicity of human atherosclerotic plaques. *Circulation*, **95**, 594–9.

9 Badimon, J.J., Lettino, M., Toschi, V. et al. (1999). Local inhibition of tissue factor reduces the thrombogenicity of disrupted human atherosclerotic plaques. Effects of TFPI on plaque thrombogenicity under flow conditions. *Circulation*, **99**, 1780–7.

10 Kruskal, J.B., Commerford, P.J., Franks, J.J. and Kirsh, R.E. (1987). Fibrin and fibrinogen-related antigens in patients with stable and unstable coronary artery disease. *New England Journal of Medicine*, **317**, 1361–5.

11 Broadhurst, P., Kelleher, C., Hughes, L. et al. (1990). Fibrinogen, factor VII clotting activity and coronary artery disease severity. *Atherosclerosis*, **85**, 169–173.

12 Francis, C.W., Connaghan, D.G., Scott, W.L. and Marder, V.J. (1987). Increased plasma concentration of cross-linked fibrin polymers in acute myocardial infarction. *Circulation*, **75**, 1170–7.

13 Bayes-Genis, A., Guindo, J., Oliver, A. et al. (1999). Elevated levels of plasmin-$\alpha 2$ antiplasmin complexes in unstable angina. *Thrombosis and Haematosis*, **81**, 865–8.

14 Bayes-Genis, A., Mateo,J., Santaló,M. et al. (2000). D-dimer is an early diagnostic marker of coronary ischemia in patients with chest pain. *American Heart Journal*, **140**, 379–84.

15 Bounameaux, H., Schneider, P.A., Reber, G. et al. (1989). Measurement of plasma D-dimer for diagnosis of deep venous thrombosis. *American Journal of Clinical Pathology*, **91**, 82–5.

16 Bounameaux, H., Cirafici, P., de Moerloose, P. et al. (1991). Measurement of D-dimer in plasma as a diagnostic aid in suspected pulmonary embolism. *Lancet*, **337**, 196–200.

17 Lee, L.V., Ewald, G.A., McKenzie, C.R. and Eisenberg, P.R. (1997). The relationship of soluble fibrin and cross-linked fibrin degradation products to the clinical course of myocardial infarction. *Arteriosclerosis, Thrombosis, and Vascular Biology*, **17**, 628–33.

18 Gurfinkel, E., Bozovich, G., Cerdá, M. et al. (1995). Time significance of acute thrombotic reactant markers in patients with and without silent myocardial ischemia and overt unstable angina pectoris. *American Journal of Cardiology*, **76**, 121–4.

19 Puleo, P.R., Meyer, D., Wathen, C. et al. (1994). Use of a rapid assay of subforms of creatine kinase MB to diagnose or rule out acute myocardial infarction. *New England Journal of Medicine*, **331**, 561–6.

20 Larue, C., Calzolari, C., Bertinchant, J.P. et al. (1993). Cardiac-specific immunoenzymometric assay of troponin I in the early phase of acute myocardial infarction. *Clinical Chemistry*, **39**, 972–9.

21 Salomaa, V., Stinson, V., Folsom, A.R. et al. (1995). Coronary heart disease/myocardial infarction: association of fibrinolytic parameters with early atherosclerosis. The ARIC study. *Circulation*, **91**, 284–90.

22 Ridker, P.M., Hennekens, C.H., Cerskus, A. and Stampfer, M.J. (1994). Plasma concentration of cross-linked fibrin degradation product (D-dimer) and the risk of future myocardial infarction among apparently healthy men. *Circulation*, **90**, 2236–40.

23 Moss, A.J., Goldstein, R.E., Marder, V.J. et al. (1999). Thrombogenic factors and recurrent coronary events. *Circulation*, **99**, 2517–22.

24 Bauer, K.A. and Rosenberg, R.D. (1987). The pathophysiology of the prethrombotic state in humans: insights gained from studies using markers of hemostatic system activation. *Blood*, **70**, 343–50.

35

# Endothelin: what does it tell us about myocardial and endothelial dysfunction?

John Pernow

Karolinska Hospital, Stockholm, Sweden

## Introduction

Endothelin-1 is a 21-amino acid peptide that was identified in 1988 as a potent constrictor substance produced by vascular endothelial cells [1]. Endothelin-1 belongs to a family of peptides which also includes endothelin-2 and endothelin-3. These isopeptides are encoded by separate genes. It was originally thought that endothelin-1 was produced exclusively by endothelial cells, but it was demonstrated subsequently that it can be produced by several different cell types, such as vascular smooth muscle cells and cardiac myocytes. The expression of endothelin-1 mRNA is stimulated by different growth factors, cytokines, vasoactive substances such as angiotensin II and catecholamines, shear stress, hypoxia and oxidized low-density lipoproteins (Figure 35.1). The expression is inhibited by nitric oxide and atrial natriuretic factor.

Endothelin-1 is synthesized from a larger prepropeptide which is further processed to the 38-amino acid intermediate big endothelin-1 (Figure 35.1). Big endothelin-1 is cleaved to the 21-amino acid endothelin-1 by a family of endothelin-converting enzymes [1]. The endothelin-converting enzyme is located both intracellularly and in connection with the cell membrane of endothelial cells and vascular smooth muscle cells (Figure 35.1). The conversion of big endothelin-1 to endothelin-1 is essential for its biological activity.

The receptors for endothelin-1 are widely distributed in the cardiovascular system. Two different subtypes of ET receptors, $ET_A$ and $ET_B$, have been cloned and characterized [1]. In the vascular wall, the $ET_A$ receptor is present on the vascular smooth muscle cells and activation of this receptor leads to vasoconstriction (Figure 35.1). Activation of the $ET_B$ receptor located on vascular smooth muscle cells also results in vasoconstriction. Stimulation of $ET_B$ receptors located on endothelial cells leads to the release of nitric oxide and/or prostacyclin, resulting in vasodilation. In the heart, the $ET_A$ receptor dominates and is present on cardiac myocytes and fibroblasts in addition to the vascular wall.

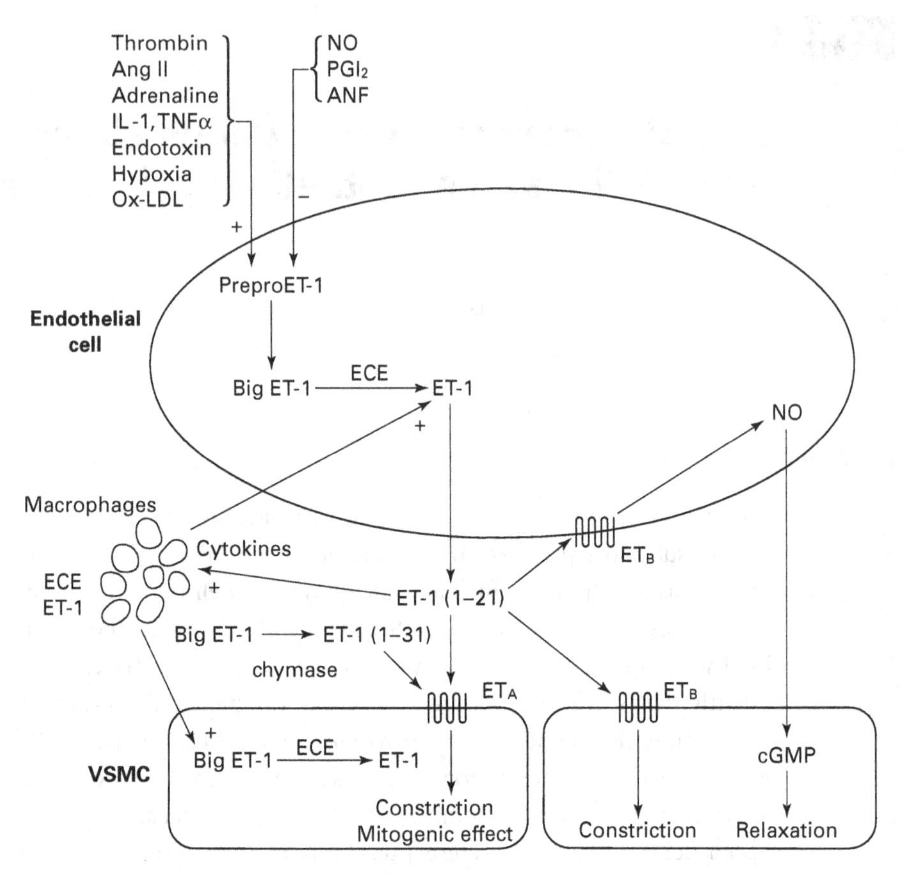


Figure 35.1 Schematic illustration of the production and biological effects of endothelin-1 (ET-1) under normal and pathological conditions. The 21-amino acid peptide ET-1 (1–21) is produced from the larger precursor peptide big endothelin-1 by endothelin-converting enzyme (ECE) in endothelial cells. The expression of preproET-1 mRNA and the production of ET-1 is stimulated by different growth factors, cytokines, vasoactive substances and oxidized low-density lipoprotein (ox-LDL). The production is inhibited by nitric oxide (NO), prostacyclin ($PGI_2$) and atrial natriuretic factor (ANF). ET-1 produces its effects via stimulation of the $ET_A$ or $ET_B$ receptors. These effects include vasoconstriction, mitogenic effects and stimulation of inflammatory cells. ET-1 may also produce vasorelaxation via release of NO from endothelial cells. Under pathophysiological conditions, such as congestive heart failure, atherosclerosis or ischaemic heart disease, the production of ET-1 is increased and occurs in vascular smooth muscle cells (VSMC) and infiltrating inflammatory cells. ET-1 is a chemoattractant for macrophages. The infiltrating macrophages express both ECE and ET-1. Cytokines produced by macrophages further enhance ET-1 production in various cells. A 31-amino acid form of ET-1 (1–31) may be formed from big ET-1 by chymase in the atherosclerotic vascular wall.

Endothelin-1 exerts a variety of potent biological effects on different cell types. Besides being one of the most potent vasoconstrictors known, endothelin-1 exerts positive myocardial inotropic effects and mitogenic effects on cardiac myocytes and vascular smooth muscle cells. Endothelin-1 also regulates water balance via reduction in renal plasma flow and direct renal tubular effects. In addition, endothelin-1 can act as a chemoattractant for monocytes, macrophages and neutrophils. It has also been demonstrated that endothelin-1 stimulates expression of adhesion molecules on leucocytes and endothelial cells.

Circulating plasma levels of endothelin are usually very low. The biological effects of endothelin-1 are mainly mediated via an autocrine and/or paracrine action. The endothelin system is markedly upregulated during several different pathological cardiovascular conditions. This upregulation results in an increased production of endothelin-1 in several cell types and elevated circulating plasma levels of endothelin-1 in many cardiovascular diseases. This chapter will focus on the potential involvement of endothelin-1 in some pathological cardiovascular conditions, associated with increased production and release of endothelin-1, and on how elevated endothelin levels relate to the progression and prognosis of these conditions.

## Congestive heart failure

Impairment of ventricular contractility in congestive heart failure (CHF) is associated with neuronal and humoral activation which results in marked cardiovascular responses. This neurohormonal activation is aimed at preserving cardiac output and organ perfusion. However, the reflex response results in increased peripheral resistance, which further impairs left ventricular function and leads to a vicious cycle of CHF. Accordingly, effective treatment of CHF aims at inhibiting the reflex neurohormonal activation. An example of this is the well-documented treatment with inhibitors of angiotensin-converting enzyme.

Endothelin-1 typically induces slowly developing and extremely long-lasting vasoconstrictor responses. Endothelin-1 also augments the vasoconstrictor action of other vasoactive agents of importance in CHF such as angiotensin II and catecholamines. Conversely, the production of endothelin-1 is stimulated by angiotensin II and catecholamines. Thus, endothelin-1 and renin–angiotensin as well as the sympathetic nervous systems seem to potentiate each other. In addition to its vasoactive effects, endothelin-1 stimulates sodium retention. Endothelin-1 also exerts mitogenic effects, which may contribute to the myocardial and vascular remodelling in CHF. All these biological actions of endothelin-1 may be of importance for the progression of CHF.

There are several studies demonstrating that plasma levels of endothelin-1 and

**Table 35.1.** Univariate relationship between various clinical and biochemical variables and mortality in 120 heart failure patients

| Variables | *P* |
|---|---|
| Endothelin-1 | 0.0001 |
| NYHA | 0.0007 |
| Atrial natriuretic peptide | 0.006 |
| LVED diameter | 0.007 |
| Noradrenaline | 0.01 |
| Age | 0.02 |
| Ejection fraction | 0.13 |

*Notes:*
The median follow up was 361 days (15 days–3.8 years).
NYHA: New York Heart Association functional class; LVED: left ventricular end diastolic
*Source:* From Pousset et al. [4].

big endothelin-1 are elevated in experimental animal models of heart failure and in patients with CHF [2]. Endothelin-1 immunoreactivity is usually elevated 2–5-fold in patients with CHF. The increase in endothelin-1 levels is independent of the aetiology of the heart failure. The plasma concentration of endothelin-1 correlates with the severity of heart failure. Thus, plasma endothelin-1 concentration correlates positively with the New York Heart Association functional class and inversely with the left ventricular ejection fraction. In addition, plasma endothelin-1 concentrations correlate closely with pulmonary artery pressure and pulmonary vascular resistance [2].

The plasma concentrations of endothelin-1 and big endothelin-1 have significant prognostic value. Pacher et al. [3] found that the plasma concentration of big endothelin-1 was significantly higher in patients with CHF who died within 1 year, and in transplant recipients, than in 1-year survivors. The survival rate was significantly lower in patients with big endothelin-1 levels >4.3 fmol/ml. Plasma big endothelin-1 and functional class provided independent prognostic information using a stepwise Cox regression multivariate analysis. In another study of 120 patients with CHF and a mean left ventricular ejection fraction of 28%, Pousset et al. [4] found that plasma levels of endothelin-1 constituted the most powerful prognostic marker among the variables tested, including plasma atrial natriuretic peptide and noradrenaline, functional class, age and left ventricular ejection fraction (Table 35.1). The mortality rate was significantly higher in patients with elevated endothelin-1 levels. These findings clearly demonstrate that circulating plasma levels of both big endothelin-1 and endothelin-1 provide prognostic information in patients with moderate to severe CHF. However, in patients with asymptomatic or minimally symptomatic left ventricular dysfunction, plasma endothelin-1 does not seem to be a good predictor of mortality and morbidity [5].

The origin of circulating endothelin-1 in CHF is most likely from sources other than the vascular endothelium alone. In CHF, the production of endothelin-1 is markedly increased in cardiac myocytes and vascular smooth muscle cells [1]. The upregulated endothelin system may therefore act via an autocrine and/or paracrine mechanism in the myocardium and vascular wall. Prolonged upregulation of the endothelin system may have a detrimental effect in CHF. It is also possible that the elevated circulating plasma levels of endothelin-1 in CHF are due to reduced clearance of the peptide. In fact, the clearance of exogenous endothelin-1 is reduced in dogs with CHF. In addition, it has been noted that the pulmonary clearance of endothelin-1 was reduced in patients with pulmonary hypertension [2].

The pathophysiological role of endothelin-1 in CHF has been investigated using endothelin receptor antagonists. In experimental models of heart failure in rats, both a selective $ET_A$ receptor antagonist (BQ 123) and a mixed $ET_A/ET_B$ receptor antagonist, bosentan, improve haemodynamics and survival [6, 7]. In a study on patients with CHF, intravenous bolus administration of bosentan substantially reduced pulmonary and systemic vascular resistance and pulmonary capillary wedge pressure. Cardiac index was increased, but heart rate was unaffected [2]. Similar results were obtained during a 2-week period of oral administration of bosentan [8].

## Atherosclerosis

Development and progression of atherosclerosis involve endothelial cell injury, release of cytokines, accumulation of monocytes, macrophages and lipids, as well as proliferation and migration of smooth muscle cells. Endothelin-1 production in the vascular wall is increased in atherosclerosis (Figure 35.1). In the atherosclerotic vessel, the production of endothelin-1 is induced in vascular smooth muscle cells and in macrophages, in addition to endothelial cells [9]. The production of endothelin-1 is increased in human atherosclerotic arteries and the accumulation seems to be especially high in unstable atherosclerotic plaques. The enhanced production of endothelin-1 is related to an increased expression of endothelin-1 mRNA in atherosclerotic arteries. Furthermore, there are reports of increased expression of endothelin-converting enzyme in endothelial cells, vascular smooth muscle cells and macrophages of the atherosclerotic artery. There is also evidence that endothelin (1–31) may be formed by chymase from big endothelin-1. Since chymase is accumulated in atherosclerotic arteries, this pathway may be of importance in atherosclerotic vessels (Figure 35.1). The production of endothelin-1 may be stimulated by several proatherogenic factors such as cytokines and oxidized low-density lipoproteins. Increased production and release of endothelin-1 in atherosclerosis results in elevated circulating plasma levels. It has been reported that circulating plasma levels are increased 2.3-fold in patients with atherosclerosis in comparison with a control group [10]. The plasma levels of endothelin-1 correlated significantly with the numbers of atherosclerotic lesions. Plasma levels of endothelin-1 are also elevated in patients and experimental animals with hypercholesterolaemia.

The potential involvement of endothelin-1 in the pathogenesis of atherosclerosis and endothelial dysfunction has been investigated in various animal models of atherosclerosis using endothelin receptor antagonists. In hamsters fed a high cholesterol diet, administration of a selective $ET_A$ receptor antagonist reduced the development of fatty streaks [9]. Both selective $ET_A$ receptor blockade and mixed $ET_A/ET_B$ receptor blockade were found to preserve coronary endothelial function in hypercholesterolaemic pigs [11]. In a study on atherosclerotic mice lacking apolipoprotein E, treatment with the $ET_A$ receptor antagonist LU 135252 for 30 weeks resulted in improved endothelium-dependent relaxation and reduction of atherosclerotic lesion development [12]. The author's own unpublished observations suggest that an $ET_A$ receptor antagonist improves endothelium-dependent relaxation in patients with atherosclerosis. These important findings suggest that endothelin-1, via activation of the $ET_A$ receptor, is involved in the development of atherosclerosis and that $ET_A$ receptor blockade may have therapeutic potential in atherosclerotic disease.

## Ischaemic heart disease

There are several studies demonstrating that the production and release of endothelin-1 are increased during myocardial ischaemia and reperfusion, both in experimental animals and in patients [13]. Patients with unstable angina and nonQ-wave myocardial infarction (MI) have elevated systemic plasma levels of endothelin-1. Furthermore, the plasma endothelin-1 levels are significantly higher and remain elevated in patients who have subsequent clinical events, such as acute MI or refractory angina, in comparison with patients who have no complications [14]. In a study of patients with both Q-wave and nonQ-wave MIs, plasma levels of endothelin-1 were significantly related to 1-year mortality (Table 35.2). Age, presence of heart failure and plasma atrial natriuretic factor levels provided no additional prognostic information. The increase in plasma endothelin-1 concentrations following acute MI seems to be inversely related to left ventricular function [16].

The role of endothelin-1 during myocardial ischaemia and reperfusion has been

**Table 35.2.** Risk ratio estimates based on univariate Cox models of 1-year mortality in 142 patients following acute myocardial infarction

| Variable | Risk ratio estimate | *P* |
|---|---|---|
| Endothelin | | 0.0002 |
| <4.9 pg/ml | 1.0 | |
| 4.9–6.5 pg/ml | 2.8 | |
| >6.5 pg/ml | 12.6 | |
| Atrial natriuretic factor | | 0.0032 |
| <93.6 pg/ml | 1.0 | |
| 93.6–128.1 pg/ml | 6.2 | |
| >128.1 pg/ml | 8.9 | |
| No heart failure | 1.0 | 0.0009 |
| Heart failure | 2.4 | |
| Age (years) | | 0.0708 |
| <70.0 | 1.0 | |
| 70.0–75.3 | 3.4 | |
| >75.3 | 3.3 | |

*Note:*
Cut-off points represent median and 75th percentile values.
*Source:* From Omland et al. [15].

extensively investigated in various experimental models [13]. Myocardial ischaemia followed by reperfusion results in increased expression of endothelin-1 in cardiac myocytes. Endothelin-1 is released from the jeopardized myocardial area during reperfusion of a previously ischaemic myocardium. Ischaemia and reperfusion also results in an enhanced vasoconstrictor activity of endothelin-1, which seems to be related to an upregulation of $ET_A$ receptors. Administration of endothelin receptor antagonists has been shown to ameliorate the ischaemia–reperfusion injury. Thus, both selective $ET_A$ receptor antagonists and mixed $ET_A/ET_B$ receptor antagonists reduce the infarct size and improve myocardial performance following ischaemia and reperfusion. Endothelin receptor antagonists also preserve endothelial function following ischaemia and reperfusion. These data indicate that endogenous endothelin-1 is involved in the development of ischaemia–reperfusion injury and that administration of endothelin receptor antagonists may be a useful tool to limit myocardial and endothelial damage in this situation. The mechanism behind the cardioprotective effect of endothelin receptor antagonists during ischaemia–reperfusion remains to be established. The endothelin receptor antagonists may improve coronary blood flow and prevent development of the 'no reflow phenomenon'. They may also inhibit the endothelin-1-induced increase in intracellular calcium or activation of the renin–angiotensin system. Furthermore, it has been suggested that endothelin receptor blockade attenuates neutrophil-mediated injury during reperfusion [17].

## Conclusions

The endothelin system is activated in a number of different cardiovascular diseases. This activation involves the induction of production of endothelin-1 in several different cell types in the heart and in the vascular wall, as well as release into plasma. The increased activity of the endothelin system seems to be deleterious in several situations. The deleterious effects may include enhanced vasoconstriction, mitogenic effects and proinflammatory effects. In such situations, administration of endothelin receptor antagonists may prove useful. The plasma levels of endothelin-1 in moderate to severe CHF, atherosclerosis and following acute MI relate to the severity of the disease. Furthermore, high levels of endothelin-1 in patients with CHF and following MI correlate highly significantly with mortality. However, it should also be emphasized that plasma levels of endothelin-1 are usually low and studies in experimental animals indicate that endogenous endothelin-1 may also be of pathophysiological significance even when plasma endothelin-1 levels are low.

## REFERENCES


1 Miyauchi, T. and Masaki, T. (1999). Pathophysiology of endothelin in the cardiovascular system. *Annual Review of Physiology*, **61**, 391–415.

2 Love, M. and McMurray, J. (1996). Endothelin in chronic heart failure: current position and future prospects. *Cardiovascular Research*, **31**, 665–74.

3 Pacher, R., Stanek, B., Hülsmann, M. et al. (1996). Prognostic impact of big endothelin-1 plasma concentrations compared with invasive hemodynamic evaluation in severe heart failure. *Journal of the American College of Cardiology*, **27**, 633–41.

4 Pousset, F., Isnard, R., Lechat, P. et al. (1997). Prognostic value of plasma endothelin-1 in patients with chronic heart failure. *European Heart Journal*, **18**, 254–8.

5 Tsutamoto, T., Wada, A., Maeda, K. et al. (1999). Plasma brain natriuretic peptide levels as a biochemical marker of morbidity and mortality in patients with asymptomatic or minimally symptomatic left ventricular dysfunction. Comparison with plasma angiotensin II and endothelin-1. *European Heart Journal*, **20**, 1799–807.

6 Sakai, S., Miyauchi, T., Kobayashi, M., Yamaguchi, I., Goto, K. and Sugishita, Y. (1996). Inhibition of myocardial endothelin pathway improves long-term survival in heart failure. *Nature*, **384**, 353–5.

7 Mulder, P., Richard, V., Derumeaux, G. et al. (1997). Role of endogenous endothelin in chronic heart failure. Effect of long-term treatment with an endothelin antagonist on survival, hemodynamics, and cardiac remodeling. *Circulation*, **96**, 1976–82.

8 Sütsch, G., Kiowski, W., Xiao-Wei, Y. et al. (1998). Short-term oral endothelin-receptor antagonist therapy in conventionally treated patients with symptomatic severe chronic heart failure. *Circulation*, **98**, 2262–8.

9 Kowala, M.C. (1997). The role of endothelin in the pathogenesis of atherosclerosis. *Advances in Pharmacology*, **37**, 299–318.

10 Lerman, A., Edwards, B.S., Hallett, J.W., Heublein, D., Sandberg, S.M. and Burnett, J.C. (1991). Circulating and tissue endothelin immunoreactivity in advanced atherosclerosis. *New England Journal of Medicine*, **325**, 997–1001.

11 Best, J., McKenna, C., Hasdai, D., Holmes, D. and Lerman, A. (1999). Chronic endothelin receptor antagonism preserves coronary endothelial function in experimental hypercholesterolemia. *Circulation*, **99**, 1747–52.

12 Barton, M., Haudenschild, C., D'Uscio, L., Shaw, S., Münter, K. and Lüscher, T. (1998). Endothelin $ET_A$ receptor blockade restores NO-mediated endothelial function and inhibits atherosclerosis in apolipoprotein E-deficient mice. *Proceedings of the National Academy of Sciences USA*, **95**, 14367–72.

13 Pernow, J. and Wang, Q-D. (1997). Endothelin in myocardial ischaemia and reperfusion. *Cardiovascular Research*, **33**, 518–26.

14 Wieczorek, I., Haynes, W.G., Webb, D.J., Ludlam, C.A. and Fox, K.A.A. (1994). Raised plasma endothelin in unstable angina and non-Q wave myocardial infarction: relation to cardiovascular outcome. *British Heart Journal*, **72**, 436–41.

15 Omland, T., Lie, R., Aakvaag, A., Aarsland, T. and Dickstein, K. (1994). Plasma endothelin determination as a prognostic indicator of 1-year mortality after acute myocardial infarction. *Circulation*, **89**, 1573–9.

16 Battistelli, S., Billi, M., Manasse, G., Vittoria, A., Roviello, F. and Forconi, S. (1999). Behavior of circulating endothelin-1 in a group of patients with acute myocardial infarction. *Angiology*, **50**, 629–38.

17 Gonon, A.T., Wang, Q.-D. and Pernow, J. (1998). The endothelin A receptor antagonist LU 135252 protects the myocardium from neutrophil injury during ischemia/reperfusion. *Cardiovascular Research*, **39**, 674–82.

36

# Homocysteine: a reversible risk factor for coronary heart disease

John C Chambers

Hammersmith Hospital, London, UK

## Homocysteine metabolism

Homocysteine is a sulphur-containing amino acid, derived from the metabolism of methionine. Homocysteine concentrations are determined by genetic and nutritional factors; mutations in the genes for enzymes involved in homocysteine metabolism, such as the common 5,10-methylenetetrahydrofolate reductase (MTHFR) 677→T mutation, and deficiencies of vitamins $B_6$, $B_{12}$ and folic acid are associated with hyperhomocysteinaemia.

## Homocysteine and the risk of coronary heart disease

The clinical importance of homocysteine was first recognized in the early 1960s with the recognition of the rare inherited metabolic disorder homocystinuria. Affected children, who have severe hyperhomocysteinaemia (plasma homocysteine >100 μmol/l), develop widespread, premature atherosclerosis affecting large- and medium-sized arteries and, if untreated, die from vascular complications in their teens or 20s. These observations led to the hypothesis that mild–moderate elevations in plasma homocysteine (>15 μmol/l) might contribute to the development of vascular disease in adults. This possibility has now been evaluated in more than 30 cross-sectional and prospective studies, involving over 10000 subjects [1]. The results have been remarkably consistent, with few exceptions. Elevated plasma homocysteine is an independent risk factor for peripheral vascular disease, cerebrovascular disease and coronary heart disease (CHD); in patients with CHD, homocysteine concentrations of 9, 15 and 20 μmol/l predict total mortality ratios of 1.9, 2.8 and 4.5, respectively [2]. Elevated homocysteine is common, and concentrations exceeding the upper limit of normal (15 μmol/l) are found in almost 30% of patients with vascular disease. In North American and European populations, it is estimated that elevated homocysteine may contribute to 10% of the population CHD risk [1].

## National variations in homocysteine concentrations

Previous studies in healthy men have shown that mean homocysteine concentrations vary from 7 to 11 μmol/l between European countries, and that homocysteine concentrations correlate closely with the national standardized mortality rates [3]. Homocysteine concentrations are also higher in Indian Asians, compared with European whites, and homocysteine may contribute to the increased CHD mortality in Asians [4]. Patients with chronic renal failure, and with renal or cardiac transplants, have high homocysteine concentrations. These observations suggest that the contribution of homocysteine to coronary risk may be greater in some populations and in some patient groups.

The precise reasons for the variation in homocysteine concentrations between populations remain to be determined. Previous studies have shown that elevated homocysteine in Indian Asians is accounted for by their reduced levels of vitamins $B_{12}$ and folate, implying that dietary B vitamin deficiency may underlie elevated homocysteine and increased CHD risk in Indian Asians, and perhaps other populations. Observations that the MTHFR 677→T mutation is present in 17% of North American patients with CHD, but 7% of patients from the Netherlands with premature vascular disease, suggest that MTHFR allele frequencies may underlie differences in homocysteine between populations. The MTHFR 677T mutation is rare in South African Blacks, who have more effective homocysteine metabolism and lower homocysteine levels than South African whites. This may explain their relative resistance to CHD, despite a high prevalence of obesity, hypertension and smoking.

## Mechanisms linking homocysteine to vascular disease

It remains unknown whether homocysteine is directly involved in the pathogenesis of vascular disease or is simply a marker for increased risk. Evidence to support a causal role for homocysteine has emerged from in vitro, animal and human experimental studies demonstrating that elevated plasma homocysteine concentrations induce abnormalities of vascular endothelial function. Exposure of cultured endothelial cells to homocysteine impairs nitric oxide-mediated inhibition of platelet aggregation, promotes leucocyte adhesion, inhibits thrombomodulin-dependent activation of the anticoagulant protein C and decreases endothelial cell binding of tissue plasminogen activator (t-PA) and expression of anticoagulant heparin sulphate [5–8].

Studies in nonhuman primates show that diet-induced increments in plasma homocysteine concentration are associated with impaired endothelium-dependent vasodilatation, a sensitive marker of the atherosclerotic process, which occurs early during the development of atherosclerosis, and improves during regression of the atherosclerotic lesion. Impaired endothelium-dependent dilatation has been described in children with cystathionine-beta-synthase deficiency, and in adults with moderate hyperhomocysteinaemia. However, in both groups, homocysteine concentrations were chronically elevated and it has been difficult to distinguish whether impaired endothelium-dependent dilatation results from elevated homocysteine or is a consequence of subclinical atherosclerotic disease. More recent studies show that an acute elevation in total plasma homocysteine concentrations after oral methionine is accompanied by a rapid and reciprocal fall in endothelium-dependent dilatation [9, 10]. Vascular endothelial dysfunction can be demonstrated at homocysteine concentrations similar to those associated with an increased risk of myocardial infarction and stroke, and can also be induced by physiological increments (2–3 μmol/l) in plasma homocysteine following dietary animal protein. These findings suggest that diet-related increments in plasma homocysteine may contribute to the development and progression of atherosclerosis, and may help to explain the incremental risk of vascular events with increasing homocysteine concentrations.

## Mechanisms linking homocysteine to endothelial dysfunction

The mechanisms linking homocysteine to endothelial dysfunction are not clear. There is growing evidence that homocysteine exerts its effects by promoting oxidative damage in endothelial cells. Initial exposure of cultured endothelial cells to homocysteine leads to the formation and release of nitric oxide, S-nitrosothiols and S-nitrosohomocysteine, substances with potent vasodilator and platelet inhibitor properties. However, with continued exposure, the oxidative effects of homocysteine predominate, with the resultant generation of superoxide anion radicals ($O_{2-}$) and hydrogen peroxide ($H_2O_2$) [5]. Generation of free radical superoxide anions may promote oxidation of low-density lipoprotein, and deactivation of nitric oxide. Deactivation of nitric oxide, the major endothelium-derived vasodilator, may lead to vasoconstriction, platelet aggregation and monocyte adhesion, all of which promote atherosclerosis. In vivo studies have confirmed the presence of oxidant stress during hyperhomocysteinaemia, using measurement of lipid peroxides as indicators of oxidant stress. Recent studies also show that vascular endothelial dysfunction associated with hyperhomocysteinaemia can be prevented by pretreatment with vitamin C [10], lending further support to the view that the adverse effects of homocysteine on vascular endothelial cells are mediated through oxidative stress mechanisms.

## Potential beneficial effects of homocysteine lowering

B vitamin supplementation lowers homocysteine concentrations by 30% [11]. Based on this, and the reported relationship between homocysteine and vascular disease, there are presently at least seven large-scale, randomized, placebo-controlled intervention trials underway in order to determine whether lowering homocysteine concentrations through B vitamin supplementation can improve survival in patients with CHD [11]. However, these studies are not expected to report for at least 2 years and, at present, little is known about the beneficial vascular effects of lowering homocysteine concentrations.

In physiological studies, homocysteine lowering through dietary folate supplementation in healthy volunteers, and in patients with CHD, is associated with a reduction in homocysteine concentrations and an improvement in vascular endothelial function [12]. More recent studies have examined the effect of homocysteine-lowering treatment on the progression of subclinical atherosclerosis. In healthy siblings of patients with premature atherosclerosis, the effect of folate and vitamin $B_6$ on the development or progression of subclinical atherosclerosis was estimated from exercise ECG, the ankle-brachial pressure index, and carotid and femoral ultrasonography [13]. B vitamin therapy lowered homocysteine concentrations by 50%, and decreased the rate of development of abnormal exercise tests. However, the other indices of subclinical atherosclerosis improved in both placebo and vitamin groups over the study period. A specific beneficial effect of homocysteine lowering on the progression of subclinical vascular disease remains, therefore, to be demonstrated.

## Conclusions

Epidemiological studies provide convincing and consistent evidence that elevated homocysteine is a risk factor for vascular disease, including CHD, in adults. Physiological studies show that elevated homocysteine induces vascular endothelial dysfunction, which may be mediated by oxidation stress mechanisms, supporting the view that the relationship between homocysteine and vascular disease may be causal. However, at present, there are no data to show that lowering homocysteine will reduce major cardiovascular end-points. The results of large-scale intervention studies with hard end-points are keenly awaited.

## REFERENCES

1 Boushey, C.J., Beresford, S.A., Omenn, G.S. and Motulsky, A.G. (1995). A quantitative assessment of plasma homocysteine as a risk factor for vascular disease. Probable benefits of increasing folic acid intakes. *Journal of the American Medical Association*, **274**, 1049–57.

2 Nygard, O., Nordrehaug, J.E., Refsum, H., Ueland, P.M., Farstad, M. and Vollset SE. (1997). Plasma homocysteine levels and mortality in patients with coronary artery disease. *New England Journal of Medicine*, **337**, 230–6.

3 Alfthan, G., Aro, A. and Gey K.F. (1997). Plasma homocysteine and cardiovascular disease mortality. *Lancet*, **349**, 397.

4 Chambers, J.C., Obeid, O.A., Refsum, H. et al. (2000). Plasma homocysteine concentrations and risk of coronary heart disease in UK Indian Asian and European men. *Lancet*, **355**, 523–7.

5 Stamler, J.S., Osborne, J.A., Jaraki, O. et al. (1993). Adverse vascular effects of homocysteine are modulated by endothelium-derived relaxing factor and related oxides of nitrogen. *Journal of Clinical Investigation*, **91**, 308–18.

6 Dudman, N.P., Temple, S.E., Guo, X.W., Fu, W. and Perry, M.A. (1999). Homocysteine enhances neutrophil-endothelial interactions in both cultured human cells and rats in vivo. *Circulation Research*, **84**, 409–16.

7 Rodgers, G.M. and Kane, W.H. (1986). Activation of endogenous factor V by a homocysteine-induced vascular endothelial cell activator. *Journal of Clinical Investigation*, **77**, 1909–16.

8 Lentz, S.R. and Sadler, J.E. (1991). Inhibition of thrombomodulin surface expression and protein C activation by the thrombogenic agent homocysteine. *Journal of Clinical Investigation*, **88**, 1906–14.

9 Bellamy, M.F., McDowell, I.F., Ramsey, M.W. et al. (1998). Hyperhomocysteinemia after an oral methionine load acutely impairs endothelial function in healthy adults. *Circulation*, **98**, 1848–52.

10 Chambers, J.C., McGregor, A., Jean Marie, J., Obeid, O.A. and Kooner, J.S. (1999). Demonstration of rapid onset vascular endothelial dysfunction after hyperhomocysteinemia: an effect reversible with vitamin C therapy. *Circulation*, **99**, 1156–60.

11 Clarke, R. and Collins, R. (1998). Can dietary supplements with folic acid or vitamin B6 reduce cardiovascular risk? Design of clinical trials to test the homocysteine hypothesis of vascular disease. *Journal of Cardiovascular Risk*, **5**, 249–55.

12 Bellamy, M.F., McDowell, I.F., Ramsey, M.W., Brownlee, M., Newcombe, R.G. and Lewis, M.J. (1999). Oral folate enhances endothelial function in hyperhomocysteinaemic subjects. *European Journal of Clinical Investigation*, **29**, 659–62.

13 Vermeulen, E.G., Stehouwer, C.D., Twisk, J.W. et al. (2000). Effect of homocysteine-lowering treatment with folic acid plus vitamin B6 on progression of subclinical atherosclerosis: a randomised, placebo-controlled trial. *Lancet*, **355**, 517–22.

37

# Matrix metalloproteinases and their tissue inhibitors

Peter M Timms, Stewart Campbell, Vinijar Srikanthan,
Christopher P Price
St Bartholomew's and The Royal London Hospitals, London, UK

## Introduction

The extracellular matrix (ECM) contains a variety of structural proteins which include collagens, proteoglycans and glycoproteins. These proteins act as a cellular 'skeleton' allowing cell to cell interactions.

Physiologically, matrix remodelling requires the synthesis of matrix which is balanced by degradation mediated via the production of active matrix metalloproteinases (MMPs). When this balance of synthesis and degradation is uncontrolled, inadequate or increased amounts of ECM are deposited. Reduced levels of matrix are associated with unstable angina whereas elevated levels of matrix are found in alcoholic cirrhosis, pulmonary fibrosis and left ventricular hypertrophy (LVH).

To date, 17 MMPs have been characterized (Table 37.1). MMPs were originally classified in terms of their substrate specificity, but this view has now been superseded and a numbering system introduced when it was appreciated that each MMP could hydrolyse a variety of substrates, including the degradation of other MMPs.

MMPs have a variety of domains, but all MMPs share three: (i) a predomain which targets the MMP for extracellular excretion; (ii) a prodomain which maintains the MMP in an inactive form called the proMMP; and (iii) a zinc-dependent catalytic domain which becomes activated on hydrolysis of the prodomain. With the exception of MMP7, all the MMPs have a haemopexin domain which enhances the binding of substrates and inhibitors. MMPs contain two conserved motifs: the prodomain and catalytic domain. MMP14, 15, 16 and 17 have a transmembrane domain and MMP14 activates proMMP2 into its biologically active form [1]. The fibronectin and collagen domains help in substrate recognition and the furin domain is an alternative site for activation of MMPs.

MMP1 hydrolyses the triple helix of collagen types I, II and III, and gelatin. In the *C* terminally truncated form, MMP1 can still hydrolyse gelatin but cannot hydrolyse collagen. MMP3 binds tissue inhibitor of metalloproteinase-1 (TIMP-1) more tightly than does its *C* terminal truncated form, but MMP7, which lacks the

**Table 37.1.** The matrix metalloproteinases

| MMP | Common Name | Substrate |
|---|---|---|
| 1 | Collagenase 1<br>Fibroblast collagenase<br>Interstitial collagenase | Collagens, IGF-BP3, IGF-BP5 MMP3, 9<br>Proteoglycan link protein, gelatin, aggrecan |
| 2 | Gelatinase A<br>72 kD gelatinase | Collagens, gelatin, laminin, MMP1, 9, 13<br>Proteoglycan link protein |
| 3 | Stromelysin 1 | Collagens III, IV, V, IX, MMP2/TIMP-2, MMP7, 8, 9, 13 and activation of MMP1 |
| 7 | Matrilysin | Collagen IV, X, MMP2, 3, 9, MMP/TIMP-1 |
| 8 | Collagenase 2<br>Neutrophil collagenase | Collagens, aggrecan, gelatin |
| 9 | Gelatinase B<br>92 kD gelatinase | Collagens III, IV and V, elastin |
| 10 | Stromelysin 2 | Collagens III, IV, V, gelatin, casein |
| 11 | Stromelysin 3 | Laminin, collagen IV, fibronectin |
| 12 | Macrophage elastase | Collagen IV, gelatin, elastin |
| 13 | Collagenase 3<br>Rat osteoblast collagenase | Collagens, gelatin, plasminogen activator inhibitor, MMP9, aggrecan |
| 14 | MT1-MMP | Collagen I, II, III, MMP2 and 13 |
| 15 | MT2-MMP | Fibronectin, large tenascin C, laminin |
| 16 | MT3-MMP | Collagen III |
| 17 | MT4-MMP | |
| 18 | Collagenase 4 | |
| 19 | No trivial name | Gelatin |
| 20 | Enamelysin | Amelogenin |

haemopexin domain, has similar binding affinities for TIMP-1 as the *C* truncated MMP3. These and other observations showed that both the *N* and *C* terminal portions of MMP1 are required for the hydrolysis of native collagens whereas the binding of TIMP-1 requires interaction with the *C* terminal end of the MMP.

## Nonmatrix hydrolytic actions of MMPs

MMPs not only degrade matrix, but are also involved in the degradation of nonmatrix proteins. Interleukin 1$\beta$ is activated to its biologically active 18 kD form

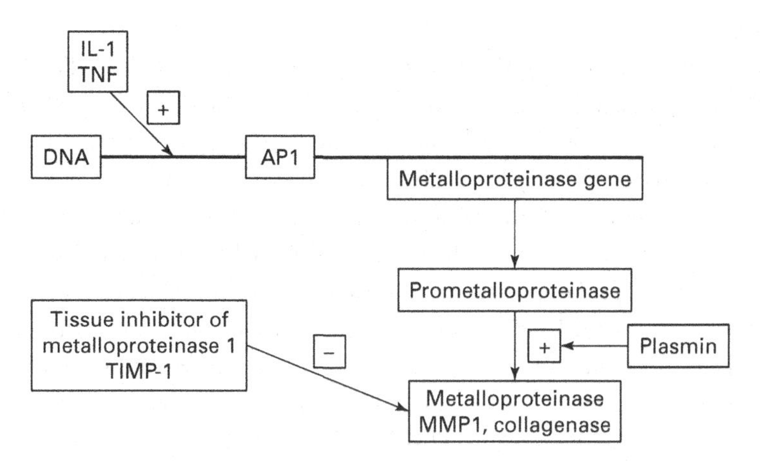


Figure 37.1 Control of matrix metalloproteinase activity.

from an inactive precursor by MMP2, 3 and, within minutes, MMP9. Insulin-like growth factors (IGFs) are mitogenic and their effects are expressed via an IGF receptor. IGF is bound to insulin growth factor-binding proteins (IGFBPs) and this binding reduces the interaction of IGF with its receptor. MMP1, 2 and 3 can hydrolyse this IGF/IGFBP complex, increasing IGF bioavailability [2].

## Control of MMPs

Because of their potency and importance in ECM remodelling, MMP levels are tightly controlled. This control occurs at several different steps in the production of the active MMPs (Figure 37.1). MMP activity is dependent on activation of the proMMP to the MMP, inhibition of activity of the MMP by a family of inhibitors (TIMPs) and transcriptional upregulation of the gene.

### Activation of proMMPs

Plasminogen-activating factor increases the concentration of plasmin, which converts (with the exception of MMP-2) the inactive prometalloproteinase to the active metalloproteinase by hydrolysis of the prodomain. The removal of the prodomain eliminates cysteine chelation of the zinc atom at the active site, permitting hydrolysis.

### Inhibition of MMPs by TIMPs

α2-macroglobulin accounts for 95% of plasma inactivation of MMPs in vitro. However, at the tissue level, TIMPs inhibit MMP activity resulting in MMP/TIMP complexes appearing in blood, synovial fluid, bronchoalveolar lavage, vitreous humour and blood.

### Structure of TIMPs

To date, four members of the TIMP family have been described – TIMP-1, TIMP-2, TIMP-3 [3, 4] and, most recently, TIMP-4 [5]. All of these TIMPs share a similar structure. However, there are significant differences in their amino acid sequence and the degree of carbohydrate substitution. TIMPs are low molecular weight proteins containing 12 sulphydryl groups that are conserved throughout the TIMP family. The sulphydryl groups form six disulphide bonds, which divide the molecule into six loops. Loops 1, 2 and 3 are associated with the *N* terminal region and are necessary for MMP inhibition, whereas loops 4, 5 and 6 are at the *C* terminal end of the molecule and are important in determining the rate at which inhibition occurs. TIMP-1, molecular weight 28 Kd, contains mannose and sialic acid sugars and is glycoslyated at two sites. TIMP-2, like TIMP-4, is not glycoslyated and has a molecular weight of around 21 Kd. TIMP-1, -2 and -4 are found in plasma, but TIMP-3 is not, as it is strongly bound to the extracellular matrix.

### Structural functionality and inhibitory function of TIMPs

By definition, each TIMP will inhibit all the MMPs, but at differing rates. At the *N* terminal region, TIMP-1 and -2 bind tightly to MMPs with a very high affinity ($K_I$ of $10^{-9}$ M) and with a 1:1 molar ratio of enzyme to inhibitor. The mechanism for inhibition of collagenase by TIMP-1 is a noncompetitive two-step process. During the first reversible stage, there is an initial rapid inactivation of MMP activity, while, in the second step, a tighter MMP/TIMP complex forms slowly [6]. Both TIMP-1, which binds specifically to proMMP9, and proMMP2, which binds to TIMP-2, do so at the *C* terminal end of the enzyme. The TIMP-2:proMMP2 complex is hydrolysed by a membrane-bound matrix metalloproteinase, MMP14, which activates proMMP2.

### Noninhibitory functions of TIMPs

TIMPs are multifunctional proteins and have roles other than their well-known inhibition of MMP activity which include a growth factor-like activity, antiangiogenesis, reduction in formation of new blood vessels and an increase in steroidogenesis [3].

TIMP-1 and -2 have been shown to stimulate a T lymphoblast cell line [7] and erythropoiesis by a mechanism independent of metalloproteinase inhibition. Hayakawa et al. [8] demonstrated that TIMP-2, at less than 10 times the concentration of TIMP-1, increased the cell growth of Raji cells which do not secrete MMPs. Alkylated TIMPs which have no metalloproteinase inhibitory activity stimulate Raji cell growth. TIMP depleted fetal calf serum or the addition of TIMP antibodies resulted in a marked reduction of growth stimulation which was reversed on the addition of TIMP-1 or TIMP-2 to the medium. TIMP-1 bound to proMMP-9 or TIMP-2 bound to proMMP-2 had no growth stimulating activity.

Chesler et al. [9] studied growth potentiation, binding to MMP and inhibitory activity of TIMP-1 using modified and truncated forms of the protein. Altered amino acids in positions 3–13 had minimal MMP inhibitory activity but normal binding to MMP and growth stimulation. Substitution of histidine for tyrosine at amino acid 35 markedly reduced growth stimulation but had no effect on the binding or inhibition of TIMPs on MMPs suggesting that the varying functions of TIMP-1 are located in different parts of the molecule.

## Transcriptional control of MMPs and TIMPs

Transcriptional control is one of the most important control points in the synthesis of MMPs and TIMPs. Several cytokines and growth factors upregulate both MMPs and TIMPs. The MMPs have different, although overlapping, substrate specificities and the TIMPs have different roles because they are differentially expressed by the same cytokine or steroid. In normal skin, there is little MMP activity but injury produces a differential increase in MMPs in different cells of the dermis. MMP1 and MMP10 are increased postinjury within the migrating keratinocytes whereas MMP3 is expressed in the proliferating basal keratinocytes. TIMP-1 expression is increased after cells are treated with oncostatin, but TIMP-3 levels fall. The synthetic steroid dexamethasone suppresses TIMP-1 expression while increasing levels of TIMP-3.

MMPs and TIMPs can be subdivided into constitutive or responsive types. MMP2, MMP10 and TIMP-2 are generally constitutively expressed in that they are not upregulated by 12-O-tetradecanoylphorbol-13-acetate (TPA) and cytokines [10]. Although there is a marked difference in the responses of MMPs and TIMPs to different cytokines, the transcription factors controlling the expression of the MMPs and TIMP genes are similar.

Most MMPs and TIMPs contain a TPA site which binds to a response element (TRE). Activated protein-1 (AP-1) is a transcription factor binding to TRE. AP-1 is a dimer consisting of several oncoproteins including c-fos and c-jun . TPA-induced dephosphorylation of c-jun increases AP-1 activity [11]. Transcriptional upregulation is affected by a second AP-1 site as well as polyoma enhancer activator (PEA3) which binds the Ets family of oncoproteins. The genes controlling transcription of the responsive MMPs and TIMPs have an AP-1 site about 30 bases upstream of their initiation site [10]. In the constitutively expressed MMP2 and TIMP-2, the AP-1 site is missing or is not close to the initiation site. There is crosstalk between the PEA3 and AP-1 sites which both act in tandem to modulate transcription.

The TIMP genes are located on different chromosomes and, like the MMPs, genes can be divided into constitutive or responsive types. TIMP-1 is located on the X chromosome and, like TIMP-3, has no TATA box, but, unlike TIMP-3, TIMP-1 has multiple transcription sites.

TIMP-1 and TIMP-3 can be upregulated whereas TIMP-2 is constitutively expressed and generally is unaffected by cytokines, making the plasma level of TIMP-2 relatively constant [11]. Currently, it is unclear how TIMP-4 is regulated.

## Pathology

The pathological role of abnormal matrix deposition with the exception of inherited metabolic defects has, until the last 15 years, been largely ignored. The study of MMPs and TIMPs in relation to disease is stimulating because it provides a focus for understanding morbidity and mortality in terms of deficient or excessive matrix deposition. Two examples of defective matrix deposition will be discussed: alcoholic fibrosis, which leads to an increase in collagen deposition and to cirrhosis, and cardiac diseases including hypertensive fibrosis and unstable angina.

### Liver

Cirrhotic liver contains approximately six times the amount of collagen (mainly types I and III) of normal liver. TIMPs and MMPs have been studied to understand the underlying mechanism(s) of hepatic fibrosis, their possible use as clinical markers of disease and their therapeutic potential.

The activity of hepatic MMP1 decreases as hepatic fibrosis increases in severity. Liver injury, including inflammation, stimulates Kupffer cells to release various factors, including transforming growth factor-$\beta$ (TGF$\beta$), which causes activation, proliferation and transformation of the normally quiescent stellate cells. Hepatic damage without inflammation (e.g. in haemochromatosis) may also activate hepatic stellate cells, probably via the production of reactive oxygen species.

Activated stellate cells secrete both collagenous and noncollagenous extracellular matrix proteins as well as regulating levels of collagenous matrix via production of MMP2 and MMP3 and TIMP-1 [12]. The correlation between TIMP-1 and TIMP-2 mRNA and tissue hydroxyproline content was 0.65 and 0.80 ($p<0.05$ and $p<0.01$ for TIMP-1 and -2), respectively, for adult liver samples.

There is now persuasive data to suggest that excess hepatic collagen deposition arises both as the result of decreased collagen degradation mediated by increased TIMPs, and increased collagen synthesis.

Recently, interest has focused on circulating TIMP levels. There is a good correlation between the serum and hepatic levels of TIMP-1 in a range of liver diseases, encouraging studies on the role of plasma TIMP-1 as a noninvasive marker of histological damage in alcoholic liver disease, hepatitis C [13] and other chronic liver diseases.

Li et al. [14] studied 44 alcoholics versus eight controls. Alcoholics had elevated

TIMP-1 levels, which could distinguish patients with septal fibrosis from controls. TIMP-1 could not, however, discriminate steatosis from perivenular fibrosis in cirrhosis. These studies suggest a correlation between the increasing histological evidence of damage and increasing levels of plasma TIMP-1.

Liver biopsy is an invasive intervention with an associated mortality and morbidity. Perhaps the most promising potential role for plasma TIMP-1 is as a minimally invasive marker of fibrosis and cirrhosis. Such a marker would reduce the necessity for multiple biopsies or could define stages at which biopsy was necessary; however, there are no published studies addressing this longitudinal role at present.

## Cardiac disease

Currently, LVH, most commonly caused by hypertension, is diagnosed by echocardiography (ECG). LVH is associated with a very poor prognosis, particularly if diagnosed using ECG criteria. Broadly, LVH can be subdivided into physiological LVH, which occurs in athletes and in which collagen deposition in the left ventricle is normal, and pathological LVH which is associated with an increase in tissue collagen. The increase in collagen within the left ventricle is associated with a reduction in its elasticity, an increase in the number of ectopics and a reduction in coronary arteriole dilatation due to collagen deposition within the vessel. Two mechanisms are involved in the pathophysiology of LVH. Firstly, the myocyte, which is a terminally differentiated cell, hypertrophies due to increased blood pressure. Secondly, Brilla and coworkers [15] showed in a series of experimental models of hypertension that upregulation of angiotensin II (the one clip model) was associated with increased fibrosis in the left ventricle. Rats rendered hypertensive by infra-aorta banding, although having similar elevations in blood pressures to the one clip model, had normal angiotensin II levels and no evidence of ventricular fibrosis, suggesting that the fibrosis associated with LVH is angiotensin II dependent. Further experiments showed that treatment of spontaneously hypertensive rats (SHR) with suppressor doses of the angiotensin-converting enzyme inhibitor (ACEI) lisinopril had no effect either on blood pressure or the development of LVH but did reduce ventricular fibrosis. In contrast, treatment of SHR with a diuretic or alpha blocker reduced blood pressure but had no effect on cardiac fibrosis, whereas treatment with an ACEI at pressor doses reduced blood pressure, ventricular stiffness and ventricular fibrosis. In the left ventricle of the hypertrophied heart, metalloproteinase activity was reduced. This observation led the authors to measure plasma TIMP-1 levels in patients with untreated hypertension. The median TIMP-1 in hypertensive patients was more than twice the level found in controls (Figure 37.2) [16] and the plasma TIMP-1 level correlated with left ventricular mass index as measured by ECG. However, since the TIMP-1 levels in hypertensive patients were elevated

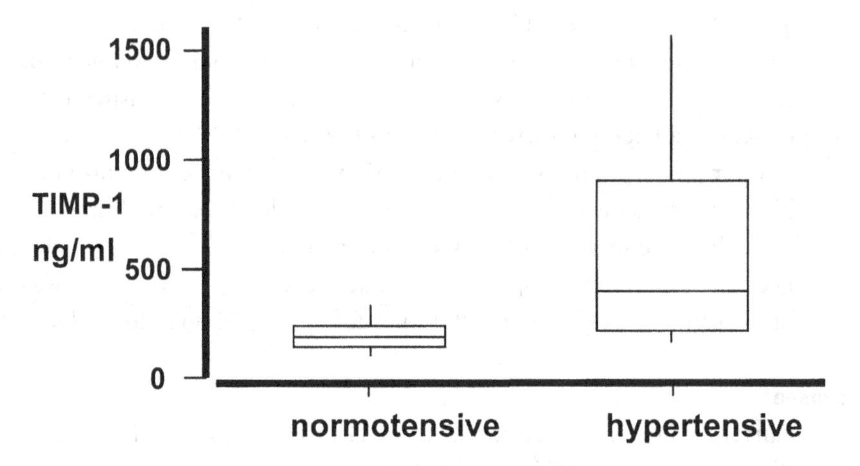


Figure 37.2 Box and Whisker plots of plasma TIMP-1 in normotensive and hypertensive patients. The bar represents median, the box represents the first and third quartiles and the whiskers represents the range of values.

without evidence of LVH, and were as high as those measured in liver disease, it is likely that TIMP-1 also arose from a noncardiac source, possibly the endothelium.

Chua et al. [17] showed that TIMP-1 could be upregulated by angiotensin II. Cardiac remodelling may occur by the following mechanism: ACEI reduces angiotensin II levels, although these rebound after treatment with ACEI; and ACEI reduces TIMP-1 concentration, with an increase in plasma levels of MMP1 assessed immunologically [18].

Angiotensin II poses multiple profibrotic effects which include the upregulation of TGF-$\beta$, c-fos and mitogen-activated protein (MAP) kinase [19]. Angiotensin II increases plasminogen activator inhibitor-1 (PAI-1) and, together with the increase in TIMP-1, this is likely to reduce metalloproteinase activity – consequently increasing matrix deposition.

In the pathogenesis of unstable angina, Libby [20] and others have demonstrated that MMP1, 2, 3, 9 and 13 were elevated in unstable plaques and Kai et al. [21] made the important observation that plasma MMP2 and 9 were elevated in the coronary syndromes acute myocardial infarction and unstable angina.

Bellosta et al. [22] showed, in tissue culture, that the addition of a statin to monocyte-derived macrophages reduced MMP9 levels in the culture medium. This work suggests that the protective role of statins was not only due to their cholesterol-lowering properties but also due to a statin-induced reduction of MMP9 secretion which, in theory, would stabilize plaque.

Cardiovascular risk factors associated with polymorphisms of MMP3 and 12 have been studied in an effort to relate genotype with morbidity and mortality. MMP12 transcription, like that of most MMPs, is controlled by the transcriptional protein AP-1. Insulin increases the binding of AP-1 as well as increasing MMP12 expression. In the promoter region of MMP12 at position −82, a substitution of adenine (A) for guanine (G) has been detected. This A allele was associated with smaller coronary vessels in diabetics suggesting a potential mechanism of coronary artery narrowing in diabetics [23].

## Conclusion

MMPs and TIMPs are involved in many disease states including malignancy, rheumatoid arthritis, chronic wounds and aortic aneurysms. However, because MMPs and TIMPs are found in a wide variety of cells, they may not be useful as specific disease markers, but may become important in staging disease progression.

The clinical promise of TIMPs and MMPs would be the development of new therapies specifically altering levels of MMPs and TIMPs – thus modulating tissue matrix. This may require a greater clinical emphasis on the latter part of the disease process, with the ideal treatment addressing both the removal of the toxin as well as attempting to modulate levels of collagen. For example, in alcoholic cirrhosis, a dual pronged attack might be the abolition of alcohol as well as treatment to prevent the excessive collagen deposition. Currently, there are few data on the modification of MMP and TIMP levels in disease.

## REFERENCES

1 Murphy, G., Knäuper, V. and Cowell, S. et al. (1999). Evaluation of some newer matrix metalloproteinases. *Annals of the New York Academy of Sciences*, **878**, 25–39.
2 Fowlkes, J., Enghild, J., Suzuki, K. et al. (1994). Matrix metalloproteinases degrade insulin-like growth factor-binding protein-3 in dermal fibroblast cultures. *Journal of Biological Chemistry*, **269**, 2572–6.
3 Gomez, D., Alonso, D., Yoshiji, H. et al. (1997). Tissue inhibitors of metalloproteinases: structure, regulation and biological function. *European Journal of Cell Biology*, **74**, 111–22.
4 Woessner, J. (1991). Matrix metalloproteinases and their inhibitors in connective tissue remodeling. *FASEB Journal*, **5**, 2146–54.
5 Greene, J., Wang, M., Liu, E. et al. (1996). Molecular cloning and characterization of human tissue inhibitor of metalloproteinase-4. *Journal of Biological Chemistry*, **271**, 30375–80.
6 Taylor, K., Windsor, J., Cateriana, N. et al. (1996). The mechanism of inhibition of collagenase by TIMP-1. *Journal of Biological Chemistry*, **271**, 23938–45.
7 Stetler-Stevenson, W., Bersch, N. and Golde, D. (1992). Tissue inhibitor of metalloproteinase-2 (TIMP-2). *FEBS Letters*, **296**, 231–4.

8 Hayakawa, T., Yamashita, K., Tanzawa, K. et al. (1992). Growth promoting activity of tissue inhibitor of metalloproteinase-1 (TIMP-1) for a wide range of cells. *FEBS Letters*, **298**, 29–32.

9 Chesler, L., Golde, D., Bersch, N. et al. (1995). Metalloproteinase inhibition and euthroid potentiation are independent activities of tissue inhibitor of metalloproteinases-1. *Blood*, **12**, 4506–15.

10 Crawford, H. and Matrisian, L. (1996). Mechanisms controlling the transcription of matrix metalloproteinase genes in normal and neoplastic cells. *Enzyme Protein*, **49**, 20–37.

11 Angel, P. and Karin, M. (1991). The role of Jun, Fos, and the AP-1 complex in cell proliferation and transformation. *Biochemica et Biophysica Acta*, **1072**, 129–57.

12 Iredale, J., Murphy, G., Hembry, R. et al. (1992). Human hepatic lipocytes synthesize tissue inhibitor of metalloproteinase-1 (TIMP-1): implication for regulation of matrix degradation in the liver. *Journal of Clinical Investigation*, **90**, 282–7.

13 Walsh, K., Timms, P., Campbell, S. et al. (1999). Plasma levels of matrix metalloproteinase-2 (MMP-2) and tissue inhibitors of metalloproteinase 1 and 2 (TIMP-1 and -2) as non-invasive markers of liver disease in chronic hepatitis C. Comparison using ROC analysis. *Digestive Diseases and Sciences*, **44**, 624–30.

14 Li, J., Rosman, A., Loe, M. et al. (1994). Tissue inhibitor of metalloproteinase is increased in the serum of precirrhotic and cirrhotic alcoholic patients and can serve as a marker of fibrosis. *Hepatology*, **6**, 1418–23.

15 Brilla, C., Maisch, B., Rupp, H. et al. (1995). Pharmacological modulation of cardiac fibroblast function. *Herz*, **20**, 127–34.

16 Timms, P.M., Srikanthan, V., Maxwell, P. et al. (1998). Hypertension causes an increase in tissue inhibitor of metalloproteinase-1. *American Journal of Hypertension*, **11** (4 Part 2), 1A.

17 Chua, C., Hamdy, R., Chua, B.H. (1996). Angiotensin II induces TIMP-1 production in rat heart endothelial cells. *Biochimica et Biophysica Acta*, **1311**, 175–80.

18 Lavaides, C., Varo, N., Fernandes, J. et al. (1998). Abnormalities of the extracellular degradation of collagen type 1 in essential hypertension *Circulation*, **98**, 535–40.

19 Kawano, H., Yung, S., Kawano, Y. et al. (2000). Angiotensin II has multiple profibrotic effects in human cardiac fibroblasts. *Circulation*, **101**, 1130–7.

20 Libby, P. (1995). Molecular bases of the acute coronary syndromes. *Circulation*, **91**, 2844–50.

21 Kai, H., Ikeda, H. and Yasukawa, H. (1998). Peripheral blood levels of matrix metalloproteinases-2 and -9 are elevated in patients with acute coronary syndromes. *Journal of Acute Coronary Care*, **32**, 368–72.

22 Bellosta, S., Via, D., Canavesi, M. et al. (1998). HMG-CoA reductase inhibitors reduce MMP-9 secretion by macrophages. *Arteriosclerosis, Thrombosis, and Vascular Biology*, **18**, 1671–8.

23 Jormsjo, S., Ye Shu, Moritz, J. et al. (2000). Allele-specific regulation of matrix metalloproteinase-12 gene activity is associated with coronary artery luminal dimensions in diabetic patients with manifest coronary artery disease. *Circulation Research*, **86**, 998–1003.

Part 8

# Biomarkers of neurological disease and dysfunction

38

# Biomarkers of neurodegenerative disorders

John M Land
Institute of Neurology and National Hospital, London, UK

Biomarkers of disease may reflect tissue damage or disease mechanisms – that is, the pathophysiology of the condition. For many years, neurological diseases could only be described in clinical or neuropathological terms. However, in the last decade, we have come to understand some of the mechanisms underlying some of those diseases associated with the most severe morbidity and mortality – particularly Alzheimer's disease, multiple sclerosis (MS) and Parkinson's disease. In addition, we are increasingly recognizing substances which appear to be specific to the cells found in the brain which could act as biomarkers. Examples include N-acetylaspartate (in neurones), neurone-specific enolase (NSE) and an isoform of S-100 protein found in glial cells.

However, despite these advances, the application of biomarkers to neurodegenerative diseases has not been as easy as it has been for disorders of other tissues – notably, the heart, liver and skeletal muscle. Two major reasons exist for this: firstly, the presence of the blood–brain barrier and, secondly, the relatively small size of the adult brain (2% of body weight) as compared with the rest of the body and the systemic circulation. The former, therefore, impedes the loss of markers from the brain into the bloodstream while the relatively large size of the systemic circulation acts to dilute any marker released. In (pre)term infants, the brain as a proportion of total body weight (13%) is much larger than that of adults and the blood–brain barrier less well developed. This suggests that the development of plasma/serum biomarkers for neurological insults in neonates may be a productive area for future research.

## Alzheimer's disease

In part, the development of good biomarkers for neurodegenerative diseases has been impeded by the imprecision of clinical diagnoses. One striking example of this is Alzheimer's disease. Diagnosis, of course, can be only confirmed at autopsy, although Khachaturian in 1985 devised a series of clinical criteria. Recently, these have been improved upon and the suggestion added that an ideal biomarker for

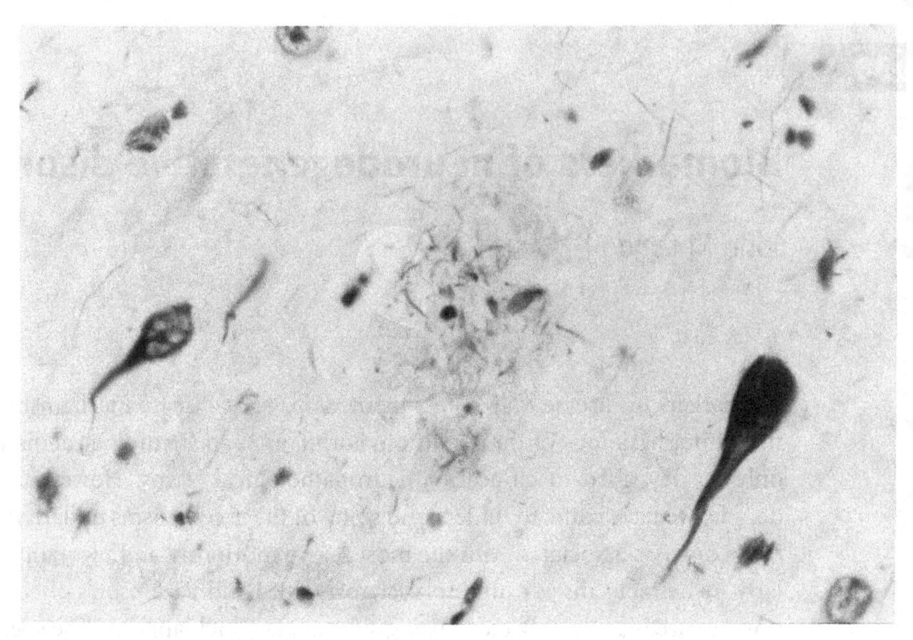

Figure 38.1 A hippocampal section from a patient with Alzheimer's disease showing an amyloid plaque and neurofibrillary tangles stained with silver. Courtesy of Dr T Revesz, Institute of Neurology.

Alzheimer's should have a sensitivity of ≥80% for distinguishing it from other dementias [1].

At autopsy, in the brains of patients with Alzheimer's disease, two major pathological hallmarks are noted: intracellular neurofibrillary tangles and extracellular senile or amyloid plaques (Figure 38.1). Tangles consist of paired helical filaments comprising hyperphosphorylated tau protein, a molecule which, when in its native state, promotes the assembly of the microtubules required for intracellular transport processes. Typically, in Alzheimer's, neurofibrillary tangles are found in the axonal and somato-dendritic domains of the cholinergic neurones. The altered phosphorylation may be easily demonstrated using polyacrylamide electrophoresis where, in place of six bands, hyperphosphorylated tau is detected as three slower moving isoforms.

Recently, it has been reported that the cerebrospinal fluid (CSF) tau levels are increased in Alzheimer's patients compared with controls. Moreover, it is observed that CSF tau levels are higher in the carriers of common amyloid precursor protein mutations. Although the intraassay variation was quite large, the sensitivity of this test was in excess of 90%, with a specificity against all controls of 93% and versus neurological controls of 87% [2]. Early antibodies did not discriminate between native and hyperphosphorylated tau protein. However, such antibodies are now available and it has now been shown conclusively that hyperphosphorylated tau is found in significantly higher concentrations in the CSF of patients with Alzheimer's disease, as compared with those in nonAlzheimer's, and that hyperphosphorylated tau discriminates far better than total tau concentration [3].

The second and perhaps most well-known pathological hallmark of Alzheimer's disease is the amyloid plaque. Typically found in the neocortex and hippocampus, the central amyloid core is associated with dystrophic neurites and abnormal dendrites. The plaque is composed of A$\beta$ peptide, an amyloidogenic fragment some 43 amino acids long derived from the transmembrane section of amyloid precursor protein. The plaques tend to be surrounded by activated microglia and reactive astrocytes. It is now recognized that CSF A$\beta$ is decreased by as much as 50% as compared with controls. Moreover, the sensitivity of CSF A$\beta$ as a marker for Alzheimer's disease is high ($\sim$92%, using the lower 0.95 fractile of control values as a cut-off), while individual biological variation is low [4]. Just why A$\beta$ is reduced remains unclear, but probably reflects the fact that, in Alzheimer's A$\beta$ is hydrophobic and tends to aggregate to a greater extent than in healthy individuals.

The previously described studies of CSF tau and amyloid protein were undertaken in single centre research institution settings. For biomarkers to be truly useful, the assay used needs to be robust and shown to be applicable in a range of clinical settings. Such a study in 10 centres in Europe and the USA examined the discriminatory power of using CSF A$\beta$ and tau in 150 CSF samples drawn from Alzheimer's patients and 100 control CSF samples from healthy volunteers or patients free of central nervous system disease [5]. In this situation, with discrimination of Alzheimer's from controls set at 85% sensitivity, the combined A$\beta$ and tau test gave a specificity of 86% (95% confidence interval [CI]: 81–91%) as against 55% (95% CI: 47–62%) for A$\beta$ and 65% (95% CI: 58–72%) for tau alone. These figures, albeit using CSF, are in line with the recommendation in the consensus statement on diagnostic markers of Alzheimer's disease [1]. They give hope, therefore, that, for a disease which is so difficult to distinguish clinically from age-associated memory impairment or depressive pseudodementia, a biochemical test is now available, which in future years will no doubt become still more refined.

## Stroke

Stroke is one of the major causes of death in the developed world and arguably the most common cause of disability. The development of acute therapies, especially thrombolytics, combined with the need to predict outcome, so that appropriate use may be made of limited resources, as well as for reasons of clinical governance, has given an impetus to the quest for biomarkers of the condition. One early candidate was S-100 protein.

S-100 is an acidic calcium-binding protein first described in 1965 and named as such because it required 100% ammonium sulphate to achieve solubility. It is now known that there are at least 14 different homologous proteins, although their exact function is unknown. Each molecule is a dimer made up of either the homo- or heterodimer of $\alpha$ and $\beta$ subunits. The $\beta$ homodimer, S-100$\beta$, is present in glial cells and the Schwann cells of the peripheral nervous system. The heterodimer, $\alpha\beta$ S100, is found in glia, while the $\alpha$ homodimer is found in neurones. The $\alpha$ homodimer is also found in skeletal muscle, heart, adipose tissue and lymphocytes, albeit at much lower concentrations than those found in neurones.

Several early studies showed that CSF S-100$\beta$ was elevated in a range of conditions including MS, acute encephalomyelitis, cerebral tumours, Guillain–Barré syndrome, and dementias, as well as cerebrovascular disorders. Cerebrovascular disorders, however, appear to give the most consistent changes and, in the case of subarachnoid haemorrhage, CSF S-100$\beta$ levels appear to correlate with the level of damage to the brain as well as having some degree of prognostic value [6]. Interestingly, no correlation between changes in another calcium-binding protein, calmodulin, and the preoperative grade of subarachnoid haemorrhage or clinical outcome has been noted. CSF sampling is not always possible or appropriate in patients with cerebrovascular disease. The testing of plasma or serum samples would be more convenient. In the original studies, serum S-100 tended to be raised in patients with cerebral tumours, meningioma, glioblastoma and secondary melanomas as well as in cerebrovascular disease. Increases were not found, however, in MS, the dementias or in meningitis. However, all early studies were performed using assays that were technically difficult, lacked specificity and sensitivity, and which were slow to perform. They were therefore ill-suited to the task of informing clinical decision making.

The introduction into clinical practice of stroke therapies such as thrombolysis or cerebroprotective agents has once again stimulated interest into means of identifying those patients most likely to benefit from these agents. New, rapid and ultrasensitive enzyme-linked immunosorbent assays are being developed for S-100$\beta$ which are applicable to serum samples. Takahashi et al. [7] describe one such assay and illustrate its potential utility in diagnosing stroke in computed tomography-negative patients. Serum-100$\beta$ measurements can also be related to cerebral infarct sizes, as well as clinical outcome, using a variety of disability and handicap scales. Typically, serum S-100$\beta$ levels rise during the first 2 days after a stroke, before returning to normal in the next 7 days.

The presence of serum S-100$\beta$ post-ischaemic stroke appears to be due to the combined loss from necrotic glial cells as well as leakage through a damaged blood–brain barrier. It is of interest to note that serial measurements of S-100$\beta$, despite being primarily glial in origin, appear to be better than NSE for indicating infarct volume and possibly clinical outcome [8]. However, as S-100$\beta$ is not specific for cerebral infarction, increased levels are indicative of a range of types of cell damage within the brain.

## Multiple sclerosis

A molecule which does appear to be almost exclusively localized to neurones in high concentrations is N-acetylaspartate (NAA). NAA is synthesized within neuronal mitochondria and exported to the cytoplasm and then to the dendrite and axonal processes of the neurone. Its exact function is unknown, although it has been postulated to be an acetyl donor for the synthesis of N-acetylglucosamine and acetylcholine and for the elongation of free fatty acids to form very long chain fatty acids. Whatever its exact role is, it has several important characteristics. Firstly, NAA is present in neurones in high concentration (8–12 mmoles/l). Secondly, its production and steady state concentration appears to be highly sensitive to a continual supply of ATP. Thirdly, and in part related to its high concentration within the brain, it is visible to proton magnetic resonance imaging (MRI) and spectroscopy (MRS). Accordingly, whole brain NAA as assessed by proton MRI is a potentially powerful tool for assessing disease pathology, progression and responses to treatment where neuronal cell bioenergetics are compromised or neuronal cells are damaged.

One condition in which assessment of whole brain NAA (WBNAA) is being found to be informative is MS. For many years, T2-weighted MRI (T2-WMRI) has been used to diagnose MS. However, correlations between disability and lesion volumes are, on occasions, poor, not least because T2-WMRI appears to be influenced in a similar way by a spectrum of pathological processes, including oedema, gliosis and demyelination. Recently, it has become apparent that WBNAA is significantly reduced in the brains of patients with MS as compared with controls and that the loss of WBNAA correlates more closely with disability scores than with T2-WMRI. Moreover, these observations cast new light on the pathological processes occurring in MS, by providing evidence that axonal and dendritic loss occurs in otherwise normal-looking white matter. In addition, by taking serial scans, it has been possible to show that in patients with relapsing, remitting MS there is a 10-fold faster decline in NAA (~0.8% per year of age) as compared with controls – suggesting that chronic neuronal loss is an important facet of the disease [9].

NAA is not specific to MS, reflecting as it does the bioenergetic status of neurones. Falls in WBNAA concentrations are therefore observed in a range of conditions in which either neurones die or in which there is mitochondrial dysfunction. These include the mitochondrial encephalomyopathies, Alzheimer's disease, acute intermittent porphyria, hypothyroidism and stroke. A range of other markers have therefore been used to assess disease progression in MS. These have included adhesion molecules (soluble intercellular adhesion molecule, soluble vascular intercellular adhesion molecule and sE-selectin), costimulatory molecules and cytokines (CD30, CD95, neopterin and interleukin-10) and the nitric oxide breakdown metabolites (nitrite and nitrate). Many show changes in serum/plasma or CSF concentrations which, in population studies, correlate with disease activity. Furthermore, some such as nitrite and nitrate throw new light on the mechanisms of the disease process in MS. Thus, the enhanced levels of nitrite and nitrate taken with the reduced NAA levels suggest nitric oxide- or peroxynitrite-mediated damage to the mitochondrial electron transport chain as being of fundamental importance in the pathophysiology of MS. However, as Sorensen observed [10]: 'Currently no single body fluid marker is sufficiently correlated to disease activity to be used in the individual patient (with MS) to monitor disease activity, progression or therapeutic effects.'

## Creutzfeldt–Jakob disease

In recent years, much attention has focused on Creuzfeldt–Jakob disease (CJD) and the new variant of CJD, vCJD, which is linked to bovine spongiform encephalopathy. Both are rapidly progressive and fatal disorders of the nervous system, and belong to the transmissible spongiform encephalopathies which are caused by prions. Definitive diagnosis of both vCJD and CJD may only be made at autopsy, so a search for a suitable biomarker has been undertaken. CSF S-100$\beta$ and tau protein are both known to be raised in CJD but neither are specific for the disease. By searching CSF protein electrophoresis patterns taken from patients, a constant constituent has been noted by several groups and identified as a 14–3–3 protein isoform.

The 14–3–3 family of proteins is a group of highly conserved proteins found in a wide range of eukaryotic organisms. Their role is still debated but it is possible that they act as chaperone proteins modulating the interactions between signal transduction proteins and intracellular kinase systems. At least seven isoforms are known, of which four, $\beta$, $\gamma$, $\varepsilon$ and $\eta$, appear in the CSF of patients with CJD. Excellent figures for both sensitivity (90–96%) and specificity (94–100%) have been reported for 14–3–3 in patients with CJD, particularly if one looks for the $\eta$ isoform. However, the figures for vCJD are less good, with as few as 50% of patients being reported as being positive [11]. This appears to be especially true when only the 14–3–3 $\gamma$ isoform is looked for (A. Green, Institute of Neurology, personal communication).

Finally, it has been suggested that 14–3–3 proteins are possibly both surrogate markers of CJD as well as being implicated in the pathophysiology of CJD. If this were the case, then 14–3–3 could be considered in the same light as A$\beta$ peptide and tau are in Alzheimer's disease. Perhaps more importantly, the different specificity and sensitivity of 14-3-3 observed in CJD and vCJD would tend to suggest that these conditions have differing pathophysiological mechanisms.

## Conclusion

As neurodegenerative diseases are increasingly understood, and as patients and their families rightly ask for more information about their conditions, the need for biomarkers for neurodegenerative diseases becomes ever more urgent. Biomarkers should allow the rapid diagnosis and prediction of outcome. At least in Alzheimer's, MS and stroke, all conditions in which new therapies are being introduced, there is hope that sensitive and specific markers are being developed that may also inform us about pathophysiology.

## REFERENCES

1 The Ronald and Nancy Reagan Research Institute of the Alzheimer's Association and the National Institute of Aging Working Group (1998). Consensus Report of the Working Group on Molecular and Biochemical Markers of Alzheimer's disease. *Neurobiology of Aging*, **19**, 109–16.
2 Jensen, M., Basun, H., Lanfert, L. et al. (1995). Increased cerebrospinal fluid tau in patients with Alzheimer's disease. *Neuroscience Letters*, **186**, 189–91.
3 Ishiguro, K., Ohno, H., Arai H. et al. (1999). Phosphorylated tau in human cerebrospinal fluid is a diagnostic marker in Alzheimer's disease. *Neuroscience Letters*, **270**, 91–4.
4 Andreason, N., Herse, C., Davidsson, P. et al. (1999). Cerebrospinal fluid $\beta$-amyloid$_{(1-42)}$ in Alzheimer's disease. *Archives of Neurology*, **56**, 673–80.
5 Hulstaert, F., Blennow, K., Ivanoiu, A. et al. (1999). Improved discrimination of AD patients using $\beta$-amyloid$_{(1-42)}$ and tau levels in CSF. *Neurology*, **52**, 1555–62.
6 Takayasu, M., Shibya, M., Kanamari, M. et al. (1985). S-100 protein and calmodulin levels in cerebrospinal fluid after subarachnoid haemorrhage. *Journal of Neurosurgery*, **63**, 417–20.
7 Takahashi, M., Chamczuk, A., Hong, Y. and Jackowski, G. (1999). Rapid and sensitive immunoassay for the measurement of serum S100$\beta$ using isoform-specific monoclonal antibody. *Clinical Chemistry*, **45**, 1307–11.
8 Missler, V., Weissman, M., Friedrich, C. and Kaps, M. (1997). S-100 protein and neurone-specific enolase concentrations in blood as indicators of infarction volume and prognosis in acute ischaemic stroke. *Stroke*, **28**, 1956–60.
9 Gonen, O., Catalaa, I., Babb, J.S. et al. (2000). Total brain N-acetylaspartate. A new measure of disease load in MS. *Neurology*, **54**, 15–19.
10 Sorensen, P.S. (1999). Biological markers in body fluids for activity and progression in multiple sclerosis. *Multiple Sclerosis*, **5**, 287–90.
11 Wiltfang, J., Otto, M., Baxter, M.C. et al. (1999). Isoform pattern of 14–3–3 proteins in the cerebrospinal fluid of patients with Creuzfeld-Jakob disease. *Journal of Neurochemistry*, **73**, 2485–90.

39

# Traumatic brain injury: assessment by biochemical serum markers

Tor Ingebrigtsen[1], Bertil Romner[2]

[1]Tromsø University Hospital, Tromsø, Norway. [2]Lund University Hospital, Lund, Sweden

## Introduction

The annual incidence rates of hospital-admitted head injuries vary between 90 and 400/100000 population in different Western countries [1]. Most of the patients with head injuries are fully conscious when evaluated. The management of such minor head injuries (MHI) is focused on the early detection of patients who deteriorate due to a post-traumatic intracranial haematoma. Expensive screening methods such as hospital admission for overnight observation or computed tomographic (CT) scanning are used to detect the relatively few individuals who develop this life-threatening complication [2]. After the injury, many patients experience postconcussion symptoms, even after uneventful recoveries in the acute stage.

In most western countries, injuries are the leading cause of death among individuals under 45 years of age [1]. Traumatic brain injury (TBI) accounts for one-half of the deaths and most cases of permanent disability after injury. Accurate evaluation of the primary injury and prevention of secondary ischaemic injury is essential in the clinical management. TBI is, however, difficult to assess. Clinical examination is of limited value in the first hours and days after a severe head injury (SHI). Most of the diagnostic process is based on modern neuroimaging techniques, such as CT or magnetic resonance imaging (MRI), but CT has a relatively low sensitivity for diffuse brain damage and the availability of MRI is limited.

A biochemical marker in the serum with the ability to both detect intracranial pathology and to predict postconcussion symptoms would be highly desirable in the management of MHI. There is also a major need for biochemical markers in SHI, both to assess the severity of the primary TBI and to detect secondary injury. Finally, such biochemical markers may be of value as surrogate end-points in clinical trials of neuroprotective treatments.

**Table 39.1.** Potential biochemical markers of traumatic brain injury

| |
|---|
| Alpha hydroxybutyric acid dehydrogenase (HBD) |
| Creatine kinase isoenzyme BB (CK-BB) |
| Fructose 1,6-diphosphate aldolase (ALD) |
| Glial fibrillary acidic protein (GFAP) |
| Glutamic oxaloacetic transaminase (GOT) |
| Glutamic pyruvic transaminase (GPT) |
| Lactate dehydrogenase (LDH) |
| Malate dehydrogenase (MDH) |
| Myelin basic protein (MBP) |
| Neurone-specific enolase (NSE) |
| S-100$\beta$ protein |

## Potential markers of traumatic brain injury

Numerous substances synthetized in astroglial cells or neurones have been proposed as biochemical markers of neurological injury (see Table 39.1), but, until recently, none has been found sufficiently useful in routine clinical practice. Bakay et al. [3] suggested that an ideal biochemical marker should fulfil the following prerequisites:

- high specificity for the brain;
- high sensitivity for brain injury;
- release following irreversible destruction of brain tissue;
- a rapid appearance in serum or cerebrospinal fluid (CSF);
- release in a time-locked sequence with the injury; and
- show low age- and sex-related variability.

Two other important properties may be added to this list. Firstly, it should be possible to measure the marker in serum since CSF sampling is impractical, especially in MHI. Secondly, reliable assays for urgent analyses should be commercially available.

## Early enzyme studies

Numerous studies performed in the 1970s included serum measurements of enzymes such as alpha hydroxybutyric acid dehydrogenase (HBD), fructose 1,6-diphosphate aldolase (ALD), glutamic oxaloacetic transaminase (GOT), glutamic pyruvic transaminase (GPT), lactate dehydrogenase (LDH) and malate dehydrogenase

**Table 39.2.** Relative concentrations (%) of CK-BB, NSE and S-100$\beta$ in human tissues

| | CK-BB | NSE | S-100$\beta$ |
|---|---|---|---|
| Brain cortex | 100 | 100 | 100 |
| Rectum | 49.1 | 1.9 | 2.5 |
| Stomach | 35.3 | 2.6 | 0.7 |
| Urinary bladder | 35.3 | 2.6 | 0.7 |
| Prostate gland | 31.9 | 2.0 | 0.1 |
| Small intestine | 19.2 | 1.9 | 2.1 |
| Uterus | 22.1 | 1.1 | 0.2 |
| Vein | 12.1 | 1.4 | 0.2 |
| Thyroid gland | 11.3 | 2.6 | 0.2 |
| Gall bladder | 5.4 | 0.9 | 1.7 |
| Kidney | 5.7 | 0.1 | 0.3 |
| Lung | 3.5 | 1.5 | 0.2 |
| Mammary gland | 0.5 | 0.1 | 1.8 |
| Spleen | 0.7 | 2.5 | 1.8 |
| Aorta | 0.8 | 0.5 | 0.1 |
| Liver | 0.3 | 0.2 | 0.1 |
| Skeletal muscle | 0.3 | 0.2 | 0.7 |
| Heart | nr | nr | 0.2 |

*Notes:*

Abbreviation: nr is no result.

(MDH). These early reports demonstrated increased enzyme activity in serum in many head-injured patients, but these markers did not have sufficient specificity for brain injury [4]. In the 1980s, promising results were observed in studies of the creatine kinase isoenzyme BB (CK-BB). Recent research has been focused on neurone-specific enolase (NSE) and S-100$\beta$ protein. These are specific markers of damage to neurones and astroglial cells, respectively. Interesting reports on two other markers of astroglial damage, myelin basic protein (MBP) and glial fibrillary acidic protein (GFAP), have also been published.

## CK-BB

In the central nervous system, CK-BB is located in the astrocytes. CK-BB is, however, also present in other organs, such as the large intestine and prostate. The concentration in these organs is one-third to one-quarter of that found in the brain (see Table 39.2). Several reports published in the 1980s showed correlations between serum levels of CK-BB and the severity of brain injury.

Skogseid et al. [5] performed both CK-BB measurements and CT scanning in patients with MHI, but there was no correlation between CK-BB serum levels and CT findings. A weak tendency towards more postconcussion symptoms and mild neuropsychological dysfunction was, however, observed among patients with increased CK-BB levels, indicating more pronounced TBI in these patients.

CK-BB has been extensively studied in patients with SHI. It is clear that enzyme levels are highest on the first day after injury and that serum levels usually decline to normal levels within a few days in surviving patients [4, 5]. The level of the enzyme in serum is, however, not consistently increased even in patients with SHI and its presence is not consistently associated with poor outcome. Therefore, Bakay and Ward [4] concluded that the sensitivity and specificity of CK-BB is inadequate for use as an indicator of neurological trauma.

## NSE

At present, NSE is the only marker which specifically indicates traumatic damage to neurones. It is located in the cytoplasm of neurones and is probably involved in increasing neuronal chloride levels during the onset of neural activity. NSE is also found in peripheral neuroendocrine tissues and tumours and, therefore, is currently used as a biochemical marker of tumours such as neuroblastoma, melanoma and small cell lung cancer. The biological half-life of this protein is probably more than 20 hours.

Increased serum levels of NSE have been observed in patients with head injury [5], but NSE levels are neither significantly different between MHI patients and controls nor between MHI patients and patients with SHI [6]. Several studies have correlated NSE serum levels with both clinical and neuroradiological measures of injury severity and with outcome, but the results have been disappointing. Some authors have reported a weak correlation between NSE and CT findings, as well as between NSE and Glasgow Coma Scale (GCS) scores [5], but these findings have not been confirmed by others. No correlation has been found between serum levels of NSE and long-term outcome after SHI [7].

## S-100β protein

The S-100 proteins are involved in the $Ca^{2+}$-dependent regulation of a variety of intracellular processes. The isoform S-100$\beta\beta$ is predominantly present in astroglial cells of the central nervous system (see Table 39.2) and is commonly referred to as S-100$\beta$. The biological half-life is significantly shorter than that of NSE, probably well below 60 minutes. The median S-100$\beta$ level in healthy individuals is 0.05 μg/l and there is no significant variation with sex or age.

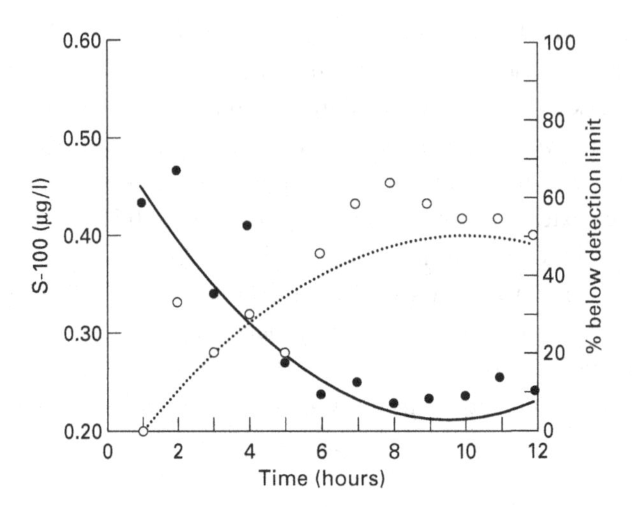


Figure 39.1 Smoothed graph (solid line) depicting mean serum level of S-100β protein (solid circles) during the first hours after MHI in 14 patients with detectable serum levels (detection limit 0.2 μg/l) [8]. Dotted line: the proportion of patients (open circles) with initially detectable serum levels whose levels had declined to below the detection limit at each point in time.

The authors introduced S-100β as a marker of traumatic brain damage in MHI. In a detailed study of 50 MHI patients with normal CT scans, S-100β was measured hourly and the serum levels were correlated with MRI and neuropsychological examinations, both in the acute stage and after 3 months [8]. Increased serum levels of S-100β were observed in 28% of patients (mean 0.4 μg/l, standard deviation ± 0.3). Figure 39.1 shows that the levels were highest immediately after the trauma in all patients, and declined each hour thereafter. The proportion of patients with detectable serum levels was significantly higher when MRI revealed a brain contusion. A trend towards impaired neuropsychological functioning was also observed, as assessed from measures of attention, memory and speed of information processing, among patients with high serum levels of S-100β. In a further study of 182 patients with MHI, S-100β was detected in 38%. The proportion of patients with increased serum S-100β levels was significantly greater among those with intracranial pathology revealed by CT (90%) when compared with those without this pathology (35%). The accuracy for CT-evident injury was 66% and the negative predictive value of an undetectable serum level was 99%. Results that are consistent with these have recently been published by others [9].

Several studies of patients with SHI agree that there is a correlation between early S-100β measurements and GCS scores, CT findings, intracranial pressure and long-term outcome [7, 9–11]. All patients with intracranial pathology revealed by

CT scan show increased serum S-100$\beta$ levels. A cut-off value of about 2.0 $\mu$g/l seems to distinguish between those patients with good and bad clinical outcomes, with a specificity of well over 90% [7, 11].

## MBP and GFAP

MBP is found in growing oligodendroglial cells and is specific to myelin. GFAP constitutes the major part of the cytoskeleton of astrocytes and is found only in astroglial cells of the central nervous system. There are few studies on the serum levels of these proteins in head-injured patients. GFAP is, however, of great interest since its brain specificity may be even better than that of NSE and S-100$\beta$. Recently, Missler et al. [12] reported increased serum GFAP levels in 12 of 25 head-injured patients. After the first 6 hours, serum levels declined rapidly, including in those patients with SHI. Hence, GFAP may be a very specific marker of the primary TBI.

## Summary and discussion

The early studies of different enzymes showed that substances may be released from the brain and detected in serum after a brain injury. Such enzyme release requires that there is cell damage with release of the enzyme into the interstitium, and that there is concomitant injury to the blood–brain barrier. This is probably the case in most brain injuries. Experimental studies show that even mild diffuse axonal injury is accompanied by endothelial disruption in cerebral capillaries with astrocyte swelling. Therefore, the appearance in serum of substances released from damaged astrocytes and neurones would be expected in most cases of brain injury. Today, it is clear that the main drawback with the measurement of enzymes such as LDH, GPT and GOT has been their lack of specificity for brain injury.

The studies of CK-BB showed more clearly that the amount and severity of TBI may correlate with serum levels of a substance released from the brain. The results obtained with CK-BB measurements were encouraging, probably because the brain specificity of this enzyme is better than that of the enzymes studied earlier. The specificity of this test is not, however, adequate for the accurate assessment of injury to the brain (Table 39.2) [4].

CK-BB and the other enzymes mentioned above are released from astroglial cells. Direct interpretation of injury to neurones by measurements of a substance released from these cells may be considered more relevant. NSE is the only marker available for such studies, but the experience with this marker is disappointing despite high specificity for the brain (Table 39.2). The reason for this may be related to difficulties in defining the optimal point of time for NSE measurements since the biological half-life of the enzyme is more than 20 hours. This slow elimination of NSE from the blood probably makes it difficult to assess the amount of primary damage and impossible to distinguish between the primary injury and secondary injuries.

At present, S-100$\beta$ seems to be the marker that best fulfils the prerequisites suggested by Bakay et al. [3] for assessing brain damage. The specificity is good (see Table 39.2) and the authors' studies of MHI clearly demonstrate a high sensitivity for brain injury. The authors have also shown that S-100$\beta$ appears in serum immediately after a TBI, and that it is rapidly eliminated over a few hours (see Figure 39.1). Others have documented that the age- and sex-related variability in S-100$\beta$ concentration is insignificant. It is, however, unclear whether S-100$\beta$ release requires irreversible cell damage or if it can be released during less severe injury, such as that which might occur in ischaemia. Present knowledge of the release mechanisms and routes of elimination of S-100$\beta$ is also incomplete. These issues should be studied in detail in experimental models.

It is a paradox that the correlation between precise measures of the severity of an injury, such as those made by MRI or CT, with S-100$\beta$ levels is much better than that between sensitive outcome measures, such as neuropsychological examinations, and S-100$\beta$ levels. This is probably attributable to the complex functions of the brain and the fact that injuries of similar size to different anatomical regions may cause sequelae of varying severity. There is, for example, strong evidence that the postconcussion syndrome is a multifactorial entity influenced by both organic, psychosocial and behavioural factors. Accordingly, a strong association would not necessarily be anticipated between a specific measure of TBI, such as S-100$\beta$ protein levels in serum, and subjective ratings of postconcussion symptoms.

The assessment of the primary injury and the evaluation of neurointensive care may, therefore, be the most promising applications for serological markers of TBI. The authors' findings in studies of patients with MHI indicate that serum S-100$\beta$ measurements can be used to select patients for CT scanning. In the future, more detailed studies of GFAP are warranted.

Finally, one major limitation of the routine use of serological markers of brain injury is their restricted availability in clinical practice. For the purposes mentioned above, results of the analyses have to be available within an hour after blood sampling. Simple assays for urgent analyses are not currently commercially available. This is a major challenge for the diagnostics industry.

## REFERENCES

1 Jennet, B. (1996). Epidemiology of head injury. *Journal of Neurology, Neurosurgery and Psychiatry*, **60**, 362–9.

2 Ingebrigtsen, T., Romner, B. and Kock-Jensen, C. (2000). Scandinavian guidelines for initial management of minimal, mild and moderate head injuries. *Journal of Trauma*, **48**, 760–6.

3 Bakay, R.A.E., Sweeney, K. M. and Wood, J.H. (1986). Pathophysiology of cerebrospinal fluid in head injury: part 2. Biochemical markers for central nervous system trauma. *Neurosurgery*, **18**, 376–82.

4 Bakay, R.A.E. and Ward, A.A. (1983). Enzymatic changes in serum and cerebrospinal fluid in neurological injury. *Journal of Neurosurgery*, **58**, 27–37.

5 Skogseid, I.M., Nordby, H.K., Urdal, P., Paus, E. and Lilleaas, F. (1992). Increased serum creatine kinase BB and neuron specific enolase following head injury indicates brain damage. *Acta Neurochirurgica (Wien)*, **115**, 106–11.

6 Ross, S.A., Cunningham, R.T., Johnston, C.F. and Rowlands, B.J. (1996). Neuron specific enolase as an aid to outcome in head injury. *British Journal of Neurosurgery*, **10**, 471–6.

7 Raabe, A., Grolms, C. and Seifert, V. (1999). Serum markers of brain damage and outcome prediction in patients after severe head injury. *British Journal of Neurosurgery*, **13**, 56–9.

8 Ingebrigtsen, T., Waterloo, K., Jacobsen, E.A., Langbakk, B. and Romner, B. (1999). Traumatic brain damage in minor head injury: relation of serum S-100 protein measurements to magnetic resonance imaging and neurobehavioral outcome. *Neurosurgery*, **45**, 468–76.

9 Herrmann, M., Curio, N., Jost, S., Wunderlich, M.T., Synowitz, H. and Wallesch, C.-W. (1999). Protein S-100$\beta$ and neuron specific enolase as early neurobiochemical markers of the severity of traumatic brain injury. *Restorative Neurology and Neuroscience*, **14**, 109–14.

10 Raabe, A., Grolms, C., Sorge, O., Zimmermann, M. and Seifert, V. (1999). Serum S-100$\beta$ in severe head injury. *Neurosurgery*, **45**, 477–83.

11 Woertgen, C., Rothoerl, R.D., Metz, C. and Brawanski, A. (1999). Comparison of clinical, radiological, and serum marker as prognostic factors after severe head injury. *Journal of Trauma*, **47**, 1126–30.

12 Missler, U., Wiesmann, M., Wittmann, G., Magerkurth, O. and Hagenström, H. (1999). Measurement of glial fibrillary acidic protein in human blood: analytic method and preliminary clinical results. *Clinical Chemistry*, **45**, 138–41.

40

# An overview of S-100$\beta$ as a clinically useful biomarker of brain tissue damage

John Azami, Basil F Matta

Addenbrooke's Hospital, Cambridge, UK

The ideal marker of organ damage or dysfunction must be specific and sensitive. Of paramount importance is also that it should be easily and reliably measured with clearly defined thresholds, and preferably independent of age, sex and other concurrent systemic disorders. In addition, the serum levels of the marker should correlate with the severity of disease so that it is possible to predict accurately recovery, further damage and the likelihood of permanent injury. The diagnosis, prognosis and natural history of central nervous system (CNS) injury relies on repeated clinical neurological examination and radiological techniques such as computed tomography (CT) and/or magnetic resonance imaging (MRI). A number of clinical scenarios exist in which the application of such assessments is not practical or feasible. Therefore, the benefits of having a reliable, sensitive and specific serum biomarker for diagnosing and predicting prognosis after CNS injury are self-evident [1]. This chapter will discuss the merits and the potential pitfalls of protein S-100$\beta$ as a useful marker of CNS damage and dysfunction.

## S-100 proteins

The S-100 protein family consists of 17 members [2]. These proteins are made of two subunit amino acid chains, $\alpha$ and $\beta$. S-100 $\beta\beta$ is found in high concentrations in glial and Schwann cells, S-100 $\alpha_1\beta$ in glial cells and S-100 $\alpha_1\alpha_1$ in kidney, and striated and cardiac muscles. The isoform of main interest in brain damage is the homodimer which consists of two monomers (S-100 $\beta\beta$) and which is usually abbreviated to S-100$\beta$. This isoform is only synthesized in glial cells and, therefore, is only present in significant amounts in brain tissue. Although this acidic calcium-binding protein with a molecular weight of 21 kD is a neurotrophic factor involved in CNS healing and maturation, it may also be a trigger for apoptosis via nitric oxide synthase and lipid peroxidation. S-100$\beta$ is metabolized in the kidney and excreted in the urine, and has a half-life ranging between 25 and 120 mins [3]. Although serum concentrations of S-100$\beta$ are normally very low (0.03–0.12 $\mu$g/l), higher extracellular and intracellular brain tissue values have been reported. The serum concentration can be easily measured using luminescence immunoassay (Sangtec 100, AB Sangtec, Bromma, Sweden). S-100$\beta$ levels increase with age, and are higher in females. Many questions regarding the prognostic value of S-100$\beta$ as a marker of CNS damage remain unanswered. Is there a correlation between the serum concentration of S-100$\beta$ and brain injury? Are higher levels always associated with poorer prognosis despite the underlying disease process? Is an increase in the serum concentration of S-100$\beta$ following head injury as significant as the increase observed in patients with liver coma?

## Clinical uses of S-100$\beta$

It is generally accepted that, in adults, S-100$\beta$ serum levels in excess of 0.2 μg/l are associated with some degree of brain injury. Such increases in S-100$\beta$ serum levels have been observed in patients who have suffered head trauma, cerebrovascular accidents, subarachnoid haemorrhage or hypoxic brain injury. Similar increases have also been detected during cardiopulmonary bypass, liver transplantation, pregnancy and labour (e.g. fetal intrapartum hypoxia and neonatal intraventricular haemorrhage), and in patients with neurodegenerative diseases.

### Head trauma

Although the majority of patients with minor head injury (MHI) who present to accident and emergency departments have a good outcome, up to 50% of these patients develop postconcussional symptoms [4]. The severity of head injury is often difficult to assess as this may not correlate with radiological findings. Furthermore, it is not uncommon for a normal CT scan to be followed by subsequent late injury due to the evolution of minor contusions. S-100$\beta$ levels have been shown to correlate with the degree of brain injury [5]. Increased levels are associated with prolonged hospital stay, increased incidence of abnormalities on MRI or CT scans and increased risk of delayed neurobehavioural deficit. However, S-100$\beta$ is a predictor of global rather than focal brain injury. Damage to small highly 'functional' areas, such as the internal capsule, may result in disability out of proportion to the increase in S-100$\beta$ levels observed.

### Subarachnoid haemorrhage

Serum levels of S-100$\beta$ have been shown to increase for several days following subarachnoid haemorrhage (SAH) as a result of a ruptured cerebral aneurysm [6]. This increase may predict early neurological deficit with similar sensitivity to the Hunt and Hess scores in predicting neurological outcome. It seems that serum S-100$\beta$ levels constitute a reliable noninvasive method of monitoring progress and predicting outcome following SAH.

**Table 40.1.** Comparison of concentrations of S-100 (μg/l) on day 1 and day 2 in patients deceased or alive within 14 days after cardiac arrest

| | Number of patients at day 14 | | |
|---|---|---|---|
| S-100 concentration by day (μg/l) | Deceased | Alive | Total |
| Day 1 | | | |
| ≥0.2 | 10 | 4 | 14 |
| <0.2 | 3 | 17 | 20 |
| Total | 13 | 21 | 34 |
| Day 2 | | | |
| ≥0.2 | 11 | 0 | 11 |
| <0.2 | 3 | 24 | 27 |
| Total | 14 | 24 | 38 |

*Source:* From Rosen et al. [7].

## Cardiac arrest

Mortality of patients who suffer an out-of-hospital cardiac arrest remains high despite successful resuscitation and hospitalization. In a large proportion of patients, this unfavourable final outcome is due to global hypoxic injury suffered during the cardiac arrest period. Predicting survival based on clinical assessment is difficult but may be improved by the use of S-100$\beta$ as a marker of brain injury. Rosen et al. [7] were able to link outcome to the serum levels of S-100$\beta$ in this population of patients. S-100$\beta$ levels increased (and were highest) on day 1 postcardiac arrest. The levels on days 1 and 2 correlated with the degree of coma. While all patients with S-100$\beta$ levels greater than 0.2 μg/l on day 2 died after 14 days, patients in whom the S-100$\beta$ levels remained below 0.2 μg/l had an 89% survival rate (Table 40.1) [7]. These findings suggest that the use of S-100$\beta$ in these patients will supplement the clinical assessment and may be a useful guide to therapeutic intervention.

## Cardiopulmonary bypass

The effects of cardiac surgery on life expectancy and the neurobehavioural complications associated with cardiopulmonary bypass (CPB) are well documented. The incidence and severity of these complications have been correlated with postoperative S-100$\beta$ levels in both adults and children [3], and seem to be related to the duration of CPB. Raised levels of S-100$\beta$ have also been reported to correlate with the duration of aortic cross-clamping and circulatory arrest, preoperative arteriosclerosis of the ascending aorta, and the number and size of cerebral emboli.

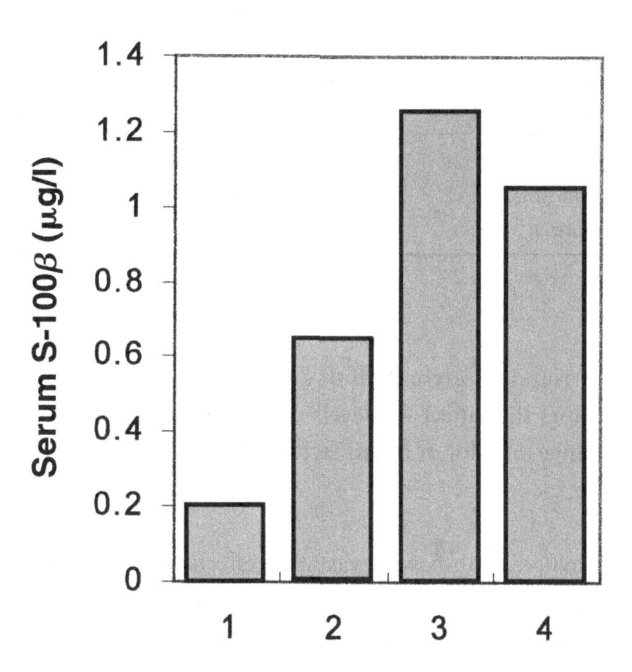


Figure 40.1 Mean serum levels of S-100$\beta$ ($\mu$g/l) in 10 patients: (1) postinduction; (2) at the beginning of the anhepatic phase; (3) at the end of the anhepatic phase; and (4) at skin closure.

Although a disruption of the blood–brain barrier has been postulated as being responsible for this rise in S-100$\beta$, the detection of extracranial sources of S-100$\beta$ such as fat and mediastinal tissue further complicate the picture. While contamination is expected if cardiotomy suckers are used and subsequent autotransfusions are made, modifying CPB methods and duration, surgical technique and the use of neuroprotective and anti-inflammatory drugs can significantly reduce serum level of S-100$\beta$ with clinically measurable improvements in outcome.

## Hepatic disease and liver transplantation

Neurological complications are common in patients with severe hepatic dysfunction and are frequent after orthotopic liver transplantation. In a pilot study, the authors measured the serum levels of S-100$\beta$ in 10 patients at different stages of the liver transplant procedure (Figure 40.1)[8].

The S-100$\beta$ levels were above the normal reported level preoperatively in several of the patients but were markedly elevated in patients with hepatic encephalopathy. S-100$\beta$ levels further increased during the anhepatic phase of the procedures and began to fall again at skin closure. These results suggest either a disruption in

**Table 40.2.** Intrapartum hypoxia and umbilical vein (UV) levels of S-100$\beta$ at delivery

| Group | Intrapartum CTG | Actual UV pH | S-100$\beta$ (μg/l) |
|---|---|---|---|
| A | Normal | $7.32 \pm 0.01$ | $0.94 \pm 0.06$ |
| B | Abnormal | $7.31 \pm 0.01$ | $1.20 \pm 0.14$ |
| C | Normal | $7.16 \pm 0.01$ | $1.09 \pm 0.13$ |

the blood–brain barrier or a cerebral insult during the procedure. However, there are many other factors that affect the levels of S-100$\beta$, such as altered clearance, rapid volume exchange and donor organ factors.

## Pregnancy and labour

Fetal intrapartum hypoxia can have devastating effects, resulting in neurological deficits in the neonate. The methods of assessment currently used, such as fetal pH and fetal heart rate, are sometimes unreliable. It has been postulated that S-100$\beta$ may be useful in detecting brain insult as a result of intrapartum hypoxia much earlier than pH changes, thereby allowing earlier intervention and early delivery. In a pilot study, the authors examined the potential use of S-100$\beta$ in three groups of mothers during labour: group A, 10 mothers with normal intrapartum cardiotocograph (CTG) and umbilical vein pH; group B, 10 mothers with abnormal intrapartum CTG and normal umbilical vein pH; and group C, 10 mothers with abnormal intrapartum CTG and umbilical vein pH $<7.2$ [9]. The main findings of the study are shown in Table 40.2.

There was no statistically significant difference in the level of S-100$\beta$ between the groups. While there may be many other factors that affect the level of S-100$\beta$ in the pregnant mother, the authors were unable to identify a benefit for using S-100$\beta$ as an indicator for the need of earlier intervention during labour in this study.

## Neonatal intraventricular haemorrhage

Preterm infants are at risk of neonatal intraventricular haemorrhage (IVH), which often leads to developmental abnormalities. In a study of 25 preterm infants, Gazzolo et al. [10] were able to demonstrate that a rise in the serum levels of S-100 in those babies who developed IVH, diagnosed by ultrasound scan and confirmed by neurological examination, correlated with the grade of haemorrhage and the change in middle cerebral artery pulsatility index. This led them to conclude that S-100 may be a useful indicator in the early diagnosis and management of neonatal IVH.

### Cerebrovascular accident

S-100 protein and neurone-specific enolase concentrations have been investigated as indicators of the size of brain infarction and prognosis following a cerebrovascular accident (CVA) [11]. Although the increase in the levels of S-100 seen in the majority of patients who have suffered a CVA correlated with the size of the infarct, it did not correlate with the degree of functional injury. It is doubtful whether serum levels of S-100 have a prognostic value in focal brain injuries.

### Neurodegenerative diseases

Recently, there has been increasing interest in the search for reliable biomarkers of neurodegenerative diseases. Research has focused on Alzheimer's disease, multiple sclerosis, Creutzfeldt–Jakob disease, Parkinson's disease and stroke [12]. Biomarkers such as S-100$\beta$ may be useful research tools in the diagnosis of some of these diseases and to measure the effectiveness of new therapies. Furthermore, S-100$\beta$ may also be used to predict prognosis and survival times following the diagnosis of a neurodegenerative condition.

## Summary

The reliability and clinical utility of S-100$\beta$ as a marker of brain pathology has been intensively researched in recent years. Because it can be assayed easily using a variety of commercially available kits, serum levels of S-100$\beta$ are increasingly being measured in many situations in which there is a need for a sensitive marker of CNS damage. The available evidence suggests that S-100$\beta$ is most beneficial in the detection of global brain injury but is probably of limited value in detecting damage from focal insults. Nevertheless, many issues remain unresolved. These include the temporal relationship between elevations of serum S-100$\beta$ and the level of injury, as well as the exact specificity and sensitivity of S-100$\beta$ as a marker of CNS damage. More work is also needed to define clearly the diagnostic and prognostic thresholds above which serum levels of S-100$\beta$ indicate brain damage.

## REFERENCES


1 Abraha, H.D., Grubb, N.R., Simpson, C. et al. (2000). Prediction of death and cognitive impairment after resuscitated cardiac arrest using serum NSE and S-100β. *European Meeting on Biomarkers of Organ Damage and Dysfunction*, April, Robinson College, pp. 81–2.

2 Zimmer, D.B., Cornwall, E.H., Landar, A. and Song, W. (1995). The S100 protein family: history, function and expression. *Brain Research Bulletin*, **37**, 417–29.

3 Johnson, P. (2000). Markers of brain cell damage related to cardiac surgery. *European Meeting on Biomarkers of Organ Damage and Dysfunction*, April, Robinson College, pp. 56–7.

4 Middelboe, T., Anderson, H.S., Birket-Smith, M. and Friis, M.L. (1992). Minor head injury: impact on general health after 1 year. *Acta Neurologica Scandinavica*, **85**, 5–9.
5 Ingebrigtsen, T., Romner, B. and Trumpy, J.H. (1997). Management of minor head injury: the value of early computed tomography and serum protein S-100 measurements. *Journal of Clinical Neuroscience*, **4**(1), 29–33.
6 Wiesmann, M., Missler, U., Hagenstrom, H. and Gottmann, D. (1997). S-100 protein plasma levels after aneurysmal subarachnoid haemorrhage. *Acta Neurochirurgica (Wien)*, **139**, 1155–60.
7 Rosen, H., Rosengren, L., Herlitz, J. and Blomstrand, C. (1998). Increased serum levels of the S-100 protein are associated with hypoxic brain damage after cardiac arrest. *Stroke*, **29**, 473–7.
8 Matta, B.F., Bakewell, S.E., Klinck, J. and Gunning, K.E. (1998). Neurologic injury associated with liver transplantation. *Journal of Neurosurgical Anesthesiology*, **10**, A277.
9 Matta, B.F., Bland, E., Trull, A. and Thornton, S. (1999). The effect of foetal acidosis on umbilical vein plasma S 100. *British Journal of Anaesthesia*, **82**, A520.
10 Gazzolo, D., Vinesi, P., Bartocci, M. et al. (1999). Elevated S-100 blood levels as an early indicator of intraventricular haemorrhage in preterm infants. Correlation with cerebral Doppler velocimetry. *Journal of the Neurological Sciences*, **170**, 32–5.
11 Missler, U., Wiesmann, M., Christine, F. and Kaps, M. (1997). S-100 protein and neuron-specific enolase concentrations in blood as indicators of infarction volume and prognosis in acute ischemic stroke. *Stroke*, **28**, 1956–60.
12 Land, J.M. (2000). Biomarkers of neurodegenerative diseases. *European Meeting on Biomarkers of Organ Damage and Dysfunction*, April, Robinson College, p. 55.

# Part 9

# Biomarkers in transplantation

41

# Monitoring heart and lung transplant patients

Marlene L Rose
Harefield Hospital, Harefield, UK

## Introduction

Approximately 10 000 hearts and 2000 lungs are transplanted every year. Rejection remains the most common complication following transplantation and is the major source of morbidity and mortality. Constant vigilance is required to monitor the immune response to the grafted organ in the first 3 months, when acute rejection is most likely to occur. In contrast to the management of kidney transplant recipients, in whom raised levels of serum creatinine and urea can be used to monitor graft function, monitoring the function of transplanted hearts and lungs relies entirely on histological or clinical parameters.

Thus, for patients who have undergone cardiac transplantation, surveillance endomyocardial biopsies are taken at weekly intervals for the first 6 weeks and then at 2-weekly intervals until the end of the third postoperative month. In addition, any positive biopsy is followed up by a repeat biopsy 1 week later to ensure that antirejection therapy has been successful. Patients also undergo further biopsies when clinically indicated. Thus, every heart transplant patient has a minimum of nine biopsy procedures within the first postoperative year. Lung allograft function is monitored daily by the patients themselves by means of a spirometer. Any unexplained persistent fall in the forced expiratory volume will be followed up by transbronchial biopsy to confirm the diagnosis histologically. It is especially important to obtain a differential diagnosis between rejection and infection after lung transplantation. For this reason, the transbronchial biopsy procedure is usually accompanied by bronchoalveolar lavage, which is sent for culture and microbiological analysis.

Routine histological analysis of cardiac biopsies remains the gold standard technique for detecting and grading heart allograft rejection and any new diagnostic methods are compared with this. A standardized histopathological nomenclature for the grading of both heart and lung allograft rejection was suggested in 1990 [1, 2] and is now used by the majority of transplant centres. However, the endomyocardial biopsy procedure is unpleasant for the patient, is associated with a small chance of complications and is highly labour intensive and expensive. It would be of huge benefit to the patient and the hospital to have a noninvasive method to replace the endomyocardial biopsy. In theory, there are many possibilities of noninvasively detecting rejection including noninvasive monitoring of heart function such as magnetic resonance imaging [3], signal averaged electrocardiogram [4], specialised echocardiographic indices [5] and the use of markers in peripheral blood. This chapter will focus on blood markers. There are two major approaches: one is to exploit what is known about activation of the allograft recipient's immune system, the second is to look for markers of graft damage.

## Markers of immune activation

Over the last 10 years, there has been an explosion of knowledge regarding the effect of the allograft on the immune system (see Figure 41.1). Rejection is initiated by CD4+ recipient T lymphocytes recognizing foreign major histocompatibility complex (MHC) class II molecules on antigen-presenting cells in the donor graft. This initiates a cascade of cytokines which may be acting directly to damage graft parenchymal cells or which may be acting to recruit and amplify further effector mechanisms such as those mediated by CD8+ T cells, macrophages and B cells. In heart transplant recipients, where there can be dissociation between the size of the graft infiltrate and the extent of cardiac haemodynamic compromise, it is thought that tumour necrosis factor-$\alpha$ (TNF$\alpha$) and nitric oxide may have negative inotropic effects on contracting myocytes.

Over the years, there have been numerous attempts to find signs of immune activation in peripheral blood. These have included studies of the blood levels of interleukin (IL)-2, soluble IL-2 receptor (IL-2R), IL-6, IL-7, IL-8, TNF$\alpha$, interferon-$\gamma$, soluble intercellular adhesion molecule-1 (ICAM-1), soluble MHC antigens, activated T cells and T-cell populations as well as cytoimmunological monitoring [6]. Often, these are cross-sectional studies and, when results are pooled (i.e. a comparison made between rejection and nonrejection periods), significant differences have been obtained. However, when one performs longitudinal studies of individual patients, the values for a particular marker vary so widely on a daily or weekly basis that sensitivities and specificities so derived are inadequate for practical use. It is clear that the immune system is highly labile during the first 3 postoperative months, when most rejection episodes occur. This early propensity to rejection will certainly be modified by augmented immunosuppression, both directly (e.g. antithymocyte globulin binds to soluble human leucocyte antigen and adhesion molecules) and indirectly, by altering the balance of cell subpopulations as they recover from depletion. Taken as a whole, these immune activation markers are always elevated after transplantation if they are compared with levels found in nontransplant

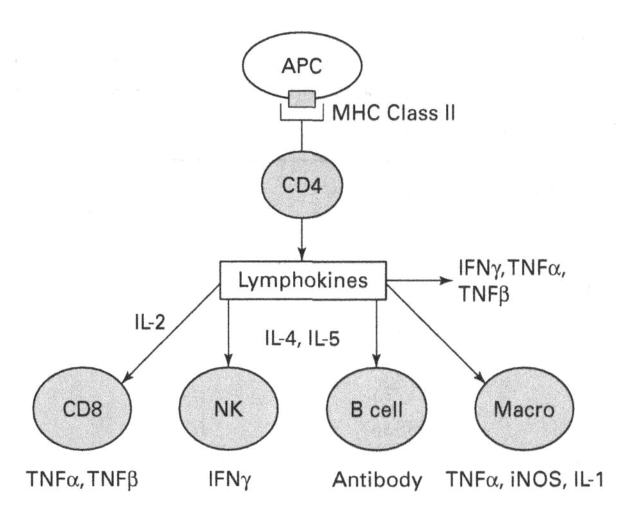


Figure 41.1 Diagrammatic representation of the initiation of allorejection and subsequent maturation of possible effector mechanisms. CD4+ T cells are activated by recognition of foreign MHC class II antigens in donor antigen-presenting cells (APC). Activated CD4+ T cells release various cytokines (including IL-2, IL-4, IL-5), some of which (e.g. IFN$\gamma$, TNF$\alpha$, iNOS) may be directly toxic to beating cardiac myocytes. IFN$\gamma$, interferon-gamma; TNF$\alpha$, tumour necrosis factor-alpha; TNF$\beta$, tumour necrosis factor beta; iNOS, inducible nitric oxide synthase; NK, natural killer cells.

patients, but they are not reliable indicators of rejection within the individual transplant recipient.

One of the reasons for the low specificity and sensitivity of these markers is that they do not distinguish between rejection and infection. In order to circumvent this criticism, the author's group [7] and others [8] have distinguished between donor-specific and third-party T-cell responses as a way of assessing the state of the patient's immune system. The author's group also addressed the issue of where one would expect to find donor-specific lymphocytes – in the peripheral circulation or in the graft. To this end, lymphocytes from patients' endomyocardial biopsies were cultured and a limiting dilution analysis performed to quantify numbers of cytotoxic precursor cells with donor-specific or third-party specificity. At the same time, lymphocytes were cultured from patients' blood and a comparison made of the precursor frequencies of donor-specific cells found in blood and in the graft. The results (Table 41.1) showed the presence of donor-reactive CD8+ T cells during rejection, but they were found almost exclusively in the graft, not in the blood. This diminishes the chances of finding specific reactivity in the peripheral circulation unless a particularly sensitive assay is used. A similar argument can be used for the detection of circulating cytokines. Thus, it has been shown that high levels of IL-6 and soluble

**Table 41.1.** Cytotoxic CD8+ T-cell frequencies in biopsy-derived lymphocytes for donor-specific and third-party cells: comparison with peripheral blood lymphocytes (PBL) taken at the same time as the biopsy

| | Biopsy-derived lymphocytes | | PBL | |
|---|---|---|---|---|
| Patient | Donor | Third party | Donor | Third party |
| 1 | 1/4978 | 1/17273 | — | — |
| 2 | 1/781 | 1/84773 | 1/29821 | 1/24113 |
| 3 | 1/317 | 1/13313 | 1/28589 | Undetectable |
| 4 | 1/528 | 1/3725 | 1/4215 | 1/6243 |
| 5 | 1/1318 | 1/38361 | 1/37259 | Undetectable |
| 6 | 1/9561 | Undetectable | 1/19439 | 1/164967 |
| 7 | 1/11818 | 1/31397 | 1/134248 | 1/52565 |

*Source:* Reproduced from reference [7] with permission from Williams and Wilkins.

TNF-R1 (TNF receptor) in coronary sinus, but not in aortic blood, correlated with poor coronary vasomotor tone during rejection episodes [9].

Interestingly, donor-specific alloantibody is produced during cell-mediated acute rejection episodes in some patients [10] but it is unlikely that this response will be sufficiently rapid to exploit as a means of monitoring patients. An association between blood eosinophil counts and acute cardiac and pulmonary allograft rejection has been recently reported [11] and it has been demonstrated that eosinophil count monitoring can be used as a guide to steroid dosage, and can result in a reduction in the rate of early allogaft rejection [12].

## Markers of graft damage

In the early days of heart transplantation in the 1970s, before the advent of cyclosporine, conventional serological markers of cardiac damage, such as lactate dehydrogenase and creatine kinase, were used as markers of graft failure. However, they lacked sensitivity and were often found to be elevated too late to serve as an effective prompt to the treatment of cardiac allograft rejection. Troponin is a contractile regulatory complex found in striated and cardiac muscle. It consists of three distinct polypeptide components: troponin C (the calcium-binding element), troponin I (the actinomyosin ATPase inhibitory element) and troponin T (the tropomyosin-binding element). The complex serves to regulate the calcium-dependent interaction of myosin and actin, and, thus, plays an integral role in muscle contraction. In the 1990s, specific enzyme immunoassays were developed against cardiac-specific isoforms of troponin T and I, which show little cross-reactivity with the isoforms from skeletal muscle [13]. With the current commercially available kits, circulating troponin T or I is only detectable in the circulation of patients with severe cardiac muscle damage such as that caused by myocardial infarction [14] or cardiac surgery [15]. Katus' group first reported that the use of troponin T to monitor heart allograft rejection was limited by the observation that high levels are usually found in the first few days after transplantation, and that levels can remain well above normal for 2–3 months [16]. This was not related to the surgical ischaemic time and the reasons for these elevated troponin T levels are still unclear. They probably reflect low-level immunological damage caused by humoral factors (antibodies or cytokines). For this reason, the assay cannot be used to monitor rejection during the first 3 months, when rejection is most likely to occur. After this period, the assay does detect grade 3 (moderate) or 4 (moderate to severe) rejection with a high sensitivity of 80.4% and a strong negative predictive value of 96.2% [17]. It has also been used in patients at 6 months after transplantation, when rising levels are said to predict chronic rejection [18]. An interesting adaptation of this assay to transplantation has been to measure levels of serum troponin T in donors; high levels correlate with the occurrence of rejection in the recipients of such hearts [19], presumably reflecting damage and release of donor graft antigens which are then recognized by the recipient's immune system.

## Proteomics and metabonomics

The approaches described thus far assume a prior understanding of the mechanism of transplant rejection (e.g. Figure 41.1) and utilize this knowledge to select prospective markers. Another, more global, approach assumes no prior knowledge of the pathophysiology of allograft rejection and simply seeks to identify any changes in the composition of serum that coincide with rejection. The technology now exists for the relatively rapid analysis of global metabolites (metabonomics) or proteins (proteomics) in tissues and serum.

The first stage in proteome analysis involves the separation of complex mixtures of proteins obtained from whole cells or tissues. The best method is two-dimensional polyacrylamide gel electrophoresis. This uses a combination of a first dimensional separation by isoelectric focusing under denaturing conditions with a second dimensional separation using sodium dodecyl sulphate polyacrylamide gel electrophoresis (SDS-PAGE). This separation by molecular charge (isoelectric point) in combination with size separation (relative molecular mass) results in the sample proteins being distributed across the two-dimensional gel profile (see Figure 41.2). Interlaboratory studies of heart proteins have demonstrated that excellent reproducibility can be achieved through the use of this method. Identification and characterization of proteins can be achieved in various ways [20], but currently the

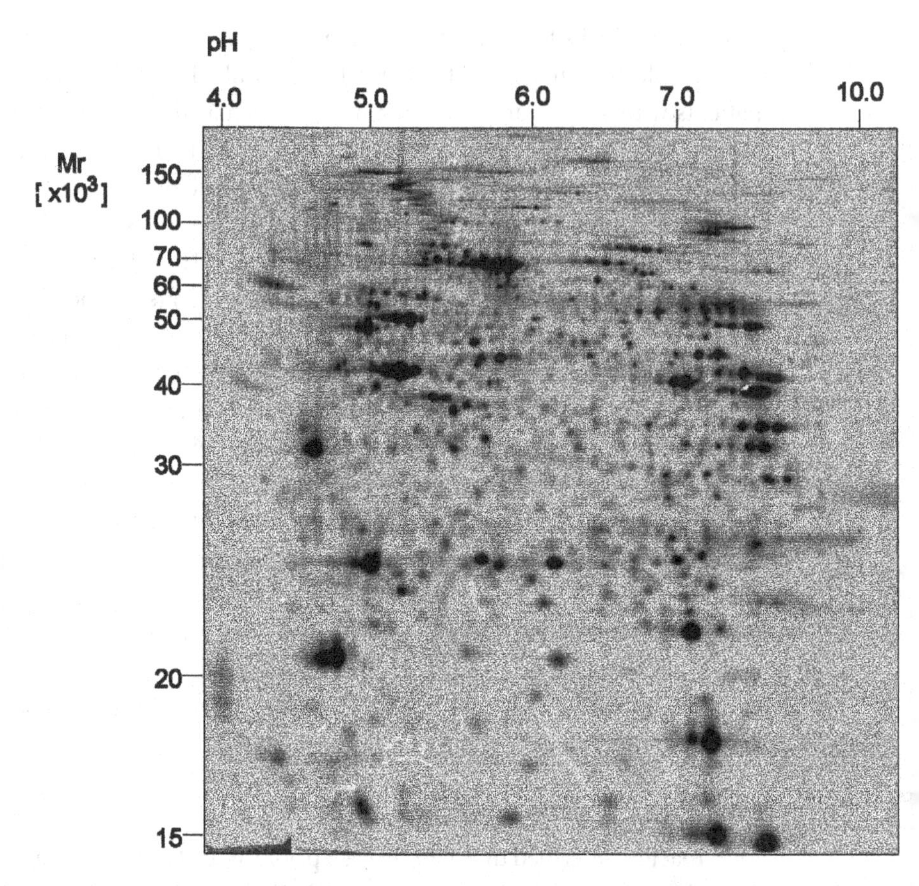


Figure 41.2 A two-dimensional electrophoresis separation of 80 $\mu$g of heart (ventricle) proteins. An 18-cm nonlinear pH 3–10 immobilized pH gradient isoelectric focusing gel was used in the first dimension. The second dimension was a 21-cm 12% sodium dodecyl sulphate-polyacrylamide gel electrophoresis gel. The proteins were detected by silver staining. The scale at the top indicates the nonlinear pH range of the first dimension strip from which the apparent isoelectric points of the separated proteins can be estimated. The relative molecular mass ($M_r$) scale can be used to estimate the molecular weights of the separated proteins. Reproduced from reference [19] with permission from Elsevier Science London.

most important method is mass spectrometry because of its high sensitivity and capacity for high sample throughput. Three groups have established databases of human cardiac proteins accessible via the World Wide Web [20]. The author's group is currently compiling a database of proteins expressed in endomyocardial biopsies from heart transplant patients and using computer analysis to identify changes associated with rejection episodes. Should any coincident changes be found, the information could be used to develop noninvasive assays if the proteins or markers are also found to be released into the blood in detectable quantities.

Metabonomics has arisen from work on the application of $^{1}$H-nuclear magnetic resonance (NMR) spectroscopy to study the multicomponent metabolic composition of biofluids, cells and tissues over the past two decades [21]. Also, studies utilizing pattern recognition, expert systems and related bioinformatic tools are used to interpret and classify complex NMR-generated metabolic data sets. There seems to be no reason why both of these approaches should not be used to discover novel markers of transplant rejection.

## Conclusions

Despite our understanding of the cellular and humoral factors involved in acute cardiac and lung allograft rejection, monitoring the histological changes in the graft directly remains the only reliable way of detecting rejection after heart and lung transplantation. Markers of rejection have to differentiate this event from nonspecific immune activation occurring in a labile immune system. Similarly, changes in markers of graft damage are not necessarily the consequence of an immune response against the graft. Modern methods for identifying global changes associated with rejection, and the subsequent development of appropriate assay systems, may, however, succeed in producing a noninvasive way of monitoring patients after thoracic organ transplantation.

## REFERENCES

1 Billingham, M.E., Cary, N.R.B., Hammond, M.E. et al. (1999). A working formulation for the standardisation of nomenclature in the diagnosis of heart and lung rejection: heart rejection study group. *Journal of Heart and Lung Transplantation*, **9**, 587–93.

2 Yousem, S.A., Berry, G.J., Brunt, E.M. et al. (1990). A working formulation for the standardisation of nomenclature in the diagnosis of heart and lung rejection: lung rejection study group. *Journal of Heart and Lung Transplantation*, **9**, 593–8.

3 Smart, F.W., Young, J.B., Weilbaecher, D., Kleiman, N.S., Wendt, R.E. and Johnston, D.L. (1993). Magnetic resonance imaging for assessment of tissue rejection after heterotopic heart transplantation. *Journal of Heart and Lung Transplantation*, **12**, 403–10.

4 Lacroix, D., Kacet, S., Savard, P., et al. (1992). Signal averaged electrocardiography and detection of heart transplant rejection: comparison of time and frequency domain analysis. *Journal of the American College of Cardiology*, **19**, 553–8.

5 Dodd, D.A., Brady, L.D., Carden, K.A., Frist, W.H., Boucek, M.M. and Boucek, R.J. (1993). Pattern of echocardiographic abnormalities with acute cardiac allograft rejection in adults: correlation with endomyocardial biopsy. *Journal of Heart and Lung Transplantation*, **12**, 1009–17.

6 Koelman, C.A., Vaessen, L.M., Balk, A.H., Weimar, W., Doxiadis, I.I. and Class, F.H. (2000).

Donor derived soluble HLA plasma levels cannot be used to monitor graft rejection in heart transplant recipients. *Transplant Immunology*, **8**, 57–64.

7 Suitters, A.J., Rose, M.L., Dominguez, M.J. and Yacoub, M.H. (1990). Selection for donor-specific cytotoxic T lymphocytes within the allografted human heart. *Transplantation*, **49**, 1105–9.

8 Jutte, N.H., Knoop, C.J., Heijse, P. et al. (1996). Human heart endothelial-restricted allorecognition. *Transplantation*, **62**, 403–6.

9 Weis, M., Wildhirt, S.M., Schulz, C. et al. (1999). Modulation of coronary vasomotor tone by cytokines in cardiac transplant recipients. *Transplantation*, **68**, 1263–7.

10 Smith, J.D., Danskine, A.J., Rose, M.L. and Yacoub, M.H. (1992). Specificity of lymphocytotoxic antibodies formed after cardiac transplantation and correlation with rejection episodes. *Transplantation*, **53**, 1358–62.

11 Trull, A., Steel, L., Cornelissen, J. et al. (1998). Association between blood eosinophil counts and acute cardiac and pulmonary allograft rejection. *Journal of Heart and Lung Transplantation*, **17**, 517–24.

12 Trull, A.K., Steel, L.A., Sharples, L.D. et al. (2000). Randomized trial of blood eosinophil count monitoring as a guide to corticosteroid dosage adjustment after heart transplantation. *Transplantation*, **70**, 802–9.

13 Katus, H.A., Looser, S., Hallermayer, K. et al. (1992). Development and in vitro characterisation of a new immunoassay of cardiac troponin T. *Clinical Chemistry*, **38**, 386–93.

14 Katus, H.A., Remppis, A., Neumann, F.J. et al. (1991). Diagnostic efficiency of troponin T measurements in acute myocardial infarction. *Circulation*, **83**, 902–12.

15 Katus, H.A., Schoeppenthau, M., Tanzeem, A., et al. (1991). Non-invasive assessment of perioperative myocardial cell damage by circulating cardiac troponin T. *British Heart Journal*, **65**, 259–64.

16 Zimmermann, R., Baki, S., Dengler, T.J. et al. (1993). Troponin T release after heart transplantation. *British Heart Journal*, **69**, 395–8.

17 Dengler, T.J., Zimmerman, R., Braun, K. et al. (1998). Elevated serum concentrations of cardiac troponin T in acute allograft rejection after human heart transplantation. *Journal of the American College of Cardiology*, **32**, 405–12.

18 Faulk, W.P., Labarrere, C.A., Torry, R.J. and Nelson, D.R. (1998). Serum cardiac troponin-T concentrations predict development of coronary artery disease in heart transplant recipients. *Transplantation*, **66**, 1335–9.

19 Vijay, P., Scavo, V.A., Morelock, R.J., Sharp, T.G. and Brown, J.W. (1998). Donor cardiac troponin-T: a marker to predict heart transplant rejection. *Annals of Thoracic Surgery*, **66**, 1034–9.

20 Dunn, M.J. (2000). Studying heart disease using the proteomic approach. *Drug Discovery Today*, **5**, 76–84.

21 Nicholson, J.K., Lindon, J.C. and Holmes, E. (1999). Metabonomics: understanding the metabolic response of living systems to pathophysiological stimuli via multivariate analysis of biological NMR spectroscopic data. *Xenobiotica*, **29**, 1181–9.

42

# Monitoring liver transplant recipients

Andrew Trull
Addenbrooke's and Papworth Hospitals, Cambridge, UK

A complex array of serious medical complications can influence recovery of the allograft after liver transplantation. In the peri- and immediate postoperative periods, complications that affect the allograft commonly include primary non-function, preservation or ischaemia reperfusion injury, thrombotic (particularly arterial) and nonthrombotic infarction, biliary obstruction and sepsis. The risk of acute (cellular) rejection is greatest in the second postoperative week and can be clinically indistinguishable from other causes of graft dysfunction at this time. Furthermore, the especially high levels of immunosuppression required in the first postoperative month to control acute rejection later predispose the transplant recipient to widespread opportunistic infections. Other potentially serious complications of immunosuppression include renal dysfunction, hypertension, hyperglycaemia, hypercholesterolaemia, hyperuricaemia, central and peripheral neuropathies, osteoporosis and lymphoproliferative disease. Chronic ductopenic rejection can become manifest as early as the second postoperative month and is the greatest obstacle to morbidity-free, long-term survival.

The differential diagnosis of the complications that affect the function and/or histological integrity of the graft may be facilitated by their chronologically distinct pattern of clinical presentation, but other clinical and laboratory investigations are usually required to confirm a diagnosis. Laboratory tests form an essential component of the diagnostic tools available to the transplant surgeon and physician, and changes in such tests may serve to prompt biopsy or a modification of therapy. It is essential to select the optimum test panel for monitoring patients at different stages after transplantation – particularly when their interpretation is crucial, as may occur in the patient with impaired coagulation who cannot be subjected to core needle biopsy. Critical factors in the selection of a laboratory test menu for a liver transplant programme include turnaround times that support clinical decision making and real cost-effective diagnostic or prognostic value. In the last two decades, a number of new laboratory tests have been evaluated in the liver transplant population and some may genuinely offer supplementary or improved diagnostic value when compared with the conventional 'liver function tests' that have been in use for over 30 years. Whether they have offered a pragmatic and cost-effective alternative is more debatable, as few have been adopted routinely.

## Tests of graft metabolic function, functioning mass and blood flow

Rational assessment of liver function by the measurement of bilirubin, albumin and the prothrombin time is limited by the relative lack of sensitivity of these tests and their inability to quantify the functional reserve of the liver. Simple, cheap and reliable tests of liver metabolic function, functioning hepatic mass or liver blood flow have long been the 'Holy Grail' of liver transplant units. Methods that showed early promise involve the intravenous administration of model drugs that are metabolized exclusively by the hepatic cytochrome P450 cyclo-oxygenase (CYP 450) enzyme systems. These permit the quantitative assessment of any impairment in either the rate of clearance of the parent drug (e.g. antipyrine and caffeine clearance tests) or formation of one of its metabolites (e.g. formation of monoethylglycinexylidide [MEGX] from lignocaine).

The MEGX test has been proposed as a measure of metabolic function, but, since lignocaine is a high extraction drug, its metabolism will be predominantly influenced by blood flow. Thus, the total systemic clearance of lignocaine ($CL_s$) is essentially equivalent to the drug's hepatic clearance ($CL_H$) which equates with the product of the hepatic blood flow ($Q_H$) and the hepatic extraction ratio (E):

$$CL_s = CL_H = Q_H E$$

In standard pharmacokinetic modelling, this product is related to the unbound fraction of the drug ($f_{ub}$) and the intrinsic hepatic clearance of the unbound drug ($CLu_{int}$) according to the following equation:

$$Q_H E = \frac{Q_H \times f_{ub} \times CLu_{int}}{Q_H + f_{ub} \times CLu_{int}}$$

With very high extraction drugs, intrinsic clearance is not limiting and the clearance of the drug is predominantly influenced by delivery of the drug to the liver. Indeed, it has been shown that MEGX concentrations in patients with congestive heart failure were significantly lower than those measured in healthy subjects [1].

Measurement of antipyrine clearance is still regarded by many as the ideal test of hepatic CYP 450 function. The drug is well tolerated, it is almost completely absorbed after an oral dose, it is not bound to plasma proteins to any great extent, its volume of distribution is equivalent to total body water and it is mainly eliminated by hepatic metabolism. Unlike lignocaine, antipyrine is a poorly extracted compound and so its clearance is independent of liver blood flow – its clearance being largely constrained by the availability of CYP 450 enzymes. Furthermore, the

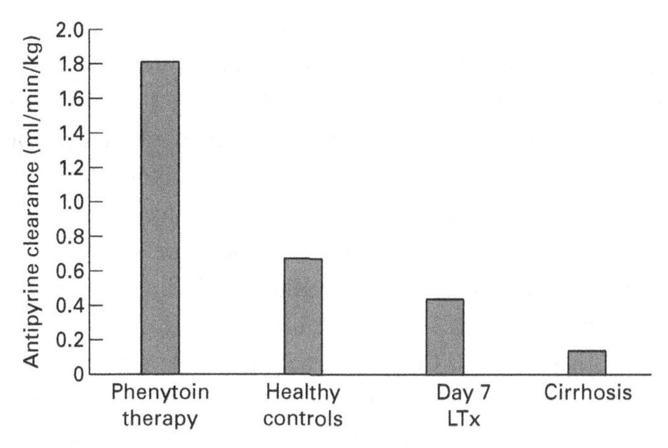


Figure 42.1 Antipyrine clearance estimated in patients receiving phenytoin therapy, in healthy subjects, in patients 7 days after liver transplantation (LTx) and in patients with severe liver disease (cirrhosis). Data summarized from population distributions derived from multiple sources including unpublished personal observations in the liver transplant group [2, 3]. Clearance measurements were significantly different between all groups ($p<0.001$).

three major metabolites are each formed via different CYP 450 enzyme pathways and antipyrine can provide a ‘broad spectrum’ model substrate of hepatic metabolism. However, enzyme-inducing (e.g. phenytoin) or -inhibiting factors can have a profound influence on antipyrine clearance in humans (Figure 42.1) and must be controlled for in any comparisons between populations of patients.

The elimination of an intravenous dose of galactose from the blood by the liver is linear with time and can be used to estimate the galactose elimination capacity. This correlates well with other measures of hepatocellular function such as the aminopyrine breath test – a measure of cytochrome P450-dependent demethylation of isotopically labelled ($^{13}$C or $^{14}$C) aminopyrine to carbon dioxide – but has the advantage that it measures a single aspect of hepatocellular function. Hepatic functional imaging with a $^{99m}$technitium-labelled ligand of the asialoglycoprotein receptor (where internalization of galactose-terminated glycoproteins occurs) has been shown to correlate with conventional markers and scoring systems of liver dysfunction and could provide a sensitive measure of functioning hepatic mass. Removal of intravenous doses of bromosulphathalein, and the safer indocyanin green, by hepatic parenchymal cells has been used to measure both hepatic uptake and blood flow, but the latter, potentially very useful, application, unfortunately involves hepatic venous catheterization. The arterial ketone body ratio (AKBR) has been proposed to indicate hepatic mitochondrial redox state and changes in AKBR following revascularization are claimed to reflect the viability and subsequent clinical outcome of living related donor grafts.

Most of these functional tests have been applied in the context of liver transplantation, but usually for assessing pretransplant residual liver function, the donor graft or function immediately after transplantation. The MEGX test has probably received the most attention because an automated assay is available for the measurement of the lignocaine metabolite and only a single blood sample is required. In a prospective multivariate analysis of 102 potential liver donors, it was shown that MEGX testing offered limited incremental value in predicting early donor graft function when used in conjunction with traditional clinical, laboratory and histological criteria but that it may be of value in predicting graft survival after liver transplantation [4]. Similarly, a Cox proportional hazard model has shown that the indocyanin green elimination constant on day 1 after 50 liver transplants was a better predictor of liver-related graft outcome than conventional liver function tests. The elimination constant correlated with the severity of preservation injury, intensive care unit and hospital stay, prolonged liver dysfunction and septic complications [5]. The galactose elimination capacity has been used in a prospective study to predict the survival of 194 patients with cirrhosis and to identify those who might benefit most from urgent transplantation. Only Child–Pugh score, creatinine, varices and galactose elimination capacity were independent predictors of mortality in a multivariate analysis [6].

In most institutions, the application of such functional tests for the longitudinal monitoring of liver transplant recipients has not been a practical option as they usually involve expensive and complicated procedures, and reference ranges are difficult to establish in patients receiving polydrug therapy and who differ in sex and age, as well as dietary and smoking habits. Consequently, very few serial studies have been done to compare prospectively the clinical value of these quantitative liver function tests after transplantation and such studies have been largely confined to comparisons of new and conventional markers of biliary epithelial or hepatocyte damage. The biliary, endothelial and hepatocyte damage incurred by acute allograft rejection has probably received most attention. Although the influence of isolated rejection episodes on long-term outcome is debatable, a prospective study of 170 consecutive liver transplant recipients with a mean follow up of $3.7 \pm 0.2$ years showed that, in comparison with those who experienced only one or no rejection episodes, the 30 patients who experienced recurrent acute rejection had greater impairment of liver function tests, including lower indocyanin green clearance, and more severe histological damage [7]. It was proposed that the 50% of patients who experience rejection should receive heavier immunosuppression, leaving the remainder on a lifelong light maintenance immunosuppressive regimen.

Much attention has been paid to the development of improved methods for detecting the onset of acute rejection on the unsubstantiated premise that if it could

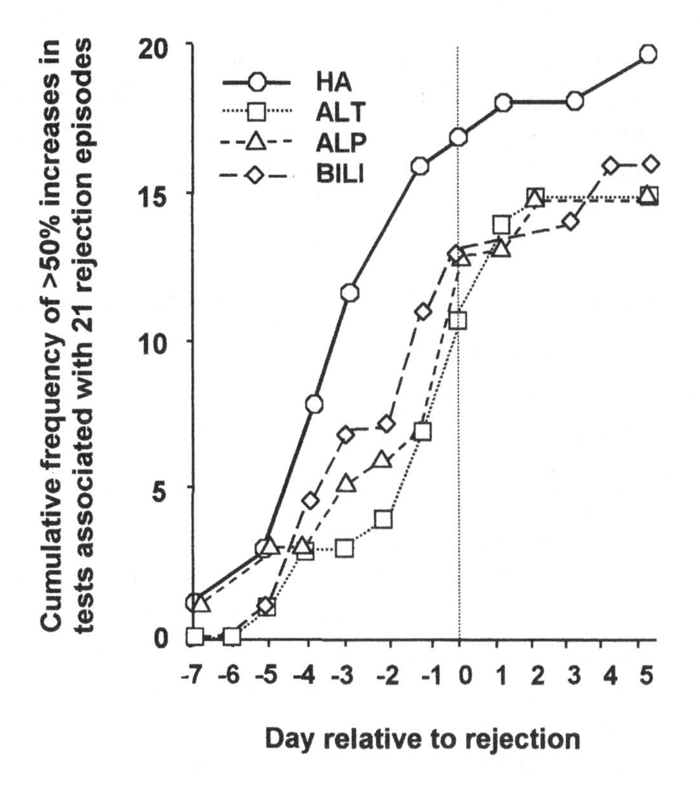


Figure 42.2 The cumulative frequency of >50% increases in plasma hyaluronic acid (HA), alanine transaminase (ALT), alkaline phosphatase (ALP) and bilirubin (BILI) in the 7 days before and 5 days after the clinical diagnosis (vertical broken line) of 21 episodes of acute liver allograft rejection. The median time to the first >50% increase in HA was 2.5 days (95% confidence interval −3.5 to −1.0 days) prior to the diagnosis of rejection – 1 or more days earlier than the conventional liver function tests.

be treated earlier there would be less sequellar damage to the allograft. Reflecting the characteristic histopathological features of rejection, surrogate markers of rejection have conventionally included the transaminases and bilirubin. More novel markers of vascular injury have included increased nitric oxide production [8] and reduced clearance of plasma hyaluronic acid [9]. In the author's experience, increases in plasma hyaluronate are a very early and sensitive indicator of acute rejection (Figure 42.2) – a greater than 50% increase in concentration occurring in association with 20 of 21 acute rejection episodes. This probably does largely reflect endothelial damage as the hepatic endothelium has specific receptors for hyaluronic acid and this is the major route of its elimination. However, changes in production of hyaluronate in extrahepatic tissues and changes in liver blood flow could also influence the plasma concentration. Although there is a simple enzyme-linked immunosorbent assay method for quantitating hyaluronic acid (Chugai Diagnostics, Japan), the method has not been widely adopted for monitoring transplant recipients. Significant decreases in aminopyrine clearance have also been found coinciding with acute rejection [10], but tests of this type are not a practical option for routine monitoring.

## Markers of hepatocellular damage, inflammation and immune activation

Improved methods for detecting the hepatocellular damage associated with allograft rejection may be a more pragmatic option and objective evidence that the earlier detection and treatment of acute rejection can influence subsequent clinical outcome has emerged from longitudinal studies of $\alpha$-glutathione *S*-transferase ($\alpha$-GST) monitoring. This enzyme has a much shorter in vivo plasma half-life than conventional transaminases and can provide a more accurate indication of ongoing liver damage. In probably the first randomized controlled trial of a laboratory diagnostic test, the author and colleagues found that increases in $\alpha$-GST occurred at a median of 1 day earlier than transaminases before rejection in 60 de novo liver transplant recipients [11]. Monitoring $\alpha$-GST improved survival and halved the risk of graft loss during the first 3 postoperative months. It also reduced hospitalization, requirements for biopsy, the severity of rejection and the incidence of infection. One practical problem with the measurement of $\alpha$-GST is that it cannot be measured enzymatically because of endogenous interferents and it requires time-consuming immunoassay. The author and colleagues have now identified an alternative enzyme, fructose 1,6 bisphosphatase (FBPase), which shares the short plasma half-life of $\alpha$-GST but which can be quantitated enzymatically [12]. This substantially reduces assay turnaround time as well as costs, making the method much more adaptable to clinical application as it could be requested outside of routine working hours. This new method is currently at an advanced stage of commercial development (Cambridge Life Sciences, Ely).

The clinical value of a wide variety of inflammatory cytokines or their receptors (e.g. tumour necrosis factor-$\alpha$, interleukin (IL)6, IL2R) and acute phase reactants (C-reactive protein [CRP] and serum amyloid A [SAA]) have been the subject of studies of variable quality and it is difficult to assess their clinical merits in the differential diagnosis of rejection as most lack specificity and are particularly influenced by infection. Intriguingly, among the potential immunological markers of rejection, increases in the absolute blood eosinophil count appear to have a generic relationship with the early stages in the development of rejection of all solid organ allografts. Indeed, in the liver transplant group, the presence of an eosinophil infiltrate in the graft itself is one of the strongest pathognomonic features of more severe rejection. The author's own group found that blood eosinophil counts increased at a median of 4 days before the clinical diagnosis of rejection and these changes preceded increases in $\alpha$-GST and transaminases [13]. Receiver–operating characteristics analysis showed that the optimum combination of eosinophil and $\alpha$-GST monitoring had a predictive efficiency of 84% for acute rejection.

## Markers of infection

The differential diagnosis of serious opportunistic infection and rejection after liver transplantation is particularly difficult. For example, the clinical manifestations of bacterial cholangitis and acute rejection can be identical, but the treatments for these events are diametrically opposed. It is beyond the scope of this chapter to cover the microbiological aspects of monitoring infection but biochemical diagnostic tests should be considered as they can provide a first line of investigation and can be used to prompt more sophisticated analyses to identify the offending microorganism. Currently, procalcitonin is showing some promise and Professor Hammer provides a more extensive review of this development in chapter 48. The author's group's retrospective longitudinal study of 150 liver, heart and lung transplant recipients has suggested that procalcitonin has a greater predictive efficiency in the 7 days preceding the diagnosis of serious bacterial or fungal infection (versus viral infection, acute rejection or uneventful periods) than either CRP or SAA[14].

## Monitoring immunosuppression

The propensity to rejection or infection after liver transplantation is, to a variable extent, influenced by maintenance immunosuppressive regimens. Regular measurement of the blood concentration of calcineurin inhibitors has become a routine component of monitoring after transplantation, although the presumed clinical benefits of therapeutic monitoring have rarely been subjected to rigorous scrutiny in this population. In a retrospective review of data from three clinical trials, it was shown by logistic regression that trough tacrolimus concentrations, measured within a 7-day window before the onset of rejection or toxicity, correlated with the incidence of toxicity but not with rejection (although the latter relationship has been found in the kidney, heart and lung transplant groups) [15]. By the use of similar methods to analyse the pharmacodynamics of cyclosporin in the early post-operative period, the author and colleagues similarly found only a very weak relationship between the median blood cyclosporin concentration and either rejection or infection (unpublished data). Indeed, the risk of rejection was more closely associated with the dose of cyclosporin. No relationship between cyclosporin concentrations and renal dysfunction was found, although it did correlate weakly with the mean arterial pressure.

One explanation for the weak pharmacodynamic relationship between the total blood concentrations of these drugs and both efficacy and toxicity in the liver transplant population may relate to the influence of binding proteins on their distribution characteristics, particularly in the context of any impairment of hepatic drug metabolism after transplantation. It is already known that very small changes in the unbound, pharmacologically active fraction of highly protein-bound drugs can have profound effects and such changes are not registered in conventional measurements of the total drug concentration. The author's group recently completed a randomized trial in which the influence of replacing any deficit in plasma albumin after liver transplantation with either human albumin solution or the artificial plasma expander, Gelofusine®, on the efficacy and toxicity of tacrolimus (an albumin-binding drug) and cyclosporin (a nonalbumin-binding drug) was compared (unpublished data). Patients who received human albumin solution had significantly greater plasma albumin concentrations than those who received Gelofusine® during the first 2 postoperative weeks, but this predominantly influenced the pharmacodynamics of tacrolimus. It was always anticipated that monitoring immunosuppressive drug therapy would be particularly complex in this population of patients but we are only just beginning to understand why. Hopefully, this will lead to the development of improved pharmacodynamic monitoring techniques.

Nonallograft-related conditions that affect the liver transplant recipient are frequently either direct complications of immunosuppressive drugs or indirect complications of this therapy, such as those associated with the administration of powerful antibiotics for the prevention or treatment of opportunistic infections. The comorbidities of immunosuppression also require careful monitoring and some of the latest procedures will be reviewed in Chapter 45.

## Markers of chronic allograft rejection

Chronic allograft rejection typically occurs several months to a year post-transplantation and is characterized by the histological manifestations of ductopenia and a decrease in the number of hepatic arteries in portal tracts in the presence of foam cell (obliterative) arteriolopathy. In contrast to kidney or heart transplantation, this does not initially present as a vascular process but as a biliary phenomenon and is manifested clinically and biochemically as cholestatic jaundice. The incidence in adults is decreasing, and is currently 4%, but remains at more than double this in children. Some of the clinical risk factors could be amenable to modification by improvements in management. These risk factors include: donor–recipient human leucocyte antigen and sex matching, positive lymphocytotoxic cross-match, cytomegalovirus infection, frequency and intensity of acute rejection episodes and low blood cyclosporin concentrations. Improved methods for monitoring immunosuppression as well as the earlier detection, and more effective treatment, of infection or rejection are likely to have a beneficial impact in delaying or preventing the onset of chronic rejection. Crucially, in contrast to other solid organ allografts, chronic rejection of the liver is not always an irreversible process and 20–30% of patients may respond to augmented immunosuppression – especially if it is diagnosed in its early histological stages. This quality has been generally attributed to its unique immunobiological properties and the regenerative capacity of one of the main targets in chronic rejection – the bile ducts.

Since early chronic rejection can be treated, differential diagnosis is crucial. Chronic rejection is usually not difficult to recognize on the basis of the histopathological findings, but establishing the diagnosis with certainty often requires a thorough review of previous biopsies, which usually show acute rejection with significant bile duct damage. The major problem areas are distinguishing: (i) early chronic rejection from a normal or near normal biopsy; (ii) early chronic rejection from chronic hepatitis with bile duct damage; and (iii) whether chronic rejection or some other insult is responsible for the ductopenia or perivenular fibrosis, if present. Liver function tests are probably most helpful in the first problem area and selective elevation of the 'canalicular' enzyme, alkaline phosphatase, in an otherwise healthy long-term liver allograft survivor should prompt the pathologist to look more closely at the integrity of the bile ducts and the number of portal tracts sampled. If there are less than 6–8 portal tracts sampled, and there is no other explanation for the elevated alkaline phosphatase, rebiopsy should be suggested.

## REFERENCES


1 Halkin, H., Meffin, P., Melmon, K.L. and Rowland M. (1975). Influence of congestive heart failure on blood vessels of lidocaine and its active monodeethylated metabolite. *Journal of Clinical Pharmacy and Therapeutics*, **17**, 669–76.

2 St. Peter, J.V. and Awni, W.M. (1991). Quantifying hepatic dysfunction in the presence of liver disease with phenazone (antipyrine) and its metabolites. *Clinical Pharmacokinetics*, **20**, 50–65.

3 Farell, G.C. and Zaluzny, L. (1984). Accuracy and clinical utility of simplified tests of antipyrine metabolism. *British Journal of Clinical Pharmacology*, **18**, 559–65.

4 Woodside, K.J., Merion, R.M. and Williams, T.C. (1998). Prospective multivariate analysis of donor monoethylglycinexylidide (MEGX) testing in liver transplantation. *Clinical Transplantation*, **12**, 43–8.

5 Tsubono, T., Todo, A., Jabbour, N. et al. (1996). Indocyanine green elimination test in orthotopic liver recipients. *Hepatology*, **24**, 1165–71.

6 Salerno, F., Borroni, G., Moser, P. et al. (1996). Prognostic value of the galactose test in

predicting survival of patients with cirrhosis evaluated for liver transplantation. A prospective multicentre Italian study. AISF Group for the Study of Liver Transplantation. *Journal of Hepatology*, **25**, 474–80.

7 Dousset, B., Conti, F., Cherruau, B. et al. (1998). Is acute rejection deleterious to long-term liver allograft function? *Journal of Hepatology*, **29**, 660–8.

8 Devlin, J., Palmer, R.M., Gonde, C.E. et al. (1994). Nitric oxide generation. A predictive parameter of acute allograft rejection. *Transplantation*, **58**, 592–5.

9 Adams, D.H., Wang, L. and Neuberger, J.M. (1989). Serum hyaluronic acid following liver transplantation: evidence of hepatic endothelial damage. *Transplantation Proceedings*, **21**, 2274.

10 Adolf, J., Martin, W.G., Muller, D.F. et al. (1992). The effect of acute cellular rejection on liver function following orthotopic liver transplantation. Quantitative functional studies with the 14C-aminopyrine breath test. *Deutsche Medizinische Wochenschrift*, **117**, 1123–8.

11 Hughes, V.F., Trull, A.K., Gimson, A. et al. (1997). Randomized trial to evaluate the clinical benefits of serum $\alpha$-glutathione S-transferase concentration monitoring after liver transplantation. *Transplantation*, **64**, 1446–52.

12 Morovat, A., Trull, A. and Maguire, G.A. (1999). Serum fructose-1,6-bisphosphonate measurements for the early detection of acute hepatocellular rejection in liver transplant recipients. In *Advances in Critical Care Testing*, eds. W.F. List, M.M. Muller and A. St John. Schafthausen: AVL Medical Instruments, pp. 3–17.

13 Hughes, V.F., Trull, A.K., Joshi, O. and Alexander, G.J. (1998). Monitoring eosinophil activation and liver function after liver transplantation. *Transplantation*, **65**, 1334–9.

14 Cooper, D., Sharples, L., Cornelissen, J., Wallwork, J., Alexander, G. and Trull, A. (2001). Comparison between procalcitonin, serum amyloid A, and C-reactive protein as markers of serious bacterial and fungal infections after solid organ transplantation. *Transplantation Proceedings*, **33**, 1808–10.

15 Kershner, R.P. and Fitzsimmons, W.E. (1996). Relationship of FK506 whole blood concentrations and efficacy and toxicity after liver and kidney transplantation. *Transplantation*, **62**, 920–6.

43

# Chronic allograft damage index as a surrogate marker for chronic allograft rejection

Serdar Yilmaz[1], Mark Nutley[1], Eero Taskinen[2], Timo Paavonen[2], Pekka Hayry[2]

[1] University of Calgary and Foothills Medical Centre, Calgary, Alberta, Canada
[2] University of Helsinki and University of Helsinki Central Hospital, Helsinki, Finland

## Introduction

In the last two decades, existing immunosuppressive agents have significantly improved short-term graft survival in renal transplant recipients. However, this improvement has failed to translate into long-term survival [1].

Apart from death with a functioning transplant, chronic rejection has emerged as a leading cause of renal allograft loss [1]. There is currently no specific therapy for chronic rejection. Identification of treatment strategies that result in enduring benefits in the long term has therefore become essential. Any such study on long-term treatment strategies must evaluate the effect of treatment over several years' post-transplantation and will take between 5 and 10 years to complete. This is hardly feasible and, therefore, it has become important to identify and characterize appropriate near-term surrogate end-points [2]. Such a surrogate end-point should be visible between 1 or 2 years' post-transplantation and should predict graft loss 5–10 years' post-transplantation, even if the function of the graft at the time of measurement is normal.

In this chapter, we will briefly describe some silent features of chronic allograft rejection, particularly in kidney transplants, and propose the use of protocol core needle biopsy as a surrogate end-point in clinical studies.

## Clinical manifestations and prevention of chronic rejection

Chronic kidney allograft rejection is both a clinical and a histological diagnosis. Clinically, it manifests as progressive deterioration of renal function in the absence of other causes [3]. However, the natural history of chronic rejection is quite variable in terms of the time of onset and speed of progression [4]. Therefore, the diagnosis still depends largely on biopsy-proven confirmation for typical histological alterations.

Chronic kidney allograft rejection results from a multitude of various interacting immunological, nonimmunological and infectious insults with a similar final common pathway [5–7]. Identified risk factors include acute rejection episodes, delayed initial graft function, histoincompatibility, high donor and recipient age, black race, hyperlipidaemia, regraft, insufficient immunosuppression, hypertension, diabetes mellitus and cytomegalovirus (CMV) infection.

Most clinical series suggest, however, that the most important risk factor is acute rejection, particularly an acute rejection episode with any level of arteritis (acute vascular rejection) [8, 9]. Experimental studies in the rat suggest, furthermore, that some of the risk factors referred to above – for example, delayed graft function, insufficient immunosuppression and CMV infection – may represent confounding variables reflecting an increased proinflammatory state in the transplant and exposing the graft to acute episodes of rejection [10].

Exclusion of identified risk factors currently offers the only possibility of prophylaxis; however, a lack of suitable numbers of donors means that compromises must be made also in this respect.

## Histopathology of chronic rejection

In light microscopy, the pathological changes of chronic kidney allograft rejection are myointimal proliferation of arteries and arterioles, interstitial fibrosis with mild-to-moderate mononuclear cell infiltration, tubular atrophy and glomerular changes, including increased mesangial matrix, glomerular basement membrane and Bowman capsular thickening and segmental focal glomerulosclerosis [11, 12]. The histological findings of increased interstitial fibrosis and tubular changes, in the absence of vascular and glomerular changes, are not sufficient for the diagnosis of chronic allograft rejection [11].

The differentiation of chronic cyclosporin (CsA) nephrotoxicity from chronic kidney allograft rejection may present a problem in renal allograft biopsy interpretation. The vascular lesions in chronic rejection occur in arteries, whereas CsA nephrotoxicity mainly affects arterioles [11]. Isometric vacuolization, tubular calcification, giant mitochondria and focal or striped forms of interstitial fibrosis, which were useful diagnostic indicators of CsA nephrotoxicity during the high-dose CsA administration era, are not seen today and hyaline arteriolar thickening appears to be the only specific finding that distinguishes CsA nephrotoxicity from other entities [13].

## Quantitative histology as a surrogate marker for chronic rejection

There is substantial evidence deriving from several centres that the incipient changes of chronic rejection can be detected histologically in protocol biopsies long before the manifestation of clinical symptoms, and that these changes correlate with the duration of subsequent allograft function and survival. In order to use this information in clinical trials, the descriptive pathology reports must be given numerical values, enabling statistical treatment of the data.

In the study of Kasiske et al. [14], the histopathological findings that were most closely associated with chronic deteriorating graft function were glomerular mesangial expansion and sclerosis, interstitial fibrosis, basement membrane reduplication, arterial intimal occlusion and tubular atrophy. The sum score of these six parameters was expressed as a single numerical figure, called the 'chronic rejection score' (CRS). Kasiske and his colleagues [14] also found that this score correlated with the duration of subsequent survival ($r=0.65$, $p<0.001$).

In Isoniemi's study in Helsinki [15], eight quantitative histological variables were found to correlate significantly with impairment of kidney function, as measured by the level of serum creatinine at the time of protocol biopsy, at 2 years after transplantation (Table 43.1). These variables include interstitial inflammation and fibrosis, glomerular mesangial matrix increase and sclerosis, vascular intimal proliferation and tubular atrophy [16]. A 'Chronic Allograft Damage Index' (CADI) was constructed from the sum score of these, with each parameter quantitated separately from 0 to 3. This group also investigated the predictive value of the CADI index and correlated the CADI index at 2 years to transplant function during the subsequent years of follow up. The correlation coefficient (*r* value) between the CADI index at 2 years and graft function at 6 years post-transplantation was 0.717 and highly significant ($p=0.0001$) [16].

This association was further supported by an independent study from Fellström's group in Uppsala [17]. They performed protocol transplant biopsies at 6 months and used a 'chronic graft damage' score (CGD score), based on vascular intimal hyperplasia, glomerular mesangial changes, focal lymphocytic infiltration, interstitial fibrosis and tubular atrophy, to predict graft survival beyond 2 years. Patients with a CGD score of $>6$ had a significantly higher graft loss rate at 2 years than those with a score of $<6$ ($p=0.037$).

Nickerson et al. showed that Banff scoring [18] of chronic allograft pathology at 6 months correlated with the 12- and 24-months' serum creatinine [19]. They also suggested that chronic pathology independently predicts long-term decline of renal allograft function. Since then, other clinical studies have also confirmed the association between quantitative histopathology and a subsequent decline in allograft function.

**Table 43.1.** Parameters evaluated in the histological specimens of human and rat renal allografts

| *Glomeruli* | *Interstitium* |
|---|---|
| Number of glomeruli | Inflammation* |
| Mesangial cell proliferation | Pyroninophilic cells (%) |
| Mesangial matrix increase* | Oedema |
| Capillary basement membrane thickening | Haemorrhage |
| Capillary basement membrane duplication | Fibrin deposits |
| Capillary thrombosis | Fibrosis* |
| Bowman capsular thickening | |
| Glomerular inflammation | |
| Glomerular sclerosis* | |
| Glomerular necrosis | |
| *Tubuli* | *Vessels* |
| Epithelial swelling | Endothelial swelling |
| Epithelial vacuolation | Endothelial proliferation |
| Epithelial desquamation | Intimal proliferation* |
| Epithelial atrophy* | Inflammation |
| Epithelial necrosis | Sclerosis |
| Basement membrane thickening | Obliteration |
| Inflammation | |
| Dilatation | |
| Casts | |

*Notes:*
The parameters marked with (*) contribute to CADI.
CADI represents the sum score of interstitial inflammation and fibrosis, glomerular mesangial matrix increase and sclerosis, vascular intimal proliferation and tubular atrophy – each one scored separately from 0 to 3.
*Source:* After Isoniemi *et al.* [15, 16].

Taken together, several independent single centre studies have demonstrated that blinded quantitative human renal allograft histology is predictive for subsequent chronic rejection well before any alteration in transplant function is observed. Whether these observations also hold in multicentre clinical trials remains to be seen.

## CADI as a surrogate marker in multicentre trials

The multicentre study of renal histology in post-transplantation patients was performed as part of the US and tricontinental multicentre clinical trials of mycophenolate mofetil (MMF) in cadaver renal allograft recipients. The clinical results of the study have been reported [20, 21]. The primary objective of the biopsy study was to investigate how histological parameters at 1 year relate to long-term renal function and graft and patient survival. The secondary objectives were to investigate which parameters contribute to high 1-year CADI score and to explore the progression of CADI scores over a 3-year period after transplantation.

The biopsies included in this study were either 'protocol biopsies', obtained exclusively for study analysis, or 'for-cause biopsies' obtained in conjunction with a clinical event and substituting for a protocol biopsy if obtained at a corresponding time point. A protocol biopsy was preferred to a 'for-cause' biopsy. For quantitation of the CADI parameters in each biopsy, at least three stained slides were used, one stained with haematoxylin–eosin, one with Masson's trichrome and one with periodic acid schiff (D-PAS).

The centralized reading was performed in Helsinki. All samples were coded and evaluated blindly. Each histological sample was reviewed and quantitated by two pathologists and the CADI score obtained was based on the sum of individual component scores for diffuse or focal inflammation and fibrosis in the interstitium, mesangial matrix increase and sclerosis in glomeruli, intimal proliferation of vessels and tubular atrophy [22, 23], each one scored separately from 0 to 3. Thus, the theoretical minimum 'CADI' was 0 and the maximum 18. If the total CADI obtained independently by the two pathologists differed by more than one standard deviation unit from the mean, a consensus reading was performed. The data were filed using FileMaker Pro (Claris, California) software and the Helsinki Quantitative Renal Transplant Histopathology Database (for details, see [2]), and transferred to appropriate software for further handling and statistical analysis.

The US multicentre study [22] demonstrated that graft loss or patient death at 3 years in the 214 patients included in the biopsy study were significantly correlated with high CADI scores at 1 year ($p=0.0269$) and with acute rejection ($p=0.005$) during the first year (Table 43.2). Factors predicting high 1-year CADI, identified using multiple regression analyses, included high donor age ($p=0.0005$), long cold ischaemia time ($p=0.0330$) and rejection ($p=0.0012$), or CMV tissue-invasive disease ($p=0.0002$) during the first post-transplantation year. There also appeared to be a direct relationship between 1-year CADI score and renal function, as shown by a linear increase in serum creatinine with increasing CADI score ($p=0.0001$). Histopathological damage to the transplant appeared to be progressive, with an average increase in CADI score of 1.75 units during the first year and 2.37 units during the 3 years after transplantation.

In the tricontinental trial [23], graft loss or patient death at 3 years in the 174 patients included in the biopsy study were significantly correlated with high CADI scores at 1 year ($p=0.008$, Table 43.2). Besides the 1-year CADI, none of the risk

**Table 43.2.** Parameters correlating (*p* values) with subsequent graft loss in the US, tricontinental and combined studies

| | US | Tricontinental | Combined |
|---|---|---|---|
| Number of patients | 214 | 174 | 388 |
| CADI score at 1 year | 0.0269* | 0.0080* | 0.0014* |
| Number (%) with rejection at 1st year | 0.0049* | ns | 0.0257* |
| Number (%) of black patients | ns | ns | ns |
| Number (%) with two DR mismatches | ns | ns | ns |
| Number (%) with CMV infection at 1st year | ns | ns | ns |
| Number (%) with second allograft | N/A | ns | ns |
| Mean cold ischaemia time | ns | ns | ns |
| Mean donor age | ns | ns | ns |
| Number (%) with delayed function | ns | ns | ns |

*Notes:*
*P* values calculated from logistic regression model including all explanatory variables.
*P* values $<0.05$ are considered nonsignificant (ns). Delayed graft function is defined as the need for dialysis during the first week post-transplant. CMV infection is tissue-invasive disease and/or viraemia. N/A is not applicable.
*Source:* For details, see separate reports [22, 23].

factors (i.e. acute rejection during first year, previous transplant, donor age, cold ischaemia time, number of HLA-DR mismatches, delayed graft function and presence of CMV tissue-invasive disease or viraemia syndrome during the first year post-transplant) was found to be associated with 3-year graft loss or patient death. There was a significant linear relationship between the 1-year CADI score and the 1-year serum creatinine level ($p=0.0001$). High donor age ($p=0.0001$) and rejection during the first year ($p=0.0308$) were associated with high 1-year CADI scores. In virtually all patients, CADI scores inexorably progressed, with average changes of 1.87 units from baseline to year 1 and 2.64 units from baseline to 3 years.

These two analyses were also pooled – now consisting of 388 patients [24]. A relationship between 1-year CADI and 3-year graft and patient survival was identified using a logistic regression model for 3-year graft loss (including patient death). At 3 years, 30 of the 388 patients with a biopsy at 1 year had lost their graft. Graft loss or patient death at 3 years' post-transplant was significantly associated with 1-year CADI score ($p=0.0014$) and rejection during the first year ($p=0.0257$) (Table 43.2). With each unit increase in CADI score, the odds of graft loss or death increased by almost 150%. Multiple regression identified significant predictors for high 1-year CADI score. A high 1-year CADI score correlated with graft from an older donor ($p=0.0001$), prolonged cold ischaemia time ($p=0.0190$) and the pres-

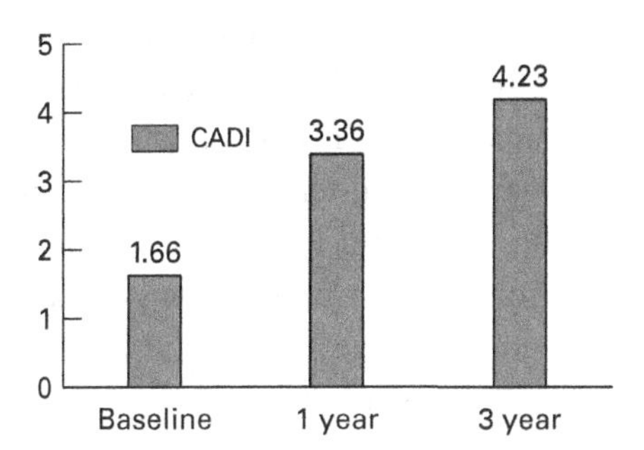


Figure 43.1 Progression of the CADI values from 0 to 3 years in the combined MMF study. Baseline CADI score obtained from earliest biopsy within 100 days post transplantation ($n=102$)

ence of an acute rejection episode during the first year ($p=0.003$). In the subset of 197 patients where baseline CADI score was available, the mean CADI score increased from 1.63 during the early period after transplantation to 3.45 at the first year post-transplant. In the subset of patients with serial biopsies from 0 to 3 years ($n=102$), the mean CADI score increased from 1.66 at baseline to 4.23 at 3 years' post-transplantation (Figure 43.1).

## Is it dangerous to perform a protocol biopsy?

In the multicentre study referred to above, the Uppsala spring-loaded biopsy device (Biopty) was introduced to all participating centres. This device, when used together with ultrasound guidance, has previously been demonstrated to be safe and virtually free of complications [15]. The authors are not aware of any fatal biopsy complication occurring with this equipment during the two multicentre studies or in the authors' experience in Helsinki (unpublished data). Since the introduction of modern transplant biopsy technology – fine needle aspiration biopsy (FNAB) with a 0.8 mm OD needle in the 1970s for kidney and liver and the Uppsala Needle Biopsy (Biopty) technology (NB) in the 1990s with a 1.2 mm outernal diameter (OD) needle for kidney and 1.2–1.6 mm OD needle for the liver – over 17 165 kidney and 3950 liver FNABs, and 1675 kidney and 1437 liver NBs have been performed. This represents a total of 24 227 needle biopsies with no patient or graft losses (H Isoniemi, personal communication).

## Conclusion

Several lines of research indicate that the typical signs of chronic histopathology appear in the transplant long before a decline in graft function. Different centres have performed protocol biopsies in stable allografts to characterize the prognostic value of incipient histopathology to subsequent chronic allograft rejection. Both single centre studies and two large multicentre trials have shown that incipient histopathological changes in protocol biopsy predict long-term graft survival.

Current immunosuppressive therapies have only marginally improved long-term graft outcome. Therapies to improve long-term success rate are currently impractical, as a 5–10-year follow up study and large numbers of patients are needed. The authors suggest that protocol core needle biopsy and centralized, blinded quantitation of 'Chronic Allograft Damage Index', may be employed as surrogate end-points in future chronic rejection studies.

## ACKNOWLEDGEMENTS

The authors are grateful to Ms Eerika Wasenius, RN, Helsinki Transplantation R&D Ltd, for efficient handling of biopsy material and filing of the CADI data.

## REFERENCES

1 Paul, L.C. and Fellström, B. (1992). Chronic vascular rejection of the heart and the kidney – have rational treatment options emerged? *Transplantation*, **53**, 1169–79.

2 Häyry, P., Yilmaz, S., Taskinen, E. and Isoniemi, H. (1996). Intermediate efficacy end-points for chronic kidney allograft rejection: implications of protocol core biopsy for early registration of new agents. In *Principles of Drug Development in Transplantation and Autoimmunity*, eds. R. Lieberman and A. Mukherjee, Austin, TX: RG Landes Company, pp. 557–64.

3 Paul, L.C., Häyry, P., Foegh, M. *et al.*. (1993). Diagnostic criteria for chronic rejection/accelerated graft atherosclerosis in heart and kidney transplants: joint proposal from the Fourth Alexis Carrel Conference on Chronic Rejection and Accelerated Arteriosclerosis in Transplanted Organs. *Transplantation*, **25**, 2022–3.

4 Kasiske, B.L., Heim-Duthoy, K.L., Tortorice, K.L. and Rao, K.V. (1991). The variable nature of chronic declines in renal allograft function. *Transplantation*, **51**, 330–4.

5 Tilney, N.L. (1999). Chronic rejection and its risk factors. *Transplantation Proceedings*, **31**(1–2A), 41S–4.

6 Hayry, P., Aavik, E. and Savolainen, H. (1999). Mechanisms of chronic rejection. *Transplantation Proceedings*, **31**(7A), 5S–8.

7 Paul, L.C. and Solez, K. (1992). Chronic rejection of vascularized organ allografts. In *Organ Transplantation: Long-Term Results*, eds. L. C. Paul and K. Solez, New York: Marcel Dekker, pp. 99–134.

8 Van Saase, J.L., van der Woude, F.J., Thorogood, J. et al. (1995). The relation between acute vascular and interstitial renal allograft rejection and subsequent chronic rejection. *Transplantation*, **59**, 1280–5.

9 Nickeleit, V., Vamvakas, E.C., Pascual, M., Poletti, B.J. and Colvin, R.B. (1998). The prognostic significance of specific arterial lesions in acute renal allograft rejection. *Journal of the American Society of Nephrology*, **9**, 1301–8.

10 Yilmaz, S., Paavonen, T. and Häyry, P. (1992). Chronic rejection of rat renal allografts. II. The impact of prolonged ischaemia time on transplant histology. *Transplantation*, **53**, 823–7.

11 Mihatsch, M.J., Ryffel, B. and Gudat, F. (1993). Morphological criteria of chronic rejection: differential diagnosis, including cyclosporine nephropathy. *Transplantation Proceedings*, **25**, 2031–7.

12 Sanfilippo, F. (1990). Renal transplantation. In *The Pathology of Organ Transplantation*, ed. G.E. Sale, Boston, MA: Butterworths, pp. 51–101.

13 Solez, K., Racusen, L.C., Marcussen, N. et al. (1993). Morphology of ischemic acute renal failure, normal function, and cyclosporine toxicity in cyclosporine-treated renal allograft recipients. *Kidney International*, **43**, 1058–67.

14 Kasiske, B.L., Kalil, R.S.N., Soon, Lee H. and Rao, K.V. (1991). Histopathologic findings associated with a chronic, progressive decline in renal allograft function. *Kidney International*, **40**, 514–24.

15 Isoniemi, H.M., Krogerus, L., von Willebrand, E., Taskinen, E., Ahonen, J. and Häyry, P. (1992). Histopathological findings in well-functioning, long-term renal allografts. *Kidney International*, **41**, 155–60.

16 Isoniemi, H., Taskinen, E. and Häyry, P. (1994). Histological chronic allograft damage index accurately predicts chronic renal allograft rejection. *Transplantation*, **58**, 1195–8.

17 Dimeny, E., Wahlberg, J., Larsson, E. and Fellström, B. (1995). Can histopathological findings in early renal allograft biopsies identify patients at risk for chronic vascular rejection? *Clinical Transplantation*, **9**, 79–84.

18 Solez, K., Axelsen, R.A., Benediktsson, H. et al. (1993). International standardization of criteria for the histologic diagnosis of renal allograft rejection: the Banff working classification of kidney transplant pathology. *Kidney International*, **44**, 411–22.

19 Nickerson, P., Jeffery, J., Gough, J. et al. (1998). Identification of clinical and histopathologic risk factors for diminished renal function 2 years' posttransplant. *Journal of the American Society of Nephrology*, **9**, 482–7

20 Mathew, T.H. (1998). Tricontinental Mycophenolate Mofetil Renal Transplantation Study Group. A blinded, long-term, randomized multicentre study of mycophenolate mofetil in cadaveric renal transplantation: results at three years. *Transplantation*, **65**, 1450–4.

21 US Renal Transplant Mycophenolate Mofetil Study Group. (1999). Mycophenolate mofetil in cadaveric renal transplantation. *American Journal of Kidney Disease*, **34**, 296–303.

22 Tomlanovich, S.J., Häyry, P., Yilmaz, S., Navarro, M. and Ramos, E. (1999). Correlation of serial biopsy data with long term function and graft and patient survival in the US renal transplant mycophenolate mofetil trial. American Society of Nephrology, 32nd Annual Scientific Meeting in Miami, Fl, November 5–8.

23 Mathew, T., Häyry, P., Yilmaz, S., Navarro, M. and Romos, E. (2000). High 1–year chronic allograft damage index (CADI) significantly predicts 3–year graft loss and patient death. 4th International Conference on New Trends in Clinical and Experimental Immunosuppression in Geneva, Switzerland, February 17–20.

24 Yilmaz, S., Nutley, M., Taskinen, E., Paavonen, T. and Hayry, P. (2000). Post-transplantation histology as a surrogate marker for long-term kidney allograft outcome. *Annals of Transplantation*, **5**, 37–43.

44

# Advances in pharmacodynamic biomarkers for monitoring the response to immunosuppressive drug therapy

Victor W Armstrong

Georg-August-Universität Göttingen, Göttingen, Germany

## Introduction

In recent years, improved knowledge of the immunological mechanisms underlying transplant rejection, as well as the discovery of novel biological and pharmaceutical agents, which selectively block various steps of the immune response, has enabled the effective control of graft rejection by the use of combined immunosuppressive therapy. Because of their complex and variable pharmacokinetics, as well as their narrow therapeutic indices, therapeutic drug monitoring (TDM) is an essential part of patient care, in order to reduce the risks of toxicity or rejection. Most transplant centres using cyclosporin or tacrolimus as primary immunosuppressants for the prophylaxis of graft function rely on trough concentration measurements of these drugs to help guide dosage. Despite its widespread use, this approach has its limitations in that the apparent concentrations of the drug may not accurately reflect the concentrations of the pharmacologically active moieties or may not correlate with the actual concentrations at the site of action of the drug. An alternative approach is pharmacodynamic monitoring in which the biological effect of the drug of interest is determined to assess the state of immunosuppression [1]. This chapter will present a critical appraisal of recent advances in the development of pharmacodynamic markers for the immunosuppressive drugs cyclosporin, tacrolimus, sirolimus, azathioprine and mycophenolate mofetil.

## Calcineurin inhibitors (cyclosporin, tacrolimus)

Although structurally quite different, cyclosporin and tacrolimus inhibit the same early events in T-lymphocyte activation, thus preventing T-helper cells from progressing from the $G_0$ to the $G_1$ phase of the cell cycle. Both drugs bind with high affinity to cytoplasmic proteins, termed immunophilins. The major immunophilins which bind cyclosporin and tacrolimus are cyclophilin and FK-binding protein

(FKBP12), respectively. Several minor immunophilins have also been identified [2]. The immunophilin-ligand dimers are able to engage the intracellular trimeric complex calcineurin–calmodulin–$Ca^{2+}$ to form a pentameric complex. This leads to inhibition of calcineurin phosphatase activity, thereby blocking activation of the nuclear factor of activated T cells (NF-AT) and ultimately inhibiting activation of the genes encoding interleukin-2 (IL-2) and other cytokines. In the search for pharmacodynamic markers which might better guide immunosuppressive therapy with either cyclosporin or tacrolimus, analytical procedures have been developed based on the various steps involved in this inhibitory pathway.

### Immunophilin receptor assays

Radiolabelled immunophilin-binding assays have been described for both cyclosporin and tacrolimus [2]. In one clinical study [3], the results of a cyclophilin-binding assay for cyclosporin were compared with those of a specific monoclonal radioimmunoassay. A close correlation was observed between the two methods, but the mean cyclosporin concentrations obtained with the cyclophilin-binding assay were 2-fold higher than those found with the radioimmunoassay. No information was given relating cyclosporin concentrations to therapeutic outcome. The higher cyclosporin concentrations obtained with the cyclophilin-binding assay are presumably due to the presence of cyclosporin metabolites which also bind to cyclophilin but do not cross-react in the radioimmunoassay [2]. Since these metabolites do not possess any appreciable immunosuppressive activity, the cyclophilin-binding assay would not appear to offer any advantage over specific immunoassays for monitoring cyclosporin therapy. In the case of tacrolimus, several metabolites are capable of binding to FKBP12 with similar efficiency to tacrolimus. However, not all metabolite–FKBP12 complexes can form the pentamer complex [4], which is a prerequisite for the inhibition of calcineurin phosphatase activity.

### Pentamer complex formation

An enzyme-linked immunosorbent assay (ELISA) technique using recombinant FKBP12 coated to a microtitre plate was used to study the ability of tacrolimus and seven of its metabolites to sustain pentamer formation [4]. A good correlation was observed between the ability of the metabolites to form the pentamer complex and their activities in the mixed lymphocyte response assay. These data suggested that a pentamer formation assay (PFA) would better reflect the total immunosuppressive activity of tacrolimus and its metabolites in whole blood samples from transplant recipients. The author and coworkers therefore developed a pentamer formation assay capable of quantifying both the parent drug and its immunologically active metabolites in whole blood specimens [5]. Tacrolimus trough concentrations, as estimated with the pentamer formation assay, were compared with the results obtained using an LC/MS/MS procedure specific for the parent drug and a microparticle enzyme immunossay (MEIA-II) that is used by many transplant centres to individualize tacrolimus dosage. The MEIA-II is known to possess substantial cross-reactivity [4] with the three tacrolimus metabolites 31–O-demethyl (M-II), 15–O-demethyl (M-III) and 15,31–O-didemethyl-tacrolimus (M-V). Of these three metabolites, only M-II is capable of sustaining pentamer complex formation. The results obtained with the MEIA were on average higher (median difference 2.0 μg/l) than those determined with the specific LC/MS/MS procedure, whereas the median difference for the pentamer formation assay was only 1.1 μg/l compared with the specific procedure. Inspection of the difference plot for MEIA-II and PFA revealed that 25 specimens exhibited a difference of 3 μg/l or greater between the two procedures. These 25 specimens, however, displayed good agreement between PFA and LC/MS/MS. These data suggest that these specimens contained relatively large amounts of pharmacologically inactive (not active) tacrolimus metabolites which cross-reacted with the immunoassay, but which were not capable of pentamer formation. In line with this supposition was the finding that these specimens tended to come from patients with evidence of liver dysfunction in the early post-transplant period. The results of this investigation suggest that the PFA could provide a more accurate estimate of the state of immunosuppression in patients receiving tacrolimus as their primary immunosuppressant, particularly during the early post-transplant period.

## Calcineurin activity

The measurement of calcineurin phosphatase activity in peripheral blood lymphocytes (PBLs) of patients has been proposed as a better guide to the degree of immunosuppression with cyclosporin than trough concentration measurements [6]. The most frequently used substrate is a 19 amino acid peptide, labelled with $^{32}$P, that mimics the *N*-terminal region of NF-AT – a natural substrate for calcineurin. In addition to calcineurin, several other phosphatases are present in PBLs. The release of $^{32}$P from the phosphorylated peptide is therefore determined in the presence of okadaic acid which inhibits protein phosphatases 1 and 2A (PP1 and PP2A). The remaining phosphatase activity is due to calcineurin and PP2C. The latter is cyclosporin resistant and so its activity can be determined by assaying PBLs after exposure of the cells to 1000 μg/l cyclosporin. In an investigation of clinically stable adult renal transplant recipients with a mean trough cyclosporine whole blood concentration of 180 μg/l, calcineurin activity was reduced by around 50% compared with untreated controls [6]. Similar results were observed in a small collective of eight stable paediatric transplant recipients, with a mean cyclosporin trough concentration of 148 μg/l being associated with an inhibition of calcineurin activity of around 50% [7]. In a recent publication [8], it was demonstrated that calcineurin inhibition by cyclosporin could be determined directly in whole blood. Peak cyclosporin levels of 800–2285 μg/l at 1–2 hours postdose produced a calcineurin inhibition of 70–96%. There was no observable lag-time between the rise and fall of cyclosporin levels and calcineurin inhibition. According to Quien et al. [7] the current assay still possesses an intrinsic variability which needs to be overcome before it can be made available for routine monitoring.

### IL-2 production

Cyclosporin inhibition ultimately leads to a suppression of IL-2 production. A pharmacodynamic procedure has been developed to measure the effect of cyclosporin determined in whole blood both in vitro and ex vivo [9]. This procedure involves the ex vivo stimulation of lymphocytes in whole blood with phytohaemagglutinin followed by measurement of the IL-2 production with an ELISA. In subjects who had received cyclosporin, the peak inhibition of IL-2 production (*c.* 85%) was found to occur 90–120 min after drug administration and coincided with peak cyclosporin concentrations of around 950 μg/l. From these ex vivo studies, the IC50 concentration of cyclosporin was found to be 407 μg/l which was similar to the IC50 concentration of 301 mg/l derived from in vitro experiments. Interestingly, these values are close to the upper limit of the therapeutic range for trough whole blood cyclosporin concentrations that are used by most transplant centres [10].

## Sirolimus

Sirolimus (rapamycin) is a macrocyclic triene antibiotic with immunosuppressive properties. As with cyclosporin and tacrolimus, its cellular activity depends on binding to immunophilins. Due to structural similarity in their binding domains, sirolimus and tacrolimus both bind to the same family of immunophilins. Whereas tacrolimus and cyclosporin block lymphokine gene transcription, sirolimus acts at a later stage in the cell cycle by blocking and inhibiting several cytokine- or growth factor-induced signal transduction pathways through inhibition of the P70 S6 kinase.

### Immunophilin binding

A radioreceptor assay based on a minor 52 kD immunophilin has been described for sirolimus. This assay displayed only low or minimal (<10%–26%) cross-reactivity for sirolimus metabolites and a good correlation was observed with a high performance liquid chromatography procedure specific for the parent drug [2].

### P70 S6 kinase assay

The activity of P70 S6 kinase in peripheral blood mononuclear cells has been suggested as a potential target for pharmacodynamic monitoring of sirolimus. In a preliminary in vitro study, a dose-dependent inhibition of P70 S6 kinase occurred on incubation of white blood cells with sirolimus, whereas cyclosporin had no effect [11]. As yet, no data have been published on clinical samples from transplant recipients receiving sirolimus as part of their immunosuppressive regimen.

## Mycophenolate mofetil

This drug was approved in 1995 as an adjunctive therapy with corticosteroids and cyclosporin in renal transplantation. Mycophenolate mofetil (MMF) is rapidly metabolized in vivo to its active constituent mycophenolic acid (MPA). The latter compound reversibly inhibits inosine monophosphate dehydrogenase (IMPDH) by an uncompetitive mechanism. This enzyme occupies a pivotal position in purine metabolism since it controls the first committed step in GTP biosynthesis through the de novo synthetic pathway in which inosine monophosphate is oxidized to xanthine monophosphate. Two isoenzymes of IMPDH have been identified in human tissues: IMPDH-1 is widely expressed whereas IMPDH-2 is induced during cellular proliferation and its activity is high in malignant cells and in stimulated lymphocytes. The inhibition of IMPDH-2 in activated lymphocytes by MPA causes a reduction in intracellular guanine nucleotide pools and leads to an arrest of lymphocyte proliferation.

### IMPDH inhibition

A radiolabelled assay has been described for the measurement of IMPDH activity in whole blood [1, 12]. The assay involves incubation of whole blood with [2,8–$^{3}$H]-hypoxanthine which is taken up by the cells and converted to [2,8–$^{3}$H]-inosine monophosphate. The latter compound is then oxidized by IMPDH to xanthine monophosphate with the release of $^{3}$H. Aliquots of the cell suspension are removed at timed intervals and mixed with a cold suspension of activated charcoal in 50 g/l trichloroacetic acid. After centrifugation, the radioactivity in the supernatant is determined by scintillation counting. An alternative nonradioactive HPLC procedure has also been developed to follow IMPDH activity [13].

In a clinical investigation, IMPDH activity was measured using the radiolabelled assay in two groups of stable kidney transplant recipients [14]. The first group of nine patients had been on MMF for 1 year or less, while the second group had received MMF for over 2 years. Whole blood samples were collected before, and 0.5, 1, 2, 4 and 6 hours after, dosing for the determination of IMPDH activity and plasma MPA concentrations. Patients treated with MMF for less than 1 year showed a similar IMPDH activity profile to that reported by Langman et al. [12], with a peak inhibition of IMPDH activity (range: 40%–90%) coinciding with the peak plasma MPA concentration. Subsequently, there was a gradual restoration of IMPDH activity to predose levels. The IMPDH activity profiles of the eight patients on MMF for more than 2 years were different to those of the previous nine patients. Three patients displayed a substantial inhibition of IMPDH activity coinciding with a mean peak MPA concentration of around 30 mg/l. Only a weak inhibition of IMPDH activity occurred in the remaining five patients, consistent with the much lower mean peak MPA concentration of around 10 mg/l. However, in all eight patients, there was a subsequent increase in IMPDH activity to levels that were 3-fold higher and more than the original predose values. In three patients who could be tested at 1 and 2 years, the IMPDH activity profile was found to alter in accordance with the profiles observed in the two different groups. The investigators concluded that long-term treatment (2 years) with MMF is associated with induction of IMPDH activity, which might have implications for the role of MMF in long-term maintenance therapy. The interpretation of these intriguing observations must, however, take into consideration an induction of erythrocyte IMPDH due to the long-term treatment with MPA. The activity of IMPDH is normally extremely low in erythrocytes from healthy subjects. Grossly elevated IMPDH activities have been observed in immunodeficient children treated with the IMPDH inhibitor ribovarin [15]. In a further study, the ratio of whole blood IMPDH activity to the activity in mononuclear cells was found to be substantially higher in two patients receiving MMF than in healthy controls [13]. It was concluded that the high activity observed in the whole blood of the renal transplant recipients on MMF was mainly due to the induction of erythrocyte IMPDH. The interesting data of Sanquer et al. [14] now need to be confirmed in isolated mononuclear cells, and the inhibition assay will have to be carried out in the plasma from the corresponding time point to ensure that the MPA concentration in the cell is not depleted.

### Lymphocyte function

A whole blood culture technique has been used to study the in vivo pharmacodynamics of MPA in a rat model by monitoring lymphocyte proliferation and activation [16]. This procedure involved incubation of diluted blood for 72–96 hours with the mitogen concanavalin A. The cells were then analysed for lymphocyte proliferation or for lymphocyte activation. The proliferative response was measured by [$^3$H]TdR incorporation and by bivariate analysis of proliferating nuclear cell antigen expression and DNA content with two-colour flow cytometry. Lymphocyte activation was measured by flow cytometric detection of the expression of CD25 and CD134. A dose-dependent effect of MPA on cell proliferation and on the expression of cell surface antigens was documented. The IC50 concentrations for inhibition of lymphocyte proliferation (0.83 mg/l) as determined from the cellular DNA content and for the inhibition of CD25 (0.81 mg/l) and CD134 (0.74 mg/l) expression were all very similar. Interestingly, these values are close to the predose MPA concentration of 1 mg/l, below which an increased incidence of acute rejection was observed in paediatric renal transplant recipients early after transplantation [17]. The assessment of lymphocyte function should be a useful pharmacodynamic tool for monitoring the efficacy of immunosuppressive therapy.

## Azathioprine

Azathioprine has been used in transplantation for more than 25 years. It rapidly undergoes metabolism to 6-mercaptopurine (6–MP) which is further metabolized to the pharmacologically active 6-thioguanine nucleotides. Two further metabolic pathways for 6–MP include conversion to 6-thiouric acid by xanthine oxidase and methylation to 6-methylmercaptopurine, catalysed by the enzyme thiopurine-methyltransferase (TPMT). The latter exhibits a wide range of activity in the normal population. One person in 300 from the Caucasian population has a complete deficiency of this enzyme and such individuals are at extremely high risk for myelotoxicity if treated with azathioprine. Measurement of red blood cell TPMT activity is a convenient assay for identifying those individuals at high risk [18].

In a prospective study [19], red blood cell TPMT activity was serially monitored in renal transplant recipients on therapy with azathioprine. Three patterns of TPMT activity variation were observed. In one group of patients, TPMT activity was induced early after transplantation and rose steadily until month 3. In the second group, TPMT activity rose at month 1 after transplantation, while, in the third group, TPMT activity remained unchanged. There was a significant difference in the incidence of acute rejection between the three groups: the lowest incidence occurred in group 1 and the highest incidence in group 3. The authors of this study reasoned that the azathioprine-induced increase in TPMT activity reflects conversion of azathioprine to 6–MP and thus also reflects the immunosuppressive potency of the drug.

## Conclusions

Currently, none of the pharmacodynamic markers so far investigated has proven its clinical utility for the routine monitoring of immunosuppression in transplant recipients. The assays are still rather laborious and time consuming and generally have a high intrinsic variability. Nonetheless, they are useful research tools in assessing the extent of immunosuppression in transplant recipients, in comparing the immunosuppressive efficacy of different agents acting on the same mechanism and evaluating the relationship between immunosuppressive drug levels and immunosuppressive efficacy in order to help develop better monitoring strategies.

## REFERENCES

1 Yatscoff, R.W., Aspeslet, L.J. and Gallant, H.L. (1998). Pharmacodynamic monitoring of immunosuppressive drugs. *Clinical Chemistry*, **44**, 428–32.

2 Soldin, S. (2000). Immunophilins: their properties and their potential role in therapeutic drug monitoring. *Therapeutic Durg Monitoring*, **22**, 44–6.

3 Huupponen, R., Hirvisalo, E.L. and Neuvonen, P. (1997). Comparison of cyclophilin binding assay and radioimmunoassay in monitoring of blood cyclosporin. *Therapeutic Drug Monitoring*, **19**, 446–9.

4 Tamura, K., Fujimura, T., Iwasaki, K. et al. (1994). Interaction of tacrolimus (FK506) and its metabolites with FKBP and calcineurin. *Biochemical and Biophysical Research Communications*, **202**, 437–43.

5 Armstrong, V.W., Schuetz, E., Zhang, Q. et al. (1998). Modified pentamer formation assay for measurement of tacrolimus and its active metabolites: comparison with liquid chromatography-tandem mass spectrometry and microparticle enzyme-linked immunoassay (MEIA-II). *Clinical Chemistry*, **44**, 2516–23.

6 Batiuk, T.D., Pazderka, F. and Halloran, P.F. (1995). Calcineurin activity is only partially inhibited in leukocytes of cyclosporine-treated patients. *Transplantation*, **59**, 1400–4.

7 Quien, R.M., Kaiser, B.A., Dunn, S.P. et al. (1997). Calcineurin activity in children with renal transplants receiving cyclosporin. *Transplantation*, **64**, 1486–9.

8 Halloran, P.F., Helms, L.M.H., Kung, L. and Noujaim, J. (1999). The temporal profile of calcineurin inhibition by cyclosporin in vivo. *Transplantation*, **68**, 1356–61.

9 Stein, C.M., Murray, J.J. and Wood, A.J.J. (1999). Inhibition of stimulated interleukin-2 production in whole blood: a practical measure of cyclosporin effect. *Clinical Chemistry*, **45**, 1477–84.

10 Oellerich, M., Armstrong, V.W., Kahan, B. et al. (1995). Lake Louise Consensus Conference on cyclosporin monitoring in organ transplantation: report of the consensus panel. *Therapeutic Drug Monitoring*, **17**, 642–54.

11 Gallant, H.L. and Yatscoff, R.W. (1996). P70 S6 kinase assay: a pharmacodynamic monitoring strategy for rapamycin; assay development. *Transplantation Proceedings*, **28**, 3058–61.

12 Langman, L.J., LeGatt, D.F., Halloran, P.F. et al. (1996). Pharmacodynamic assessment of mycophenolic acid-induced immunosuppression in renal transplant recipients. *Transplantation*, **62**, 666–72.

13 Storck, M., Abendroth, D., Albrecht, W. et al. (1999). IMPDH activity in whole blood and isolated blood cell fraction for monitoring of CellCept-mediated immunosuppression. *Transplantation Proceedings*, **31**, 1115–16.

14 Sanquer, S., Breil, M., Baron, C. et al. (1999). Induction of inosine monophosphate dehydrogenase activity after long-term treatment with mycophenolate mofetil. *Clinical Pharmacology and Therapeutics*, **65**, 640–8.

15 Montero, C., Duley, C.A., Fairbanks, L.D. et al. (1995). Demonstration of erythrocyte inosine monophosphate dehydrogenase activity in ribavirin-treated patients using a high performance liquid chromatography linked method. *Clinica Chimica Acta*, **238**, 169–78.

16 Gummert, J.F., Barten, M.J., Sherwood, S.W. et al. (1999). Pharmacodynamics of immunosuppression by mycophenolic acid. Inhibition of both lymphocyte proliferation and activation correlates with pharmacokinetics. *Journal of Pharmacology and Experimental Therapeutics*, **291**, 1100–12.

17 Oellerich, M., Shipkova, M., Schutz, E. et al. (2000). Pharmacokinetic and metabolic investigations of mycophenolic acid in pediatric renal transplant recipients: implications for therapeutic drug monitoring. *Therapeutic Drug Monitoring*, **22**, 20–6.

18 Schutz, E., Gummert, J., Armstrong, V.W. et al. (1996). Azathioprine pharmacogenetics: the relationship between 6-thioguanine nucleotides and thiopurine methyltransferase in patients after heart and kidney transplantation. *European Journal of Clinical Chemistry and Clinical Biochemistry*, **34**, 199–205.

19 Mircheva, J., Legendre, C., Soria-Royer, C. et al. (1995). Monitoring of azathioprine-induced immunosuppression with thiopurine methyltransferase activity in kidney transplant recipients. *Transplantation*, **60**, 639–42.

45

# The use of biomarkers for monitoring the response to immunosuppressive drug therapy

Atholl Johnston

St Bartholomew's and the Royal London School of Medicine and Dentistry, London, UK

## ntroduction

Immunosuppressive drug therapy is used to prevent acute or chronic rejection of organ allografts in patients who have received a transplant. As with most prophylactic treatments, it only becomes apparent that the treatment has failed when the unwanted event happens. If there is a large safety margin between the dose of drug that prevents the unwanted event and the dose at which unacceptable drug toxicity occurs then the drug dose given can be sufficiently high to ensure treatment effectiveness. However, in transplant patients, the consequences of under- or over-immunosuppression can be equally catastrophic. On the one hand, underimmunosuppression can result in organ rejection that may, in turn, lead to graft loss and even death, while, on the other hand, overimmunosuppression can result in life-threatening infections, carcinoma and a range of drug-related adverse events such as nephrotoxicity. At present, we use plasma or blood drug concentrations as a surrogate measure of immunosuppression in the belief that there is a tight relationship between drug concentration and effect [1]. However, interindividual variation in pharmacological response can make drug concentration data difficult to interpret. For this reason, the use of biomarkers for monitoring the response to immunosuppressive drug therapy has been investigated.

## :hoice of biomarker

The choice of biomarkers for monitoring the response to immunosuppressive drug therapy is wide. Monitoring the effect of therapy on the primary intracellular drug targets offers the most promise, but these targets are also the most difficult to measure (Table 45.1). In any case, within- and between-subject variation in signal transduction may result in differences in immunosuppressive efficacy 'downstream' of the primary drug target. For this reason, the majority of studies have

**Table 45.1.** The primary intracellular drug targets of the currently used immunosuppressive drugs

| Drug | Target |
|---|---|
| Azathioprine | Thiopurine methyltransferase (TPMT) [2] |
| Antilymphocyte, antithymocyte globulin | Lymphocytes [3] |
| Cyclosporin | Calcineurin [4] |
| Basiliximab and daclizumab | IL-2 receptor of activated T cells [5] |
| Muromononab-CD3 (OKT3) | CD3 complex on T cells [6] |
| Mycophenolic acid | Inosine monophosphate dehydrogenase (IMPDH) [7] |
| Steroids | Transcription factors such as AP-1 and NF-KB |
| Sirolimus | P70 S6 kinase [8] |
| Tacrolimus | Calcineurin [4] |

focused on the measurement of secondary events such as cytokine production [9], soluble adhesion molecules [10], DNA binding [11], expression of antibodies [12], markers of cellular damage [13], p-glycoprotein expression [14] or the number (absolute or relative) of specific lymphocyte subtypes [3].

Few trials have examined the use of biomarker measurements in a prospective manner. A good example was the use of serum $\alpha$-glutathione *S*-transferase (GST) concentration in clinical practice reported by Hughes et al. [13]. These authors tested the benefits of measuring GST in liver transplant patients in the first 100 days' post-transplant using a randomized prospective trial. Patients were randomized to a reporting or a nonreporting group. In the former, the GST concentrations were reported to the clinical staff, while, in the latter, the results were recorded but not reported by the laboratory. The study showed clearly the benefits of the GST measurement. Compared with the nonreporting group patients, the patients for whom the GST was available to the clinicians had significantly less risk of graft loss, spent less time in hospital, had fewer and less severe rejection episodes and a lower incidence of rejection.

However, most studies have been designed to generate rather than test hypotheses [15]. For example, in a very thorough study by Daniel et al. [16], predictive indicators of rejection or infection in renal transplant patients were examined by measuring relative and absolute numbers of CD3-, CD4-, CD8-, CD8/CD56-, CD19-, CD16-, CD3/CD25-, CD3/DR- and OKM1-positive blood cells and plasma concentrations of neopterin, interleukin-1$\alpha$ (IL-1$\alpha$), IL-1$\beta$, soluble IL-1 receptor antagonist, IL-2, soluble IL-2 receptor, IL-3, IL-4, IL-6, soluble IL-6 receptor, IL-8, IL-10, tumour necrosis factor-$\alpha$, transforming growth factor-$\beta$2, interferon-$\gamma$ and soluble intercellular adhesion molecule 1 (sICAM1). The samples studied were 153 blood samples from 28 patients with an uncomplicated postoperative course, 103 samples from 19 patients during rejection, 92 samples from 20 patients during viral infection, 17 samples from nine patients with rejection and infection, 241 samples from 25 patients before or after rejection or infection, 86 samples from 22 patients during acute tubular necrosis, 30 samples from 16 patients after operative interventions and 118 samples from 12 patients in the symptom-free period after operative intervention or acute tubular necrosis. In summary, this study examined 21 parameters in 840 blood samples from 145 patients in eight different groups.

## Statistical issues

The study by Daniel et al. [16] outlined above generated over 17 000 data points. The authors defined cut-offs for each of the parameters in terms of median + 1 standard deviation (a strange mixture of nonparametric and parametric statistics) and the statistical analysis was carried out using Fisher's exact test. The authors quoted the cut-offs used and the probability values for comparisons between groups of patients. Not surprisingly, given the number of parameters and comparisons made, they found significant differences between the groups for many of the measured parameters. However, they made no mention of correcting the type I statistical error rate for the number of multiple comparisons made. Although the authors could make broad statements, such as 'our data suggest that increased CD4%, CD4/CD8, SIL2RA, SIL6R and IL-10 may define patients with impending rejection . . .', they gave no indication of how the results of this study could be used in a quantitative manner to discriminate between the eight groups of patients and they did not give or discuss the number of false-positive or false-negative results.

Given the richness and complexity of the data, a better approach might have been to use a multivariate statistical technique such as discriminant analysis or nominal logistic regression. The linear combination of variables used in multivariate procedures is more powerful than a series of pairwise comparisons. Many authors avoid these techniques because they are considered arcane and difficult to implement. However, modern statistical software such as MINITAB™ (www.minitab.com) and SPSS™ (www.spss.com) contain many multivariate procedures that are easy both to understand and use. These programs are available in Windows™-based, stand-alone, desktop computer environments.

If you have a series of biomarker measurements in patients who can be classified into groups – for example into those who are clinically stable, experiencing rejection or nephrotoxicity – then discriminant analysis can be used to investigate how the variables contribute to group separation. This allows one to determine which combinations of biomarkers are important diagnostic tools and which are not.

**Table 45.2.** Simulated data (mean ± sd) for three biomarkers in three groups of renal transplant patients

| Patient group | Renal function Marker A | Renal damage Marker B | Immune function Marker C |
|---|---|---|---|
| Stable | 498 ± 101 | 84 ± 16 | 4364 ± 894 |
| Rejection | 1008 ± 198 | 83 ± 16 | 2166 ± 428 |
| Nephrotoxicity | 996 ± 204 | 160 ± 33 | 4351 ± 821 |

Once the important biomarkers have been chosen and the discriminant function has been determined using the data from the initial series of patients, the use of these markers to predict clinical events can be tested prospectively.

The process can be illustrated with an example using simulated data for a series of renal transplant patients. Three hypothetical biomarkers were measured: one was a measure of kidney function, one was specific for renal damage and the third was a marker of immune function. The data were simulated using random, normally distributed, parameters with means and standard deviations (sd) as shown in Table 45.2. A discriminant function was derived to differentiate between those patients who were stable and those who were experiencing either rejection or nephrotoxicity.

According to current clinical practice, a measure of kidney function (for example, serum creatinine) is routinely used for monitoring renal allograft recipients. This correctly identifies 99% of the stable patients but fails to diagnose renal impairment in about 10% of those patients experiencing rejection and nephrotoxicity (see Table 45.3, Step 1). In addition, this single biomarker cannot accurately differentiate between the renal dysfunction attributable to rejection or nephrotoxicity. By the use of either an additional biomarker of renal damage or of immune function, it is possible to increase the proportion of patients diagnosed correctly (see Table 45.3, Step 2). However, it requires a third biomarker to decrease the proportion of patients misclassified to below 8% in each group (see Table 45.3, Step 3).

This example is simplistic and, for this reason, the procedure may appear to be self-evident. In real life, there is no single biomarker that clearly and specifically measures renal damage or immune function and, if there were, it is unlikely that the differentiation between groups of patients would be so clear cut. Nevertheless, discriminant analysis and other multivariate procedures can be used to set objective decision criteria in order to rule in or rule out specific diagnoses. The use of other techniques, such as artificial neural networks [17, 18], expert systems [19] or a Bayesian belief network [20], should also be considered.

**Table 45.3.** The prediction accuracy of combining diagnostic tests for renal function, renal damage and immune function

Step 1 – Test renal function

| Patient group ⇒ | Stable | Rejection | Nephrotoxicity |
|---|---|---|---|
| Stable | **98.9%** | 9.4% | 11.1% |
| Rejection | 0% | **51.3%** | 46.9% |
| Nephrotoxicity | 1.1% | 39.3% | **42.0%** |

Step 2 – Test for renal function and renal damage

| Patient group ⇒ | Stable | Rejection | Nephrotoxicity |
|---|---|---|---|
| Stable | **99.0%** | 9.6% | 3.0% |
| Rejection | 0.8% | **90.4%** | 10.2% |
| Nephrotoxicity | 0.2% | 0% | **86.8%** |

*or* Test for renal function and immune function

| Patient group ⇒ | Stable | Rejection | Nephrotoxicity |
|---|---|---|---|
| Stable | **97.7%** | 1.3% | 10.8% |
| Rejection | 1.2% | **97.6%** | 8.6% |
| Nephrotoxicity | 1.1% | 1.1% | **80.6%** |

Step 3 – Test for renal function, renal damage and immune function

| Patient group ⇒ | Stable | Rejection | Nephrotoxicity |
|---|---|---|---|
| Stable | **98.7%** | 1.3% | 4.9% |
| Rejection | 1.2% | **98.7%** | 2.7% |
| Nephrotoxicity | 0.1% | 0% | **92.4%** |

*Notes:*

The figures in bold show agreement – that is, the percentage of patients in whom the actual and predicted coincide.

## Summative or outcome markers

When prophylactic drug therapy is successful, there is often very little out of the ordinary to measure since the patient is stable. However, during periods of instability, large numbers of parameters are changing and it may be difficult to make sense of the complex patterns exhibited. Given that for immunosuppressive drug therapy the activity of the main pharmacological target or receptor can be difficult to measure and the intermediate responses are too complex, a simple outcome measure that combines or summarizes the process would be a useful biomarker. One such candidate is nitric oxide.

## Nitric oxide

Nitric oxide (NO) plays a critical role in a diverse range of biological processes, including control of vascular tone and blood pressure, neurotransmission, inflammation and immunity [21]. In the kidney, NO participates in the regulation of glomerular and medullary haemodynamics, the tubuloglomerular feedback response and renin release [22]. While NO is beneficial in the immune response system, excessive NO is cytotoxic as it gives rise to reactive oxygen and nitrogen species, causing cellular damage. The dual role of NO as both an agent of good and ill within the body makes it, potentially, an ideal biomarker to monitor immunosuppressive drug therapy [23]. However, NO is an extremely reactive molecule and is difficult to measure in vivo. Therefore, its stable metabolites, $NO_2$, $NO_3$ or $NO_x$, are usually measured in serum and urine samples [24].

## Nephrotoxicity

In animals, experimental nephrotoxicity produced by cyclosporin can be enhanced by simultaneous NO blockade, suggesting that NO has a protective effect in drug-induced nephropathy [25, 26]. Further evidence to support this hypothesis was provided by Yang et al. [27] who showed that, when cyclosporin-treated rats were given oral supplements of L-arginine, the precursor to NO, nephrotoxicity was prevented. In humans, NO has been shown to be an important regulating mechanism that protects against cyclosporin-associated vasoconstriction in vivo [28]. After prolonged exposure of rats to high concentrations of cyclosporin, there was a significant rise in arterial blood pressure coupled with a steady decline in urinary nitrate/nitrite excretion, suggesting depressed NO production [29]. This was attributed to a reduction in inducible NO synthase (INOS) in the kidney and thoracic aorta. Thus, reduced urinary excretion of nitrate or low nitrate concentrations in serum may be indicative of impending renal damage.

## Rejection

Experimental work in animals suggests that the cytotoxic action of NO may play an important role in the in vivo response to allogeneic tissue. In rats, following cardiac transplantation, INOS and increased NO production occurred during the early stages of acute rejection and was localized to infiltrating mononuclear inflammatory cells [30]. In dogs with lung transplants, plasma nitrate rose following removal of immunosuppressive drug treatment when compared with a group of animals who were maintained on drug therapy [31].

Rejection of a variety of human organ allografts is also associated with increased NO production and, hence, raised serum nitrate concentrations and urinary nitrate excretion. In a study of liver transplant patients, Fábrega and coworkers [32] showed a more than 2-fold greater concentration of serum nitrate in patients with biopsy-proven rejection (50 ± 15 μmol/l; mean ± sd) than in those without rejection (19 ± 13 μmol/l). When treated with a 3-day course of 1 g/day methylprednisolone, the serum nitrate returned to normal (19 ± 6 μmol/l). In cardiac transplant recipients, both urinary nitrate excretion [33] and serum nitrate are increased [34] during periods of biopsy-proven acute rejection. A similar pattern is seen during acute rejection in kidney transplant patients [35, 36]. However, there is at least one study (of children) in which decreased nitrate excretion was found, followed by a return to basal levels with antirejection therapy [37]. In bone marrow transplantation, high nitrate levels were associated with graft versus host disease [38].

Lung transplantation provides a unique opportunity to measure NO directly in expired air. Exhaled NO was measured in a study of 108 lung transplant recipients [39]. In stable patients, the amount of exhaled NO was not different from that found in healthy controls, but was more than two and a half times greater in patients during acute rejection. In contrast to a study by Fisher and colleagues [40], they also found no increase in NO exhalation in patients with acute infection or bronchiolitis obliterans. During acute vascular rejection, exhaled NO was not increased. Based on the data presented by Vora et al. [38], it can be also postulated that NO is a common proximate regulator of the immune response in host versus graft and graft versus host reactions.

## Conclusions and future directions

It is now over 40 years since the first successful human kidney transplant [41] and 35 years since the first use of azathioprine for immunosuppression by Calne [42]. Yet, we seem to have no better biomarker of kidney function than that which was available then – the serum creatinine concentration. Indeed, for all transplanted solid organs, biopsy remains the gold standard for the definitive diagnosis of rejection. If we are to progress towards better and noninvasive measures of the effectiveness of immunosuppressive drug therapy, we need to use not only the data we collect more effectively but also be prepared to test our conclusions prospectively.

## REFERENCES

1 Johnston, A. and Holt, D.W. (1999). Therapeutic drug monitoring of immunosuppressant drugs. *British Journal of Clinical Pharmacology*, **47**, 339–50.

2 Gummert, J.F., Schutz, E., Oellerich, M., Mohr, F.W. and Dalichau, H. (1995). Monitoring of TPMT in heart transplant recipients under immunosuppressive therapy with azathioprine. *Artificial Organs*, **19**, 918–920.

3 Clark, K. (1996). Monitoring antithymocyte globulin in renal transplantation. *Annals of the Royal College of Surgeons of England*, **78**, 536–40.

4 Halloran, P.F., Kung, L. and Noujaim, J. (1998). Calcineurin and the biological effect of cyclosporine and tacrolimus. *Transplantation Proceedings*, **30**, 2167–70.

5 Amlot, P.L., Rawlings, E., Fernando, O.N. et al. (1995). Prolonged action of a chimeric interleukin-2 receptor (CD25) monoclonal antibody used in cadaveric renal transplantation. *Transplantation*, **60**, 748–56.

6 Cinti, P., Cocciolo, P., Evangelista, B. et al. (1996). OKT3 prophylaxis in kidney transplant recipients: drug monitoring by flow cytometry. *Transplantation Proceedings*, **28**, 3214–16.

7 Shaw, L.M. and Nowak, I. (1995). Mycophenolic acid: measurement and relationship to pharmacologic effects. *Therapeutic Drug Monitoring*, **17**, 685–9.

8 Yatscoff, R.W., LeGatt, D.F. and Kneteman, N.M. (1993). Therapeutic monitoring of rapamycin: a new immunosuppressive drug. *Therapeutic Drug Monitoring*, **15**, 478–82.

9 Masri, M.A., Stephan, A., Barbari, A., Kamel, G., Kelany, H. and Karam, A. (1997). Correlation between cyclosporine pharmacokinetics and immunologic parameters in dialyzed patients awaiting transplantation. *Transplantation Proceedings*, **29**, 2953–4.

10 Salmaggi, A., Corsini, E., La, M.L. et al. (1997). Immunological monitoring of azathioprine treatment in multiple sclerosis patients. *Journal of Neurology*, **244**, 167–74.

11 Akioka, K., Nakajima, H., Fujiwara, I. et al. (1997). The DNA-binding activities of NF-AT and AP-1 can predict levels of immunosuppression of kidney transplant recipients. *Transplantation Proceedings*, **29**, 2621–3.

12 Kocher, A.A., Dockal, M., Weigel, G. et al. (1997). Immune monitoring in cardiac transplant recipients. *Transplantation Proceedings*, **29**, 2895–8.

13 Hughes, V.F., Trull, A.K., Gimson, A. et al. (1997). Randomized trial to evaluate the clinical benefits of serum alpha-glutathione S-transferase concentration monitoring after liver transplantation. *Transplantation*, **64**, 1446–52.

14 Melk, A., Daniel, V., Weimer, R. et al. (1999). P-glycoprotein expression in patients before and after kidney transplantation. *Transplantation Proceedings*, **31**, 299–300.

15 Weimer, R., Zipperle, S., Daniel, V., Carl, S., Staehler, G. and Opelz, G. (1998). Superior 3-year kidney graft function in patients with impaired pretransplant Th2 responses. *Transplant International*, **11**(Suppl 1), S350–6.

16 Daniel, V., Arzberger, J., Melk, A. et al. (1999). Predictive indicators of rejection or infection in renal transplant patients. *Transplantation Proceedings*, **31**, 1364–5.

17 Furness, P.N., Levesley, J., Luo, Z. et al. (1999). A neural network approach to the biopsy diagnosis of early acute renal transplant rejection. *Histopathology*, **35**, 461–7.

18 Abdolmaleki, P., Movhead, M., Taniguchi, R.I., Masuda, K. and Buadu, L.D. (1997). Evaluation of complications of kidney transplantation using artificial neural networks. *Nuclear Medicine Communications*, **18**, 623–30.

19 Ivandic, M., Hofmann, W. and Guder, W.G. (1996). Development and evaluation of a urine protein expert system. *Clinical Chemistry*, **14**, 1214–22.

20 Kazi, J.I., Furness, P.N. and Nicholson, M. (1998). Diagnosis of early acute renal allograft rejection by evaluation of multiple histological features using a Bayesian belief network. *Journal of Clinical Pathology*, **51**, 108–13.

21 Gross, S. and Wolin, M. (1995). Nitric oxide: pathophysiological mechanisms. *Annual Reviews of Physiology*, **57**, 737–69.

22 Kone, B.C. (1997). Nitric oxide in renal health and disease. *American Journal of Kidney Diseases*, **30**, 311–33.

23 Takahashi, N., Suzuki, T., Yamaya, K. and Funyu, T. (1998). Nitric oxide generation in renal allograft recipients. *Transplantation Proceedings*, **30**, 2960–2.

24 Marzinzig, M., Nussler, A.K., Stadler, J. et al. (1997). Improved methods to measure end products of nitric oxide in biological fluids: nitrite, nitrate, and S-nitrosothiols. *Nitric Oxide*, **1**, 177–89.

25 Gardner, M.P., Houghton, D.C., Andoh, T.F., Lindsley, J. and Bennett, W.M. (1996). Clinically relevant doses and blood levels produce experimental cyclosporine nephrotoxicity when combined with nitric oxide inhibition. *Transplantation*, **61**, 1506–12.

26 Bobadilla, N.A., Gamba, G., Tapia, E. et al. (1998). Role of NO in cyclosporin nephrotoxicity: effects of chronic NO inhibition and NO synthase gene expression. *American Journal of Physiology*, **274**, F791–8.

27 Yang, C.W., Kim, Y.S., Kim, J. et al. (1998). Oral supplementation of L-arginine prevents chronic cyclosporine nephrotoxicity in rats. *Experimental Nephrology*, **6**, 50–6.

28 Stroes, E.S., Luscher, T.F., de Groot, F.G., Koomans, H.A. and Rabelink, T.J. (1997). Cyclosporin A increases nitric oxide activity in vivo. *Hypertension*, **29**, 570–5.

29 Vaziri, N.D., Ni, Z., Zhang, Y.P., Ruzics, E.P., Maleki, P. and Ding, Y. (1998). Depressed renal and vascular nitric oxide synthase expression in cyclosporine-induced hypertension. *Kidney International*, **54**, 482–91.

30 Worrall, N.K., Misko, T.P., Botney, M.D. et al. (1999). Time course and cellular localization of inducible nitric oxide synthases expression during cardiac allograft rejection. *Annals of Thoracic Surgery*, **67**, 716–22.

31 Wiklund, L., Lewis, D.H., Sjoquist, P.O. et al. (1997). Increased levels of circulating nitrates and impaired endothelium-mediated vasodilation suggest multiple roles of nitric oxide during acute rejection of pulmonary allografts. *Journal of Heart and Lung Transplantation*, **16**, 517–23.

32 Fábrega, E., Casafont, F., La Mantia, L. et al. (1997). Nitric oxide production in hepatic cell rejection of liver transplant patients. *Transplantation Proceedings*, **29**, 505–6.

33 Mugge, A., Kurucay, S., Boger, R.H. et al. (1996). Urinary nitrate excretion is increased in cardiac transplanted patients with acute graft rejection. *Clinical Transplantation*, **10**, 298–305.

34 Benvenuti, C., Bories, P.N. and Loisance, D. (1996). Increased serum nitrate concentration in cardiac transplant patients. A marker for acute allograft cellular rejection. *Transplantation*, **61**, 745–9.

35 Smith, S.D., Wheeler, M.A., Zhang, R. et al. (1996). Nitric oxide synthase induction with renal transplant rejection or infection. *Kidney International*, **50**, 2088–93.

36 Castillo, J., Berrazueta, J.R., Herrera, L. et al. (1996). Nitric oxide production during the hyperacute vascular rejection. *Journal of Surgical Research*, **63**, 375–80.

37 Dedeoglu, I.O. and Feld, L.G. (1996). Decreased urinary excretion of nitric oxide in acute rejection episodes in pediatric renal allograft recipients. *Transplantation*, **62**, 1936–8.

38 Vora, A., Monaghan, J., Nuttall, P. and Crowther, D. Cytokine-mediated nitric oxide release – a common cytotoxic pathway in host-versus-graft and graft-versus-host reactions? *Bone Marrow Transplantation*, **20**, 385–9.

39 Silkoff, P.E., Caramori, M., Tremblay, L. et al. (1998). Exhaled nitric oxide in human lung transplantation. A noninvasive marker of acute rejection. *American Journal of Respiratory and Critical Care Medicine*, **157**, 1822–8.

40 Fisher, A.J., Gabbay, E., Small, T., Doig, S., Dark, J.H. and Corris, P.A. (1998). Cross sectional study of exhaled nitric oxide levels following lung transplantation. *Thorax*, **53**, 454–8.

41 Murray, J.E. (1999). Reminiscences for the '50-year retrospective' of transplantation. *Transplantation Proceedings*, **31**, 34.

42 Calne, R.Y. (1968). The present position and future prospects of organ transplantation. *Annals of the Royal College of Surgeons, England*, **42**, 283–306.

46

# Post-transplantation bone disease

Juliet E Compston
University of Cambridge School of Clinical Medicine, Cambridge, UK

## Introduction

The development of techniques for solid organ transplantation represents a major advance in the management of patients with end-stage disease. However, in spite of the increasing survival rates achieved after transplantation, a number of serious medical complications may occur, one of which is bone disease. This chapter provides an overview of the prevalence, pathogenesis, pathophysiology and management of post-transplantation bone disease and reviews the use of biochemical markers of bone turnover in the diagnosis and monitoring of treatment of this condition.

## Bone loss and fracture incidence after transplantation

Increased rates of bone loss have been documented after transplantation in many studies and certain characteristics have emerged. The rate of bone loss appears to be most rapid in the first 3–6 months and both cancellous and cortical bone are affected. Reported rates of bone loss vary considerably and have generally been higher for patients undergoing liver, lung or cardiac transplant than in those undergoing renal transplantation [1]. Even within a single transplant category, however, there are large variations in reported rates, reflecting both differences in the selection of patients and in the immunosuppressive regimens used. Some studies have demonstrated a tendency for bone mineral density (BMD) to recover after 12 months' or so post-transplantation[2], but this has not been a consistent observation. Thus, data from a recent study suggest that, in patients with renal transplants, recovery of BMD may be less prominent. Measurements of BMD in 54 patients, studied a mean of 10 years after transplantation, demonstrated reduced BMD at several skeletal sites [3]. There is also some evidence that rates of bone loss have decreased in recent years, possibly as a result of lower doses of glucocorticoids than were formerly used.

The important clinical consequence of bone loss is fracture and, in prospective studies, fracture incidence rates of around 35–40% have been reported after liver or cardiac transplantation [1]. The risk of fracture after lung and renal transplantation

**Table 46.1.** Pathogenetic factors in post-transplantation bone disease

- Pre-existing bone disease
- Drugs – glucocorticoids, cyclosporin and other immunosuppressive agents
- Calcium and vitamin D deficiency
- Hypogonadism
- Reduced physical activity
- Malnutrition

has not been clearly established, although prevalence rates of 37% and 7–11%, respectively, have been reported in cross-sectional studies. In a recent cohort study set in tertiary care centres, the incidence of symptomatic fractures after solid organ transplant performed between 1986 and 1996 was investigated in 600 patients, most of whom had undergone renal transplantation. The overall fracture incidence in this cohort was 9.3%, the highest rates being observed in women undergoing heart and kidney transplantation [4].

## Pathogenesis of post-transplantation bone disease

The pathogenesis of post-transplantation bone disease is multifactorial (Table 46.1). Many patients have low bone mass prior to transplantation. Other pathogenetic factors include glucocorticoid therapy, cyclosporin and other immunosuppressive agents, hypogonadism, calcium and vitamin D deficiency and reduced physical activity, some of which may also contribute to bone disease prior to transplantation. Several studies have documented a high prevalence of osteoporosis prior to transplantation; in 243 consecutive patients with chronic liver disease undergoing assessment prior to transplantation, 37% were shown to have osteoporosis, defined by World Health Organization criteria, at the hip and/or spine and only 15% had normal BMD, i.e. a T score $>-1$ at both sites [5]. Regression analysis demonstrated that age was a significant predictor of low BMD in women but not men; the only other independent risk factor, again in women, was low body weight. Other groups have reported a similar or even higher prevalence of osteoporosis and osteopenia in patients with end-stage pulmonary or cardiac disease prior to transplantation [6].

## Pathophysiology of bone loss

There have been very few histomorphometric studies of the pathophysiology of bone loss following transplantation. One study in patients undergoing renal transplantation demonstrated a tendency for resolution of hyperparathyroid bone disease and some increase in bone turnover in those with adynamic disease [7]. The author's group recently reported the changes in bone turnover and remodelling in patients with chronic liver disease undergoing liver transplantation in a prospective study in which iliac crest biopsies were obtained, before and at 3 months postoperatively in 21 patients, and assessed by the use of histomorphometric techniques. Prior to transplantation, there was evidence of low bone turnover, with reduced bone formation rate at tissue level. At 3 months postoperatively, there was a highly significant increase in activation frequency and bone formation rate [8]. These data thus indicate that the major pathophysiological mechanism underlying early bone loss after liver transplantation is increased bone turnover.

Pathophysiological mechanisms in patients undergoing renal transplantation may differ because of the presence, in many, of renal osteodystrophy prior to transplantation. Histomorphometric data are sparse, but, in one prospective study, the changes of increased bone turnover and secondary hyperparathyroidism which were present preoperatively showed a tendency to resolve 6 months postoperatively.

## Biochemical changes associated with post-transplantation bone disease

There have been conflicting reports of the changes in biochemical markers of bone turnover following solid organ transplantation. This is likely to reflect, at least in part, the heterogeneity of the populations of patients studied, the use of different immunosuppressive regimens and the different markers studied. In general, where reliable markers of bone resorption have been studied, these have been shown to increase following transplantation [9]. Thus, in a prospective study of patients undergoing cardiac transplantation, Shane et al. [6] reported a significant increase in urinary excretion of deoxypyridinoline in the first 3 months postoperatively, followed by a return to preoperative values by 6 months. In a cross-sectional study of men undergoing cardiac transplantation, increased urinary excretion of N-telopeptide was demonstrated [10]. Serum osteocalcin has been the most widely studied bone formation marker. In some studies, this has been shown to decrease transiently following transplantation but recovery is apparent by 6 months and thereafter. Other studies in patients undergoing liver or cardiac transplantation indicate that osteocalcin may increase to abnormally high levels.

The available evidence is thus consistent with an increase in bone turnover in the early postoperative months, with resorption exceeding formation. Whether bone turnover returns to normal at a later stage has not been clearly defined and further studies are required.

An increase in plasma parathyroid hormone (PTH) levels has also been reported after both liver and cardiac transplantation [11], although this finding has not been universal. Glucocorticoid therapy, when administered in large doses, leads to a decrease in intestinal calcium absorption and the observed increase in plasma PTH concentrations may thus reflect a resulting secondary hyperparathyroidism which, in turn, may contribute to bone loss early after transplantation. A high prevalence of vitamin D deficiency or insufficiency has been reported in patients prior to lung, heart and liver transplantation, although serum PTH levels have been normal in the majority of such patients preoperatively.

## Management of post-transplantation bone disease

Prior to transplantation, patients should undergo thorough assessment for the presence of bone disease. Bone densitometry should be performed and lateral X-rays obtained of the thoracic and lumbar spine. Serum calcium, phosphate, alkaline phosphatase, 25-hydroxyvitamin D, PTH and thyroid function tests should be routinely measured. In men, sex hormone status should also be assessed by the measurement of serum testosterone and sex hormone-binding globulin.

After transplantation, bone densitometry should ideally be repeated at 6 and 12 months and yearly thereafter. X-rays should be performed as indicated by clinical events. However, since two-thirds or more of vertebral fractures are asymptomatic, spinal X-rays should be performed in all patients at 1 year.

Management of post-transplantation bone disease consists of measures to optimize BMD preoperatively and prophylaxis of bone loss during and after transplantation. Known risk factors for osteoporosis should be avoided where possible, vitamin D deficiency corrected and hormone replacement therapy advised for hypogonadal patients, both men and women. Transdermal preparations are preferable since they avoid first pass hepatic metabolism and both oestrogen and testosterone can be administered by this route. In those patients who receive glucocorticoids for their underlying disease, treatment with a bisphosphonate should be advised. In nonglucocorticoid-treated patients with evidence of osteoporosis prior to transplantation, appropriate treatment should be given according to age, gender and clinical circumstances.

At present, there are no adequately powered randomized controlled trials in transplant cohorts with fracture as the primary end-point that have been reported. Because of the complex and probably specific pathogenesis of post-transplantation bone disease, extrapolation from studies in women with postmenopausal osteoporosis or in subjects with glucocorticoid-induced osteoporosis cannot be made with confidence. Interpretation of many of the existing prevention and treatment studies of post-transplantation bone disease is hindered by the small numbers studied, different timing of the intervention with respect to transplantation, non-randomization of treatment and lack of appropriate control data. Thus, conflicting results have been reported for a number of agents including calcitonin, vitamin D metabolites and bisphosphonates, alone or in combination.

The time course of bone loss after transplantation, and evidence that it is due predominantly to increased bone turnover, provides a strong rationale for prophylaxis with antiresorptive therapy in the peri- and early postoperative period. Several groups have assessed the effects of intravenous pamidronate infusions given preoperatively and at various times after transplantation. So far, reported studies have been small and definitive results have not emerged. However, a randomized controlled trial of intravenous pamidronate, calcium and vitamin D in patients with cystic fibrosis demonstrated significant treatment benefits in both the hip and spine. Other, larger, studies are currently in progress and the results are awaited.

## Conclusions

In conclusion, osteoporosis is a serious complication of solid organ transplantation. The pathogenesis is multifactorial, but pre-existing bone disease and immunosuppressive therapy are likely to be major contributory factors. Management of post-transplantation bone disease requires optimization of bone mass prior to transplantation and measures to prevent postoperative bone loss; with respect to the latter, effective therapeutic regimens have not been established and further studies are required.

In recent years, lower rates of bone loss following transplantation have been reported than formerly and some groups have reported much lower morbidity from fractures. This is supported by the author's own experience and suggests that modification of immunosuppressive regimens and transplantation earlier in the course of disease may significantly reduce the prevalence of post-transplantation bone disease. The challenge for the future will be to eradicate osteoporosis as a complication of transplantation. Current trends are encouraging and indicate that improvements in immunosuppressive regimens and optimization of bone health prior to transplantation may eventually obviate the need for routine prophylactic and therapeutic measures at the time of transplantation.

## REFERENCES

1 Rodino, M.A. and Shane, E. (1998). Osteoporosis after organ transplantation. *American Journal of Medicine*, **104**, 459–69.
2 Sambrook, P.N., Kelly, P.J., Keogh, A.M. et al. (1994). Bone loss after heart transplantation: a prospective study. *Journal of Heart and Lung Transplantation*, **13**, 116–20.
3 Parker, C.R., Freemont, A.J., Blackwell, P.J., Grainge, M.J. and Hosking, D.J. (1999). Cross sec-

tional analysis of renal transplantation osteoporosis. *Journal of Bone and Mineral Research*, **14**, 1943–51.

4 Ramsey-Goldman, R., Dunn, J.E., Dunlop, D.D. et al. (1999). Increased risk of fracture in patients receiving solid organ transplants. *Journal of Bone and Mineral Research*, **14**, 456–63.

5 Ninkovic, M., Bearcroft, P.W.P., Skingle, S.J., Love, S., Alexander, G.A. and Compston, J.E. (1998). Vertebral fractures after liver transplantation: incidence and relationship to bone loss. *Journal of Bone and Mineral Research*, **13**, 521.

6 Shane, E., Thys-Jacobs, S., Papadopoulos, A. et al. (1996). Antiresorptive therapy prevents bone loss after cardiac transplantation. *Journal of Bone and Mineral Research*, **11**, S340.

7 Julian, B.A., Laskow, D.A., Dubowsky, J., Dubovsky, E.V., Curtis, J.J. and Quarles, L.D. (1991). Rapid loss of vertebral mineral density after renal transplantation. *New England Journal of Medicine*, **325**, 544–50.

8 Vedi, S., Greer, S., Skingle, S.J. et al. (1999). Mechanism of bone loss after liver transplantation: a histomorphometric analysis. *Journal of Bone and Mineral Research*, **14**, 281–7.

9 Kulak, C.A.M. and Shane, E. (1999). Transplantation osteoporosis. In *Dynamics of Bone and Cartilage Metabolism*, eds. M.J. Seibel, S.P. Robins and J.P. Bilezikian. New York: Academic Press, pp. 515–26.

10 Guo, C.-Y., Johnson, A., Locke, T.J. and Eastell, R. (1998). Mechanisms of bone loss after cardiac transplantation. *Bone*, **22**, 267–71.

11 Compston, J.E., Greer, S., Skingle, S.J. et al. (1996). Early increase in plasma parathyroid hormone levels following liver transplantation. *Journal of Hepatology*, **25**, 715–18.

47

# Molecular diagnosis of cytomegalovirus disease

Aycan F Hassan-Walker, Vincent C Emery and Paul D Griffiths

Royal Free and University College Medical School, London, UK

## Introduction

Cytomegalovirus (CMV) causes a variety of end-organ diseases in the immunocompromised host (see Table 47.1). CMV disease is diagnosed when patients meet an internationally agreed case definition [1], which essentially consists of symptoms of the affected organ and compatible clinical signs, plus detection of CMV in the affected organ (the only exception being the retina, where biopsy is not required). The pathogenesis of these conditions is distinct, probably involving contributions from the host as well as the virus. However, it is important to note that all of the CMV diseases have an underlying common pathogenetic factor – that is, CMV appears in the bloodstream before the onset of disease and rises to a high value to cause disease. It is this measurement of viral load that represents the biomarker being discussed in this chapter.

For more than two decades, a series of investigators defined several risk factors for CMV disease in different populations of patients (see Table 47.2). Thus, it is clear that primary infection represents a major risk factor in pregnant women and in recipients of solid organ transplants, whereas this is not the case in bone marrow transplant or acquired immune deficiency syndrome (AIDS) patients. Viraemia has been shown to be a risk factor in all the transplant patients and AIDS patients, but has not systematically been looked for in pregnancy. High CMV load was first identified by Stagno and colleagues in 1975 when they showed that neonates with congenital infection plus symptoms had, on average, a 1 log higher viral load in their urine than those with congenital infection who remained symptom free [2]. It was this observation using the bioassay available 25 years ago – namely, serial tube dilution of the clinical sample – which provided the impetus for a series of studies that have defined the value of modern molecular biological approaches to measuring CMV load.

**Table 47.1.** CMV diseases in the immunocompromised patient

| Symptoms | Solid $T_x$ | $BMT_x$ | AIDS |
|---|---|---|---|
| Fever/hepatitis | ++ | + | + |
| Gastrointestinal | + | + | + |
| Retinitis | + | + | ++ |
| Pneumonitis | + | ++ | |
| Myelosuppression | | ++ | |
| Encephalopathy | | | + |
| Polyradiculopathy | | | + |
| Addisonian | | | + |
| Immunosuppression | + | | |
| Rejection/graft versus host disease | + | ? | |

**Table 47.2.** Risk factors for CMV disease

| Factor | Pregnancy | Solid $T_x$ | $BMT_x$ | AIDS |
|---|---|---|---|---|
| Primary infection | + | + | — | — |
| Viraemia | ? | + | + | + |
| CMV load | + | + | + | + |

## Studies in transplant recipients

The authors' studies began with the same clinical analyte used by Stagno et al. (urine) and used quantitative–competitive polymerase chain reaction (QCPCR) to define the viral load in renal transplant recipients. The viral load in urine was significantly higher in those patients who developed disease [3]. Formal clinicopathological analysis showed three factors to be associated with CMV disease: raised viral load, presence of viraemia and recipient-positive serostatus (which was protective). Importantly, multivariate statistical analyses showed that the latter two risk factors were no longer statistically significant once viral load had been controlled for. Thus, viral load is the determinant of CMV disease and the other risk factors have prognostic value simply because they correlate statistically with high viral loads. It was, therefore, possible to construct a probability curve relating viral load to CMV disease and demonstrate the threshold concept of CMV disease (Figure 47.1). Subsequently, this work was taken forward to measure CMV load in

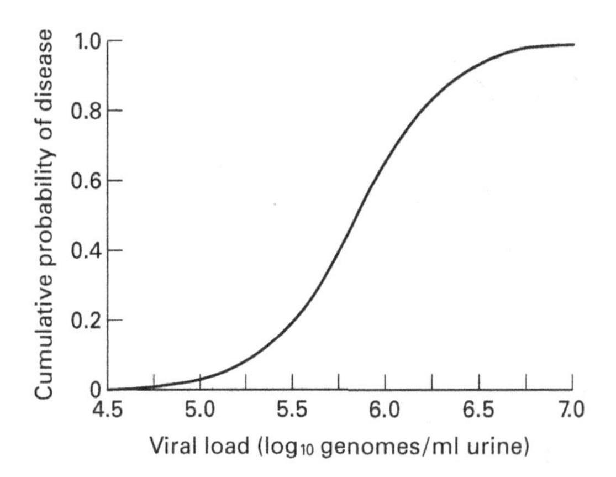


igure 47.1 Viral load and probability of human cytomegalovirus (HCMV) disease in renal transplant patients. From Cope et al. [3].

the blood of renal allograft recipients, with very similar conclusions [4]. In addition, it was possible to show that receipt of antithymocyte globulin increased CMV disease risk by increasing the peak CMV load. Similar results were also described for bone marrow transplant patients [5].

In liver transplant patients, similar results were obtained in that the viral load data explained the association of donor/recipient serostatus with CMV disease. However, the multivariate model showed an additional feature because receipt of augmented methylprednisolone was a risk factor for CMV disease which was statistically independent of high viral load (odds ratio: 1.61/1 g increase; $p=0.03$) [6]. Thus, it is possible to construct the viral load : probability-of-disease curve for these patients and show that augmented methylprednisolone modulates the risk associated with a given CMV load (Figure 47.2).

## AIDS patients

The authors' proposal that the viral load threshold pathogenesis concept described above would apply also to CMV disease in AIDS patients was treated sceptically by the AIDS community who felt that the predominant retinitis involvement of CMV disease in AIDS patients indicated a different pathogenetic process. This controversy ended in 1997 when three groups simultaneously published results showing that CMV viraemia preceded CMV disease in AIDS patients [7–9] and that the predictive values associated with the detection of viraemia were very similar to those found in transplant patients [10]. Furthermore, the authors' quantitative studies showed that an increase in viral load was an additional risk factor for CMV disease over and above the qualitative presence of CMV viraemia [9]. The work by Dodt et

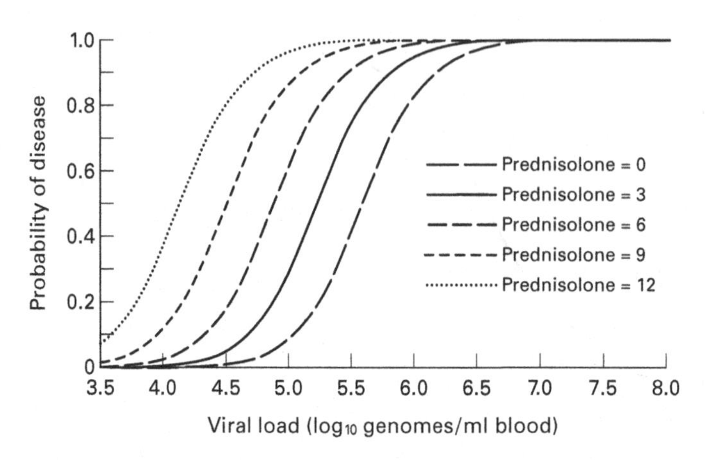


Figure 47.2 Cumulative probability of symptomatic disease according to viral load and methylprednisolone usage. From Cope et al. [6].

al. showed that antigenaemia also performed well in these analyses, although PCR had the advantage of detecting infection at earlier times before disease became apparent [8].

It could be shown that high viral loads in AIDS patients responded to treatment with intravenous ganciclovir, oral valaciclovir or oral ganciclovir [11, 12]. It was interesting to note in the valaciclovir prophylaxis study [11] that control of CMV disease was achieved with an average reduction in viral load of only 1.3 logs, thus confirming the implications of the viral threshold concept – that is, dramatic control of CMV disease can be obtained by relatively modest reductions in viral load by shifting the probability of disease curve in Figure 47.1 to the left. In addition, it was shown that patients with first episode retinitis, who had viral loads in blood higher than the median of the group, had a significantly higher mortality [12]. These results were subsequently confirmed and expounded by Spector and colleagues, who showed that mortality in AIDS patients is driven more by the CMV load than by the HIV load [13].

## CMV dynamics

All of these results could best be explained by a two-compartment dynamic model, with CMV replicating initially at a peripheral site and subsequently transmitting through viraemia episodes to the target organ. To address the dynamics of the flux from one compartment to the other, the authors have been making serial measurements of CMV viraemia. Results from a series of informative clinical situations led to the conclusion that the half-life of CMV in the blood is approximately 1 day [14].

By the use of serial measurement of the appearance of CMV viraemia post-transplant, and then following the trajectory of the quantity:time curve, it was possible to show that the doubling time was approximately 2 days in bone marrow transplant patients ($n=18$). The half-life of decline from a high viral load following treatment with intravenous ganciclovir was shown to be 2.5 days in AIDS patients ($n=35$), 1.5 days in bone marrow transplant patients ($n=11$) and 2.4 days in liver transplant patients ($n=12$). These differences were not attributable to differences between the groups of patients but were a manifestation of the timing of sample collection. Specifically, when samples were obtained on alternate days from AIDS patients, it was possible to reduce these upper limit estimates of the half-life down to approximately 1 day, whether measured in blood or in urine [14]. Finally, in patients where resistant strains were evolving under the selective pressure of ganciclovir, it was possible to monitor the trajectory of both resistant and wild-type strains on the quantity:time curve during continued selective pressure with oral ganciclovir and then to follow the reversal in viral competition when an alternative drug, offering no selective advantage to the mutant resistant strain, was given. In summary, the calculations of the fitness gained under these two distinct circumstances again led to the conclusion that the doubling time of CMV or its half-life was approximately 1 day [14].

## Applications to rapid viral diagnosis

Armed with the concept that CMV is a rapidly replicating virus, it is possible to determine whether quantification of the viral load in the first positive sample from a transplant patient could provide prognostic value. Specifically, the authors looked at whether the absolute value of the first positive sample or the rate of increase to reach this value identified patients at risk of CMV disease. In multivariate statistical models, each of these parameters was significantly associated with CMV disease and independent of each other (odds ratios 1.28 and 1.52, respectively). It was, therefore, possible to use a combination of these parameters to identify patients with a statistically increased risk of CMV disease [15]. This information could be used in several different ways by setting distinct cut-off values for the two parameters. Perhaps the most useful application will be to set the parameters to capture all individuals at risk of CMV disease. Any cut-off will necessarily include a small number of others who will be given pre-emptive therapy unnecessarily, but in a population of patients with a 15% incidence of CMV disease, for example, pre-emptive therapy would need to be given to 25%, leaving 75% of patients untreated. Such an approach will involve some overtreatment, yet should be a significant advance over using qualitative PCR alone and will be greatly superior to giving prophylaxis – that is, treating all patients in a transplant population.

## Summary

In summary, CMV causes several distinct diseases in the immunocompromised host including hepatitis, gastrointestinal disease, pneumonitis, retinitis and bone marrow suppression. Although these diseases are clinically distinct, they share the common pathogenetic feature of following a period of CMV viraemia. Recent prospective studies have demonstrated that increasing levels of viraemia represent a risk factor for CMV disease, with end-organ disease developing once the viral load threshold reaches a critical value. Multivariate statistical analyses also show that, once viral load has been controlled for, most of the other clinically recognized risk factors, such as donor/recipient serostatus, are no longer significant. Thus, high viral load is the major determinant of disease and methods for diagnosis should be focused on identifying patients destined to develop high viral load. Recent studies of the serial investigation of viral load changes with time have also defined the dynamics of CMV replication. CMV is a rapidly replicating virus, with a half-life of approximately 1 day in vivo, which contrasts with its reputation as a slowly growing virus in the laboratory. This new information will have a direct practical application in identifying patients in need of *pre-emptive therapy* (where a drug is administered expectantly to patients at imminent risk of developing high viral loads).

## REFERENCES

1 Ljungman, P. and Plotkin, S.A. (1995). Workshop of CMV disease: definitions, clinical severity scores, and new syndromes. *Scandinavian Journal of Infectious Diseases – Supplementum* **99**, 87–9.

2 Stagno, S., Reynolds, D.W., Tsiantos, A., Fuccillo, D.A., Long, W. and Alford, C.A. (1975) Comparative serial virologic and serologic studies of symptomatic and subclinical congenitally and natally acquired cytomegalovirus infections. *Journal of Infectious Diseases*, **132** 568–77.

3 Cope, A.V., Sweny, P., Sabin, C., Rees, L., Griffiths, P.D. and Emery, V.C. (1997). Quantity of cytomegalovirus viruria is a major risk factor for cytomegalovirus disease after renal transplantation. *Journal of Medical Virology*, **52**, 200–5.

4 Hassan-Walker, A.F., Kidd, I.M., Sabin, C., Sweny, P., Griffiths, P.D. and Emery, V.C. (1999) Quantity of human cytomegalovirus (CMV) DNAemia as a risk factor for CMV disease in renal allograft recipients: relationship with donor/recipient CMV serostatus, receipt of aug-

mented methylprednisolone and anti-thymocyte globulin (ATG). *Journal of Medical Virology*, **58**, 182–7.

5 Gor, D., Sabin, C., Prentice, H.G. et al. (1998). Longitudinal fluctuations between peak virus load, donor/recipient serostatus, acute GvHD and CMV disease. *Bone Marrow Transplantation*, **21**, 597–605.

6 Cope, A.V., Sabin, C., Burroughs, A., Rolles, K., Griffiths, P.D. and Emery, V.C. (1997). Interrelationships among quantity of human cytomegalovirus (HCMV) DNA in blood, donor-recipient serostatus, and administration of methylprednisolone as risk factors for HCMV disease following liver transplantation. *Journal of Infectious Diseases*, **176**, 1484–90.

7 Shinkai, M., Bozzette, S.A., Powderly, W., Frame, P. and Spector, S.A. (1997). Utility of urine and leukocyte cultures and plasma DNA polymerase chain reaction for identification of AIDS patients at risk for developing human cytomegalovirus disease. *Journal of Infectious Diseases*, **175**, 302–8.

8 Dodt, K.K., Jacobsen, P.H., Hofmann, B. et al. (1997). Development of cytomegalovirus (CMV) disease may be predicted in HIV-infected patients by CMV polymerase chain reaction and the antigenemia test. *AIDS*, **11**, F21–8.

9 Bowen, E.F., Sabin, C.A., Wilson, P. et al. (1997). Cytomegalovirus (CMV) viraemia detected by polymerase chain reaction identifies a group of HIV-positive patients at high risk of CMV disease. *AIDS*, **11**, 889–93.

10 Kidd, I.M., Fox, J.C., Pillay, D., Charman, H., Griffiths, P.D. and Emery, V.C. (1993). Provision of prognostic information in immunocompromised patients by routine application of the polymerase chain reaction for cytomegalovirus. *Transplantation*, **56**, 867–71.

11 Emery, V.C., Sabin, C., Feinberg, J.E. et al. (1999). Quantitative effects of valaciclovir on the replication of cytomegalovirus in patients with advanced human immunodeficiency virus disease: baseline cytomegalovirus load dictates time to disease and survival. *Journal of Infectious Diseases*, **180**, 695–701.

12 Bowen, F., Wilson, P., Cope, A. et al. (1996). Cytomegalovirus retinitis in AIDS patients: influence of cytomegaloviral load on response to ganciclovir, time to recurrence and survival. *AIDS*, **10**, 1515–20.

13 Spector, S.A., Hsia, K., Crager, M., Pilcher, M., Cabral, S. and Stempien, M.J. (1999). Cytomegalovirus (CMV) DNA load is an independent predictor of CMV disease and survival in advanced AIDS. *Journal of Virology*, **73**, 7027–30.

14 Emery, V.C., Cope, A.V., Bowen, E.F., Gor, D. and Griffiths, P.D. (1999). The dynamics of human cytomegalovirus replication in vivo. *Journal of Experimental Medicine*, **190**, 177–82.

15 Emery, V.C., Sabin, C.A., Cope, A.V., Gor, D., Hassan-Walker, A.F. and Griffiths, P.D. (2000). Application of viral load kinetics to identify patients destined to develop cytomegalovirus disease following transplantation. *Lancet*, **355**, 2052–6.

48

# Diagnosis and monitoring of inflammatory events in transplant patients

Claus Hammer, Gudrun Höbel, Stefanie Hammer, Peter Fraunberger, Bruno Meiser

Klinikum Grosshadern, LM-University of Munich, Munich, Germany

## Introduction

The history and management of an intensive care patient suffering from systemic infection or sepsis or a patient who is in danger of becoming septic is totally different from that of a patient who will receive, or has received, a transplant. A patient who is registered on the waiting list for heart and/or lung transplantation has been chronically ill for months and even years. Patients waiting for a heart have circulatory and respiratory problems, multiple organ dysfunction and often show signs of immune depression. Lung transplant patients, however, have a long history of cystic fibrosis with all its associated infections, or have suffered from chronic bronchitis, emphysema or idiopathic fibrosis. All such patients have been under intensive medical care and observation and their underlying disease known and heavily and specifically treated. All transplant patients have to undergo major surgery with a long period of anaesthesia and extended trauma; they often receive multiple blood transfusions from foreign donors. After surgery, the patient experiences aggressive and chronic immunosuppression. Azathioprine, cyclosporin, tacrolimus and steroids are given within the first 7 days and remain at relatively high doses for another month. A rejection episode, which occurs in more than 60% of all transplant patients, needs an immediate boost of immunosuppression. Under this antirejection treatment, patients are at a significant risk of developing opportunistic infections.

According to the European Transplant Registry, more than 50% of transplanted patients were found to have infections in the first postoperative year, 40% of which were of viral, 42% of bacterial, 10% of fungal and 8% of protozoal origin. Their early and rapid detection is vital for the effective implementation of appropriate antimicrobial therapy. Improved methods for the early diagnosis of infection would facilitate post-transplant monitoring and minimize unnecessary antibiotic treatment that often has toxic side effects. For those individuals who become septic after sudden and unexpected severe trauma or insult, monitoring and managing is very different. Procalcitonin (PCT) seems to be a new and excellent early marker for identifying nonviral infections in both groups of patients. PCT helps to differentiate nonviral infections from rejection episodes in transplanted patients; in these circumstances, PCT is much more specific than acute phase proteins, leucocyte counts and even interleukin-6 [1].

## Biochemistry and characterisics of PCT

PCT was first described by Bohoun in 1992 as the precursor molecule of human calcitonin (molecular weight of 13 kD). The polypeptide of 116 amino acids comprises an N-terminal region, the midregional calcitonin and the C-terminal region of katacalcin.

In healthy persons, PCT is quickly processed by proteolysis into calcitonin and its other regional fragments. The normal plasma concentrations are under the detection limit of the PCT assay of 0.01 ng/ml. In contrast, very high levels of PCT are typical in patients with systemic bacterial infections, sepsis or multiorgan failure. In these cases, PCT concentrations of over 100 ng/ml have been measured without any changes in plasma calcitonin [2]. The hormonally active region of calcitonin has a very short half-life of 10–20 mins. In contrast, PCT has an ideal half-life of about 24 hours and lacks hormonal activity [3].

In comparison with interleukins, PCT is a very stable protein in vivo and in vitro. At room temperature, PCT values remain constant and measurable for several hours. Stored frozen, the activity remains unchanged for 6 months. Repeated thawing and freezing has been found to have no influence on the PCT molecule [4].

### Measurement of PCT

PCT is analysed using the immunoluminometric assay LUMItest® PCT (BRAHMS Diagnostica, Berlin, Germany). This test system determines PCT quantitatively in human serum or plasma. It is easy to handle and can be performed in less than 3–4 hours. Inter- and intra-assay variations at both low and high concentrations are less than 7%.

### Origin and effect

The biological function of PCT and the major organs/tissues of origin are still uncertain. PCT cannot be induced in vitro, by allogeneic, lectin or endotoxin stimulation of peripheral blood or isolated cell populations [5]. Equally as unknown as the cellular origin of PCT is the function of PCT.

## PCT – indicator of bacterial and fungal infections

In the last 7 years, numerous studies have shown a correlation between PCT increase and bacterial or fungal infections. As PCT reacts very specifically to nonviral infections, it can be used to differentiate these from non-specific inflammation.

Interestingly enough, PCT correlates with the severity and activity of the infection. Assicot et al. [1] describe values up to 53 ng/ml in patients with severe systemic bacterial/fungal infections. Local bacterial/fungal infections resulted only in a moderate, or no, rise in PCT (0.3–1.5 ng/ml). In the case of an encapsulated process or infections limited to one organ – such as pneumonia or peritonitis – PCT levels were only slightly increased (1–3 ng/ml) [1]. The authors' own studies with heart, lung or liver transplant patients have confirmed these results (Tables 48.1 and 48.2). In liver transplant patients, PCT was found to reach slightly higher values (local infection: 2.2 ng/ml; systemic infection: 11.9 ng/ml) than in patients after heart or lung transplantation (local infection: 0.6 ng/ml; systemic infection: 10.5 ng/ml). PCT not only reacts sensitively to severe bacterial infections but also to those infections of parasites and fungi. As a consequence of the immunosuppressive therapy, transplanted patients often suffer from generalized fungal and opportunistic infections. In the authors' study, the maximal PCT values of 82 ng/ml were recorded in patients with *Aspergillus* sepsis. In the case of localized fungal infection – like candidiasis of the oral mucosa – PCT remained within normal limits.

## PCT and viral infections

In comparison with bacterial infections, PCT was found not to rise in response to viral infections. Assicot et al. [1] describe marginal concentrations between 0.1 ng/ml and 1.5 ng/ml in patients with different kinds of viral infections [1]. Even children with severe viral meningitis showed only a slight increase in PCT of 0.32 ng/ml and, therefore, differed clearly from children with bacterial meningitis [6].

After transplantation, cytomegalovirus (CMV) infections occur frequently. Normal PCT levels of 0.1 ng/ml were found in heart transplant patients with CMV infection [5]. Even HIV-infected patients in the final stage of AIDS did not differ from healthy persons with PCT levels of 0.5 ng/ml [7].

## PCT and different types of germs

PCT concentrations correlate well with the severity (Table 48.2) of the infection, but are not influenced by the type of organisms. In fact, the highest levels of PCT are measured in patients with fungal infections – especially those taken ill with

Table 48.1. Values of PCT, IL-6, CRP and leucocytes in patients with and without rejection episodes or infection after heart and/or lung transplantation

| | PCT (ng/ml) | | IL-6 (pg/ml) | | CRP (mg/l) | | WBC (g/l) | |
|---|---|---|---|---|---|---|---|---|
| | mean v. ± sd | *n* | mean v. ± sd | *n* | mean v. ± sd | *n* | mean v. ± sd | *n* |
| No AR | 0.3 ± 0.3 | 59 | 33.1 ± 25.9 | 30 | 3.8 ± 5.1 | 38 | 13.4 ± 4.8 | 38 |
| AR | 0.2 ± 0.2 | 19 | 29.0 ± 17.0 | 9 | 1.9 ± 1.4 | 9 | 12.1 ± 3.5 | 9 |
| AR + infection | 0.6 ± 0.6 | 7 | 69.4 ± 41.8 | 4 | 6.5 ± 3.0 | 5 | 12.0 ± 6.5 | 5 |
| Local infection | 0.6 ± 0.3 | 25 | 58.3 ± 53.5 | 22 | 6.8 ± 6.2 | 14 | 13.8 ± 6.2 | 14 |
| Systemic infection | 10.5 ± 18.2 | 28 | 254.1 ± 430.1 | 27 | 6.8 ± 5.4 | 17 | 16.0 ± 7.9 | 17 |

Table 48.2. Values of PCT, IL-6, CRP and leucocytes in relation to the severity of infections

| | PCT (ng/ml) | | IL-6 (pg/ml) | | CRP (mg/l) | | WBC (g/l) | |
|---|---|---|---|---|---|---|---|---|
| | mean v. ± sd | *n* | mean v. ± sd | *n* | mean v. ± sd | *n* | mean v. ± sd | *n* |
| Discharge | 0.3 ± 0.5 | 32 | 29.2 ± 21.3 | 25 | 2.8 ± 4.0 | 9 | 9.5 ± 2.6 | 9 |
| Local infection | 0.8 ± 0.9 | 30 | 55.7 ± 50.8 | 27 | 6.0 ± 5.7 | 18 | 13.3 ± 7.4 | 18 |
| General multi-infection | 6.5 ± 15.0 | 29 | 170.7 ± 324.6 | 11 | 5.9 ± 4.1 | 21 | 12.7 ± 7.6 | 21 |
| Sepsis | 20.9 ± 14.5 | 13 | 633.7 ± 813.4 | 22 | 11.7 ± 7.7 | 9 | 27.0 ± 16.8 | 9 |
| Day of death | 38.1 ± 57.3 | 12 | 1583 ± 1279 | 6 | 11.4 ± 7.9 | 10 | 19.2 ± 16.2 | 10 |

*Aspergillus* infection. The mean PCT values did not differ significantly in patients infected with different bacteria or fungi [8]. This was confirmed by studies from Al-Nawas et al. in septic patients [9].

## Trigger of induction and release of PCT

PCT concentrations have been shown to be elevated in patients with gram-negative infections. Based on this information, Dadona et al. [10] investigated the effect of endotoxin on PCT release and production in healthy volunteers. Four hours after intravenous injection of endotoxin from *Escherichia coli*, the PCT level began to increase and peaked after 6 hours; then, the PCT concentration stayed at a plateau for 8–24 hours. PCT concentrations also reached peak values in patients with fungal or parasitic infections, which are not accompanied by endotoxin release [3].

Thus, endotoxin seems to be only one possible mediator responsible for PCT release. Endotoxin, however, could be the trigger of a kind of cytokine cascade which, as a consequence, induces the release of PCT. This was suggested in the study of Dadona et al. in which tumour necrosis factor-$\alpha$ and IL-6 peaked after 1 hour and 3 hours, respectively, followed by PCT [10].

## PCT and rejection

In transplanted patients with acute or chronic rejection, no PCT increase has been measured [11]. In 63% of the renal transplant patients with acute rejection (AR) PCT remained lower than 0.5 ng/ml and did not correlate with the degree of the rejection. Yet, PCT showed a significant difference between rejection and infection [12].

This observation was also found in heart- and lung-transplanted patients with AR where PCT remained within the normal limit. In these recipients, a mean value of 0.2 ng/ml was measured during AR and 0.5 ng/ml after live transplantation. Even very severe rejections (grade 3 according to the International Society of Heart and Lung Transplantation) were not accompanied by a PCT increase. A correlation between the degree of the rejection and PCT values was not found. With a sensitivity, specificity and positive predictive value of 89%, the authors were able to differentiate a systemic infection from a pure rejection in heart or lung transplant patients when the PCT concentration was above a cut-off point of 0.5 ng/ml [8].

## PCT and immunosuppression/deficiency

To avoid an acute rejection after transplantation, patients receive massive immunosuppression for weeks. In the case of AR, a maximum dose of steroids as a bolus treatment is also administered. The influence of various immunosuppressive agents on PCT has been explored in different studies.

In septic patients, with and without immunosuppression, PCT does not differ during the first 3 days of the disease. However, immunocompromised patients already show low PCT values between days 3 and 5, while, in patients with a normal immune system, PCT levels do not drop until day 5 [9]. These observations are in complete contrast to reports of exceedingly high PCT values of 103 ng/ml in a 4-year-old liver transplanted girl with disseminated candidiasis who received a combined immunosuppressive therapy consisting of cyclosporin, azathioprine and corticosteroids [12]. PCT was not influenced by these immunosuppressive agents. The only exception appeared to be immunosuppressive therapy with OKT 3, which increased PCT release.

## PCT and antibiotic therapy

In the same way as PCT increases in association with the severity of an infection, it also decreases again under an effective antibiotic or antifungal therapy. Due to the half-life of PCT of 24 hours, a suitable therapy halves the PCT value day by day.

This observation has also been confirmed in patients who were successfully treated because of sepsis. In children with bacterial meningitis receiving antibiotic therapy, a similarly typical fall of PCT values was observed [6]. PCT, therefore, proved to be a suitable parameter for monitoring the efficacy of the treatment of bacterial, fungal and parasitic infections.

## PCT after trauma/operation

In a number of heart-, lung- and liver-transplanted patients, moderate levels of PCT were reported in the first days after operation. In the case of an uncomplicated course of the transplant, PCT values halve every day. The persistence of high levels of PCT or a new increase heralds a postoperative infection. After small, aseptic operations, PCT increases only in 32% of patients and only seldom is higher than 1 ng/ml. Extended abdominal surgery raises PCT in 95% of cases, but rarely over 10 ng/ml. An abnormal postoperative course is usually connected with significantly higher PCT values than found in uncomplicated courses. Translocation of intestinal bacteria would explain the increased PCT values after abdominal operations.

## PCT and premortal increase

PCT levels correlate clearly with the severity of an infection. Consistent with this observation, PCT increases continuously in patients with lethal infections and systemic inflammatory response syndrome (SIRS) and reaches maximum values on the day of death. In patients with SIRS, mortality can be predicted from the PCT concentration. There is a fatality ratio of 65% in the case of PCT values above 2.1 ng/ml. Below this limit, no patient died in the authors' study.

Besides the assessment of the severity of an infection, the PCT concentration also has prognostic value following infection. If PCT increases continuously and reaches maximum values, the mortality rate also rises (Table 48.2) [13].

## Sensitivity, specificity and positive predictive value of PCT

Sensitivity and specificity are especially important for evaluating a new possible marker of infection like PCT. At a cut-off point of 0.5 ng/ml in heart- and lung-transplanted patients, a sensitivity, specificity and positive predictive value of 89% is achieved. Thus, PCT proves to be a reliable marker to monitor such patients [8]. Similar results of PCT above 1.0 ng/ml exclude rejection without infection in 98% of patients. In the case of an infection, the sensitivity is 77% and specificity 100% at a cut-off point of above 1.0 ng/ml [5]. After renal transplantation, there is a sensitivity of 87% and specificity of 70% in the diagnosis of an invasive bacterial infection [11].

## Summary

PCT seems to be an elegant predictor in transplant patients not only of severe infections but also to distinguish infections from acute rejection episodes. One of the advantages of PCT monitoring is the rapid detection of an infection within 3 hours. PCT levels are correlated with the activity and dynamics of the infection, as well as with the efficacy of therapy and prognostic outcome. Critical care patients, neonates and unconscious patients can be monitored reliably and rapidly. Differentiation between infectious and noninfectious clinical events is possible and valuable. However, PCT does not replace tests to identify specifically the different types of infectious organisms involved, or those which test the sensitivity or the resistance of these organisms to antibiotics. Thus, PCT represents a new marker of serious infections which, in combination with other parameters like C-reactive protein, leucocyte numbers and interleukins, can be used to accelerate the introduction of appropriate treatment – hopefully leading to improved clinical outcome, quality of life and prolonged survival.

## REFERENCES

1 Assicot, M., Gendrel, D., Carsin, H., Raymond, J., Guilbaud, J. and Bohuon, C. (1993). High serum procalcitonin concentrations in patients with sepsis and infection. *Lancet*, **341**, 515–18.

2 Nylen, E.S., O'Neill, W., Jordan, M.H. et al. (1992). Serum procalcitonin as an index of inhalation injury in burns. *Hormone and Metabolic Research*, **24**, 439–42.

3 LeMoullec, J.M., Julliene, A., Chenais, J. et al. (1984). The complete sequence of human preprocalcitonin. *FEBS Letters*, **167**, 93–7.

4 Meisner, M., Tschaikowsky, K., Schnabel, S. et al. (1997). Procalcitonin – influence of temperature, storage, anticoagulation and arterial or venous asservation of blood samples on procalcitonin concentrations. *European Journal of Clinical Chemistry and Clinical Biochemistry*, **35**, 597–601.

5 Staehler, M., Ueberfuhr, P., Reichart, B., Hammer, C. (1997). Differential diagnostik der Abstossungsreaktion und Infektion bei herztransplantierten Patienten: neue Wege mit

Zytokinen und Procalcitonin als Marker. *Transplantation Medicine*, **9**, 44–50.

6 Gendrel, D., Raymond, J., Assicot, M. et al. (1997). Measurement of procalcitonin levels in children with bacterial or viral meningitis. *Clinical Infectious Diseases*, **24**, 1240–42.

7 Gerard, Y., Hober, D., Assicot, M. et al. (1997). Procalcitonin as a marker of bacterial sepsis in patients infected with HIV 1. *Journal of Infection*, **35**, 41–6.

8 Hammer, S., Meisner, F., Dirschedl, P. et al. (1998). Procalcitonin: a new marker for diagnosis of acute rejection and bacterial infection in patients after heart and lung transplantation. *Transplant Immunology*, **6**, 235–43.

9 Al-Nawas, B. and Shah, P.M. (1996). Procalcitonin in patients with and without immunosuppression and sepsis. *Infection*, **6**, 434–6.

10 Dadona, P., Nix, D., Wilson, M.F. et al. (1994). Procalcitonin increase after endotoxin injection in normal subjects. *Journal of Clinical Endocrinology and Metabolism*, **79**, 1606–8.

11 Meisner, M., Tschaikowsky, K., Schmidt, J. et al. (1996). Procalcitonin (PCT) indications for a new diagnostic parameter of severe bacterial infection and sepsis in transplantation, immunosuppression and cardiac assist devices. *Cardiovascular Engineer*, **1**, 67–76.

12 Staehler, M., Hammer, C., Meiser, B. et al. (1997). Procalcitonin: a new marker for differential diagnosis of acute rejection and bacterial infection in heart transplantation. *Transplantation Proceedings*, **29**, 584–5.

13 Hergert, M., Lestin, H.G., Scherkurs, M. et al. (1998). Procalcitonin in patients with sepsis and polytrauma. *Clinical Laboratory*, **44**, 659–70.

# Index